Instructor's Resource Manual
to Accompany

Differential Equations

A Modeling Perspective

To use or not to use: that is the question:
Whether 'tis nobler in the mind to suffer
The stings and errors of outrageous computation,
Or to apply technology against a sea of questions,
And by its power resolve them?
To model: perchance to visualize: ay, there's the key:
For in that model what insights may come...

Instructor's Resource Manual
to Accompany

Differential Equations

A Modeling Perspective

Robert L. Borrelli
Courtney S. Coleman

·

Harvey Mudd College

John Wiley & Sons, Inc.

New York • Chichester • Weinheim • Brisbane • Singapore • Toronto

Copyright © 1998 by John Wiley & Sons, Inc.

Excerpts from this work may be reproduced by instructors for distribution on a not-for-profit basis for testing or instructional purposes only to students enrolled in courses for which the textbook has been adopted. *Any other reproduction or translation of this work beyond that permitted by Sections 107 or 108 of the 1976 United States Copyright Act without the permission of the copyright owner is unlawful. Requests for permission or further information should be addressed to the Permissions Department, John Wiley & Sons, Inc., 605 Third Avenue, New York, NY 10158-0012.*

ISBN 0-471-24593-3

Printed in the United States of America

10 9 8 7 6 5 4 3 2 1

Printed and bound by Hamilton Printing Company

Preface

The purpose of this manual is twofold. First, it serves the purpose of providing worked-out solutions for all the problems in the text (except the group problems) as well as all graphs asked for in the problems. A second purpose is to provide a chapter-by-chapter and section-by-section glimpse of our mind-sets when writing the text. You'll see why we arranged the material the way we did, as well as the focus of each section. We even tell you what our favorite problems are, and why. We hope that all this information will make it easier to design your own course based on this text.

The text's emphasis on modeling and visualization is made to order for the study of ODEs. The only trick is how to weave these threads together into whole cloth. The text provides enough material for a variety of introductory ODE courses taught at several different levels to students of varying interests and backgrounds. In the table below we have listed the topics we feel belong in a modern ODE course. Starting with these topics, you will need to select the models and the visualization projects that you think would appeal to your students.

Topics for an ODE Course

First and Second Order ODEs	Systems
• Solution Formulas	• Linear/Nonlinear Models
• Sensitivity to Data	• Linear Systems
• Numerical Solutions	• Stability
• Models	• Planar Systems
• Linear ODEs	• Long-term Behavior
• Basic Theory	• Bifurcations
• Long-term Behavior	• Orbital Geometry
• Geometry of Solution Curves	

We have tried to organize this selection process by providing a chart entitled "Assignment Schedule" on page viii. You can experiment with a selection of sections by arranging them in the second column of the schedule until you are comfortable. Next you might select the modeling and visualization projects in each section that you wish to cover. No harm is done in choosing some models in a section and skipping over others. Each model in a section is usually labeled as a subsection (e.g., in Section 1.5: Modeling Radioactive Decay, Galilean Approach to Vertical Motion, Newtonian Approach to Vertical Motion), so the selection process is straightforward. Some sections (at least one per chapter) are devoted mostly to one type of model, but even so, nothing is lost by doing only part of the section. Selection of visualization projects is a more intuitive process. Looking over the graphs in a section and in this "Instructor's Resource Manual", you may pick up something that appeals to you (we often build a lecture around a graph). You can list these graphs by Figure number (or Example number) in the "visualization" column of the Assignment Schedule.

Finally, you might list the problems in each section you wish to assign, using your selection of models and visualization projects as a guide. Problems related to specific models are always labeled so they are easy to spot. When assigning problems relating to visualization projects, we recommend looking over the graphs in the relevant section of this "Instructor's Manual" to see if the graphs there are what you want your students to see.

A Word about Teaching Styles
•

Having available a powerful visualization tool at your disposal, you would not be well served if all you used it for was to do what you have always done in class but faster (albeit with good visualization). With a good software tool you can make otherwise intimidating topics readily accessible to the beginning student. For example, your students can begin to test the theory around the fringes, examine the sensitivity of dynamical systems when the data change, visualize the response of systems to discontinuous driving functions, consider models that lead to nonlinear systems, compare various models for the same phenomenon, follow a system through bifurcation, visualize the behavior of chaotic dynamical systems, and in general answer what-if questions. Encourage your students to be creative in using software to examine dynamical systems—you just might be surprised with their ingenuity. We highly recommend group projects because the performance of the whole is usually much more than the sum of the individual contributions.

We advise against keyboard entry of systems on the computer in front of the class because you might look foolish when you have trouble recovering from an error. It's hard to pay attention to what you want to say at the same time you are thinking about the syntax of your software package. Just be aware that if you are going to do keyboard entry in front of a class, then be prepared for "advice" while you are fumbling around. In the process, the point you wanted to make will be lost. A better way is to prepare your graphics displays in advance and to play them out with the click of a cursor.

Restrain yourself from giving away the punchline too soon when setting up a particular application. A graphic display may be so convincing that you want to tell the world about it. In doing this you may deny your students the thrill of discovery on their own. We have tried to invent problems that will give students an opportunity to "discover" properties of dynamical systems themselves, and these are among our favorite problems.

Software
•

Before you even ask, the answer is no: the software used to do the figures in the text is not available for distribution. The reasons are involved and are not likely to change anytime soon. The road to the development of our solver, ODETOOLKIT, was long and arduous because we had so few resources at our disposal. In the end, though, we are quite pleased at the outcome and believe it to be one of the best all-around numerical ODE solvers we have seen so far. But to our regret, the software cannot be distributed. Live and learn!

That said, we now do our best in helping you look for alternative numerical ODE solvers that will meet your needs. A good place to start is the article by Andy Flint and Ron Wood in the CMJ Special Issue on Differential Equations.[1] The article lists the best known solvers then available commercially (except for ours), but there are plenty of other good (and not so good) solvers out there that didn't make the list. As indicated, some packages will run only on Mac or on PC platforms, whereas others are designed primarily for work stations (but have versions that also run on Macs and PCs). Also listed is a telephone number where you can get more information. Most of us have very little choice of platform (we use whatever is available), but you might be able to exert some influence on the choice of software. Think out how you want to use the software before making a decision. If students are going to be left more or less on their own, then the package you choose should have an easy-to-learn interface. If expense is your primary consideration, then you might

---

[1] A. Flint and R. Wood, "ODE Solvers for the Classroom," *College Mathematics Journal*, 25, No. 5 (Nov. 1994), 458–461.

want to consider MDEP because it is free. If your students are already familiar with MAPLE or MATHEMATICA or MATHCAD or DERIVE from their calculus course, then you may do well to stick with that same package. Our solution to this problem at Harvey Mudd College has been to negotiate site licenses for all the software packages on the list, and then let the students use whichever solver appeals to them. If you opt for a solver rather than a CAS package, then keep in mind that the Mac and PC solvers listed are fairly inexpensive, and many students who own one of these platforms buy their own solver and use it at home or in the dorms.

Lab Work
•

Working computer experimentation into the course is a big challenge with a big payoff. It is clear that it can't be added on top of everything you have always done in your course. (Students and your faculty colleagues would rightly perceive that the workload has been increased.) At Harvey Mudd College we assign regular computer problems along with "pencil and paper" problems, and (like homework) the computer problems are collected, graded, and returned. A group computer problem, however, replaces a "regular" problem set, and the due dates are staggered by section so as not to overwhelm the college computer facilities. This arrangement has worked well, in part because a computer tutor holds regular office hours in the lab. A better solution would be to have a one-unit, two-hour classroom lab session once a week where students can complete their computer problem assignment for the week. Of course, this is what our colleagues in Physics and Chemistry do. Not only do students get an extra unit credit for extra work, but also the faculty member gets additional teaching credit for the extra work setting up the lab projects and selecting appropriate software. Some colleges and universities have already successfully implemented this option.

Student Resource Manual
•

The Student Resource Manual (SRM) contains the solutions (and graphs) of every other part of every odd-numbered problem (except the group problems). The SRM also contains the Comments and the Background Material of this Instructor's Manual.

Conclusion
•

When all is said and done, what do you want your students to remember most when they leave your course? An arsenal of techniques for finding solution formulas? Some basic modeling principles and practice? Some important properties of systems that evolve in time? How to use numerical solvers to extract information about the behavior of solutions of ODEs?

All of these goals are valuable in and of themselves, but there is another goal that often gets lost in the shuffle: for students to come away feeling that they can use a software tool to explore the rich landscape of dynamical systems on their own. This feeling will convert them from a passive to an active participant in their own educational development. This may be all that some students will remember from the course.

And last, but not least, the use of solvers in an ODE course should encourage a sense of playfulness that can be enjoyed, even on a superficial level, just because of the interesting graphical displays. Who can resist smiling when looking at the "teddy bears" of Fig. 1.7.1, or the "wine glass" in the Chapter 5 cover graphic? And who can fail to appreciate the striking geometrical quality of the Chapter 3 cover graphic? Ultimately, it should be fun to use solvers to probe the behavior of solutions of ODEs, and let's hope that students will at least remember this about their differential equations course.

ASSIGNMENT SCHEDULE

Day	Section/Topic	Models	Visualization	Problems
1				
2				
3				
4				
5				
6				
7				
8				
9				
10				
11				
12				
13				
14				
15				
16				
17				
18				
19				
20				
21				
22				
23				
24				
25				
26				
27				
28				
29				
30				
31				
32				
33				
34				
35				
36				
37				
38				
39				

Acknowledgments

This manual was prepared in LaTeX using macros written by Dave Richards, whose programming talents and meticulous attention to detail deserve our warmest appreciation. Very special thanks go to Tony Leneis, who was the principal author of the software package ODETOOLKIT, a command script and user interface over the ODE solver MATHLIB DEQSOLVE,[2] which was used to produce all the figures in this manual. Our very special thanks also go to Jenny Switkes, who put up with daily changes in the graphs and the text with good humor and remarkable efficiency as she set much of this manual in LaTeX.

We want to thank our colleagues at Harvey Mudd College and at many other colleges and universities for their comments and suggestions. We also thank students at Harvey Mudd and Pomona Colleges, Arizona State University, and several other institutions for their suggestions.

R. L. Borrelli borrelli@hmc.edu
C. S. Coleman coleman@hmc.edu
Claremont, August 1997

---

[2]MATHLIB (Registered trademark of Innosoft International Inc., West Covina, CA.) DEQSOLVE, an ODE solver from the MATHLIB software package is based on LSODA, a descendant of C. W. Gear's DIFFSUB, and is part of the package ODEPACK developed by Alan Hindmarsh at Lawrence Livermore National Laboratories. DEQSOLVE was written by Kevin Carosso, Ned Freed, and Dan Newman.

Contents

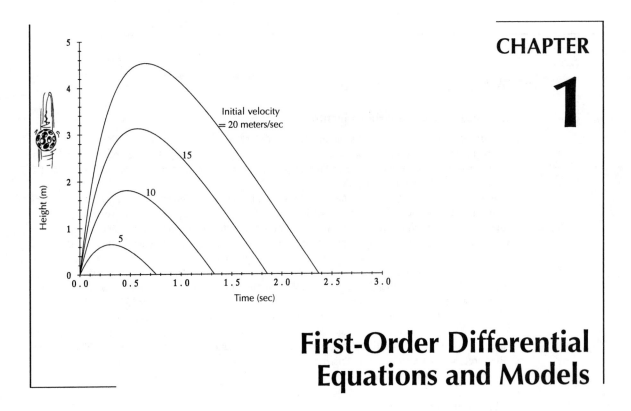

CHAPTER

1

First-Order Differential
Equations and Models

The main thrust of this chapter is to find solution formulas for first-order ODEs, discuss some ODE models, introduce the geometry of solution curves, and show the central role of numerical solvers in understanding the behavior of solutions. Solution methods are tabulated at the end of the chapter for reference purposes. Questions of the existence and uniqueness of solutions of initial value problems (IVPs) and the sensitivity of IVPs to changes in the data can be answered from these formulas. The relationships between ODEs and dynamic processes are so basic that we explore the connections right from the start. Throughout the chapter, simple processes are introduced and modeled with ODEs. We hope users of this text will use numerical solvers to approximate and graph solutions of ODEs; the computer icon designates problems for which a numerical solver is appropriate. Most CAS packages have numerical solvers, but can also produce solution formulas for many first-order ODEs (even ODEs with parameters), not to mention graphs and direction fields.

Systems of ODEs are introduced in this chapter for several reasons. Systems arise naturally in modeling. Some systems can be solved using the first-order techniques of the chapter. All numerical ODE solvers available on the market today are designed to solve systems of first-order ODEs.

Note that discontinuous driving terms (e.g., the Engineering Functions of Appendix B.1) are introduced early in a mild way and in a more substantial way later in a modeling environment (Section 1.8). If your solver does not have these functions as predefined functions, you can often find a work-around using if-then-else logical commands over a finite time interval.

1.1 A Modeling Adventure

Suggestions for preparing a lecture

Topics: Informal definitions of the terms ordinary differential equation (ODE), initial value problem (IVP), solution, solution curve, and mathematical model need to be given. Selected parts of the fish population model can be introduced and discussed in terms of solution formulas (if available) and the use of numerical solvers.

Remarks: Two of the pillars that support our approach are modeling and visualization. Numerical solvers are basic to both, and so students need to be brought up to speed as soon as possible. Several of the problems at the end of the section require the use of numerical solvers, and if they are assigned, the instructor will have to be sure that the appropriate computer facilities and guides to computing are available from the outset. Our experience is that an hour or two of instruction on a selected solver and on the use of available hardware is often enough; at least this is so for most solvers. We usually do this during the first week of the semester in order to get the point across that using a solver is an essential (and graded!) part of coursework. Students should not be discouraged from using CAS packages that they may have learned about in their calculus classes.

We have included some background material on population modeling after the problem solutions. This gives you additional material that you may want to use in your lectures. The Student Resource Manual also contains this material in case you want your students to read it. It is *not* necessary to read any of this background to do any of the problems.

Making up a problem set

Problems 1, 2(**a**)(**b**), 5, and (if you think your students are ready for problems that involve the use of a numerical solver) one of 2(**c**), 3, 4, 6, 7

Comments

In this introductory section we hit upon several of the main themes of our approach to modeling and differential equations. We construct an initial value problem that models a changing fish population. The reasoning process leading to that IVP is called *modeling* and the resulting IVP is called a *model*. In the figures we visualize the long-term behavior of solution curves of some of the differential equations used in the models. We examine the "sensitivity" of the structure of these solution curves as the initial value y_0 or the harvesting rate H is changed, although we don't yet explicitly use the term "sensitivity." This scenario is one that will be repeated many times in this book. So if you didn't grasp everything this first time around, just wait for the next opportunity. As in any field, there are a lot of terms to be defined. We have opted to include some of the definitions in this opening section, but postpone others till later.

Tips on using numerical solvers

Like any other tool, you will need a "break-in" period to use your numerical solver effectively. The documentation that comes with the solver is a good place to start, but there is no substitute for hands-on experience. Some solvers are command-line driven with a precise (and often complex) syntax for entering commands, but many solvers have a graphical interface which allows the user to select items from a menu and makes it easy to enter data and display graphs. Our advice is to get started right away in using your solver, even before you completely understand the basic concepts behind the differential equations you are studying. The reason is that the graphical displays created by the solver will give you a better feeling for solutions of ODEs and how they behave when the underlying model changes.

From time to time we will give practical tips that will help you avoid some of the frustrations in using your solver. Here is some advice you can use right now:

- Numerical solvers are designed to solve initial value problems where the ODE is in normal form. Before using your solver, write your ODE in normal form and identify the rate function. Next, examine the initial condition and identify the initial time t_0 and the initial value y_0. Finally, choose a solve-time interval for either forward or backward solving. Your solver may ask you for these items. Some solvers allow users the option to enter initial points graphically by clicking directly on the screen. Most solvers allow users to insert settings for the solver itself, but these are often set by default and not adjusted until something goes wrong.

- If your solver does not give you the option of setting your own scale on the axes at the outset, then not to worry: go ahead and let your solver set the scales automatically and then look at the result. It may be just what you wanted. If not, then at this stage your solver will let you select your own scales on the axes.

- Your solver may continue to crank away and never seem to come to an end. In that case you may want to abort and take a shorter solve-time interval. This usually happens when rate functions are very large, and so examining the rate function at the outset will give you a clue that you may run into a problem with the choice of a solve-time interval.

- Many solvers allow the user to choose the number of computed points (equally spaced in time) to display on the screen; these points are then connected with straight line segments. If the graph produced by your solver looks partly like a broken-line graph, then you should go back and increase the number of displayed points and solve again.

- Some solvers have trouble dealing with IVPs like

$$y' = ay - cy^2 - H(t), \qquad y(t_0) = y_0$$

if the harvesting function H is not continuous (for example, a step function). One way to cope with the problem is to choose a very large number of displayed points, but this doesn't work with all solvers. So you may have trouble reproducing the figures in Problems 6 and 7. In any case, your solver will have to work hard (or be clever) in order to handle on-off harvesting functions (Appendix B.1 describes these functions). Here's a way that some solvers deal with this problem: The solver computes values at the equally spaced time points displayed on the computer screen by using internal time-steps which are selected automatically in order to stay within specified error bounds (selected by default), but even this process can be defeated by discontinuous harvesting functions. The problem appears to be that internal time-steps become so large that they miss the on-off points of the harvesting function. Some solvers have a setting that allows the user to select a maximum internal step size, and if it is set low enough then the solver will be better able to "see" the on-off points of the harvesting function. Some really sophisticated solvers will handle this fairly well when on-off functions are detected.

---

1. **(a).** You can solve the IVP, $y' = ay$, $y(0) = y_0$, the easy way by replacing H by 0 in formula (9), the formula that gives the solution of initial value problem (4). So $y = y_0 e^{at}$ for $t \geq 0$. We make the restriction $t \geq 0$, not because there is any difficulty in defining the exponential for negative values of t, but because the fishing model has been formulated for future projections only. If the initial value problem (4) is valid for the past as well, then we don't need the restriction $t \geq 0$.

You could also repeat the steps given in this section for getting from IVP (4) to formula (9), replacing H by 0 every step of the way. This step-by-step derivation is an aid to understanding just why the formula $y = y_0 e^{at}$ gives the unique solution for each value of y_0. This solution formula makes sense if you remember from calculus that the exponential function e^{at} is the only function whose derivative is a multiple of itself.

(b). If the initial population y_0 is positive, then the solution formula $y = y_0 e^{at}$ tells us that the tonnage of fish grows exponentially in time since a is positive. Such rapid growth cannot be sustained for any length of time, because food and other supporting resources would soon be used up. The model is unrealistic, except possibly for short periods of time.

2. **(a).** Since the coefficient c in formula (3) is the overcrowding coefficient, the overcrowding coefficient in the fish population model IVP $y' = y - y^2/9 - 8/9$, $y(0) = y_0$, $y_0 > 0$ is $1/9$. The units for y' are tons/year, and each term in the differential equation must have the same units. Therefore, the units of c must be (ton·year)$^{-1}$, so that the death rate $y^2/9$ due to overcrowding has units of tons/year. The harvesting rate is $8/9$ tons per year.

(b). If y is to be constant, y' must be equal to 0. So the equilibrium levels are given by the values of y where y' is zero. Since $y' = y - y^2/9 - 8/9$, these are the roots of the quadratic $y - y^2/9 - 8/9$. Multiply the quadratic by -9 to obtain $y^2 - 9y + 8$, which factors to $(y - 1)(y - 8)$. The equilibrium levels are $y = 1$ and $y = 8$.

(c). Figure 2(c) was obtained by feeding the IVP into our numerical solver, using several initial values of y_0. We started with $y_0 = 1$ and then $y_0 = 8$ so that we could see the equilibrium levels. Then we started at the top with $y_0 = 15$ and worked our way down to just above $y_0 = 2$. That still left blank spaces at the right between the two lines $y = 1$ and $y = 8$. So we put solution curves into this region by taking values of t_0 and y_0 that placed the point (t_0, y_0) inside the blank region, and then we solved forward and backward in time until the solution curve reached the edges of the rectangle $0 \le t \le 10$, $0 \le y \le 15$.

We had trouble getting solution curves into the space below the line $y = 2$, because solving forward 10 units of time from some value of y_0 between 0 and 2 at time $t_0 = 0$ put us on a solution curve that was curving downward so rapidly that it exceeded our computer's capabilities, and the computer started to print on the screen rather pathetic warnings about its limitations. We got around this difficulty by choosing initial points right on the t-axis and solving backwards in time. Maybe your solver won't force you to such trickery.

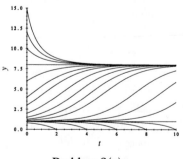

Problem 2(c).

3. **(a).** The terms in the model ODE, $y' = ay - cy^2 + R$, have the following interpretation: $y(t)$ is the fish tonnage at time t, y' is the rate of change $dy(t)/dt$ in the fish tonnage; a is the difference between the population's birth and death rates (ignoring both overcrowding and harvesting); c is the overcrowding coefficient and is measured in $(\text{ton} \cdot \text{year})^{-1}$ so that cy^2 has units of tons/year; R is the restocking rate in tons/year and is assumed to be a positive function, but not necessarily constant.

(b). See Fig. 3(b) for some solution curves for the model IVP, $y' = y - y^2/12 + 7/3$, $y(t_0) = y_0$ for various nonnegative values of t_0 and y_0. The curves at the bottom and top of the graph have positive values of t_0. In particular, the solution curves with $t_0 > 0$ and $y_0 = 0$ correspond to stocking a "dead" lake in order to rebuild the fish population.

   All of the solution curves in the first quadrant seem to approach the equilibrium population line $y = 14$. If we write the rate function $y - y^2/12 + 7/3$ as $-(1/12)(y^2 - 12y - 28) = -(1/12)(y + 2)(y - 14)$, we see that $y = 14$ is indeed an equilibrium. The other equilibrium is at $y = -2$, but this is meaningless in population terms unless we are talking about ghosts of departed fish.

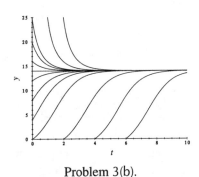

Problem 3(b).

4. **(a).** The IVP, $y' = y - y^2 + 0.3\sin(2\pi t)$, $y(0) = y_0$ models a fish population that is subject to overcrowding effects (the term $-y^2$). In the absence of overcrowding effects and of the harvesting/restocking term $0.3\sin(2\pi t)$, the population would grow exponentially [because the solution of $y' = y$, $y(0) = y_0$, is $y = y_0 e^t$]. The rate function $0.3\sin(2\pi t)$ corresponds to an annual cycle of restocking during the first half of each year [because $\sin(2\pi t)$ is positive for $0 < t < 0.5$, $1 < t < 1.5$, and so on] and harvesting during the last half [because $\sin(2\pi t)$ is negative for $0.5 < t < 1$, $1.5 < t < 2$, etc.].

   Figure 4(a) shows the rather surprising graphs of solution curves for various values of t_0 and y_0. For $t_0 = 0$, $y_0 \geq 0$, the population curves all approach some kind of sinusoidal population curve. This time-varying sinusoid seems to have period one year [which is the period of the restocking/harvesting function $0.3\sin(2\pi t)$], as we can see by measuring the distance between the tops of successive humps and comparing that distance to one unit on the time axis.

   The ODE models the restocking of an extinct population if we set $t_0 = 0, 1, \ldots, y_0 = 0$ because restocking will take place during the first six months of each year [since $\sin(2\pi t)$

is positive during that time interval]. Harvesting over the last six months when $\sin(2\pi t)$ is negative doesn't kill off the population.

(b). The solution curves starting at (t_0, y_0) and $(t_0 + 1, y_0)$ are exact translates of each other because $\sin(2\pi(t_0 + 1)) = \sin(2\pi t_0)$ for all t_0, and it is as if the clock is reset to t_0 after a year. As shown in Fig. 4(b), the solution curve of the ODE through the point $(0.5, 0)$ falls below and remains below the y-axis for $t \geq 0.5$. So it corresponds to "negative fish." This is a reminder that every mathematical model has a limited range of validity. Our model is telling us that fish are being harvested even though there aren't any fish at all!

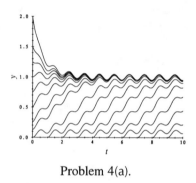 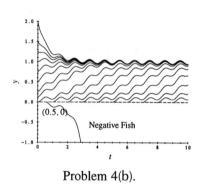

Problem 4(a).   Problem 4(b).

5. **(a).** The model IVP, $y' = ay - cy^2 - H_0 y$, $y(0) = y_0$, where a, c, H_0, and y_0 are positive constants, models a harvesting situation where the harvesting rate at time t, $H_0 y(t)$, is proportional to the tonnage $y(t)$ of fish. The term y' represents the change in fish tonnage y; a is the difference between the population's birth and death rates (ignoring overcrowding and harvesting); c is the overcrowding coefficient. The harvesting rate $H_0 y(t)$ makes a lot of sense because it means that the tonnage caught rises or falls with the amount of fish actually in the lake. This is called a constant effort model because the ratio of the harvesting rate to the population is the constant H_0, so the same amount of effort is expended whatever the size of the population. One day your nets will be full because there are a lot of fish in the lake, another day with the same effort your nets will be nearly empty because the number of fish is low.

    (b). The ODE for constant effort harvesting, $y' = ay - cy^2 - H_0 y$, can be written in the form, $y' = (a - H_0)y - cy^2$. So if the harvesting rate constant H_0 is less than a, then the harvesting may be thought of as simply decreasing the difference between the birth and death rates. The solution curves will resemble those of the unharvested population model (see, for example, Figure 1.1.3), but the equilibrium level will be lower. As in the unharvested case, the equilibrium solution line attracts all other solution curves for which $y_0 > 0$.

6. In the text we showed how a fish population can (possibly) be restored after five years of heavy harvesting by a total ban on all harvesting. A strategy that would be more acceptable to the fishing industry would be to allow harvesting to continue, but at a reduced rate. In this

problem we look at an IVP that models a particular case of this heavy harvesting/light harvesting situation: $y' = y - y^2/12 - H(t)$, where $H(t) = \begin{cases} 4 \text{ tons/year}, & 0 \le t \le 5 \\ 5/3 \text{ tons/year}, & 5 \le t \le 10 \end{cases}$,
$y(0) = y_0$, where $0 \le y \le 20$. Before feeding this IVP into a numerical solver, you should check whether your solver accepts a "piecewise-defined" function like $H(t)$. If it doesn't, maybe it will take a conditionally defined function:

$$H(t) := \text{if } 0 \le t \le 5 \text{ then } 4, \text{ else } 5/3$$

That's what we fed into our solver. For a variety of values of y_0 between 0 and 20 our solver generated Figure 6. The two dashed horizontal lines at $y = 2$ and 10 correspond to the equilibrium values of the model ODE with light harvesting, $y' = y - y^2/12 - 5/3$, since they are roots of $y - y^2/12 - 5/3 = 0$. Observe how after five years of heavy harvesting with $H = 4$, the solution curves past $t = 5$ look just like those in Figure 1.1.6. Curves that are still between the lines $y = 2$ and $y = 10$ when $t = 5$ are attracted upward toward the line $y = 10$. But population curves below $y = 2$ at $t = 5$ can't escape extinction even though the harvesting rate has been reduced from 4 to 5/3 tons per year.

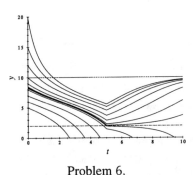

Problem 6.

7. Our solver uses the square wave function $A \operatorname{sqw}(t, d, T)$ to represent a periodic on-off function with period T, amplitude A, and "duty cycle" $d\%$. Duty cycle is an engineering term for the percentage of the total period that the function is actually on; it is assumed that the on-time is always at the beginning of each period of the cycle. For example, $4 \operatorname{sqw}(t, 50, 1)$ represents an on-off function of amplitude 4 that has value 4 for the first half (i.e., 50%) of the cycle of period 1, and value 0 for the last half. See Appendix B.1 for graphs of $\operatorname{sqw}(t, d, T)$ and other on-off functions that are widely used in engineering, science, and mathematics.

   Careful measurements of the two graphs in Figures 1.1.7 and 1.1.8 show that harvesting is "on" for the first two months of each year (since the solution curves fall during these periods) in the first graph (so $d = 100/6 = 16.66\%$), and "on" for the first eight months of the year in the second graph (since now the curves fall during the first 8/12 of each year) [so $d = (8/12)100 = 66.66\%$]. The graphs are similar, but the initial population size needed for survival starts a lot lower (somewhere slightly below $y_0 = 1.0$) in the two-month harvest season case. With an eight-month season there is a dangerously wide extinction range of values of y_0, extending all the way up to somewhere between $y_0 = 4.5$ and $y_0 = 5$.

Here's a **computer tip**: if possible, lower the maximum internal step size of your numerical solver so that your solver doesn't overlook the on-off times of the harvest season. We lowered the maximum internal step size of our solver from 1.0 to 0.05 to avoid this problem.

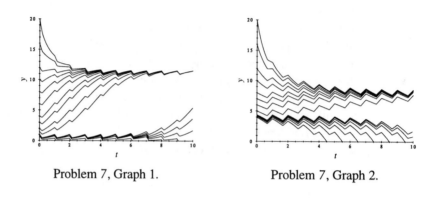

Problem 7, Graph 1.   Problem 7, Graph 2.

Background Material: Growth Processes

The U.S. Bureau of the Census frequently predicts population trends. To do this, demographers use known population data to formulate "laws" of population change. Each law is converted into a mathematical formula that can then be used to make the predictions. Since there is a good deal of uncertainty about which law best describes actual population shifts, several are derived, each proceeding from different assumptions. Similarly, ecologists make predictions about changes in the numbers of fish and animal populations, and biologists formulate laws for changing densities of bacteria growing in cultures. Let's introduce some general laws of population change, solve the corresponding mathematical models, and interpret the results in terms of the growth, stabilization, or decline of a population.

Background Material: Growth Models

Suppose that $y(t)$ denotes the population at time t of a species. The values of $y(t)$ are integers and change by integer amounts as time goes on. However, for a large population an increase by one or two over a short time span is infinitesimal relative to the total, and we may think of the population as changing continuously instead of by discrete jumps. Once we assume that $y(t)$ is continuous, we might as well smooth off any corners on the graph of $y(t)$ and assume that the function is differentiable. If we had let $y(t)$ denote the population *density* (i.e., the number per unit area or volume of habitat), the continuity and differentiability of $y(t)$ would have seemed more natural. However, we shall continue to interpret $y(t)$ as the size of the population, rather than density.

The underlying principle

$$\text{Net rate of change} = \text{Rate in} - \text{Rate out} \qquad \text{(i)}$$

applies to the changing population $y(t)$. For a population, the "Rate in" term is the sum of the birth and the immigration rates, while the "Rate out" term is the sum of the death and the emigration rates. Let's regroup the rates into an internal rate (birth minus death) and external rate (immigration minus emigration). Averaged over all classes of age, sex, and fertility, a typical individual makes a net contribution R to the internal rate of change. The internal rate of change at time t is, then, $Ry(t)$, where

$$Ry = (\text{Individual's contribution}) \times (\text{Number of individuals})$$

The *intrinsic rate coefficient* R will differ from species to species, but it always denotes the average individual's contribution to the rate. Given the rate coefficient R and the migration rate M, we see that

$$y'(t) = Ry(t) + M \qquad \text{(ii)}$$

and the study of population changes becomes the problem of solving ODE (ii) for $y(t)$.

In many cases R and M can be approximated by constants or else by simple functions of the population levels and time. If simple models suffice for observed growth processes, there is little need for more complex assumptions. In modeling, it's considered better to have fewer assumptions.

Background Material: Exponential Growth

It may safely be pronounced, therefore, that population, when unchecked, goes on doubling itself every twenty-five years.

Malthus

Thomas Robert Malthus (1766–1834) was a professor of history and political economy in England. The quotation is from "An Essay on the Principle of Population As It Affects the Future Improvement of Society." Malthus's views have had a profound effect on Western thought. Both Darwin and Wallace have said that it was reading Malthus that led them to the theory of evolution.

The Malthusian principle of explosive growth of human populations has become one of the classic laws of population change. The principle follows directly from (ii) if we set $M = 0$ and let $R = r$, a positive constant whose units are (time)$^{-1}$. ODE (ii) becomes the linear ODE $y' = ry$ which has the exponentially growing solution

$$y(t) = y_0 e^{rt}, \quad \text{for all } t \geq 0 \qquad \text{(i)}$$

where y_0 is the population at the time $t = 0$. We see from (i) that the *doubling time* of a species is given by $T = (\ln 2)/r$ since if $y(t + T) = 2y(t)$, then

$$y_0 e^{r(t+T)} = 2 y_0 e^{rt}$$

$$e^{rT} = 2$$

$$T = \frac{1}{r} \ln 2$$

Note the close connection with the half-life of a radioactive element (see Section 1.5).

Malthus claimed a doubling time of 25 years for the human population, which implies that the corresponding rate coefficient $r = (1/T) \ln 2 = (1/25) \ln 2 \cong 0.02777$. Let's use this information to calculate the annual percentage increase in a population with a 25 year doubling time. The solution formula (i) implies that

$$y(t+1)/y(t) = \frac{y_0 e^{r(t+1)}}{y_0 e^{rt}} = e^r$$

and so Malthus's value of $r = 0.02777$ gives us that $y(t+1)/y(t) \approx 1.0282$. This corresponds to a 2.8% annual increase in population. Malthus's figure for r is too high for our late-twentieth-century world. However, individual countries, for example, Mexico and Sri Lanka, have intrinsic rate coefficients which exceed 0.02777 and may be as high as 0.033 with a corresponding doubling time of 21 years. You may want to use an Almanac to calculate the intrinsic rate coefficient for the U.S.A.

Background Material: Logistic Growth

The positive checks to population are extremely various and include ... all unwholesome occupations, severe labor and exposure to the seasons, extreme poverty, bad nursing of children, great towns, excesses of all kinds, the whole train of common diseases and epidemics, wars, plague, and famine.

Malthus

The unbridled growth of a population (as predicted by the simple Malthusian law of exponential increase) cannot continue forever. Malthus claimed that resources grow at most arithmetically, i.e., the net increase in resources each year does not exceed a fixed constant. An exponential increase in the size of a population must soon outstrip available resources. The resulting hardships would surely put a damper on growth.

The simplest way to model restricted growth with no net migration is to account for overcrowding by requiring the rate coefficient R to have the form $r_0 - r_1 y$, where r_0 and r_1 are positive constants. Since $r_0 - r_1 y$ is negative when y is large, $y' = (r_0 - r_1 y)y$ is negative and the population declines. It is customary to write the rate coefficient as $r(1 - y/K)$, where r and K are positive constants, rather than as $r_0 - r_1 y$. We then have the *logistic equation* with initial condition,

$$y' = r\left(1 - \frac{y}{K}\right)y, \qquad y(0) = y_0 \tag{i}$$

Observe that $y(t) = 0$ and $y(t) = K$ are constant solutions of the logistic equation, the so-called *equilibrium* solutions. The coefficient r is the *logistic rate coefficient* and K is the *carrying capacity* of the species.

The logistic equation and its solutions were introduced in the 1840s by the Belgian statistician Pierre-François Verhulst (1804–1849). He predicted that the population of Belgium would eventually level off at 9,500,000. The estimated 1994 population of a little more than 10,000,000 is remarkably close to the predicted value.

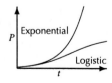

Logistic laws are well suited to laboratory or other isolated populations for which there is reason to believe that K and r are constants. Beginning with the experiments of the Soviet biologist G.F. Gauze in the early 1930s, there have been numerous experiments with colonies of protozoa growing under controlled laboratory conditions. The results of these experiments generally confirm the logistic model. See J.H. Vandermeer, "The Competitive Structure of Communities: An Experimental Approach with Protozoa," *Ecology* **50** (1969), pp. 362–371.

1.2 Visualizing Solution Curves

Suggestions for preparing a lecture

Topics: The connection between the nature of the rate function $f(t, y)$ and the behavior of solution curves of the ODE, $y' = f(t, y)$, direction fields, nullclines, equilibrium solutions; when to expect unique solutions of IVPs, and why uniqueness implies that solution curves fill up regions but don't touch one another; zooming to separate solution curves that seem to touch when viewed on a computer screen.

Remarks: Since this is all about visualizing solution curves, overheads or projected computer images are useful. Now is when the instructor will start to get across the point that you can tell something about the behavior of solution curves of $y' = f(t, y)$ without having any solution formulas. The connection with calculus and the geometric interpretation of the derivative is so strong that sometimes we as faculty think that it is obvious and doesn't need to be discussed. But for students all of this may seem quite new, and so it is worthwhile making the point several times in several different ways with several examples: the sign of f at a point tells you whether the solution curve through the point is rising or falling. Although we don't do so in this section, the instructor may also want to point out that you can even tell where solution curves curve upward or downward from the ODE: this follows from differentiating $y'(t) = f(t, y(t))$ to obtain

$$y''(t) = \partial f/\partial t + \partial f/\partial y \cdot \partial y/\partial t = \partial f/\partial t + \partial f/\partial y \cdot f$$

For example, if $y' = 3y - t$ then y'' is positive and solution curves curve upward wherever

$$y'' = \partial(3y - t)/\partial t + [\partial(3y - t)/\partial y][dy/dt]$$

$$= -1 + 3(3y - t) = -1 + 9y - 3t$$

is positive, that is, where $y > (1 + 3t)/9$, and so above the line $y = 1/9 + t/3$.

Now is also a good time to bring in the existence and uniqueness of the solution of an initial value problem where the functions f and $\partial f/\partial y$ are continuous, but without making much fuss about it (the fuss is in Chapter 2 and Appendix A). The students should begin to understand that uniqueness means that solution curves don't touch or cross, while existence means that there is a solution curve through each point of a rectangle where f and $\partial f/\partial y$ are continuous, and that solution curves don't die in the rectangle, but extend from edge to edge. These geometric features are not evident and require argument and proof (given in part in Chapter 2 and Appendix A), but the computer pictures of solution curves strongly suggest the validity of these claims. It is also helpful to say something about why a zoom on any part of most graphs of most solution curves of most first-order ODEs will show parallel lines, while the corresponding direction field shows parallel segments. The reason is that the values of the continuous rate function $f(t, y)$ don't change much if t and y vary over short ranges, and so slopes are approximately constant. Another reason to zoom is to separate apparently touching solution curves, and to note that "data compression" or "data expansion" often occurs in differential equations and needs to be watched for.

One final comment: pictures aren't proofs, but they are suggestive. Whenever we think it's appropriate to prove something, we either do so, put the proof in the Instructor's Resource Manual and in the Student Resource Manual, or else describe in words the main points of a formal proof. We have a lot of figures in this book (we think that most of them are reasonably accurate, even though they just show computer approximations), and we sometimes forget to remind the reader that a picture is not a proof. It may be a good idea to repeat that mantra once a week in lecture, but not too loudly because we do want students to use and to trust (usually) their numerical solver.

Making up a problem set

One or two parts of 2, one part of 3, one or two parts of 4, 5 or 6.

Comments

In this section we connect the form of the rate function $f(t, y)$ in the ODE $y' = f(t, y)$ to the behavior of a solution curve. For example, if $f(t, y)$ is positive at a point (t, y), then the solution curve through that point has a positive slope and is rising. The nullclines in the ty-plane divide the regions where solution curves rise from regions where they fall, so we discuss how to find nullclines by solving $f(t, y) = 0$. We discuss several other ways to visualize the behavior of solution curves.

1. The IVP is $y' = y - y^2/12 - H(t)$, $y(0) = y_0$, where

$$H(t) = \begin{cases} 4, & 0 \le t \le 5 \\ 0, & 5 \le t \le 10 \end{cases}$$

and y_0 lies between 0 and 20. The simplest way to find the smallest value of y_0 such that the population recovers after the harvesting stops is to set $y(5) = 0$, solve *backward* in t until $t = 0$, and then estimate the value of $y(0)$ from the graph of the approximate solution. See Fig. 1 where the points $(5, 0)$ and $(0, y(0))$ are specially marked. From the graph, $y(0) \approx 7.45$.

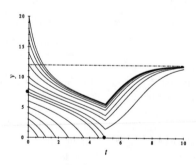

Problem 1.

2. The zeros of $f(y)$ are the equilibrium solutions of $y' = f(y)$. If f is a continuous function of y and if y_1 and y_2 are consecutive equilibrium solutions, then the sign of $f(y)$, where y is any number between y_1 and y_2, will tell us whether the solution curves are rising as t

increases in the band between y_1 and y_2 (if the sign is $+$), or falling (the sign is $-$). We describe solution behavior in words, and then back up our words with graphs of direction fields and solution curves produced by a numerical solver. In each figure we include the equilibrium solution curves (horizontal lines) and representative solution curves above and below the equilibrium lines.

(a). Here $y' = f(y) = y^2 - 11y + 10 = (y - 1)(y - 10)$. So $y_1 = 1$ and $y_2 = 10$ are the equilibrium solutions. Pick any value of y, say $y = 2$, between y_1 and y_2 and determine the sign of f: $f(2) = 4 - 22 + 10 = -8$. Since the sign is negative, the solution curves must fall as t increases in the horizontal strip in the ty-plane between y_1 and y_2. Outside this strip the value of f is always positive, e.g., $f(0) = +10$ and $f(15) = 225 - 165 + 10 = +70$, and so the solution curves are rising outside the band. See Fig. 2(a).

(b). We have $y' = f(y) = |y| - y^2 = \begin{cases} -y - y^2, & \text{if } y \le 0 \\ y - y^2, & \text{if } y \ge 0 \end{cases}$. So if $y \le 0$, then the equilibrium solutions are $y_1 = -1$, $y_2 = 0$. If $y \ge 0$, we get $y_2 = 0$ (again) and $y_3 = +1$ as equilibrium solutions. Let's look at the signs of $f(y)$ below, between, and above the equilibrium solutions. We choose any values of y in these regions and look at the sign of $f(y)$: $f(-2) = |-2| - 4 = 2 - 4 = -2$, $f(-1/2) = |-1/2| - 1/4 = +1/4$, $f(+1/2) = |+1/2| - 1/4 = +1/4$, $f(2) = |2| - 4 = -2$. Solution curves fall in the region of the ty-plane defined by $y < -1$ and in the region defined by $y > +1$ since $f(y)$ is negative in these regions. The curves rise in the horizontal strip defined by $-1 < y < 0$, and also in the strip defined by $0 < y < 1$, since $f(y)$ is positive in those regions. See Fig. 2(b).

(c). Here $y' = f(y) = \sin(2\pi y/(1 + y^2))$. Since $\sin(A)$ is zero if and only if $A = n\pi$, $n = 0, \pm 1, \pm 2, \ldots$, we see that the equilibrium solutions are given by $2\pi y/(1 + y^2) = n\pi$. Cross multiplying by $1 + y^2$, and rearranging, we see that the equilibrium solutions are given by $y_1 = 0$ (corresponding to $n = 0$) and for $n \neq 0$ by the roots of the quadratic, $n\pi y^2 - 2\pi y + n\pi = n\pi(y^2 - 2y/n + 1)$. The roots are

$$y = \frac{1}{n} \pm \frac{1}{2}\sqrt{\frac{4}{n^2} - 4}$$

If $n = -1$, then $y_2 = -1$ is an equilibrium solution, while if $n = +1$, $y_3 = +1$ is also an equilibrium solution. If $|n| > 1$, then the roots are complex and of no use to us since we are only looking at real-valued solutions. Since $f(-2) = \sin(-4\pi/5) < 0$, curves fall in the ty-region below $y_2 = -1$. Since $f(-1/2) = \sin(-4\pi/5) < 0$, curves also fall in the band $-1 < y < 0$. But curves rise in the band, $0 < y < +1$, and in the region $y > +1$ because, for example, $f(1/2) = f(2) = \sin(4\pi/5) > 0$. See Fig. 2(c).

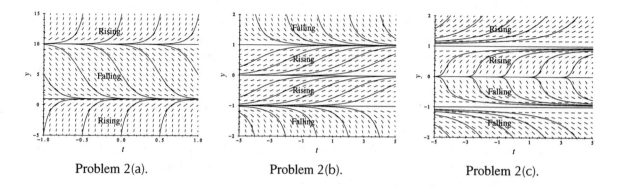

Problem 2(a).   Problem 2(b).   Problem 2(c).

3. The regions in R where a solution curve rises are precisely the regions where $y' > 0$. To find these regions, first rewrite each ODE in the form $y' = f(t, y)$. Then set $f(t, y) = 0$ so that we can find the equations of some parts of the boundary curve of the region where f is positive (i.e., where y' is positive). These boundary curves are the nullclines. In each graph we use a dashed line for the nullcline that defines the boundary.

(a). See Fig. 3(a). The region of rising solution curves for the ODE, $y' = 1 - y$, is the region below the horizontal nullcline line $y = 1$, (i.e., the region described by $y < 1$). Here the nullcline coincides with the equilibrium solution curve $y = 1$, so you can't see the dashed line.

(b). See Fig. 3(b). The region of rising solution curves for the ODE, $y' = t - y$, is the region below the line $y = t$, the region described by $y < t$.

(c). See Fig. 3(c). The region of rising solution curves for the ODE, $y' = 1 + t - y$, is the region below the nullcline line, $y = 1 + t$, the region described by $y < 1 + t$.

(d). See Fig. 3(d). Solution curves rise for the ODE, $y' = \sin(3t) - y$, wherever $\sin(3t) > y$. So curves rise below the nullcline, $y = \sin(3t)$.

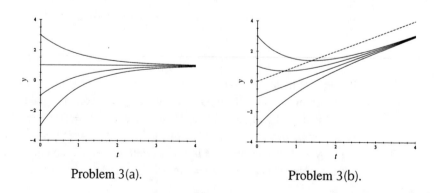

Problem 3(a).   Problem 3(b).

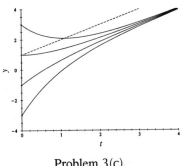

Problem 3(c).

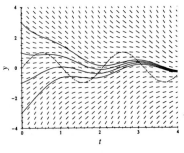

Problem 3(d).

4. See Figs. 4(a), 4(b), 4(c), 4(d), 4(e), and 4(f). To plot the nullclines (dashed) for an ODE $y' = f(t, y)$, we have graphed the curves $f(t, y) = 0$. Solution curves cross nullclines horizontally. Some of the nullclines don't enter the rectangle $|t| \leq 2$, $|y| \leq 3$. The respective nullclines are defined below for each of the ODEs. In every case the rate function $f(t, y)$ and its partial derivative $\partial f / \partial y$ is continuous for all t and y because each involves just polynomials or simple sines and cosines in t and y. So solution curves extend from edge to edge of any rectangle in the ty-plane.

(a). The nullcline for $y' = -y + 3t$ is defined by $y = 3t$. Here $f = -y + 3t$ and $\partial f / \partial y = -1$, both of which are continuous.

(b). The nullcline for $y' = y + \cos t$ is defined by $y = -\cos t$. Here $f = y + \cos t$ and $\partial f / \partial y = 1$, both of which are continuous.

(c). Since $\sin(A) = 0$ if and only if $A = n\pi$, where $n = 0, \pm 1, \pm 2, \ldots$, the nullclines for $y' = \sin(ty)$ are defined by $\sin(ty) = 0$, i.e., $ty = n\pi$, $n = 0, \pm 1, \pm 2, \ldots$. The only nullclines visible in the rectangle $|t| \leq 2$, $|y| \leq 3$ are the t and y axes (i.e., the lines $y = 0$ and $t = 0$ defined by $ty = 0$) and the hyperbolas defined by $ty = \pi$ (lower left and upper right) and $ty = -\pi$ (upper left and lower right). The nullcline $y = 0$ is also an equilibrium solution. Here $f = \sin(ty)$ and $\partial f / \partial y = t \cos(ty)$, both of which are continuous.

(d). The nullclines for $y' = \sin t \sin y$ are the horizontal and vertical lines defined by $\sin t \sin y = 0$, i.e., $t = n\pi$ or $y = k\pi$, $n, k = 0, \pm 1, \pm 2, \ldots$. The only visible nullclines in the rectangle are the two axes. The line $y = 0$ is both a nullcline and an equilibrium line. In this case $f = \sin t \sin y$ and $\partial f / \partial y = \sin t \cos y$, both of which are continuous.

(e). The nullclines for $y' = \sin t + \sin y$ are the slanted lines of slope $+1$ or -1 defined by $\sin y = -\sin t$, i.e., $y = \pm t + (2n + 1)\pi$, $n = 0, \pm 1, \pm 2, \ldots$. The only visible nullclines in the rectangle are the lines $y = -t$, $y = t + \pi$, $y = t - \pi$. Here $f = \sin t + \sin y$ and $\partial f / \partial y = \cos y$, both of which are continuous.

(f). The nullclines for $y' = \sin(t + y)$ are the lines of slope -1 defined by $\sin(t + y) = 0$, i.e., $t + y = n\pi$, or $y = n\pi - t$, $n = 0, \pm 1, \pm 2, \ldots$. The only nullclines in the rectangle are the lines $y = -t$, $y = 1 - t$, and $y = -1 - t$. In this case, $f = \sin(t + y)$ and $\partial f / \partial y = \cos(t + y)$, both of which are continuous.

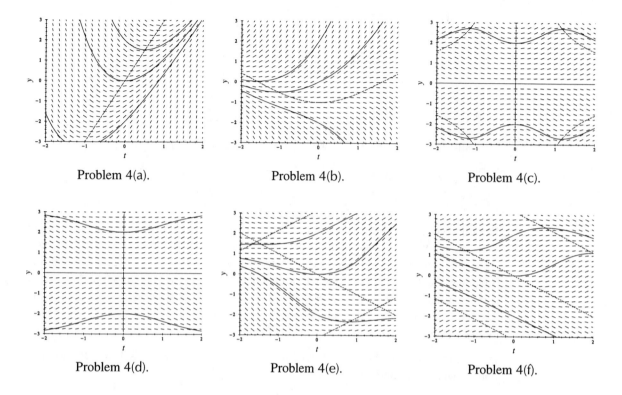

Problem 4(a).   Problem 4(b).   Problem 4(c).

Problem 4(d).   Problem 4(e).   Problem 4(f).

5. Figure (5), Graph 1 shows the solution curves of the IVP, $y' = -2y + 3e^t$, $y(0) = -5$, $-4, \ldots, 4, 5$, in the rectangle $0 \le t \le 3$, $-5 \le y \le 20$. It appears that the solution curves converge to one another as t increases and eventually curve upwards as if the corresponding solutions are exponentially increasing. The apparent merging of solution curves at the upper right of Fig. (5), Graph 1 is an illusion caused by the finite number of pixels on a computer screen. In Fig. (5), Graph 2 we zoom in on the rectangle $2.996 \le t \le 3.000$, $20.06 \le y \le 20.09$, and we see that the solution curves don't meet. Under the zoom microscope the arcs of the solution curves look like parallel lines. This is no surprise because over very short ranges for t and for y the rate function $-2y + 3e^t$ doesn't change much, and so the slopes of the curves are nearly identical.

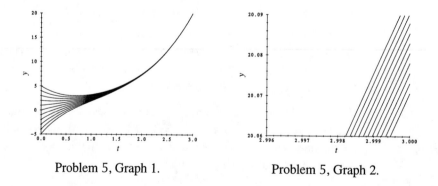

Problem 5, Graph 1.   Problem 5, Graph 2.

6.   **(a).** The nonlinear ODE $2yy' = -1$ has normal form $y' = -1/(2y)$ and we see that the rate function $f = -1/(2y)$ is discontinuous at $y = 0$. Observe that y' is positive where y is positive, negative where y is negative. So the line $y = 0$ bounds the region $y > 0$ where solution curves rise, but that line is *not* a nullcline. In fact, slopes are infinite there, solution curves can't move through the line, and numerical solvers will (usually) send error messages when you try to do so. Our solver comes to a dead stop as the solution curve through the initial point reaches the line $y = 0$. See Fig. 6(a) for a graph. Since f is discontinuous on the line $y = 0$ we should not necessarily expect solution curves to reach from one edge to another of a ty-rectangle that contains part of that line.

(b). The normal form of the nonlinear ODE $2yy' = -t$ is $y' = -t/(2y)$, and $f = -t/(2y)$ is discontinuous on the line $y = 0$. The line $t = 0$ is a nullcline, but $y = 0$ is *not* a nullcline, even though solution curves change from rising to falling as one moves across the line. As shown in Fig. 6(b), the solution curve through the point $(-2, 1.5)$ never gets near the discontinuity line of f, $y = 0$, and so this curve does extend from edge to edge of the solver screen, $|t| \le 2$, $-1 \le y \le 2$. The solution curve through $(-2, 1.4)$ approaches the line $y = 0$. It stops completely at the line with our solver sending us urgent error messages because the slope becomes infinite. We should not necessarily expect solution curves to extend from edge to edge of a rectangle that includes a portion of the discontinuity line $y = 0$.

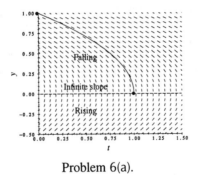
Problem 6(a).

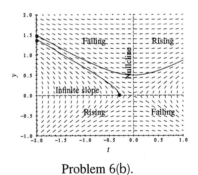
Problem 6(b).

7.   We don't usually say much about group problems, but occasionally we do. If you are looking for a rate function $f(t, y)$ that forces solution curves of $y' = f(t, y)$ to do a lot of twisting, or turning, or explosive growth, try a trigonometric function in t or y, or a polynomial in y. For example, use a numerical solver and take a look at the solution curves of $y' = y^3 \sin t$, or of $y' = 3y \sin y - t \sin t$, in the rectangle $0 \le t \le 10$, $|y| < 5$.

1.3 The Search for Solution Formulas

Suggestions for preparing a lecture

Topics: The big topic here is the integrating factor recipe for finding the general solution of a first-order linear ODE in normal form. The Antiderivative Theorem 1.3.1 supports the recipe (as well as many other methods). The needed terms can be defined as appropriate.

Remarks: This section and the next should be considered together—but it wouldn't be a good idea to cover everything in the two sections in one lecture.

Making up a problem set

Parts of 1–4, 5(**d**) or (**e**), parts of 6, several parts of 7–12.

Comments

Most methods for finding solution formulas for first-order ODEs are based on "preparing" an ODE for application of Antiderivative Theorem 1.3.1. That theorem is really "The Mother of All Solution Methods," even though it is just a slight reformulation of the Fundamental Theorem of Calculus. It is the basis for most techniques for solving first-order ODEs in closed form.

We prepare a first-order linear ODE in normal form for antidifferentiation by multiplying by an integrating factor. The separable ODE, $N(y)y' + M(x) = 0$, is already in shape for antidifferentiation (Section 1.6), as are the exact ODEs of Problem 6 in Section 1.6. The changes of variables in the last section of the chapter are applied to convert an ODE to a form that can then be prepared for antidifferentiation. So, it isn't surprising that antidifferentiation lies at the heart of finding solution formulas for ODEs.

First-order linear ODEs are the most important in applications, and that's why we tackle them first. The integrating factor technique is straightforward, but it takes practice to make the steps of the solution recipe become natural. In this section, the two antiderivatives needed for the technique can be evaluated in terms of familiar functions. Section 1.4 considers some linear ODEs for which the antiderivatives can't be easily calculated. That section takes up other features of linearity that we don't have room to do here.

---

1.   (**a**). The ODE, $y' = \sin t - t^3 y$, is first order (due to the term y'), linear in y, and has normal linear form, $y' + t^3 y = \sin t$.

(**b**). The ODE, $e^t y' + 3y = t^2$, is first order (due to the term y'), linear in y, and has normal linear form, $y' + 3e^{-t} y = t^2 e^{-t}$.

(**c**). The ODE, $(t^2 + y^2)^{1/2} = y' + t$, is first order (due to the term y') and nonlinear in y [due to the term $(t^2 + y^2)^{1/2}$].

(**d**). The ODE, $dt/dy = 1/(t^2 - ty)$, can be rewritten as $y' = dy/dt = 1/(dt/dy) = t^2 - ty$. This ODE is first order (due to y'), linear in y, and has normal linear form, $y' + ty = t^2$.

(**e**). The ODE, $y'' - t^2 y = \sin t + (y')^2$, is second order (due to y''), and nonlinear in y [due to $(y')^2$].

(**f**). The ODE, $y' = (1 + t^2)y'' - \cos t$, is second order (due to y''), linear in y, and has normal linear form, $y'' - (1 + t^2)^{-1}y' = (1 + t^2)^{-1}\cos t$.

(g). The ODE, $e^t y'' + (\sin t)y' + 3y = 5e^t$, is second order (due to y''), linear in y, and has normal linear form, $y'' + e^{-t}(\sin t)y' + 3e^{-t}y = 5$.

(h). The ODE, $(y''')^2 = y^5$, is third order (due to y'''), and nonlinear in y [due to $(y''')^2$ and also to y^5].

2. Replace y by e^{rt} in the ODE, and determine the values of the constant r for which the equation holds for all t.

   (a). Let $y = e^{rt}$. Then $y' + 3y$ becomes $re^{rt} + 3e^{rt} = (r+3)e^{rt}$. So, $y' + 3y = 0$ for all t if and only if $r = -3$ since e^{rt} is never zero. So $y = e^{-3t}$ is a solution.

   (b). Let $y = e^{rt}$. Then $y'' + 5y' + 6y$ becomes $r^2 e^{rt} + 5re^{rt} + 6e^{rt} = (r^2 + 5r + 6)e^{rt}$, which is zero for all t if and only if r is a root of $r^2 + 5r + 6 = (r+2)(r+3)$. We must have $r = -2$ or $r = -3$. So $y = e^{rt}$ is a solution of $y'' + 5y' + 6y = 0$ if and only if $r = -2$ or $r = -3$.

   (c). $y^{(5)} - 3y^{(3)} + 2y'$ becomes $(r^5 - 3r^3 + 2r)e^{rt}$, which is zero for all t if and only if r is a root of $r^5 - 3r^3 + 2r = r(r^4 - 3r^2 + 2) = r(r^2 - 2)(r^2 - 1)$. So, the values are $r = 0$, $\sqrt{2}, -\sqrt{2}, 1$, or -1. So $y = e^{rt}$ is a solution of $y^{(5)} - 3y^{(3)} + 2y' = 0$ if and only if r is one of the five values just listed.

   (d). $y'' + 2y' + 2y$ becomes $(r^2 + 2r + 2)e^{rt}$ if $y = e^{rt}$. The equation $(r^2 + 2r + 2)e^{rt} = e^{-t}$ for all t if and only if $r = -1$.

3. Replace y by rt^3 in the ODE, and find the values of r for which the equation holds for all t.

   (a). $t^2 y'' + 6ty' + 5y = 0$ becomes $t^2(6rt) + 6t(3rt^2) + 5(rt^3) = 29rt^3 = 0$, which holds for all t if and only if $r = 0$.

   (b). Since the ODE is $t^2 y'' + 6ty' + 5y = 2t^3$, the constant r must be chosen so that $t^2(6rt) + 6t(3rt^2) + 5(rt^3) = 2t^3$ for all t, that is, $29rt^3 = 2t^3$, so $r = 2/29$.

   (c). Inserting $y = rt^3$ into the ODE $t^4 y' = y^2$, we have $3rt^6 = r^2 t^6$, and so $3r = r^2$. So, $r = 0, 3$, and $y = 0$, $y = 3t^3$ are solutions.

4. Plug $y = t^r$ into the ODE and determine the values of r for which the equation holds.

   (a). $t^2 y'' + 4ty' + y = 0$ becomes $[r(r-1) + 4r + 1]t^r = 0$. The expression in the brackets is $r^2 + 3r + 1$, whose roots, by the quadratic formula, are $(-3 \pm \sqrt{5})/2$, the two values of r for which $y = t^r$ is a solution of the ODE.

   (b). $t^4 y^{(4)} + 7t^3 y''' + 3t^2 y'' - 6ty' + 6y = 0$ becomes

   $$[r(r-1)(r-2)(r-3) + 7r(r-1)(r-2) + 3r(r-1) - 6r + 6]t^r = 0$$

   The terms in the bracketed expression may be multiplied out to obtain $r^4 + r^3 - 7r^2 - r + 6$ which factors to $(r-1)(r+1)(r-2)(r+3)$. So, $r = 1, -1, 2$, or -3 are the values of r for which $y = t^r$ is a solution of the given ODE.

5. In **(a)–(f)** the Antiderivative Theorem 1.3.1 is implemented by antidifferentiating each side of the ODE. C, C_1, C_2, and C_3 are arbitrary constants in the solutions below.

   (a). Antidifferentiating both sides of $y' = 5 + \cos t$ gives $y = 5t + \sin t + C$.

(b). Antidifferentiating both sides gives $y = t^3/3 + t^2/2 - e^{-t} + C$.

(c). Antidifferentiating both sides gives $y = \int e^{-t} \cos 2t\, dt$. $\int e^{-t} \cos 2t\, dt$ is of the form $\int e^{at} \cos bt\, dt$. So from Table 1.3.1 we see that the solution is $y = (e^{-t}/5)(-\cos 2t + 2\sin 2t) + C$.

(d). Antidifferentiating $y'' = 0$ twice, we obtain $y = C_1 t + C_2$.

(e). Antidifferentiating $y'' = \sin t$ twice, we obtain $y = -\sin t + C_1 t + C_2$. Applying the initial conditions, we get the equations $C_2 = 0$ and $C_1 = 2$, so $y = 2t - \sin t$.

(f). Antidifferentiating $y''' = 2$ three times, $y = t^3/3 + C_1 t^2 + C_2 t + C_3$.

(g). Multiplying the ODE $y'' + y' = e^t$ by e^t, we obtain $(e^t y')' = e^{2t}$. Antidifferentiating, we obtain $e^t y' = e^{2t}/2 + C_1$. So, $y' = e^t/2 + C_1 e^{-t}$. Antidifferentiating again, $y = e^t/2 - C_1 e^{-t} + C_2$.

6. **(a).** Following the hint, multiply by e^t and observe that the ODE, $y' + y = 1$, becomes $e^t(y' + y) = (ye^t)' = e^t$. So, $ye^t = e^t + C$, and $y = 1 + Ce^{-t}$, C any constant.

   (b). Multiplying by e^t as in **(a)**, $y' + y = t$ becomes $(ye^t)' = te^t$. So, $ye^t = \int^t se^s\, ds = (t-1)e^t + C$. Multiply by e^{-t} to get $y = (t-1) + Ce^{-t}$, C any constant.

   (c). Multiplying by e^t as in **(a)**, $y' + y = t + 1$ becomes $(ye^t)' = (t+1)e^t$. So, $ye^t = \int^t (se^s + e^s)\,ds = (t-1)e^t + e^t + C = te^t + C$, and $y = t + Ce^{-t}$, C any constant.

   (d). Since $(y^2)' = 2yy'$, the ODE, $2yy' = 1$, can be written as $(y^2)' = 1$. By the Antiderivative Theorem 1.3.1, the solution is $y^2(t) = t + C$. Solving for y, $y = \pm(t + C)^{1/2}$, C any constant, $t > -C$.

   (e). The ODE, $2yy' = t$, may be rewritten as $(y^2)' = t$, so $y^2 = t^2/2 + C$ and $y = \pm\sqrt{t^2/2 + C}$, C any constant and $t^2/2 > -C$.

7. In each case the equation must first be written in normal linear form, $y' + p(t)y = q(t)$. Then the integrating factor $e^{P(t)}$, where $P(t) = \int p(t)\, dt$, is calculated and each side of the ODE is multiplied by $e^{P(t)}$ to obtain $(e^{P(t)}y)' = e^{P(t)}q$. Antidifferentiating and multiplying by $e^{-P(t)}$ gives the family of solutions $y = e^{-P(t)}C + e^{-P(t)}R(t)$, where $R(t) = \int e^{P(t)}q(t)\, dt$. C is an arbitrary constant, and t lies in an interval for which the ODE is normal and $p(t)$ and $q(t)$ are continuous. It is best to follow the steps of this recipe in each case, rather than memorize the solution formula. The use of integral tables will speed up the calculations. In each case, $-\infty < t < \infty$ unless otherwise noted.

   (a). For $y' - 2ty = t$, the integrating factor is $e^{\int -2t\, dt} = e^{-t^2}$. So, after multiplying by e^{-t^2}, we have that $e^{-t^2}(y' - 2ty) = (e^{-t^2}y)' = te^{-t^2}$. Integrating, we have that $e^{-t^2}y(t) = \int te^{-t^2}\, dt = -e^{-t^2}/2 + C$. The general solution is $y = Ce^{t^2} - 1/2$, where C is any constant and $-\infty < t < \infty$.

   (b). The integrating factor for $y' - y = e^{2t} - 1$ is e^{-t}. We obtain $e^{-t}(y' - y) = (e^{-t}y)' = e^t - e^{-t}$. Integrating, $e^{-t}y = e^t + e^{-t} + C$ and the general solution is given by $y = Ce^t + 1 + e^{2t}$.

   (c). The normal form is $y' + (\sin t)y = \sin t$, so the integrating factor is $e^{\int \sin t\, dt} = e^{-\cos t}$. We obtain $e^{-\cos t}(y' + (\sin t)y) = (e^{-\cos t}y)' = e^{-\cos t}\sin t$. Integrating, $e^{-\cos t}y = e^{-\cos t} +$

C and the general solution is $y = Ce^{\cos t} + 1$ since $\int e^{-\cos t} \sin t \, dt = e^{-\cos t}$.

(d). The normal form is $y' + (3/2)y = e^{-t}/2$, so the integrating factor is $e^{\int 3/2 \, dt} = e^{3t/2}$. We obtain $e^{3t/2}(y' + 3y/2) = (e^{3t/2}y)' = e^{t/2}/2$. Integrating, $e^{3t/2}y = e^{t/2} + C$ and the general solution is $y = Ce^{-3t/2} + e^{-t}$.

(e). The normal form is $y' + ty = t/2$, so the integrating factor is $e^{\int t \, dt} = e^{t^2/2}$. We obtain $e^{t^2/2}(y' + ty) = (e^{t^2/2}y)' = (t/2)e^{t^2/2}$. Integrating, $e^{t^2/2}y = (t/2)e^{t^2/2} + C$ and the general solution is $y = Ce^{-t^2/2} + 1/2$.

(f). The integrating factor for $y' + y = te^{-t} + 1$ is e^t. We obtain $e^t(y' + y) = (e^t y)' = t + e^t$. Integrating, $e^t y = t^2/2 + e^t + C$ and the general solution is $y = Ce^{-t} + 1 + t^2 e^{-t}/2$.

8. Each equation should first be written in normal linear form; then the integrating factor should be calculated, and the general solution determined. Use the initial data to find the constant of integration.

    (a). The ODE, $y' + y = e^{-t}$, is already in normal form. The integrating factor is e^t, so the ODE becomes $(e^t y)' = 1$. Integrating, $e^t y = t + C$, so $y = te^{-t} + Ce^{-t}$. Applying the initial condition, we have $1 = 0 + C$, $C = 1$, so the solution is $y(t) = (t+1)e^{-t}$, $-\infty < t < \infty$. As $t \to +\infty$, the factor e^{-t} forces $y(t) \to 0$.

    (b). The integrating factor for $y' + 2y = 3$ is e^{2t}, so the ODE becomes $(e^{2t}y)' = 3e^{2t}$. Integrating, $e^{2t}y = 3e^{2t}/2 + C$, so $y(t) = 3/2 + Ce^{-2t}$. Since $y(0) = -1$, $-1 = 3/2 + C$, so $C = -5/2$. The solution is $y(t) = 3/2 - 5e^{-2t}/2$, $-\infty < t < \infty$. As $t \to +\infty$, $y(t) \to 3/2$.

    (c). The integrating factor for $y' + 2ty = 2t$ is e^{t^2}, so the ODE becomes $(e^{t^2}y)' = 2te^{t^2}$. Integrating, $e^{t^2}y = e^{t^2} + C$, so $y(t) = 1 + Ce^{-t^2}$. Since $y(0) = 1$, $1 = 1 + C$, so $C = 0$. The solution is $y(t) = 1$, $-\infty < t < \infty$. As $t \to +\infty$, $y(t) \to 1$.

    (d). The integrating factor for $y' + (\cos t)y = \cos t$ is $e^{\sin t}$, so the ODE becomes $(e^{\sin t}y)' = (\cos t)e^{\sin t}$. Integrating, $e^{\sin t}y = e^{\sin t} + C$, so $y = 1 + Ce^{-\sin t}$. Since $y(\pi) = 2$, $2 = 1 + C$, so $C = 1$. The solution is $y = 1 + e^{-\sin t}$, $-\infty < t < \infty$. As $t \to +\infty$, $y(t)$ does not converge because $e^{-\sin t}$ oscillates.

9. You may either use graphics to plot the solutions given by the formulas derived in Problem 8, or you may apply a numerical solver directly to the appropriate IVP, as we did. See Figs. 9(a)–9(d) for the solution curves of the IVPs of Problem 8.

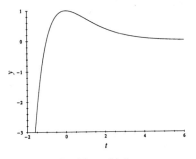

Problem 9(a).

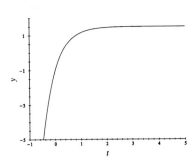

Problem 9(b).

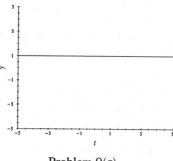

Problem 9(c).   Problem 9(d).

10.   **(a).** The normal linear form for $ty' + 2y = \sin t$ is $y' + 2t^{-1}y = t$, $t > 0$. The integrating factor is $e^{2\ln t} = t^2$, so the ODE becomes $(t^2 y)' = t^3$. Integrating, $t^2 y = t^4/4 + C$, so $y = t^2/4 + Ct^{-2}$, $t > 0$. Here, and below, C is an arbitrary constant. Note that $y = t^2/4$, all t, is a solution, even on intervals that contain $t = 0$. As $t \to 0$, $y(t) \to \pm\infty$ or zero depending on the sign of C. As $t \to +\infty$, $y(t) \to +\infty$.

(b). The normal linear form for $(3t - y) + 2ty' = 0$ is $y' - (2t)^{-1}y = -3/2$, $t > 0$. The integrating factor is $e^{-(1/2)\ln t} = t^{-1/2}$, so the ODE becomes $(t^{-1/2}y)' = -3t^{-1/2}/2$. Integrating, $t^{-1/2}y = -3t^{1/2} + C$, so $y = -3t + Ct^{1/2}$, $t > 0$. Note that although $y(t)$ is defined at $t = 0$, it has no derivative there, so $y(t)$ does not satisfy the ODE at $t = 0$. As $t \to 0$, $y(t) \to 0$. As $t \to +\infty$, $y(t) \to -\infty$.

(c). The normal linear form for $y' = (\tan t)y + t \sin 2t$ is $y' - (\tan t)y = t \sin 2t$, and we restrict t to the interval $-\pi/2 < t < \pi/2$ since $|\tan t| = \infty$ at $t = \pm\pi/2$. The integrating factor is $e^{-\int \tan t\, dt} = e^{\ln|\cos t|} = |\cos t| = \cos t$ since $\cos t > 0$ for $-\pi/2 < t < \pi/2$. Since $1/\cos t = \sec t$, the general solution is $y = \sec t \int t \sin 2t \cos t\, dt = \sec t \int 2t \sin t \cos^2 t\, dt = \sec t[-(2/3)t \cos^3 t - (2/9)\sin^3 t + (1/3)\sin t + C]$, $|t| < \pi/2$.

(d). In normal linear form, the ODE $y' - (y/t) = t^n$, $t > 0$, has integrating factor $e^{-\ln t} = 1/t$. So, $(y/t)' = t^{n-1}$. The two cases, $n = 0$ and $n \neq 0$, must be considered. If $n = 0$, $(y/t)' = 1/t$ and $y = t \ln|t| + Ct$. If $n \neq 0$, $(y/t)' = t^{n-1}$, so $y/t = t^n/n + C$ and the general solution is $y = (t^{n+1}/n) + Ct$, where $t > 0$.

$$\text{As } t \to 0^+, y \to \begin{cases} -\infty, & n < -1 \\ -1, & n = -1 \\ 0, & n > -1 \end{cases}$$

$$\text{As } t \to \infty, y \to \begin{cases} -\infty, & n < 0, C < 0 \\ -1, & n = -1, C = 0 \\ -\infty, & -1 < n < 0, C = 0 \\ 0, & n < -1, C = 0 \\ +\infty, & n < 0, C > 0 \\ +\infty, & n \geq 0 \end{cases}$$

11.   First write the ODE in normal form, then find the integrating factor and solve to get the general solution. Use the initial data to evaluate the arbitrary constant C in the general solution formula.

(a). The normal form for $ty' + 2y = \sin t$ is $y' + (2/t)y = (1/t)\sin t$, $t > 0$, and the integrating factor is $e^{\int 2t^{-1}\,dt} = e^{2\ln|t|} = t^2$. We obtain $(t^2 y)' = t\sin t$. Integrating, $t^2 y = \sin t - t\cos t + C$. So $y = (1/t^2)\sin t - (1/t)\cos t + C/t^2$. Since $y(\pi) = 1/\pi$, $1/\pi = 1/\pi + C/\pi^2$, so $C = 0$. The solution is $y = (1/t^2)\sin t - (1/t)\cos t$, $t > 0$. As $t \to +\infty$, $y(t) \to 0$.

(b). The normal form for $(\sin t)y' + (\cos t)y = 0$ is $y' + (\cot t)y = 0$, and the integrating factor is $e^{\int \cot t\,dt} = e^{\ln(\sin t)} = \sin t$. We obtain $((\sin t)y)' = 0$. Integrating, $(\sin t)y = C$, so $y = C\csc t$. Since $y(3\pi/4) = 2$, $2 = C\csc 3\pi/4 = C\sqrt{2}$, so $C = \sqrt{2}$. The solution is $y = 2^{1/2}\csc t$, $0 < t < \pi$, where the t-interval is the largest interval including $t_0 = 3\pi/4$ for which $\csc t$ is defined. As $t \to 0+$, $y(t) \to +\infty$.

(c). The integrating factor for $y' + (\cot t)y = 2\cos t$ is $e^{\int \cot t\,dt} = e^{\ln(\sin t)} = \sin t$. We obtain $((\sin t)y)' = 2\sin t\cos t$. Integrating, $(\sin t)y = \sin^2 t + C$, so $y = \sin t + C\csc t$. Since $y(\pi/2) = 3$, $3 = \sin \pi/2 + C\csc \pi/2 = 1 + C$ so $C = 2$. The solution is $y = \sin t + 2\csc t$, $0 < t < \pi$. As $t \to 0^+$, $y \to +\infty$ since $\csc t \to +\infty$.

(d). The integrating factor for $y' + 2y/t = (\cos t)/t^2$, $t > 0$, is t^2. We obtain $(t^2 y)' = \cos t$. Integrating, $t^2 y = \sin t + C$, so $y = t^{-2}\sin t + Ct^{-2}$. Since $y(\pi) = 0$, $0 = C/\pi^2$ so $C = 0$. The solution is $y = t^{-2}\sin t$, $t > 0$. As $t \to 0^+$, $y \to +\infty$ since $(\sin t)/t \to 1$ and $1/t \to +\infty$. As $t \to +\infty$, $y \to 0$.

12. You may graph the solution curves directly from the solution formulas given in Problem 11, or else do as we did and use a numerical solver. Your solver may return all sorts of error messages (ours did) when solving the IVPs of parts **(b)**, **(c)**, **(d)**. The problem is that solutions become unbounded at endpoints of the interval. This will not usually affect computed points at a distance from the endpoints. See Figs. 12(a)–12(d).

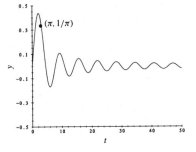

Problem 12(a).

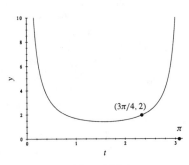

Problem 12(b).

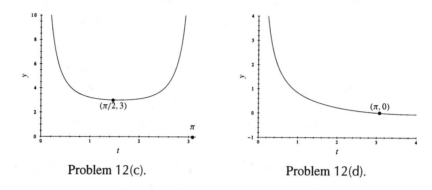

Problem 12(c).   Problem 12(d).

13.   Group problem.

1.4   Modeling with Linear ODEs

Suggestions for preparing a lecture

Topics: Spend a little time on the structure of the solution set of a linear ODE, show how and when good guessing works, but devote most of the lecture to the applications of the response interpretation and the Balance Law to mixture problems.

Remarks: We like to cover Example 1.4.3 because it gets students used to dealing with on-off rate functions and seeing their effects on solution curves (sharp corners).

If you want to take two lectures on this material (sometimes we do), you might want to take up Newton's Law of Cooling (Problem 14).

Making up a problem set

One of 1–3 [2(**e**) alone would be a good—although hard—problem], one or two parts from 4–6, one part from 7–11, one part from 14–16.

Comments

We like to focus on the total response formula (Theorem 1.4.5) and the Balance Law, because those two items have such widespread applications and interpretations. We have given some of them in this section in the context of mixture problems, and we will include others later in the book.

Other than doing a "good guessing" example, we don't make a big deal about the structure of the solution set of a first order linear ODE because it will come up again in Chapter 3 when we treat second-order linear ODEs, and again in Chapter 7, where linear systems are handled. So why have we put the structure stuff here at all? Mostly for the readers who want to see and understand the theory of a linear ODE, and because it does serve as a neat way to look at the solution formula produced by the recipe of the last section for constructing a general solution.

We always do some models with step function inputs because these inputs are so common in applications, numerical solvers (most of them) can handle these functions, and they are easy to understand and appreciate. As step functions show, a computer can easily handle some things that a solution formula has trouble with.

1. Let's denote the amount of waste (measured in gallons) in the tank at time t (measured in minutes) by $y(t)$. By the Balance Law, we have

$$y' = \text{Rate in} - \text{Rate out}$$

$$= r_{in} - \frac{y(t)}{1000} \cdot r_{out}$$

where r_{in} is the inflow rate of waste (in gal/min), $y/1000$ is the concentration of waste in the tank (in gallons of waste per gallon of mixture), and r_{out} is the outflow rate of contaminated water (in gal/min). For parts **(a)** and **(b)**, $r_{in} = 1 = r_{out}$. At time 0, the water in the tank is clean, so $y(0) = 0$.

(a). The IVP is

$$y' = 1 - \frac{y}{1000} \cdot 1 = 1 - 0.001y, \qquad y(0) = 0$$

In normal linear form the ODE is

$$y' + 0.001y = 1$$

so the integrating factor is $e^{0.001t}$. We obtain $(e^{0.001t}y)' = e^{0.001t}$. Integrating, $e^{0.001t}y = 1000e^{0.001t} + C$, so solutions of the ODE have the form

$$y = Ce^{-0.001t} + 1000$$

Imposing the initial condition $y(0) = 0$, we see that $C = -1000$, so

$$y(t) = 1000(1 - e^{-0.001t}) \text{ gallons}$$

As $t \to +\infty$, $e^{-0.001t} \to 0$, so the pollution becomes total and $y(t) \to 1000$ gallons.

(b). The concentration of waste in the tank at time t is [using the formula for $y(t)$ from **(a)**]

$$\frac{y(t)}{1000} = 1 - e^{-0.001t}$$

The concentration is 20% at that time t for which $1 - e^{-t/1000} = 0.2$, (i.e., $t = -1000 \ln 0.8 \approx$ 223 minutes).

(c). Let's model the sudden change in the waste inflow rate from 1 gal/min to 0 gal/min (the change occurs at $t = 60$ min). So

$$r_{in} = \begin{cases} 1, & 0 \leq t \leq 60 \\ 0, & t > 60 \end{cases} = \text{step}(60 - t), \quad t \geq 0$$

The modeling IVP then is

$$y' = \text{step}(60 - t) - \frac{y}{1000} \cdot 1$$

$$= \text{step}(60 - t) - 0.001y$$

$$y(0) = 0$$

Applying a numerical solver, we get the graph shown in Fig. 1(c) for the solution $y(t)$.

Since there is no inflow of waste after $t = 60$ min, we expect the amount of waste in the tank to decline exponentially to 0 as $t \to +\infty$. We see from the figure that this exponential decay is pretty slow, and we see from the answer to part **(a)** that this is expected because of the term $e^{-0.001t}$.

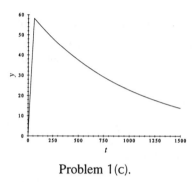

Problem 1(c).

2. **(a)**. The Balance Law applied to the amount of pollutant $y(t)$ in the tank, $t \geq 0$, gives us

$$y' = (c_0 \text{lbs/gal})(1\text{gal/min}) - (y/10)(1\text{gal/min}) = c_0 - 0.1y$$

Using the integrating factor $e^{0.1t}$ for the linear ODE, $y' + 0.1y = c_0$, we obtain $(e^{0.1t}y)' = c_0 e^{0.1t}$. Integrating, $e^{0/1t}y = 10c_0 e^{0.1t} + C$. The solution is $y = 10c_0 + Ce^{-t/10}$, where C is an arbitrary constant. Using the initial condition $y(0) = 10$, $C = 10 - 10c_0$, and we see that $y = 10c_0 + 10(1 - c_0)e^{-t/10}$. The limiting value as $t \to +\infty$ is always $10c_0$.

(b). See Fig. 2(b) for graphs of solutions $y(t)$ for various values of c_0 ranging from 0.1 lb/gal (giving the lowest curve) to 2 lb/gal (the top curve). If $c_0 = 1$ lb/gal, then the pollutant concentration in the incoming stream and the initial concentration in the tank are identical, so the amount of pollutant in the tank stays at the initial level of 10 lbs. Whatever the value of c_0, the concentration in the tank tends to c_0 as time goes on, and the pollutant's concentration in the tank begins to approach the inflow concentration. So, the total amount in the tank approaches $10c_0$.

(c). Here we have

$$c_0(t) = \begin{cases} 1, & 0 \leq t \leq 10 \\ 0, & t > 10 \end{cases} = \text{step}(10 - t), \quad t \geq 0$$

So the IVP for $t \geq 0$ is

$$y' = c_0 \cdot 1 - 0.1y = \text{step}(10 - t) - 0.1y$$

$$y(0) = 10$$

Fig. 2(c) shows the total response $y(t)$ (solid curve), the response to the initial data [long-dashed curve, the solution curve for $y' = -0.1y$, $y(0) = 10$], and the response to the input [short-dashed curve, the solution curve for $y' = \text{step}(10 - t) - 0.1y$, $y(0) = 0$].

As $t \to +\infty$ all three responses tend to 0, as we expect because from $t = 10$ min on no pollutant enters the tank.

(d). Everything is the same as in part **(c)**, except that for $t > 10$, $c_0 = 0.5$ lbs/gal instead of 0 lbs/gal. So we have

$$c_0(t) = \begin{cases} 1, & 0 \le t \le 10 \\ 0.5, & t > 10 \end{cases} = 0.5 + 0.5 \operatorname{step}(10 - t), \quad t \ge 0$$

as we see after a certain amount of playing around with step functions. There are other ways to represent c_0; here is one alternative:

$$c_0(t) = \operatorname{step}(10 - t) + 0.5 \operatorname{step}(t - 10), \quad t \ge 0.$$

Whichever form you use, the IVP is

$$y' = c_0(t) - 0.1y, \quad y(0) = 10, \quad t \ge 0.$$

Fig. 2(d) shows the total response (solid), the response to the initial data (long dashes), and the response to the input (short dashes), corresponding, respectively, to the solution of the above IVP and the two IVPs

$$y' = -0.1y, \quad y(0) = 10 \quad \text{(response to the initial data)}$$

$$y' = c_0(t) - 0.1y, \quad y(0) = 0 \quad \text{(response to the input)}$$

As $t \to +\infty$, we expect the total response and the response to the input to tend to 5 lbs since the amount of pollutant in the inflow stream is 0.5 lbs/gal and the total amount of polluted water in the tank is 10 gal. On the other hand, the response to the initial amount of 10 lbs of pollutant in the tank tends to 0 as $t \to +\infty$ because this response ignores any amount of pollutant inflow, and only models the initial amount and the outflow.

(e). Everything is as in part **(c)**, except that now the concentration in the inflow stream is given by

$$c_0 = \begin{cases} 1, & \text{the first 10 min of each hour} \\ 0.5, & \text{the remaining 50 min.} \end{cases}$$

$$= 1 - \operatorname{sqw}(t, 100/6, 60)$$

where we have used the periodic square wave function $\operatorname{sqw}(t, d, T)$ described in Appendix B.1. The parameter d is the percentage of the period when the square wave is "on" (so the first $100/6 = 16.66\%$ of each 60 minute span of time), and the parameter T is the period (60 minutes in this case). See Fig. 2(e) for the respective graphs of the total response (solid), response to the initial data (long dashes), and the response to the input (short dashes).

The graphs show that the response to the initial data tends to 0 as $t \to +\infty$ as expected since this response solves the IVP $y' = -0.1y$, $y(0) = 10$, where no pollutant enters the tank (i.e., the inflow is clean water). The responses to the periodic, on-off inflow pollutant stream where $y(0) = 0$ (short dashes) or 10 (solid) are also as expected. As t increases, the response to the inflow approaches (quite quickly) the total response. This total response seems to be periodic with period 60 minutes (expected, since the pollutant concentration in the inflow has period 60 min.), heads towards a total of 10 lbs of pollutant in the tank at the end of each hour (since the inflow concentration is 1 lb/gal), but drops sharply during the first 10 min. of each hour (while the filter removes all pollutant from the inflow).

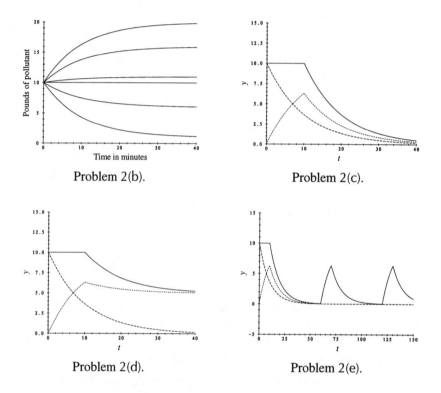

Problem 2(b).   Problem 2(c).

Problem 2(d).   Problem 2(e).

3. Two different ODEs describe the system, before, or after $t = 3$ min. If $S(t)$ denotes the amount of salt in the tank at time t, then for $t < 3$ min, $S' = (2\text{lb/gal})(3\text{gal/min}) = 6$ lb/min. So $S = 6t + C$. Since no salt is present initially, $S(0) = 0 + C$ and $C = 0$. $S = 6t$ is the amount of salt in the tank for $t < 3$ min. Note that at 3 min, $S = 6 \cdot 3 = 18$ lb, which becomes the initial condition for the next time interval ($t \geq 3$).

For $t \geq 3$ min, $S' = (2\text{lb/gal})(3\text{gal/min}) - (S/V)(3\text{gal/min}) = 6 - 3S/V$, where V is the volume of water in the tank. V is constant for $t \geq 3$ since inflow and outflow rates are equal. Using $t_0 = 3$ and an integrating factor $\exp(3t/V)$, $[S\exp(3t/V)]' = 6\exp(3t/V)$, so

$$S = 2V + C\exp(-3t/V), \quad t \geq 3.$$

Since $V = 50 + (3\text{gal/min})(3\text{min}) = 59$ gal and $S(3) = 18$ lbs, we must have $S(3) = 18 = 2(59) + C\exp(-3 \cdot 3/59) = 118 + C\exp(-9/59)$. So, $C = -100\exp(9/59)$ and

$$S = 118 - 100\exp(-3(t - t_0)/59)$$

for $t > 3$.

(a). When $t = 2$ min, $S(2) = 6 \cdot 2 = 12$ lbs salt. When $t = 25$ min, $S = 118 - 100 \exp(-3 \cdot 22/59) = 85.3$ lbs salt.

(b). As $t \to +\infty$, the exponential term approaches zero and the salt content approaches 118 lbs, which is two pounds of salt per gallon. This is expected, because in the long term the initial condition is irrelevant, and the concentration of salt in the tank is that of the inflow stream, (i.e., 2 lb of salt per gallon). Since there are 59 gallons of brine in the tank from $t = 3$ minutes on, the total amount of salt should indeed approach 118 lbs.

4.   In every ODE, guess the form of a particular solution by looking at the form of the input.

(a). Substitute $y_d = At^2 + Bt + C$ into the ODE $y' + y = t^2$ to obtain $At^2 + [2A + B]t + C + B = t^2$. Equating the coefficients of like powers of t, we must have $A = 1$, $2A + B = 0$, $C + B = 0$. The values $A = 1$, $B = -2$, $C = 2$, satisfy these conditions and so $y_d = t^2 - 2t + 2$ is a particular solution of the ODE.

Since $y_u = Ce^{-t}$ is the general solution of the undriven ODE, $y' + y = 0$, the general solution of the driven ODE is $y = y_u + y_d = Ce^{-t} + t^2 - 2t + 2$.

(b). Substitute $y_d = At + B$ into the ODE $y' + ty = t^2 - t + 1$ to find that $At^2 + Bt + A = t^2 - t + 1$. Equating the coefficients of like powers of t, we must have $A = 1$, $B = -1$, and $A = 1$. The values $A = 1$, $B = -1$ satisfy these conditions and so $y_d = t - 1$ is a solution of the ODE.

Since $y_u = Ce^{-t^2/2}$ is the general solution of the undriven ODE, $y' + ty = 0$, the general solution of the driven ODE is $y = y_u + y_d = Ce^{-t^2/2} + t - 1$.

(c). Substitute $y_d = Ate^{-2t}$ into the ODE $y' + 2y = e^{-2t}$ to obtain $te^{-2t}(-2A + 2A) + e^{-2t}(A) = e^{-2t}$, which holds only if $A = 1$, and so $y_d = te^{-2t}$ is a particular solution of the ODE.

Since $y_u = Ce^{-2t}$ is the general solution of the undriven ODE, $y' + 2y = 0$, the general solution of the driven ODE, $y' + 2y = e^{-2t}$, is $y = y_u + y_d = Ce^{-2t} + te^{-2t}$.

(d). Substitute $y_d = Ae^{-t}$ into the ODE $y' + 2y = 3e^{-t}$ to obtain $(-A + 2)e^{-t} = 3e^{-t}$, which holds only if $A = -1$, and so $y_d = -e^{-t}$ is a particular solution of the ODE.

Since $y_u = Ce^{-2t}$ is the general solution of the undriven ODE, $y' + 2y = 0$, the general solution of the driven ODE is $y = y_u + y_d = Ce^{-2t} - e^{-t}$.

(e). Guess the solution $y_d = A\cos(2t) + B\sin(2t)$. Substitute y_d into the ODE $y' + y = 5\cos 2t$ to obtain $(-2A - B)\sin(2t) + (2B - A)\cos(2t) = 5\cos(2t)$. So we must have that $-2A - B = 0$, and $2B - A = 5$. Solving, we find that $A = -1$ and $B = 2$, and so $y_d = -\cos(2t) + 2\sin(2t)$ is a particular solution of the ODE.

Since $y_u = Ce^{-t}$ is the general solution of the undriven ODE, $y' + y = 0$, the general solution of the driven ODE is $y = y_u + y_d = Ce^{-t} - \cos(2t) + 2\sin(2t)$.

(f). Guess the solution $y_d = Ae^{-t}\cos t + Be^{-t}\sin t$. Substitute y_d into the ODE $y' + y = e^{-t}\cos t$ to obtain $(-A + B + A)e^{-t}\cos t + (-A - B + B)e^{-t}\sin t = e^{-t}\cos t$. So we must have $B = 1$ and $-A = 0$, and $y_d = e^{-t}\sin t$ is a particular solution of the ODE.

Since $y_u = Ce^{-t}$ is the general solution of the undriven ODE, $y' + y = 0$, the general solution of the driven ODE is $y = y_u + y_d = Ce^{-t} - \cos(2t) + 2\sin(2t)$.

5.   We use the solution formulas found in Problem 4. In every case in this specific problem the general solution y_u of the undriven ODE tends to 0 as $t \to +\infty$ (of course, this will not always be the case). So the general solution of the driven ODE approaches the particular solution found in the corresponding part of Problem 4.

(a). As $t \to +\infty$, all solutions tend to the parabolic curve described by $y_d = t^2 - 2t + 2$.

(b). As $t \to +\infty$, all solutions tend to the straight line described by $y_d = t - 1$.

(c). As $t \to +\infty$, all solutions tend to 0, since $y_d = te^{-2t}$ has this property.

(d). As $t \to +\infty$, all solutions tend to 0, since $-e^{-t}$ does.

(e). As $t \to +\infty$, all solutions tend to the periodic function of period π, $y_d = -\cos 2t + 2\sin 2t$.

(f). As $t \to +\infty$, all solutions tend to zero since $e^{-t} \sin t$ has a decaying exponential factor.

6.   **(a).** Guess a particular solution of the form $y_d = A\cos t + B\sin t$. Substitute y_d into the ODE $y' + ay = b\cos t + c\sin t$ to obtain

$$(aA + B)\cos t + (-A + aB)\sin t = b\cos t + c\sin t$$

So we must have

$$aA + B = b$$

$$A + aB = c$$

Solving for A, B we find that

$$A = \frac{ab - c}{a^2 + 1}$$

$$B = \frac{ac + b}{a^2 + 1}$$

So a particular solution is

$$y_d = \frac{1}{a^2 + 1}[(ab - c)\cos t + (ac + b)\sin t]$$

(b). Guess a particular solution of the form $y_d = At + B$. Substitute y_d into the ODE $y' + ay = bt + c$ to obtain

$$aAt + (A + aB) = bt + c$$

So we must have

$$aA = b$$

$$A + aB = c$$

Solving for A, B, we have $A = b/a$, $B = (ac - b)/a^2$, and $y_d = bt/a + (ac - b)/a^2$.

7.   The integrating factor of the undriven linear ODE $y' + 2ty = 0$ is e^{t^2}. So the general solution of that ODE is $y_u = Ce^{-t^2}$. As $t \to +\infty$, $y_u(t) \to 0$. This means that all solutions of the driven ODE, $y' + 2ty = q(t)$, tend to some particular solution $y_d(t)$ since $y(t) = y_u + y_d$. In each case we use a numerical solver for $0 \leq t \leq 10$ with $y(0) = 0, \pm 2, \pm 4$.

(a). The general solution of $y' + 2ty = 1$ is $y = Ce^{-t^2} + e^{-t^2} \int e^{t^2} dt$, but we can't express the integral in terms of any of the elementary functions. So we turn to a numerical solver to solve the IVPs. From Fig 7(a), we conjecture that as $t \to +\infty$, all solutions tend to 0.

(b). The general solution of $y' + 2ty = 1/(1 + 2t^2)$ is $y = Ce^{-t^2} + e^{-t^2} \int e^{t^2}/(1 + 2t^2) dt$ but there is no way to express the integral in terms of elementary functions. Fig. 7(b) shows what our numerical solver did for the IVPs. As in part **(a)**, a reasonable conjecture is that as $t \to +\infty$, $y(t) \to 0$.

(c). The general solution of $y' + 2ty = 1 + t^2$ is $y = Ce^{-t^2} + e^{-t^2} \int e^{t^2}(1 + t^2) dt$. Looking at Fig. 7(c) it looks as if $y(t)$ tends to a straight line as $t \to +\infty$. Measuring the "rise" and the "run" from the figure, we guess that the straight line asymptote has a slope of $1/2$. Moreover, putting a straight-edge on the graph, it looks as if the line would go through the point $t = 0$, $y = 0$. So one might conclude that as $t \to +\infty$ all solutions tend to the line $y = t/2$, although we haven't actually proved this. Note, however, that there is no straight line solution $y(t) = A + Bt$ which solves the ODE.

(d). The general solution solution of $y' + 2ty = 1 + 2t^2$ is $y = Ce^{-t^2} + e^{-t^2} \int e^{t^2}(1 + 2t^2) dt$. Although the integral looks just as bad as those in parts **(a)**, **(b)**, **(c)**, it turns out we can guess a particular solution here: $y_d = A + Bt$. Substituting into the ODE, we have

$$B + 2At + 2Bt^2 = 1 + 2t^2$$

So $B = 1$, $A = 0$, and $y_d = t$ is a particular solution. So all solutions approach the straight line solution $y_d = t$ as $t \to +\infty$. See Fig. 7(d).

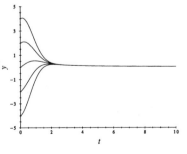

Problem 7(a).

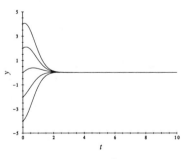

Problem 7(b).

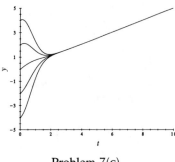

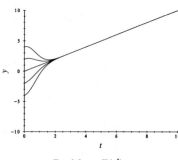

Problem 7(c).   Problem 7(d).

8. There are lots of linear ODEs with the described property. Here is one:

$$y' - y = e^{-t}$$

The integrating factor is e^{-t}. We proceed as usual to get the general solution

$$y = Ce^t - e^{-t}/2$$

Where $y_u = Ce^t$ and $y_d = -e^{-t}/2$. If C is positive then $y_u \to +\infty$ as $t \to +\infty$, while $y_d \to 0$ as $t \to +\infty$. So this ODE has a particular solution which dies out as $t \to +\infty$, while at least one solution of the undriven ODE (in fact, infinitely many) "blow-up" as $t \to +\infty$.

9. **(a).** See Fig 9(a) for the graphs of the solutions of the IVPs $y' + y = 0$, $y(-5) = 0, \pm 2, \pm 4$.

(b). See Fig 9(b) for the graph of the solution of $y' + y = t\cos(t^2)$, $y(-5) = 4$.

(c). Fig. 9(a) shows the response $y_h(t) = Ce^{-t}$ to initial data $y(-5) = 0, \pm 2, \pm 4$ with no input, and Fig. 9(b) [see part **(b)**] the total response to initial data $y(-5) = 4$ and to input $q(t) = t\cos(t^2)$. The response to any initial data dies out with increasing time. A sinusoidal response [apparently with rapidly increasing frequency like that of the input $q(t)$] due to the driving term remains. The solution curves in Fig. 9(c) show the superpositions of these two responses. As the initial response dies out, all solutions converge to the driven response.

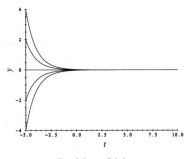

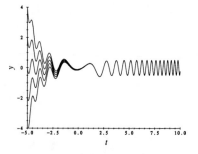

Problem 9(a).   Problem 9(b).   Problem 9(c).

10. The normal form of the ODE is $y' + (\cos t)y = \sin t$, where $p(t) = \cos t$, $q(t) = \sin t$, and the integrating factor is $e^{\sin t}$. So the general solution is $y = Ce^{-\sin t} + e^{-\sin t} \int e^{\sin t} \sin t \, dt$. Since the integral cannot be evaluated in terms of elementary functions, we cannot easily graph these solutions. So the best way to handle the IVPs is to go directly to a numerical solver. See Fig. 10 for the graphs of solutions with initial values $y(-10) = -6, -2, 0, 5$.

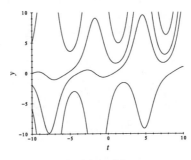

Problem 10.

11. Here you can find the solutions of the IVPs explicitly by the integrating factor technique, and then plot the graph. You can also just use a numerical solver/grapher directly without bothering to find explicit solution formulas. We give these formulas below in terms of a constant C, which must be evaluated by using the given initial condition. However, we did the graphs with our numerical solver, ignoring the solution formulas. Your solver may return error messages as ours did in parts **(c)** and **(d)** because of a singularity at the origin [part **(c)**] and at the interval's endpoints [part **(d)**]. In part **(c)** do whatever you can to get the graph of the solution $y = t^2/4$ through the origin.

(a). The integrating factor for $y' = (\sin t)(1 - y)$ is $e^{-\cos t}$, and the general solution is $y = 1 + Ce^{\cos t}$, $-\infty < t < \infty$. See Fig. 11(a) for the solution curves with $y(-\pi/2) = -1$, 0, 1.

(b). The integrating factor for $y' + y = te^{-t} + 1$ is e^t, and the general solution is $y = t^2 e^{-t}/2 + Ce^{-t} + 1$, $-\infty < t < \infty$. See Fig. 11(b) for the solution curves with $y(0) = -1$, 0, 1.

(c). The integrating factor for the ODE in normal form, $y' + 2y/t = t$, $t \neq 0$, is t^2, and the general solution is $y = t^2/4 + Ct^{-2}$, $t > 0$, or $t < 0$. See Fig. 11(c) for the solution curves with $y(\pm 2) = 0$, 1, 2. Note that $y = t^2/4$ is the only solution defined for all t. Solutions with positive C go to $+\infty$ as $t \to 0$; solutions with negative C go to $-\infty$.

(d). The integrating factor for $y' = (\tan t)y + t \sin 2t$ is $\cos t$, and the general solution is $y = \sec t(t^2/2 - \cos 2t/4 - t \sin 2t/2 + C)$, $-\pi/2 < t < \pi/2$. See Fig. 11(d) for solution curves with $y(0) = 0, \pm 1$.

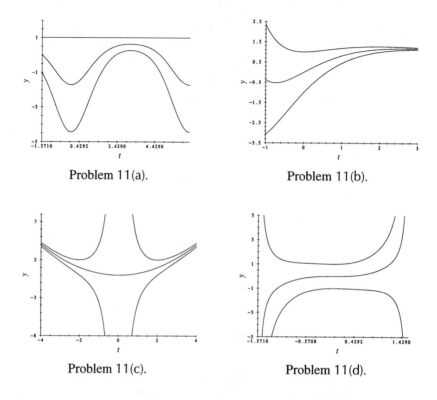

Problem 11(a).   Problem 11(b).

Problem 11(c).   Problem 11(d).

12.   We want $y(t) = y_0$, for $t \geq 0$, so $y'(t) = 0$ for $t > 0$. Plugging into the ODE $y' + p(t)y = q(t)$, we have $0 + p(t)y_0 = q(t)$. Choose the input $q(t)$ to be $y_0 p(t)$, and we are done.

13.   The general solution of the linear ODE, $y' + y = q(t)$ is $y = Ce^{-t} + y_d(t)$, where y_d is any particular solution. In this problem we assume that there is some solution $y_d(t)$ that "blows up" as $t \to +\infty$, [i.e., $y_d(t) \to +\infty$ as $t \to +\infty$]. Then, we see from the general solution formula that all solutions have this property since $Ce^{-t} \to 0$ as $t \to +\infty$, and so this term has no effect on the asymptotic behavior as $t \to +\infty$.

14.   **(a).** A direct translation into mathematical symbols of the word-description of Newton's Law of Cooling (or Warming) is given by

$$y'(t) = k[y(t) - m(t)]$$

where $y(t)$ is the body's temperature, $m(t)$ is the temperature of the medium, and k is the proportionality constant. The constant k must be negative because if the body is warmer than its surroundings (i.e., if $y > m$), then the body should cool down and so $y' < 0$. A similar argument applies if the body is cooler than the surrounding medium.

(b). The model equation is $dy/dt = k(y - m)$, where $y(t)$ is the temperature in $°F$ of the thermometer at time t measured in minutes from the time of insertion, k is a negative rate constant, and m is the temperature of the horse. We have that $y(0) = 82°F$, $y(3) = 90°F$, and $y(6) = 94°F$. The problem is to find m. The rate equation may be solved by rearranging the equation and using the fact that $(y - m)^{-1}y' = (\ln|y - m|)' = k$. So, $\ln|y - m| =$

$kt + K$, where K is a constant, or $|y(t) - m| = Ce^{kt}$, where $C = e^K$. Since $|82 - m| = C$, $|90 - m| = Ce^{3k}$, and $|94 - m| = Ce^{6k}$, we may divide the second equation by the first and the third by the second to obtain $e^{3k} = (90 - m)/(82 - m) = (94 - m)/(90 - m)$. Cross multiplying in the last equality and solving for m, we find that $m = 98°$ F. Note that the absolute value signs were removed in the quotients since the rising temperature of the thermometer implies that $m > y(t)$.

(c). The modeling IVP for the cooling egg is

$$y' = k(y - 65), \qquad y(0) = 180$$

where k is a negative constant of proportionality, the room temperature is $65° F$, and the initial temperature of the egg is $180° F$. We are given one other item of information: $y(1) = 140° F$, where "1" means "one hour," so we are measuring time in hours. Our job is to solve the IVP in terms of k, then use the extra information to find the value of k, and finally determine the time T when the egg has cooled down to $120° F$.

The linear ODE

$$y' - ky = -65k$$

has integrating factor e^{-kt} and the general solution is

$$y = Ce^{kt} + 65$$

Since $y(0) = 180$, we have that $C = 115$ and so

$$y = 115e^{kt} + 65$$

Since $y(1) = 140$, we have

$$140 = 115e^k + 65$$

and so $75 = 115e^k$, giving $k = \ln(75/115) \approx -0.427$. This gives us the egg's temperature solution formula

$$y = 115e^{-0.427t} + 65$$

If we want to find the time T when $y = 120$, we must solve

$$120 = 115e^{-0.427T} + 65$$

for T. We see that $-0.427T = -\ln(55/115) = \ln(115/55)$, so $T = 1.727$ hours, or about 1 hour and 44 minutes.

(d). This is a group problem, so we won't say much except that now there are *two* ODEs, one for the egg's temperature and the other for the water temperature. The constants k_e and k_w need not be the same.

15. **(a).** The amount $A(1)$ at the end of one year with 9% continuous compounding is $A(1) = A_0 e^{0.09}$ since the model equation is $A' = 0.09A$, $A(0) = A_0$. Setting this equal to the amount after one year with interest payable annually at a rate $r\%$, we have $A_0 e^{0.09} = A_0(1 + r/100)$, or $r/100 = e^{0.09} - 1 \approx 0.0942$, (i.e., the annual interest rate would have to be 9.42% to yield the same amount as continuous compounding at 9%).

(b). Since $A'(t) = kA$, we must have $A(8) = 2A_0 = A_0 e^{8k}$. So, $k = (\ln 2)/8 \approx 0.0866$. Thus, the interest rate is 8.66% if funds double in 8 years and interest is compounded continuously.

(c). We must find t such that $2A_0 = A_0 e^{kt}$ for $k = 0.05, \ 0.09,$ and 0.12. Canceling A_0, taking logarithms, and solving for t, the respective doubling times are approximately 13.86, 7.70, and 5.78 years.

16. For an amount A_0 to double in T years at $r\%$ interest compounded continuously, we must have $2A_0 = A_0 e^{0.01rT}$, [i.e., $T = (\ln 2)/(0.01r) = 69.315/r$]. Similarly, the span of time T for the amount to increase by 50% is $T = (\ln 1.5)/0.01r = 40.547/r$ years. The "Rules of 72 and 42" overestimate the time required, if interest is continuously compounded.

17. Group project.

1.5 Introduction to Modeling and Systems

Suggestions for preparing a lecture

Topics: The moving ball models and their solutions, equivalence with first-order systems, discussion of the question "Does it take a whiffle ball longer to rise or to fall?", a brief discussion of state variables and modeling principles, and the modeling process.

Remarks: *An alternative*: Replace the moving ball models by the models of radioactive decay and their use in dating. Sometimes, but not often, we take two lectures on all the material in this section and do both the whiffle ball and the dating models. You might do one of the whiffle ball or radioactive decay experiments mentioned below. Incidentally, many students have seen some dating problems in other courses, but few understand how the process works.

Making up a problem set

Students spend a lot of time on each modeling problem, particularly if they are asked to give explanations or do some computing. So you may want to assign no more than 3 problems from this set. Problems 5–9 deal with models of a moving ball (whiffle ball, or otherwise), and we particularly like Problem 9. Sometimes we assign only Problems 8 and 9; students find these two problems both very interesting and challenging (few can do Problem 9**(c)**).

Problems 1–4, 10–14 have to do with growth, radioactive decay, and dating problems. Any three of these would be fine for a problem set. Problem 14 gives rise to a very simple system to solve (and gives an alternative to that old standby: radiocarbon dating). If you want to assign a group problem this early in the course, give the students the snow-shoveling problem (Problem 15), but be prepared for student grumbling that the problem statement doesn't give enough information.

Comments

This section is the longest in the text. The main aim of this section is to communicate the idea that one can often create a mathematical model of a natural process if the system is described by a collection of laws or principles. These laws, which are usually based on experiment, observation, and insight, need not be expressed in mathematical terms. The mathematical modeler translates the laws into mathematical equations, ODEs and IVPs, using the methods of this chapter. In this section, the translation process, the solution of the mathematical equations, and the interpretation of the mathematical solution are explained by using the examples of motion along the local vertical and of radioactive decay. The notions of state variables and dynamical systems are introduced (in an imprecise way) so that the reader can see that a finite collection of variables that change in time often is adequate to describe the evolution of a natural process. The exact definition of a dynamical system in terms of a relation such as $y(t_2 + t_1, y_0) = y(t_2, y(t_1, y_0))$, $y(t_0) = y_0$, seems inappropriate at this early stage. In this book dynamical systems are mostly modeled with differential equations. In Chapter 2, when numerical methods for solving ODEs are introduced, discrete dynamical systems are considered (see Section 2.7). Any one-step numerical ODE solver algorithm (e.g., the Euler or Runge-Kutta methods of Section 2.5) can be considered to be a discrete dynamical simulation of the ODE. One may even treat the discrete solver algorithm as a discrete model of the original natural phenomenon.

A secondary aim of this section is to introduce the notion of a differential system in a mild way and to establish the connection between differential systems, modeling, and dynamical systems. The vertical motion model does that in a familiar environment. Section 1.7 takes up systems in a much more serious way.

Modeling is a complex process, and people become adept at it mostly by extensive practice rather than by memorizing modeling rules or schematics. For this reason, it is probably best to work through a specific modeling process rather than to think about modeling principles in the abstract.

An experiment

Throw a whiffle ball up into the air; does it take longer to go up or to come down? Ask a group this question, and we guarantee that the responses will divide four ways: longer going up, longer coming down, equal times, or it all depends. If you are in a room with a high ceiling, throw a real whiffle ball up and ask everyone the same question. You will probably get the same four responses, because it is hard to tell the times by just watching the ball. You may want to devise some way to do alternate experiments with a whiffle ball. Even in this simple setting, the proof that the fall time is longer is hard [Problem 9**(c)**], but the computer simulation (Figures 1.5.3, 1.5.4) shows the answer quite dramatically. See the article, "Projectile Motion with Arbitrary Resistance," Tilak de Alwis, *The College Mathematics Journal*, **26** (1995), 361–367, for a general discussion (with additional references) of vertical motion against resistance.

The other process modeled in the text is radioactive decay and its application to dating. You might find something radioactive and test it with a Geiger counter, but this doesn't sound very appealing. The notion of a half-life is important here. See also the background material for Section 1.1.

---

1. The rate equation is $N' = -kN$, where k is a positive constant. The ODE can be solved by multiplying by the integrating factor e^{kt} and rewriting as $e^{kt}N' + ke^{kt}N = [e^{kt}N]' = 0$. So we have that $e^{kt}N = C$, or $N(t) = Ce^{-kt}$, where C is any constant. The half-life is related to the value of k by $k\tau = \ln 2$. To see this, note that $N(0) = C$. At the time $t = \tau$, $N(0)/2 = N(0)e^{-k\tau}$ so $1/2 = e^{-k\tau}$, or $k\tau = \ln 2$. The data of the problem imply that $0.989N_0 = N_0e^{-25k}$, where $N_0 = N(0)$. Canceling N_0, taking logarithms, and solving for the constant k, we have that $k = -(\ln 0.989)/25 = 0.000442$ (years)$^{-1}$ So the half-life is $\tau = (\ln 2)/k \approx 1567$ years.

2. Since $k = (\ln 2)/\tau$ [text formula (7)], we have $k = (\ln 2)/\tau = 0.693/1000 = 0.000693$ (years)$^{-1}$. Since $N(t) = N(0)e^{-kt}$, we have that $N(100)/N(0) = e^{-100k} = e^{-0.0693} = 0.933$; approximately 93% of the original amount $N(0)$ remains after 100 years.

3. If $N(t)$ is the amount of the phosphorus at time t (in days), then $N(0) = 8$, and $N(t) = 8e^{-kt}$ [from formula (6)]. Now $k = (\ln 2)/14.2 = 0.0488$/day from text formula (7). To find the time t for which $N(t) = 1.0 \times 10^{-5}$, we have $1 \times 10^{-5} = 8e^{-0.0488t}$. Taking logarithms and solving for t, we have $t = 278.5$ days.

4. The rate constant is $k_1 = 0.03$/month for $0 \leq t \leq T$ and $k_2 = 0.05$/month for $t > T$. The basic rate equation is $N' = kN$, where $k = 0.03$, $0 \leq t \leq T$, and $k = 0.05$, $t > T$. We have that for $0 \leq t \leq T$, $N(t) = N_0 e^{k_1 t}$. We must now solve the IVP $N' = k_2 N$, $N(T) = N_0 e^{k_1 T}$, $t \geq T$. Solving, we see that $N(t) = (N_0 e^{k_1 T}) e^{k_2(t-T)}$. Since $2N_0 = N(20) = N_0 e^{.03T} e^{0.05(20-T)} = N_0 e^{1-.02T}$, we may cancel N_0, take logarithms and solve for T to obtain $\ln 2 = 1 - .02T$, or $T = 15.34$ months, the time when the rate constant changes.

5. The model equations are

$$y''(t) = -g, \qquad y(0) = 0, \quad y'(0) = 2000 \text{ cm/sec}$$

where $y(t)$ is the height of the body of mass 600 gm above the ground at time t. The model equations are valid for $0 \leq t \leq T$, where T is the positive time of impact at which $y(T) = 0$. The value of g is 980 cm/sec^2.

(a). Integrating each side of $y'' = -g$ from $t = 0$ to a general value of t, $0 \leq t \leq T$, we have that $y'(t) - y'(0) = -gt$. Since $v_0 = y'(0) = 2000$ cm/sec and since the velocity $y'(t)$ is 0 at the highest point, $0 - 2000 = -gt$ so $t = 2000/g \approx 2.04$ sec. Integrating $y'' = -g$ twice, we have $y(t) = y_0 + v_0 t - (1/2)gt^2$. With the highest point reached at time $t \approx 2.04$ sec, an initial velocity v_0 of 2000 cm/sec, and a zero initial height y_0, we find that the highest point reached is about 2041 cm.

(b). Using the equation $y(t) = y_0 + v_0 t - (1/2)gt^2$ from the solution of part **(a)** with $y_0 = 0$, $t_0 = 0$, $v_0 = 2000$, $g = 980$ cm/sec^2, and $t = 3$ sec, we have $y(3) \approx 1590$ cm. Using the equation $y'(t) - y'(0) = -gt$ in the solution of part **(a)**, we have $y'(3) = -980 \cdot 3 + 2000 = -940$ cm/sec (the body is falling). Again from **(a)**, $y(t) = y_0 + v_0 t - (1/2)gt^2$ with $y_0 = 0$, so at the time T of impact we have $y(T) = 0 = 2000T - 490T^2$, or $T = 2000/490 \approx 4.08$ sec.

6. The ODE, $y'' = -9.8$ with conditions $y(0) = y_0$, $y'(0) = v_0$ has the solution, $y = y_0 + v_0 t - (1/2)gt^2$, $0 \leq t \leq T$.

(a). Since the second derivative $y''(t)$ is the negative constant $-g$, the graph of $y(t)$ must be concave downward, and the slope y' decreases at a constant rate.

(b). Any plot should be a downward opening parabola. See Fig. 6(b) for three of these parabolas corresponding to various values of the constants y_0 and v_0.

(c). Substitute $t = 5$, $t_0 = 0$, $y_0 = 75$ and $y(5) = 0$ into the solution formula and solve for v_0. $y(5) = 0 = y_0 + v_0 t - (1/2)gt^2$ becomes $v_0 = (1/2)gt - (y_0/t)$. So, $v_0 = (1/2)(9.8)(5) - (75/5) = 9.5$ m/sec. See Fig. 6(c) for the graph of y vs t.

(d). First find the time t_1 at which $y(t)$ reaches a local maximum. Since $y'(t) = 0 = v_0 - gt$ at an extreme value, $t_1 = v_0/g$. Substitute this value of t into the solution formula for $y(t)$

and equate $y(t_1)$ to 30, the given maximum height:

$$y(t_1) = 0 + v_0\frac{v_0}{g} - \frac{1}{2}g\left(\frac{v_0}{g}\right)^2 = 30$$

and so we have that

$$v_0^2/g - (1/2)g(v_0/g)^2 = (1/2)v_0^2/g = 30$$

So, the initial velocity is given by $v_0 = \sqrt{60g} = \sqrt{(60)(9.8)} = 24.2$ m/sec. See Fig. 6(d) for the graph of y vs t.

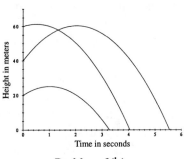

Problem 6(b).

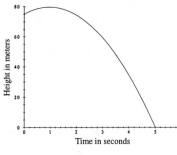

Problem 6(c).

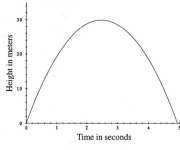

Problem 6(d).

7. The model IVPs are $s'' = -g$, $s(0) = H$, $s'(0) = 0$ for the position $s(t)$ of the stone (valid for $0 \le t \le T$), and $b'' = -g$, $b(1.5) = H$, $b'(1.5) = -20$ for the position $b(t)$ of the ball (valid for $1.5 \le t \le T$), where $s(t)$ and $b(t)$ are the heights of the stone and the ball above the ground at time t, H is the height of the building, and T is the time the stone and the ball hit the ground. In general, we have $y(t) = y_0 + v_0t - gt^2/2$ if $y'' = -g$, $y(0) = y_0$, $y'(0) = v_0$, which is the form of both model IVPs.

    (a). At impact, $s(T) = H - 4.9T^2$ since the initial velocity of the stone is 0. On the other hand, $b(T) = H - 20(T - 1.5) - 4.9(T - 1.5)^2$ since the initial velocity of the baseball is 20 meters/second and the flight time of the ball is 1.5 seconds shorter than that of the stone. Since both the ball and the stone hit the ground at time T, $s(T) = b(T)$, so $H - 4.9T^2 = H - 20(T - 1.5) - 4.9(T - 1.5)^2$. Canceling H and $-4.9T^2$ and solving for T, we have that $T \approx 3.58$ sec. Now we have that $H - S(T) = 4.9T^2 \approx 4.9(3.58)^2 \approx 62.8$, which is the height of the building in meters.

    (b). The model equations for the ball are $b'' = -g$, $b(\tau) = H$, $b'(\tau) = v_0$, where τ is the waiting time and v_0 is the initial velocity of the ball (v_0 is negative since the ball is hurled downward). We require that $s(T) = b(T)$ at the time $T > 0$ of impact, and that $0 < \tau < T$. From (a), $s(T) = H - 4.9T^2$ and $b(T) = H - 20(T - 1.5) - 4.9(T - 1.5)^2$. We have that $H - 4.9T^2 = H + v_0(T - \tau) - 4.9(T - \tau)^2$. This equation may be solved for T to obtain $T = \tau(4.9\tau + v_0)/(9.8\tau + v_0)$. Since $T/\tau > 1$, we must have $(4.9\tau + v_0)/(9.8\tau + v_0) > 1$. If both numerator and denominator are positive, we get the impossibility $4.9\tau > 9.8\tau$. So both numerator and denominator must be negative. So $9.8\tau < -v_0$. The waiting time τ must be less than $|v_0|/9.8$, regardless of the building's height!

8. Since $y''(t) = -g$, we find that the velocity is $y'(t) = v_0 - gt$. The ball has zero velocity at its maximum height, so we find the time at maximum height to be $t = v_0/g$. The ball spends as much time going down as it does going up and has an impact velocity equal in magnitude to its initial velocity. This can be revealed with a plot of $y(t) = v_0 t - (1/2)gt^2$. Since the path is parabolic, it is symmetric about the maximum height, so the slope (velocity) of the ball's trajectory has the same magnitude at the same height on the way up and the way down, and the time spent rising from the ground to height h equals the time spent falling back from the height h to the ground.

9. **(a).** With the origin at ground level and up as the positive direction, the whiffle ball is initially at $y(0) = h$ with initial velocity $y'(0) = v_0$. The whiffle ball undergoes viscous damping due to air resistance, and so we see from (10) that $my'' = -mg - ky'$. Recalling that $y' = v$, we have $y'' = v'$, so $mv' = -mg - kv$. Solving for v', $v' = -g - (k/m)v$. The initial conditions become $y(0) = h$ and $v(0) = v_0$.

(b). The curves of Fig. 9(a), Graphs 1–3 for $y_0 = 2$, $v_0 = 10, 30, 50, 70$ clearly suggest that it takes longer for the whiffle ball to fall back to the initial height of 2 meters than to rise from that height. In the case $k/m = 5$ [Fig. 9(b), Graph 2], the drag on the whiffle ball is five times what it is in the first case, $k/m = 1$ [Fig. 9(b), Graph 1], so it is reasonable to expect that the ball will not rise as high. Similarly if $k/m = 10$ [Fig. 9(b), Graph 3], the maximum height is even smaller. Visual inspection of these graphs suggests that the rise and fall times are as given below (in seconds). The first number is v_0, the second the rise time, and the third the fall time:

Graph 1 ($k/m = 1$): 10, 0.6, 1.15; 30, 1.4, 2.6; 50, 1.9, 4.35; 70, 2.25, 5.95
Graph 2 ($k/m = 5$): 10, 0.2, 0.5; 30, 0.5, 2.5; 50, 0.6, 4.65; 70, 0.7, 6.8.
Graph 3 ($k/m = 10$): 10, 0.1, 0.4; 30, 0.2, 3.3; 50, 0.25, 5.25; 70, 0.3, 7.7

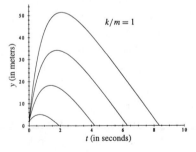

Problem 9(b), Graph 1.

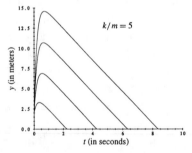

Problem 9(b), Graph 2.

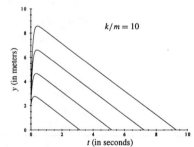

Problem 9(b), Graph 3.

(c). The strategy is to find T (the time it takes to reach maximum height) and show that $y(2T) > h$, which implies that $\tau > 2T$, where τ is the time it takes for the whiffle ball to fall back to its initial position $y = h$. Then we will have that the time τ to rise and fall is longer than twice the time T to rise to the maximum height, so the time to fall must be longer than the time to rise.

Solving the IVP, $v' + v = -g$, $v(0) = v_0$, we obtain $v(t) = -g + (v_0 + g)e^{-t}$. Setting $v(T) = 0$ and solving for T,

$$T = \ln(1 + v_0/g)$$

and so this value of T is the time it takes the whiffle ball to reach maximum height. Now since $y'(t) = v(t) = -g + (v_0 + g)e^{-t}$, and $y(0) = h$. Integrating, we see that

$$y(t) = -gt - (v_0 + g)e^{-t} + v_0 + g + h$$

So evaluating $y(t)$ at $2T$ and substituting in our expression for T, we have

$$y(2T) = -2gT - (v_0 + g)e^{-2T} + v_0 + g + h$$
$$= -2g\ln(1 + v_0/g) - g^2/(v_0 + g) + v_0 + g + h$$

Define the function $f(v_0) = y(2T) - h$, and so

$$f(v_0) = -2g\ln(1 + v_0/g) - g^2/(v_0 + g) + v_0 + g$$

If we can show that $f(v_0) > 0$ for $v_0 > 0$, then $y(2T) > h$. First, we note that $f(0) = -2g\ln 1 - g^2/g + 0 + g = 0$. To show that $f(v_0) > 0$ for $v_0 > 0$ all we need to do is to show that $df/dv_0 > 0$ for $v_0 > 0$:

$$\frac{df}{dv_0} = \frac{-2g}{v_0 + g} + \frac{g^2}{(v_0 + g)^2} + 1$$
$$= \frac{-2g(v_0 + g) + g^2 + (v_0 + g)^2}{(v_0 + g)^2} = \frac{v_0^2}{(v_0 + g)^2} > 0$$

and so $f(v_0)$ is an increasing function of v_0 which is positive if $v_0 > 0$. This shows that $y(2T) > h$ and it takes longer for the whiffle ball to fall through a given distance than it does to rise the same distance. This is reasonable because the gravitational and resistance forces oppose each other during the fall (and so the magnitude of the acceleration is lower), but these two forces act in the same direction during the rise.

10. **(a).** For $t > T$ the Radioactive Decay Law says that the ^{14}C nuclei decay via the first-order rate equation $q' = -kq$. So $q(t)$ satisfies the *backward* IVP

$$q' = -kq, \quad q(0) = q_0, \quad T \le t \le 0$$

(b). We solved the IVP $q' = -kq$, $q(0) = q_0$ of part **(a)** in Example 1.5.1; its solution is $q = q_0 e^{-kt}$. At the time T of death of the wood tissue, $q_T = q_0 e^{-kT}$. Solving for T, we obtain

$$T = -\frac{1}{k}\ln\frac{q_T}{q_0}$$

From the ODE, $q' = -kq$, we see that

$$q'(T)/q'(0) = (-kq(T))/(-kq(0)) = q(T)/q(0)$$

From Example 1.5.1, we know that $k = (\ln 2)/\tau$, so

$$T = -\frac{1}{k} \ln \frac{q_T}{q_0} = -\frac{\tau}{\ln 2} \ln \frac{q'(T)}{q'(0)}$$

(c). From the Geiger counter data we have that $q'(T)/q'(0) = 13.5/1.69$. So

$$T \approx -\frac{5568}{\ln 2} \ln \frac{13.5}{1.69} \approx -16,692 \text{ years}$$

Fig. 10 shows the graph of $q(t)/q_0$ as a function of time t, $t \le 0$. The curve begins its exponential decay at $T = -16,692$ years, and that is the age of the charcoal and, presumably, of the cave paintings.

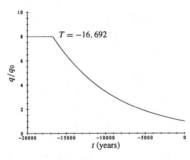

Problem 10.

11.   (a). Let $Q(t) = q(t)/q_0$. Then $Q'(t) = q'(t)/q_0$. The IVP $q' = -kq$, $q(0) = q_0$, is changed into the IVP, $Q'(t) = -kQ(t)$, $Q(0) = 1$. Also, a careful reading of Problem 10 reveals that $Q(T) = q_T/q_0 = q'(T)/q'(0) = 13.5/1.69$. Therefore, the IVP in the scaled variable Q is $Q' = -kQ$, $Q(0) = 1$, $T \le t \le 0$, and the problem is to find the value of T such that $Q(T) = 13.5/1.69 \approx 7.988$.

(b). Figure 11(b), Graph 1 shows the graph of $Q(t)$ for $-17000 \le t \le 0$. In order to estimate the time T for which $Q(T) \approx 7.988$, we zoom on the left end of the graph of $Q(t)$ [Fig. 11(b), Graph. 2] and introduce a grid. The estimated value of T is -16690 which matches the value found by the method of Problem 10. So the tree from which the charcoal in the cave was produced died about 16690 years before 1950, when the Geiger counter measurements were made.

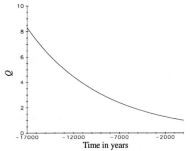

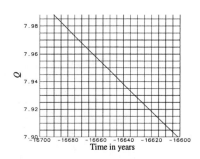

Problem 11(b), Graph 1.   Problem 11(b), Graph 2.

12. Let $q(t)$ be the amount of ^{14}C per gram of carbon in the Stonehenge charcoal at time t, where time is measured from 1977. Following the modeling techniques in the text for dating the cave painting, we see that a set of modeling equations is $q'(t) = 0$ for $t \leq T$, $q'(t) = -kq(t)$ for $T \leq t \leq 0$, and $q(0) = q_0$, where T is the time when the charcoal was living wood (T is negative since that time was long before 1977), k is to be determined, $T \leq t \leq 0$ represents the span of years up to 1977 when the residual radioactivity of the charcoal was measured, and $q(0)$ is the amount of ^{14}C per gram of charcoal in 1977. Solving the rate equation for $T \leq t \leq 0$, we have that $q(0) = q(T)e^{kT}$. Solving for T, $T = (1/k) \ln(q(0)/q(T))$. According to the text, the half-life of ^{14}C is roughly 5568 years, so $k = (\ln 2)/5568$. The numbers $q(0)$ and $q(T)$ may be determined from the rate equation and the disintegration counts: $8.2 = q'(0) = -kq(0)$ and $13.5 = q'(T) = -kq(T)$. Since $T = (1/k) \ln[q(0)/q(T)]$, we have $T = (5568/\ln 2) \ln(8.2/13.5) = -4005$ years. The tree from which the charcoal was made was living in the year 2028 B.C. The date is only approximate because of the various uncertainties mentioned in the text.

13. We have $q(T) = q_0 e^{-kT}$ where T is the time that has passed since the shell was living. Solving for T, we find $T = -(1/k) \ln(q(T)/q(0)) = -(\tau/\ln 2) \ln(q'(T)/q'(0))$, so $T = -5568 \ln(0.6 \cdot 13.5/13.5)/\ln 2 = -5568 \ln(0.6)/\ln 2 = 4103$ years, which is the approximate age of the shell.

14. **(a).** Setting $k = k_1 + k_2$, $K(0) = K_0$, $A(0) = 0$, $C(0) = 0$; solving, we have that $K(t) = K_0 e^{-kt}$. Substituting this into the second ODE, we have $A' = k_1 K_0 e^{-kt}$. Integrating and setting $A(0) = 0$, we obtain $A(t) = (K_0 k_1/k)(1 - e^{-kt})$. Similarly, $C'(t) = k_2 k_0 e^{-kt}$ and with $C(0) = 0$ we obtain $C(t) = (K_0 k_2/k)(1 - e^{-kt})$. Due to the Balance Law, total amounts are conserved, so

$$K(t) + A(t) + C(t) = K_0 e^{-kt} + (K_0 k_1/k)(1 - e^{-kt}) + (K_0 k_2/k)(1 - e^{-kt})$$
$$= K_0 e^{-kt} + K_0 - K_0 e^{-kt} = K_0$$

for all t. Another way to see this is that since $[K(t) + A(t) + C(t)]' = 0$, we have that $K(t) + A(t) + C(t) = K(0) + A(0) + C(0) = K(0) = K_0$. Equilibrium solutions may be obtained by analyzing the solution as $t \to +\infty$. As $t \to +\infty$, the exponential e^{-kt} approaches zero due to the negative exponent, and $K(t) \to 0$. Similarly, the exponential terms in $A(t)$ and $C(t)$ approach zero, and we have that $A(t) \to (K_0 k_1/k)$ and $C(t) \to$

$(K_0 k_2/k)$ as $t \to +\infty$.

(b). We see that $A/K = (K_0 k_1/k)(1 - e^{-kT})/K_0 e^{-kT} = (k_1/k)(e^{kT} - 1)$ after factoring e^{-kT} from the numerator and from the denominator. We have that $(Ak/Kk_1) = e^{kT} - 1$, and so $T = (1/k)\ln[(Ak/Kk_1) + 1]$.

(c). Substituting $k_1 = 5.76 \times 10^{-11} \text{year}^{-1}$ and $k_2 = 4.85 \times 10^{-10} \text{year}^{-1}$ into the equation $A/k = (k_1/k)(e^{kT} - 1)$ for A/K in part **(b)** and recalling that $k = k_1 + k_2$, we have

$$\frac{A}{K} = \frac{5.76 \times 10^{-11}}{5.76 \times 10^{-11} + 4.85 \times 10^{-10}} (e^{[(5.76 \times 10^{-11} + 4.85 \times 10^{-10}) \times 1.75 \times 10^{-6}]} - 1)$$

$$\approx 1.01 \times 10^{-4}$$

15. This is a hard problem because there doesn't seem to be enough information. And there isn't! The challenge is to assume just enough extra that is plausible and leads to a solution. Different people will make different assumptions and get different times for the start of the snowfall. Here is one set of assumptions and the corresponding solution. Assume that the sidewalk has constant width and the man does not turn back. Let time t in hours be measured forward from 12 noon. Since the snow falls steadily, the depth of the snow at time t in inches is $h(t) = a(t - T)$, where T is the time the snow began to fall $[T < 0]$ and a is an unknown rate constant. Let $x(t)$ be the position in blocks of the shoveler at time $t \geq 0$. Then $x(0) = 0$, $x(2) = 2$, $x(4) = 3$. The amount of snow ΔV removed in the time span from t to $t + \Delta t$ as the shoveler moves from x to $x + \Delta x$ is $\Delta V = w \Delta x h(t^*)$, where w is the width of the sidewalk, and t^* is some value of time between t and $t + \Delta t$ $[h(t)$ would be too small and $h(t + \Delta t)$ too large]. Divide ΔV by Δt and let $\Delta t \to 0$ (forcing $t^* \to t$) to obtain the snow removal rate $r = \lim_{\Delta t \to 0} \Delta V/\Delta t = w \lim_{\Delta t \to 0}(\Delta x/\Delta t)h(t^*) = wx'(t)a(t - T)$. Let $b = r/wa$ to obtain the rate equation for the shoveler's walk: $x' = b/(t - T)$. So, $x(t) = b\ln(t - T) + C$, where $t \geq 0$ and C is a constant. Our goal is to find T. We have that $0 = b\ln(-T) + C$, $2 = b\ln(2 - T) + C$, $3 = b\ln(4 - T) + C$. Divide the difference of the first two equations by the difference of the last two to obtain $2 = \ln(\frac{2-T}{-T})/\ln(\frac{4-T}{2-T})$, where we have used $\ln y - \ln z = \ln(y/z)$. Cross-multiply by $\ln\frac{4-T}{2-T}$ and use $2\ln(u) = \ln(u)^2$ for $u > 0$ to obtain $\ln(\frac{4-T}{2-T})^2 = \ln(\frac{2-T}{-T})$. So $(\frac{4-T}{2-T})^2 = \frac{2-T}{-T}$. Clearing fractions, multiplying out and combining terms, we have that $T^2 - 2T - 4 = 0$. So, $T = 1 \pm \sqrt{1 + 4}$. Since T must be negative, we conclude that the snow began to fall at $T = 1 - \sqrt{5} = -1.236$ hours before noon, in other words, at 10:46 A.M.

Background Material: Modeling Principle

A guiding principle of modeling is Occam's Razor:

> Occam's Razor. What can be accounted for by fewer assumptions is explained in vain by more.

William of Occam (1285–1349) was an English theologian and philosopher who applied the Razor to arguments of every kind. The principle is called the Razor because Occam used it so often and so sharply.

Background Material: Radioactive Decay Law

A primitive form of the decay law would go something like this:

> Radioactive Decay Law. In a sample containing a large number of radioactive nuclei the decrease in the number over time is directly proportional to the elapsed time and to the number of nuclei at the start.

Denoting the number of radioactive nuclei in the sample at time t by $N(t)$, and a time interval by Δt, the law translates to the mathematical equation

$$N(t + \Delta t) - N(t) = -kN(t)\Delta t \tag{i}$$

where k is a positive coefficient of proportionality. Observation suggests that in most decay processes k is independent of t and N. A decay process of this type is said to be of *first order* with *rate constant k*.

The mathematical model (i) of the decay law helps us to spot flaws in the law itself. $N(t)$ and $N(t + \Delta t)$ must be integers, but $k\Delta t$ need not be an integer, or even a fraction. If we want to keep the form of the law, we must transform the real phenomenon into an idealization in which a continuous rather than a discrete amount $N(t)$ undergoes decay. The number of nuclei can be calculated from the mass of the sample at any given time. So there is no need to be specific about the units for N or t at this point. Even if $N(t)$ is continuous, (i) could not hold for arbitrarily large Δt since $N(t + \Delta t) \to 0$ as $\Delta t \to \infty$. Nor does (i) make sense if Δt is so small that no nucleus decays in the time span Δt. The failure of the law for large Δt can be ignored since we are interested only in local behavior in time. The difficulty with small Δt is troublesome. We can only hope that the mathematical procedures we now introduce will lead to a mathematical model from which accurate predictions can be made.

If we divide both sides of formula (i) by Δt and let $\Delta t \to 0$ (ignoring the difficulty with small Δt mentioned above), we have the linear ODE

$$N'(t) = \lim_{\Delta t \to 0} \frac{N(t + \Delta T) - N(t)}{\Delta t} = -kN(t) \tag{ii}$$

Note that k is measured in units of reciprocal time.

Background Material: Radiocarbon Dating

The accuracy of the dating process depends on a knowledge of the exact ratio of radioactive ^{14}C to the carbon in the atmosphere. The ratio is now known to have changed over the years. Volcanic eruptions and industrial smoke dump radioactively dead ^{14}C into the atmosphere and lower the ratio. But the most drastic change in recent times has occurred through the testing of nuclear weapons, which releases radioactive ^{14}C, resulting in an increase of 100% in the ratio in some parts of the Northern Hemisphere. These events change the ratio in the atmosphere and so in living tissue. The variations are now factored into the dating process.

Since the experimental error in determining the rate constants for decay processes with long half-lives is large, we would not expect to use this process to date events in the very recent past. Recent events can be dated if a radioactive substance with a short half-life is involved, e.g., white lead contains a radioactive isotope with a half-life of only 22 years. At the other extreme, radioactive substances such as uranium, with half-lives of billions of years, can be used to date the formation of the earth itself.

1.6   Separable Differential Equations

Suggestions for preparing a lecture

Topics: The details of solving a separable ODE to get an integral, graphing an integral curve, getting the general implicit solution, finding an explicit solution, the relation between a solution curve and an integral curve.

Remarks: What do we do with exact ODEs? Part of every traditional ODE book, the topic is often omitted entirely these days. We have compromised by treating exact ODEs only in the problem set (Problem 10). If you think exact ODEs should be treated, you may want to build a second lecture around Problem 10.

Making up a problem set

Two or three parts from Problems 2–4, all or part of Problem 6 (particularly if you have mentioned the whiffle ball problem in an earlier lecture). Problem 8 is nice, but hard. You may want to assign Problems 8 and 11 together as a group problem.

Comments

People often have difficulty with separable ODEs. If you have trouble separating the ODE into the form $N(y)y' + M(x) = 0$, try to move x and y onto different sides of the equals sign to get $N(y)dy = -M(x)dx$ and then integrate each side. You may still have trouble after getting the ODE separated because you can't find the antiderivatives $G(y)$ of N and $F(x)$ of M. Because the solution is written implicitly as $G(y) + F(x) = C$, it may be hard to find $y(x)$ explicitly (sometimes impossible). These technical difficulties often lead people astray. Consequently, we encourage the use of integral tables, or a CAS (if available), and checking (and rechecking) your algebra.

In this section, we explain the difference between an integral curve [the graph of the level set $G(y) + F(x) = C_0$] and a solution curve [the graph of $y = y(x)$, x in some interval, where $G(y(x)) + F(x) = C_0$]. Because a solution curve is the graph of a differentiable function $y = y(x)$, the curve cannot double back on itself and cannot have vertical tangents.

The modeling/application involves the motion of falling bodies subject to Newtonian damping. Problem 6(d) takes up the "longer to rise or to fall" problem first encountered in Section 1.5, but now in the setting of Newtonian damping. It still takes longer to fall than to rise.

1.   The ODE $y' = 2xy^2$ evidently has the constant solution $y = 0$, for all x. In addition, note that the ODE is separable and can be written in the separable form as $y' y^{-2} = 2x$. For $y \neq 0$, we can integrate the separated ODE, obtaining $-y^{-1} = x^2 + C$, or $y = -(x^2 + C)^{-1}$ where C is any constant.

  So the general solution of $y' = 2xy^2$ is given by $y = 0$ and the family of solutions $y = -1/(x^2 + C)$, C any constant. If $C \leq 0$, x must lie in an interval on which $x^2 + C \neq 0$. So for example if $C = -1$, we have three solutions on the three intervals, $x < -1$, $|x| < 1$, and $x > 1$.

2.   In each case the variables are separated, the separated terms are integrated (using a table of integrals if necessary), and the resulting equation is solved for y in terms of x (if this is not too hard to do). Otherwise, the equation is left in an implicit form. K and C denote arbitrary constants. Be sure to check that you don't lose solutions when you divide out in order to separate variables. Note that whenever we find that $|y|$ equals the product of some positive constant and some quantity, we can solve for y simply by letting the constant be any nonzero real. If we lose a solution through division, we can sometimes recover it by allowing the constant to be zero.

(a). The ODE $y' = -4xy$ is separable: $y^{-1} dy = -4x dx$, which integrates to $\ln|y| = -2x^2 + K$. Solving for y, and setting $C = \pm e^K$, $y = Ce^{-2x^2}$, $-\infty < x < +\infty$. Setting $C = 0$ yields the solution $y = 0$, which was lost when we divided to separate the variables. The ODE is also linear in y and could be solved by the integrating factor technique of Section 1.4.

(b). The ODE $2y\,dx + 3x\,dy = 0$ separates to $2x^{-1} dx = -3y^{-1} dy$ which integrates to $2\ln|x| + K = -3\ln|y|$. Exponentiating both sides, $x^2 e^K = |y^{-3}|$. Solving for y and replacing $\pm e^{-K/3}$ with $C \neq 0$, $y = Cx^{-2/3}$, $x < 0$ or $x > 0$. Also, $y = 0$ for all x is a solution of $2y\,dx + 3x\,dy = 0$, as is $x = 0$ for all y. These two solutions were lost when the ODE was written in separated form.

(c). The separated form of $y' = -xe^{-x+y}$ is $-e^{-y}dy = xe^{-x}dx$. So, $e^{-y} = C - xe^{-x} - e^{-x}$, and solutions are given by $y = -\ln(C - xe^{-x} - e^{-x})$, where x lies in an interval for which $C - xe^{-x} - e^{-x} > 0$.

(d). The ODE $(1-x)y' = y^2$ separates to $y^{-2}dy = (1-x)^{-1}dx$, so $-y^{-1} = -\ln|1-x| - C$. Solutions are given by $y = [C + \ln|1-x|]^{-1}$, where x is restricted to an interval (not containing 1) for which $C + \ln|1-x| \neq 0$. Also, $y = 0$ for all x is a solution that was lost when the ODE was written in separated form.

(e). The ODE $y' = -y/(x^2 - 4)$ separates to $y^{-1}dy = (4-x^2)^{-1}dx$, so $\ln|y| = (1/4)\ln|(x+2)/(x-2)| + K$. Solutions are given by $y = C|(x+2)/(x-2)|^{1/4}$, where $x \neq \pm 2$ and we have set $C = e^K$. Note that while the line $y = 0$ seems to define a single solution curve of the ODE, $(4-x^2)\,dy = y\,dx$, it actually defines three solution curves [one each for $x < -2$, $-2 < x < 2$, and $x > 2$].

(f). The ODE $y' = xe^{y-x^2}$ separates to $e^{-y}dy = xe^{-x^2}dx$ which integrates to $-e^{-y} = -e^{-x^2}/2 + K$. Solving for y gives $y = \ln[2/(C + e^{-x^2})]$ after replacing $-2K$ by C; x is in an interval for which $C + e^{-x^2} > 0$.

3.   **(a).** Separating variables in $y' = (y+1)/(x+1)$, we have $(y+1)^{-1}y' - (x+1)^{-1} = 0$, which integrates to $\ln|y+1| - \ln|x+1| = K$. So, $\ln(|y+1|/|x+1|) = K$ and $|y+1| = |x+1|e^K$. That is, $y+1 = \pm e^K(x+1)$, i.e., $y = -1 + C(x+1)$, where we have renamed $\pm e^K$ by C. Since $y(1) = 1$, $1 = -1 + 2C$, so $C = 1$. The solution becomes $y = x$, $-1 < x < \infty$, since y' is undefined at $x = -1$.

(b). The ODE $y' = y^2/x$ separates to $y^{-2}y' - x^{-1} = 0$; integrating, we have $-y^{-1} - \ln|x| = C$. The initial condition $y(1) = 1$ implies that $-1 = C$. So $y = (1 - \ln|x|)^{-1}$. Since $x = 1$ must lie in the domain of definition, but $x = 0$ and $x = e$ cannot lie in the domain, the solution is defined only for $0 < x < e$. See Fig. 3(b) for the solution curve. The dashed vertical lines are at $x = 0, e$.

(c). The ODE $y' = ye^{-x}$ separates to $y^{-1}\,dy = e^{-x}\,dx$, which integrates to $\ln|y| = -e^{-x} + C$, so $y = \pm\exp(C - e^{-x})$. The initial data $y(0) = e$ imply that $C = 2$, and we take the positive solution. So, $y = \exp(2 - e^{-x})$ for all x.

(d). The ODE $y' = 3x^2/(1 + x^3)$ integrates immediately to $y(t) = \ln(1 + x^3) + C$. The initial data $y(0) = 1$ imply that $C = 1$, so $y = 1 + \ln|1 + x^3|$, $x > -1$.

(e). The ODE $y' = -x/y$ separates to $y\,dy + x\,dx = 0$, which integrates to $y^2 + x^2 = C$. Since $y(1) = 2$, $C = 5$ and solving for y the solution is $y = (5 - x^2)^{1/2}$, $|x| \le 5^{1/2}$. See Fig. 3(e) for the solution curve.

(f). The ODE separates to $2y(1 + y^2)^{-1}\,dy = x^{-1}\,dx$, which integrates to $\ln(1 + y^2) = \ln|x| + C$. The initial data $y(2) = 3$ imply that $C = \ln 5$, so $\ln(1 + y^2) = \ln(5x)$. So, $y = (5x - 1)^{1/2}$, $x > 1/5$.

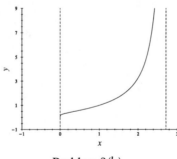   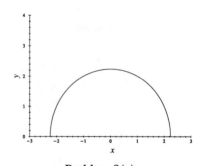

Problem 3(b).   Problem 3(e).

4.   **(a).** The ODE $y' = (x^2 + 2)(y + 1)/xy$ separates to $[y/(1 + y)]\,dy = [(x^2 + 2)/x]\,dx$, or $[1 - 1/(1 + y)]\,dy = (x + 2/x)\,dx$. Integrating, we see that the solutions are implicitly defined by $y - \ln|1 + y| = x^2/2 + 2\ln|x| + C$.

(b). The ODE $(1 + \sin x)dx + (1 + \cos y)dy = 0$ is already separated and immediately integrates to give the implicit solutions $x - \cos x + y + \sin y = C$.

(c). The solutions of the separated ODE $(\tan^2 y)dy = (\sin^3 x)dx$ are given implicitly by $\tan y - y + \cos x(\sin^2 x + 2)/3 = C$, where y is restricted to an interval of the form $k\pi - \pi/2 < y < k\pi + \pi/2$ because $\tan((2k + 1)\pi/2) = \pm\infty$.

(d). The ODE, $(3y^2 + 2y + 1)y' = x\sin(x^2)$, is already separated. It integrates to $y^3 + y^2 + y = -(1/2)\cos(x^2) + C$.

5. **(a).** By formula (15) for the solution of $y' = r(1 - y/K)y$, $y(0) = y_0$, we obtain $y = y_0 K/(y_0 + (K - y_0)e^{-rt})$. For the ODE $y' = (1 - y/20)y$, $r = 1$ and $K = 20$. We are also given initial conditions $y_0 = 5, 10, 20$, or 30, so the respective solutions are given by $y = 20y_0/(y_0 + (20 - y_0)e^{-t})$ where $y_0 = 5, 10, 20, 30$, respectively.

   (b). See Fig. 5(b). The saturation level (or carrying capacity) is $y = K = 20$.

   (c). The equilibrium (i.e., constant) solutions of $y' = 3(1 - y/12)y - 8$ are found by setting y' equal to zero and using the quadratic formula (or factoring) to find the roots. We have $3(1 - y/12)y - 8 = -\frac{1}{4}(y^2 - 12y + 32) = -\frac{1}{4}(y - 4)(y - 8)$, and so the equilibrium levels are $y = 4, 8$. This means that if y_0 lies between 0 and 4, the species falls away from the equilibrium $y = 4$ and dies out since $y' < 0$ in this region (e.g., $y_0 = 2$), while if y_0 is between 4 and 8 (e.g., $y_0 = 6$) the population follows a logistic path and is asymptotic upward to the equilibrium level $y = 8$ since $y' > 0$ in this region. If y_0 exceeds 8 (e.g., $y_0 = 10$), the population falls asymptotically toward the equilibrium level $y = 8$ since $y' < 0$ in this region.

   (d). See Fig. 5(d). The equilibrium levels are at $y = 4, 8$.

   (e). The logistic model is $y' = r(1 - y/K)y$, where $K = 5 \times 10^8$ and $r = 0.01$/day. The solution is given by formula (15) in the text: $y(t) = y_0 K/(y_0 + (K - y_0)e^{-0.01t})$. Set $t = 2$, $y_0 = 10^8$ and obtain $y(2) \approx 1.016 \times 10^8$ individuals.

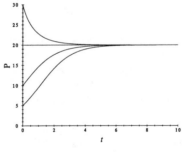
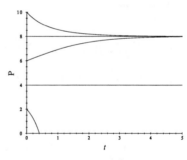

Problem 5(b).   Problem 5(d).

6. **(a).** Starting with $my'' = -mg - kv|v|$ from equation (16), divide through by m, replace y'' by v', and note that $w = mg$. So, $v' = -g - (gk/w)v|v|$. From Example 1.6.7, the limiting velocity is $(-mg/k)^{1/2}$. Solving $-400 = -(mg/k)^{1/2}$ for k, we have $k = mg/(400)^2 = 100/(400)^2 = 1/1600$ lb·sec^2/ft^2.

   (b). The system is $y' = v$, $v' = -g - (gk/w)(v|v|) = -32[1 + v|v|/160000]$.

   (c). The initial data are $y(0) = 0$, $v(0) = 500$ ft/sec. See Fig. 6(c) for the graph of position (vertical axis) against velocity (horizontal axis). The desired velocity can be read off the plot by finding where the solution curve crosses the v-axis. The value is approximately -315 ft/sec. Fig. 6(c) is produced by using a numerical solver for the system of 6(b) with initial conditions $y(0) = 0$, $v(0) = 500$.

(d). Figure 6(d), Graphs 1 and 2 show the height of the projectile above the ground as a function of time. Figure 6(d), Graph 1 shows (after a careful visual determination of the t-coordinate of the highpoint) that it takes longer for the projectile to fall than to rise. See Fig. 6(d), Graph 2 for similar graphs but with $v_0 = 100, 200, \ldots, 1000$ ft/sec (from the bottom curve upward). It is easier to see from the graph that the fall time is longer than the rise time if the initial velocity is large. On the way up, the projectile has the force of gravity and air resistance acting together in the same direction, but on the way down air resistance acts upward while gravity still acts down. The result is acceleration whose magnitude is greater for the trip up than for the trip down. We do not prove mathematically that the fall time exceeds the rise time, but rely on the visual evidence from the graphs. Graphs 1 and 2 are produced by applying a numerical solver as in 6(c), but plotting y against t instead of y against v.

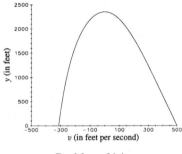

Problem 6(c).

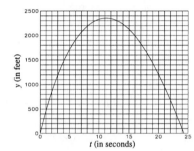

Problem 6(d), Graph 1.

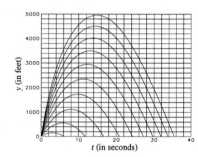

Problem 6(d), Graph 2.

7. Observe that the ODE $v' = -g + (k/m)v^2$ is separable. Separating the variables, we obtain

$$\int \frac{1}{-g + (k/m)v^2} \, dv = \int dt$$

Using partial fractions to carry out the left-hand integration, we have

$$\frac{1}{2}\left(\frac{m}{gk}\right)^{1/2} \ln\left|\frac{v - (mg/k)^{1/2}}{v + (mg/k)^{1/2}}\right| = t + C$$

Plugging in the initial condition $v(0) = 0$ gives $C = 0$. Setting $A = 2(gk/m)^{1/2}$ and taking exponentials of both sides we obtain

$$\left|\frac{v - (mg/k)^{1/2}}{v + (mg/k)^{1/2}}\right| = e^{At}$$

Note that the quantity between the absolute value bars is negative for all $t \geq 0$, and so changing signs and dropping the absolute value, we solve for v to obtain

$$v(t) = \left(\frac{mg}{k}\right)^{1/2} \frac{e^{-At} - 1}{e^{-At} + 1}$$

8. **(a).** Referring to equation (17) in the text, we see that $v' = -g + (k/m)v^2$. We know the value of all constants except the drag coefficient, k, of the closed parachute. Since $v' = 0$

at the limiting velocity $v_\infty = -(mg/k)^{1/2}$, we have from $0 = v' = -g + (k/m)v^2$ that

$$k = mg/v_\infty^2 = \frac{(240\,\text{lb})}{(250\text{ft/sec})^2} = 0.00384\,(\text{lb})(\text{sec})^2/\text{ft}^2$$

Using Equation (18), we have (after a lot of algebraic manipulation to solve for t)

$$t = \frac{1}{2}\left(\frac{m}{gk}\right)^{1/2} \ln\left|\frac{-(mg/k)^{1/2} + v}{(mg/k)^{1/2} + v}\right|$$

Noting that $(m/gk)^{1/2} = (1/g)(mg/k)^{1/2} = |v_\infty|/g$, we compute

$$t = \frac{1}{2} \cdot \frac{250}{32.2} \ln\left|\frac{-250 - 100}{250 - 100}\right| = 3.29\,\text{sec}$$

Recall that $v < 0$ because "up" is defined as positive, and the sky diver is falling.

(b). For the open parachute, $k = mg/v_\infty^2 = (240)/(17^2) = 0.8304$ (lb)(sec)^2/ft^2. Separating the variables in the differential equation $v' = -g + (k/m)v^2$ given in (17) and integrating, we have

$$\int_{100}^{25} \frac{1}{-g + (k/m)s^2}\,ds = \int_{3.29}^{t} dr$$

Using a table of integrals,

$$\frac{1}{2}\left(\frac{m}{gk}\right)^{1/2} \ln\left|\frac{s - (mg/k)^{1/2}}{s + (mg/k)^{1/2}}\right|\Bigg|_{100}^{25} = t - 3.29$$

Let $A = (m/gk)^{1/2} = (mg/k)^{1/2}/g = (1/g)v_\infty = [1/(32\,\text{ft/sec}^2)](17\,\text{ft/sec}) = 0.53$ sec. Then,

$$t = 3.29 + \frac{1}{2}(17)\left[\ln\left|\frac{25 - 17}{25 + 17}\right| - \ln\left|\frac{100 - 17}{100 + 17}\right|\right] = 3.64\,\text{sec.}$$

So, $0.35 = 3.64 - 3.29$ seconds after pulling the rip cord, the sky diver's speed has dropped to 25 ft/ sec.

(c). Use a numerical solver to solve the system

$$y' = v, \quad y(0) = 10,000$$

$$v' = -g + \frac{[k_1 + (k_2 - k_1)\,\text{step}(t - 3.29)]}{m}v^2, \quad v(0) = 0$$

where k_1 and k_2 are the damping constants of the parachutist with closed and open parachute, respectively. Note that the function step$(t - C)$ has value 0 if $t < C$ and 1 for $t \geq C$. Find the t-value where the ty-component plot crosses the t-axis. We see from Fig. 8(c), Graph 1 that this occurs at approximately $t = 580$ seconds. The region of interest is blown up in Fig 8(c), Graph 2, so we can see that the total time of descent is 581 seconds. We could plot v vs t to find the velocity on landing, but since the y vs t graph is basically linear for the time the parachute is open, we know $v(581)$ is essentially the limiting velocity of -17 ft/ sec.

One could get an excellent estimate of the duration of the jump by using the formula $y = h - (m/k)\ln\{\cosh[(gk/m)^{1/2}t]\}$. This formula can be obtained by integrating each side of (18) after replacing $(e^{-At} - 1)/(e^{-At} + 1)$ by

$$\frac{e^{-At} - 1}{e^{-At} + 1} = \frac{e^{-At/2} - e^{At/2}}{e^{At/2} + e^{-At/2}} = -\frac{\sinh At/2}{\cosh At/2} = -\tanh\frac{At}{2}$$

We find $y(3.29) = 9.83 \times 10^3$ ft when the rip cord is pulled. We see (from part **(b)**) that the terminal velocity of -178 ft/ sec is obtained very quickly, so the remainder of the jump takes about 578 sec, and the total time is approximately 581 sec, or almost 10 minutes.

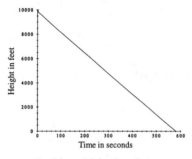

Problem 8(c), Graph 1.

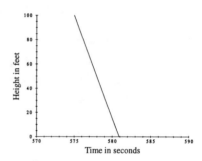

Problem 8(c), Graph 2.

9. **(a).** The rate function for the ODE $y' = y(1 - y/10) - 9/10$ is the quadratic, $-y^2/10 + y - 9/10$, which factors to yield $y' = -(y - 1)(y - 9)/10$, $y(0) = y_0$. The equilibrium populations are $y = 1$ and $y = 9$ because y' is zero if $y = 1, 9$. From the factorization of the rate function in the IVP we see that the population declines if $y > 9$ or if $0 < y < 1$ because y' is negative in those regions; $y(t)$ increases if $1 < y < 9$. Separating variables in the ODE and using partial fractions,

$$\frac{1}{y - 1} \cdot \frac{1}{y - 9} dy = -\frac{1}{10} dt, \quad y \neq 1, 9$$

$$\frac{1}{8}\left(\frac{1}{y - 9} - \frac{1}{y - 1}\right) dy = -\frac{1}{10} dt$$

$$\ln\left|\frac{y - 9}{y - 1}\right|^{1/8} = -\frac{1}{10}t + b, \quad b \text{ a constant}$$

$$\left|\frac{y - 9}{y - 1}\right|^{1/8} = e^b e^{-t/10}$$

$$\frac{y - 9}{y - 1} = Be^{-4t/5}, \quad B = \pm e^{8b}$$

Solving for y and substituting initial conditions $y(0) = y_0$, we obtain

$$y = \frac{9 - Be^{-4t/5}}{1 - Be^{-4t/5}}, \quad B = \frac{y_0 - 9}{y_0 - 1}, \quad y_0 \neq 1$$

(b). The population curves in Figure 8(b) show the expected behavior; note the horizontal lines at the equilibrium population levels of $y = 1$ and $y = 9$.

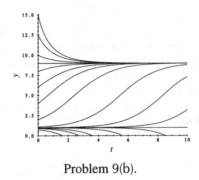

Problem 9(b).

10. Group project.

11. Group project.

1.7 Planar Systems and First-Order ODEs

Suggestions for preparing a lecture

Topics: How to go from first-order ODEs to first-order planar systems and back (with graphs), and either the combat model or reduction methods and escape velocities (but not both).

Remarks: Occasionally, we take two lectures on this section and do both combat models and escape velocities.

Making up a problem set

Two or three parts from Problems 1, 2, and Problems 3, 10 (if you talked about the combat model) or else two parts of Problems 4, 5 (if you did reduction methods) and Problem 8 or 9.

Comments

Planar systems come up in this section in a completely natural way—as a reformulation of second-order ODEs that were treated in Section 1.5. Later we show how to use planar systems to visualize solution curves of first-order ODEs; a big advantage of this approach is that one can immediately "see" the maximally-extended solution curves of the first-order ODE. Another advantage of this approach is that a contour plotter need not be used even when an integral of the first-order ODE is known. This neat trick is very useful in many applications, but the concept may seem rather tricky at first.

Another connection between second-order ODEs and first-order ODEs is made when we show how to reduce the second-order ODEs $y'' = F(t, y')$ and $y'' = F(y, y')$ to systems of first-order

ODEs that can be solved one at a time. The algebraic complications in actually constructing the solutions can be difficult. If you use a computer solver to plot solution curves of a second-order ODE, you may also want to use the solver for the corresponding system and then plot *ty*-component graphs.

For the second-order ODEs in *y* solved here, solution curves in the *ty*-plane can cross each other. This is because initial position *and* initial velocity are needed to specify a unique solution of a second-order ODE. If distinct solution curves cross, they do so with different slopes.

The section has a proof of the existence of the escape velocity from a massive body. Besides being a nice example of the use of the reduction of order technique, it also provides a striking example of what the computer cannot do i.e., prove the existence of an escape velocity. The section ends with a run-through of a Lanchester combat model, which starts with a system and ends with a separable ODE. Some people are put off by the use of mathematics in this setting. Finally, take a look at the background material on Einstein, general relativity, and the big bang.

---

1. Our basic strategy is to find integral curves of $N(x, y)y' + M(x, y) = 0$ by finding solutions of the system $dx/dt = N(x, y)$, $dy/dt = -M(x, y)$. The system can be used to generate plots on a computer solver. $\int N(y)\,dy + \int M(x)\,dx$ is an integral of the ODE when $N(x, y) = N(y)$ and $M(x, y) = M(x)$.

    (a). $y - 1/3y^3 + 1/3x^3 - x$ is an integral of the separable ODE $(1 - y^2)y' + x^2 - 1 = 0$. Plots of some integral curves are shown in Fig. 1(a).

    (b). $y - 1/3y^3 + 1/3x^3$ is an integral of $(1 - y^2)y' + x^2 = 0$. Plots of some integral curves are shown in Fig. 1(b).

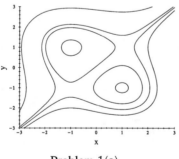
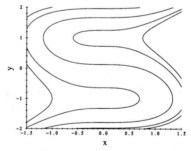

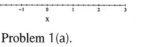

Problem 1(a).   Problem 1(b).

2. **(a)–(c).** Each solution is the largest continuous part of the integral curve that goes through the initial point and has no vertical slope. Part **(a)**'s solution is valid as $x \to -\infty$. Otherwise, in each case the solution of each IVP tends to an *x*-value where the slope becomes infinite, and so the solution must stop. See Figs. 2(a)–2(c) for the solution curves corresponding to the solutions of the given IVPs.

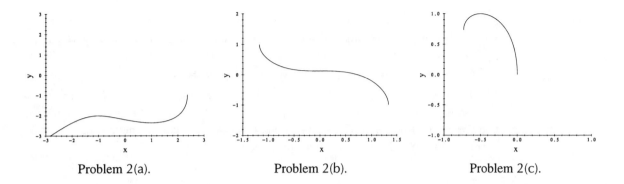

Problem 2(a).   Problem 2(b).   Problem 2(c).

3.   **(a).** From Fig. 3(a), we see that y loses if $x(0) = 10$ and $y(0) = 7$ since $y(t) \to 0$ as t increases, while $x(t) \to 4.4$.

     (b). Figure 3(b) shows the ty-plot of the orbit shown in Fig. 3(a): $y = 0$ when $t \approx 17$. So after 17 units of time (hours?, days?, years?) the struggle is over, and x has won.

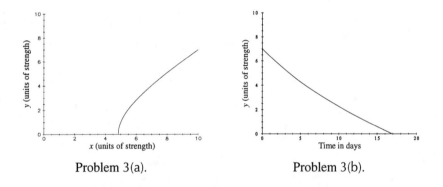

Problem 3(a).   Problem 3(b).

4.   Distinct solution curves $y = y_1(t)$ and $y = y_2(t)$ of a normal second-order ODE may intersect, but if they do, they do so at different slopes. Figures 4(a)–4(d) show a number of these crossing solution curves.

     (a). Let $v = y'$. The ODE $ty'' - y' = 3t^2$ becomes $tv' - v = 3t^2$, so $v' - t^{-1}v = 3t$, which is a first-order linear equation in v and has solutions $v = 3t^2 + Ct$. So, $y = \int^t (3s^2 + Cs)ds = t^3 + Ct^2/2 + C_2$ or $y = t^3 + C_1t^2 + C_2$, where C_1 and C_2 are any constants. See Fig. 4(a).

     (b). Let $v = y'$, $v\,dv/dy = y''$. The ODE $y'' - y = 0$ becomes $v\,dv/dy - y = 0$, or $v\,dv = y\,dy$, and so $v^2 = y^2 + C$. So, $y' = \pm(y^2 + C)^{1/2}$, which separates to $(y^2 + C)^{-1/2}\,dy = \pm dt$. Integrating, $\ln|y + (y^2 + C)^{1/2}| = \pm t + K_1$, or $y + (y^2 + C)^{1/2} = Ke^{\pm t}$, $(y - Ke^{\pm t})^2 = y^2 + C$, or $-2Kye^{\pm t} + K^2e^{\pm 2t} = C$, from which $y = C_1e^t + C_2e^{-t}$, where C_1 and C_2 are any constants. Note that we redefined the integration constants several times for convenience. See Fig. 4(b).

     (c). Let $v = y'$, $v\,dv/dy = y''$. The ODE $yy'' + (y')^2 = 1$ becomes $yv\,dv/dy + v^2 = 1$, which separates to $v(1 - v^2)^{-1}dv = y^{-1}dy$. Integrating, $-\ln|1 - v^2|/2 = \ln|y| + K$, or

$\ln(y^2|1 - v^2|) = -2K$. So, $y^2(1 - (y')^2) = C_1$, or $y' = \pm y^{-1}(-C_1 + y^2)^{1/2}$. So, $y(y^2 - C_1)^{-1/2}dy = \pm dt$. Integrating, $(y^2 - C_1)^{1/2} = \pm t + C_2$. Solving for y, $y^2 = C_1 + (C_2 \pm t)^2$. So, we have the four solution formulas, $y = \pm[C_1 + (C_2 \pm t)^2]^{1/2}$, where C_1 and C_2 are arbitrary constants and t is such that $C_1 + (C_2 \pm t)^2 \geq 0$. See Fig. 4(c).

(d). Let $y' = v$, $y'' = v'$. The ODE $y'' + 2ty' = 2t$ becomes $v' + 2tv = 2t$, a first-order linear equation with solutions $v = 1 + C_1 e^{-t^2}$. So, $dy/dt = 1 + C_1 e^{-t^2}$ and integrating we find $y = t + C_1 \int^t e^{-s^2} ds + C_2$, where C_1 and C_2 are any constants. The integral cannot be expressed in terms of elementary functions. See Fig. 4(d).

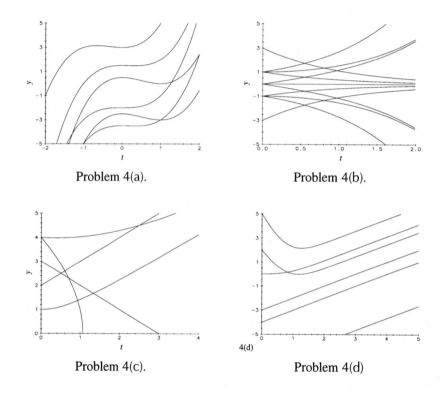

Problem 4(a).   Problem 4(b).

Problem 4(c).   Problem 4(d)

5.   **(a).** Let $y' = v$, $y'' = v\,dv/dy$. The ODE $2yy'' + (y')^2 = 0$ becomes $2yv\,dv/dy + v^2 = 0$. So $v = 0$ for all t, or $2y\,dv/dy + v = 0$. If $v = 0$ for all t, then y is a constant, but the initial condition $y'(0) = -1$ implies that y is nonconstant. So, v is not always 0. Separating variables, we have $2v^{-1}dv + y^{-1}dy = 0$. Integrating, $\ln|v^2 y| = K$, so $v^2 y = \pm e^K$. So, $v = y' = K_1 y^{-1/2}$. Separating variables, $y^{1/2}dy = K_1 dt$. Integrating, $2y^{3/2}/3 = K_1 t + K_2$ or $y = (C_1 t + C_2)^{2/3}$, where C_1 and C_2 are constants to be determined by the initial data $y(0) = 1$, $y'(0) = -1$. So, $C_2 = 1$, $C_1 = -3/2$. The solution is $y = (-3t/2 + 1)^{2/3}$.

(b). Let $y' = v$, $y'' = v\,dv/dy$. The ODE $y'' = y'(1 + 4/y^2)$ becomes $v\,dv/dy = v(1 + 4/y^2)$, which implies that $v = 0$ or $dv/dy = 1 + 4y^{-2}$. Since $y'(0) = v(0) = 3$, $v \neq 0$ and hence that alternative is discarded. Solving $dv/dy = 1 + 4y^{-2}$, we have that $v = y' =$

$y - 4y^{-1} + C_1$. Since $v = 3$ and $y = 4$ at $t = 0$, C_1 must be 0. So, $y' = y - 4y^{-1}$, which separates to $y(y^2 - 4)^{-1}dy = dt$. Integrating, $[\ln|y^2 - 4|]/2 = t + \bar{C}$, so $y^2 - 4 = Ce^{2t}$. Since $y = 4$ at $t = 0$, $C = 12$ and $y = (4 + 12e^{2t})^{1/2}$, all t.

(c). Let $y' = v$, $y'' = v'$. The ODE $y'' = -g - y'$ becomes $v' = -g - v$ or $(g + v)^{-1}dv = -dt$. Integrating, $\ln|g + v| = -t + K$, so $g + v = C_1e^{-t}$. Since $v = y'$, we have $y' = C_1e^{-t} - g$. Integrating, $y = -C_1e^{-t} - gt + C_2$. Since $y = h$ and $y' = 0$, at $t = 0$, we have $y = -ge^{-t} - gt + h + g$.

6. (a). Suppose that $z(t) \neq 0$ solves $z'' + a(t)z' + b(t)z = 0$ and $u(t)$ solves $zu'' + (2z' + az)u' = f$. We want to show that $y = uz$ solves $y'' + ay' + by = f$. Substituting $y(t) = u(t)z(t)$ into $y'' + a(t)y' + b(t)y = f(t)$ and noting that $z'' + a(t)z' + b(t)z = 0$, the ODE becomes $zu'' + (2z' + az)u' = f(t)$. So if u' solves this ODE, then $y(t) = u(t)z(t)$ solves the original ODE.

(b). Normalizing $zu'' + (2z' + az)u' = f$ as an equation in u', we have $(u')' + (2z'/z + a)(u') = f/z$. This is a linear ODE, and may be solved for u' using the integrating factor $\exp[\int 2z'(t)/z(t) + a(t)\,dt] = \exp[2\ln z + A(t)] = z^2\exp[A(t)]$, where $A(t)$ is an antiderivative of $a(t)$. We obtain $(u'z^2\exp[A(t)])' = f/2$. Integrating, $u'z^2\exp[A(t)] = \int(f/2)\,dz + C$ so

$$u' = \frac{1}{z^2\exp[A(t)]}\left[\int(f/2)\,dz + C\right]$$

(c). Normalizing the equation $tz'' - (t + 2)z' + 2z = 0$, we have $z'' - (1 + 2/t)z' + (2/t)z = 0$. Now since $z(t) = e^t$ solves the equation, $y(t) = u(t)e^t$ solves the equation, where $u(t)$ solves the ODE $e^tu'' + (2e^t - (1 + 2/t)e^t)u' = 0$. This linear equation in u' may be solved for $u' = t^2e^{-t}$. Integrating, we find $u(t) = (-t^2 - 2t - 2)e^{-t}$. So, $y(t) = u(t)e^t = -t^2 - 2t - 2$ is a second solution to the original equation.

7. The normalized ODE is $y'' + 4t^{-1}y' + 2t^{-2}y = t^{-2}\sin t, t > 0$. Following the technique described in Problem 6 and using the fact that $z = t^{-2}$ is a solution of $z'' + 4t^{-1}z' + 2t^{-2}z = 0$ (check by direct substitution), we see that $y = t^{-2}u(t)$, where $u(t)$ solves $t^{-2}u'' + (-4t^{-3} + 4t^{-3})u' = t^{-2}\sin t$, solves the normalized ODE. This last equation simplifies to $u'' = \sin t$. After two integrations, we have $u = -\sin t + C_1t + C_2$, where C_1 and C_2 are any constants. So $y = -t^{-2}\sin t + C_1t^{-1} + C_2t^{-2}$, $t > 0$, gives a family of solutions of the ODE.

8. (a). The mass of the object times its acceleration equals the net force acting on the object. The equation of motion is $my'' = -GmM(b^2 + y^2)^{-1}\sin\theta$, where G is the universal gravitational constant and M is the mass of the earth. The minus sign in front of the gravitational force is due to the fact that the acceleration is always directed towards the midpoint of the tunnel and so is negative for $\sin\theta > 0$ (above the x-axis here) and positive for $\sin\theta < 0$ (below the x-axis here). Since $\sin\theta = y(b^2 + y^2)^{-1/2}$, we have that $y'' = -GM(b^2 + y^2)^{-3/2}y$, $y(0) = (R^2 - b^2)^{1/2}$, $y'(0) = 0$. The initial data state that the object is released from rest at the surface of the earth.

(b). Let $y' = v$, $y'' = v\,dv/dy$ and divide by m to obtain the separated equation $v\,dv = -GM(b^2 + y^2)^{-3/2}y\,dy$, which integrates to $v^2 = 2GM(b^2 + y^2)^{-1/2} + C = -2GM(R^{-1} -$

$(b^2 + y^2)^{-1/2})$, where the initial data $y(0) = (R^2 - b^2)^{1/2}$, $y'(0) = 0$ have been used to evaluate C. So, $[(b^2 + y^2)^{-1/2} - R^{-1}]^{-1/2}dy = (2GM)^{1/2}dt$, or $\int_{y_0}^{y}[(b^2 + u^2)^{-1/2} - R^{-1}]^{-1/2}du = (2GM)^{1/2}t$, where $y_0 = (R^2 - b^2)^{1/2}$. The integral cannot be evaluated in term of elementary functions.

9. **(a).** The mass of the object times its acceleration equals the net force acting on the object. The equation of motion is $mz'' = -mM\tilde{G}/(z + R)^3$. Let $v = z'$, $v\,dv/dz = z''$ and separate the z and v variables to obtain $v\,dv = -M\tilde{G}(z + R)^{-3}dz$, which integrates to $v^2 = M\tilde{G}(z + R)^{-2} + v_0^2 - M\tilde{G}R^{-2}$ where it is assumed that at $t = 0$, $v = v_0$ and $z = 0$. So, v remains positive for all z if $v_0 > (M\tilde{G})^{1/2}/R$, which is the escape velocity.

(b). The ratio of the two escape velocities is given by

$$\frac{(M\tilde{G})^{1/2}/R}{(2MG/R)^{1/2}} = \left(\frac{\tilde{G}}{2GR}\right)^{1/2}$$

10. Group project.

11. Group project.

12. Group project.

Background Material: Einstein's Field Equations of General Relativity

These equations are a complicated system of nonlinear partial differential equations. In 1922, the Soviet mathematician Alexander Alexandrovich Friedmann succeeded in obtaining cosmological solutions governing the behavior of the universe as a whole. He assumed that on a broad scale the universe is homogeneous and isotropic and that pressure can be ignored (a dubious hypothesis if the universe is "small"). The PDEs reduce to a single nonlinear IVP

$$2RR'' + R'^2 + kc^2 = 0, \qquad R(t_0) = R_0 > 0, \qquad R'(t_0) = v_0 > 0 \tag{i}$$

where R is the "radius of the universe," $k = +1$ (spherical geometry), 0 (Euclidean geometry), or -1 (pseudospherical geometry), c is the speed of light, and the derivatives are with respect to time.

For each of the three cases, $k = 0, +1, -1$, decide whether the universe is born with a "big bang" [i.e., at some time t_1, as $t \to t_1^+$, $R(t) \to 0^+$, and $R'(t) \to +\infty$]. In each case, what happens to the universe as t increases? Does the universe die in a "big crunch" [i.e., for some t_2, as $t \to t_2^-$, $R(t) \to 0^+$, and $R'(t) \to -\infty$]. In carrying out this project, address the following points.

- Multiply the ODE in (i) by R' and rewrite as $[R(R')^2 + kc^2R]' = 0$. Show that $R(R')^2 + kc^2R = R_0v_0^2 + kc^2R_0 = C_0$.

- Set $k = 0$ (a Euclidean universe) and show that $R^{3/2} = R_0^{3/2} \pm 1.5C_0^{1/2}(t - t_0)$. Show that the minus sign leads to the eventual collapse of the universe (the "big crunch") at some time $T > t_0$, while the plus sign leads to perpetual expansion. What happens in both cases as $R_0 \to 0^+$?

- Set $k = 1$ (a spherical universe). Solve the ODE, $R(R')^2 + c^2 R = C_0$, in terms of a parameter u, by setting $R = C_0 c^{-2} \sin^2 u$, rewriting the ODE with u as the dependent variable, and solving to find t in terms of u. Explain why $t = t(u)$, $R = R(u)$ are the parametric equations of a cycloid in the tR-plane. Interpret each arch of the cycloid in terms of a big bang and big crunch cosmology.

- Set $k = -1$ (a pseudospherical universe). Follow the steps of the spherical case, but set $R = C_0 c^{-2} \sinh^2 u$. Interpret the solution in cosmological terms.

- Rewrite the original ODE as a system of ODEs in dimensionless variables: $dx/ds = 2y$, $dy/ds = -x^{-1}(k + y^2)$, where $x(s_0) = 1$, $y(s_0) = v_0/c$, $x = R/R_0$, $y = R'/c$, $s = ct/2R_0$. Now use an ODE solver/graphics package to plot solutions $x = x(s)$ and $y = y(s)$ as functions of s for $k = 0, +1, -1$. Interpret the graphs in cosmological terms. Discuss the advantage of scaling the original variables as indicated before computing.

Solution:

Multiply the differential equation by R' to obtain $2RR'R'' + (R')^3 + kc^2 R' = 0$, or $(R(R')^2 + kc^2 R)' = 0$. Let $R = R_0$, $R' = v_0$ at $t = t_0$. Then $R(R')^2 + kc^2 R = C_0 = R_0 v_0^2 + kc^2 R_0$.

If $k = 0$, separate variables to obtain $R^{1/2} dR = \pm C_0^{1/2} dt$, which integrates to $R^{3/2} = R_0^{3/2} \pm 3 C_0^{1/2}(t - t_0)/2$. Since $R = 0$ at $t = t_1 = 2R_0^{3/2} C_0^{-1/2}/3$, set $T = t - t_1$ to obtain $R = AT^{2/3}$, where $A = (3/2)^{2/3} C_0$. The plus sign in $R^{1/2} dR = \pm C_0^{1/2} dt$ carries down in the preceding algebra to the minus sign in the definition of t_1. At $t = t_1$, $R = 0$ and the universe is "created" in a "big bang", the latter term being used since $dR/dT = 2AT^{-1/3}/3$, which is infinite at $T = 0$, that is, at $t = t_1$. Observe that the universe subsequently expands like $T^{2/3}$ as T increases. On the other hand, if the minus sign is used in the differential equation for R, then $t_1 = t_0 + 2R_0^{3/2} C_0^{-1/2}/3$ and for $T = t - t_1 < 0$ we have that $R = AT^{2/3}$ decreases to 0 as T increases to 0. The universe expires in a "big crunch" since $|dR/dT|$ tends to infinity as $T \to 0$. Graph 1 shows the universe expanding (solid curve) from $R = 0.001$ at $t = 0.001$ (right after a big bang) if $R' = R^{-1/2}$. The dashed curve shows a future big crunch (dashed curve) if $R(0) = 5$ and $R' = -R^{-1/2}$.

Let $k = 1$. The ODE $R(R')^2 + c^2 R = C_0$ is solved parametrically by expressing both R and t in terms of a parameter u. Let $R = C_0 c^{-2} \sin^2 u = C_0(1 - \cos 2u)/2c^2$. Replacing R' by $(dR/du)(du/dt)$ in the ODE, and separating variables, we have $du/dt = \pm c^3 (2C_0 \sin^2 u)^{-1}$, or $dt/du = \pm 2C_0 \sin^2 u/c^3$. So, $t = \pm C_0(2u - \sin 2u)/(2c^3)$, where we assume $t = 0$ and $R = 0$ at $u = 0$. Since we require $t \geq 0$ here, we have for $u \geq 0$

$$t = \frac{C_0}{2c^3}(2u - \sin 2u), \qquad R = \frac{C_0}{2c^2}(1 - \cos 2u)$$

which are the parametric equations of a cycloid $\pi C_0/c^3$ since $u = \pi$ corresponds to $t =$

$\pi C_0/c^3$ and $R = 0$ at $u = n\pi$, $n = 0, 1, 2, \ldots$. Note that

$$\left|\frac{dR}{dt}\right| = \left|\frac{dR/du}{dt/du}\right| = \left|c\frac{2\sin 2u}{2 - 2\cos 2u}\right| \to \infty \quad \text{as} \quad u \to n\pi$$

This model leads to periodic expansion and collapse, that is, to alternating episodes of big bang and big crunch. Graph 2 shows the alternating bangs and crunches of a universe modeled by the spherical geometry corresponding to $k = 1$.

The ODE here is $R(R')^2 - c^2R = C_0$ since $k = -1$. Following the steps of 7(c) but using $R = C_0 c^{-2}\sinh^2 u = C_0(\cosh 2u - 1)/2c^2$ instead of $C_0 c^{-2}\sin^2 u$ and using appropriate indentities for hyperbolic functions, we have $t = C_0(\sinh 2u - 2u)/(2c^3)$, $R = C_0(\cosh 2u - 1)/(2c^2)$. As u decreases to 0 both t and R tend to 0. Creation begins from nothing and, since R and $t \to +\infty$ as $u \to +\infty$, the universe expands without bound. Graph 3 shows the future and the past of a pseudospherical universe ($k = -1$) which is at $R = 1$ when $t = 1$ with $R' = 1, 2, 3$. Note that each of these three universes starts from nothing sometime in the past, but not all start with a bang (e.g., $R'(0) \neq \infty$ if $R'(1) = 1$). In each of the three universes, $R(t)$ always increases as t increases, so each is an expanding universe.

All of these models ignore "pressure", which may be acceptable on the large scale of an essentially empty universe, but cannot be correct for small R. The scales in the figures are arbitrary.

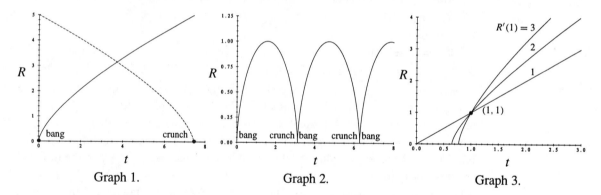

Graph 1. Graph 2. Graph 3.

1.8 Cold Pills

Suggestions for preparing a lecture

Topics: Flow of one dose, discussion (with graphs) of the repeated dose model, brief discussion of compartment models and cascades.

Remarks: Most people enjoy working with the cold pill models of this section, and we enjoy lecturing about them. Ed Spitznagel first introduced us to the models in an NSF/DUE-sponsored faculty workshop in 1992, and subsequently wrote a delightful article on compartment models in pharmacokinetics for the newsletter CODEE (Fall 1992 issue). This newsletter has short articles on models and computer experiments in elementary differential equations and is available by writing

CODEE, Dept. of Mathematics, Harvey Mudd College, Claremont, CA, 91711, or by sending an e-mail request to codee@hmc.edu.

Making up a problem set

Two parts selected from Problems 1, 2; one of Problems 5, 6, 7.

Comments

Mathematically speaking, the solution techniques of this section are all based on the integrating factor method for a first-order linear ODE (see Section 1.4). These are now applied to a cascade of two or more first-order linear ODEs, where the ODEs can be solved from the top down, one ODE at a time. Subsequent ODEs in the cascade make use of the solutions of some of the preceding ODEs. This process shows that these systems of ODEs are not at all fearsome, even though the algebra can be a little bit complicated.

On a modeling level, the first-order systems of ODEs that represent the flow of medications through the compartments of the body seem quite natural to many people, who find these systems easier to understand than the mechanical, gravitational, or electrical systems traditionally used in ODE courses. We like to introduce systems of ODEs naturally and early in the course, but informally. This approach accustoms everyone to seeing, using, and solving systems, but without all the vocabulary and theory of the full treatment of a system of n ODEs (which we do in Chapter 5). We talk about systems in other sections of Chapter 1, but this is the only section of the chapter devoted entirely to phenomena that can only be modeled using systems.

We give informal definitions of a general compartment system and a linear cascade compartment system, and note how they can be represented by labeled boxes and arrows. By careful examination of the assemblage, a system of rate equations can be written for the flow of the substance being tracked through the compartments. There is a different interpretation of a compartmental system where there is just one physical compartment containing a substance which decays over time into other substances [e.g., radioactive decay (see Problem 14 in Section 1.5 or chemical reactions (see Chapter 5)].

The sensitivity of a system to changes in a rate coefficient is shown in the cold pill models. The clearance coefficient for removing medication from the blood may be much lower for older people than for the young, and we discuss the consequences (see also Problems 5–7). The model that represents taking a fast-dissolving cold pill every four or six hours leads naturally to a system of linear ODEs with a periodic pulse input function of the form A sqw(t,d,T). Functions like this appear throughout the applications and we introduce them here in a natural setting. The system is most easily solved by using a numerical solver with piecewise continuous functions in its predefined list.

---

1. All the algebraic steps are given for part **(a)**.

   (a). Analysis of the cascade model as in Example 1.8.8 produces the linear system $x' = -3x$, $y' = x$, $z' = 2x$. The system is solved from the top down. Using the techniques of Section 1.4, we see that the general solution to the first ODE is $x = Ce^{-3t}$. Since $x(0) = 1$, $C = 1$ and the solution to the IVP is $x = e^{-3t}$. Substituting this into the other equations of the system, we have $y' = e^{-3t}$ and $z' = 2e^{-3t}$. Integrating each of these gives $y = -e^{-3t}/3 + C_1$ and $z = -2e^{-3t}/3 + C_2$. Using the initial data, $y(0) = 0 = -1/3 + C_1$, so $C_1 = 1/3$; and $z(0) = 0 = -2/3 + C_2$, so $C_2 = 2/3$. The solutions are $x = e^{-3t}$, $y = (1 - e^{-3t})/3$, and $z = 2(1 - e^{-3t})/3$. As $t \to +\infty$, $x \to 0$, $y \to 1/3$, and $z \to 2/3$.

(b). The system is similar to part **(a)**, but now includes an input and output. Here, $x' = 1 - 2.2x$, $y' = 2x - y$, $z' = 0.2x$. Solving from the top down, $x = 1/2.2 + Ce^{-2.2t}$. Using the initial condition $x(0) = 0$, $C = -1/2.2 = 5/11$, so $x = (5/11)(1 - e^{-2.2t})$. Substituting this into the next IVP with $y(0) = 0$ and solving gives $y = 10/11 + (25/33)e^{-2.2t} - (5/3)e^{-t}$. Finally, $z' = (1/11)(1 - e^{-2.2t})$, with $z(0) = 0$. Integration gives $z = (1/11)(t + (5/11)e^{-2.2t} - 5/11)$. As $t \to +\infty$, $x \to 5/11$, $y \to 10/11$, and $z \to +\infty$.

(c). The system, now with a sinusoidal input, is $x' = 1 + \sin t - 3x$, $y' = 3x - y$. We will solve from the top down. Using the integrating factor e^{3t}, $(xe^{3t})' = (1 + \sin t)e^{3t}$. Integrating, $x = 1/3 + (3\sin t - \cos t)/10 + Ce^{-3t}$. Using $x(0) = 0$, we get $C = -7/30$ and $x = 1/3 + (3\sin t - \cos t)/10 - 7e^{-3t}/30$. Since $y' = 3x - y$, another use of the integrating factor method (after replacing x by the expression in t just derived) gives $y = 1 + 3(\sin t - 2\cos t)/10 + 7e^{-3t}/20 + Ce^{-t}$. So, since $y(0) = 0$, $y = 1 + 3(\sin t - 2\cos t)/10 + 7e^{-3t}/20 - 3e^{-t}/4$. As $t \to +\infty$, x oscillates about $1/3$ and y oscillates about 1.

(d). The system is $x' = 1 - 3x$, $y' = 2 + 2x - y$, $z' = x$. Using the integrating factor e^{3t}, the solution to the first ODE is $x = (1 - e^{-3t})/3$. Substituting this for x in the next ODE, $y' + y = 2 + 2(1 - e^{-3t})/3$, we obtain the solution $y = 8/3 + e^{-3t}/3 + Ce^{-t}$. Since $y(0) = 0$, $0 = 8/3 + 1/3 + C$, so $C = -3$, and $y = 8/3 + e^{-3t}/3 - 3e^{-t}$. Another substitution into $z' = x$ gives $z' = (1 - e^{-3t})/3$ whose solution with $z(0) = 0$ is $z = -1/9 + t/3 + e^{-3t}/9$. As $t \to +\infty$, $x \to 1/3$, $y \to 8/3$, and $z \to +\infty$.

2.   **(a).** The system is $x' = 5 - x$, $y' = x - 5y$. The first compartment (i.e., the top of the cascade) is x, which has a constant rate input of 5 units of substance per unit of time. The substance leaves the x compartment at the rate of x units of substance per unit of time and enters the y compartment at the same rate. The substance leaves the y compartment at the rate of $5y$ units of substance per time unit. The corresponding compartment diagram is

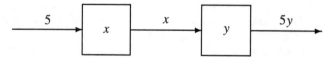

(b). The system is $x' = -x/2$, $y' = 1 - y/3$, $z' = x/2 + y/3$. Here the x and the y compartments are at the top of the cascade and the y compartment has an inflow rate of 1 unit of substance per time unit. The substance flows from the x and y compartments into the z compartment at respective rates $x/2$ and $x/3$ units of substance per time unit. So the compartment diagram is

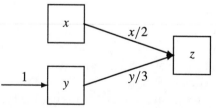

(c). The system is $x' = -x$, $y' = x/2 - 3y$, $z' = x/2 + 3y - 2z$. The substance exits the top x compartment and enters the y compartment at rate $x/2$ and the z compartment also at the rate $x/2$ units of substance per unit of time. The substance also leaves the y compartment and enters the z compartment at the rate $3y$, and leaves the z compartment at the rate $2z$. The diagram is

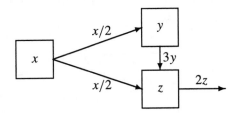

3. **(a).** Using integrating factors, solve the first ODE of system (3), and then the other. The IVP, $dx/dt = -k_1 x$, $x(0) = A$, gives $x = Ae^{-k_1 t}$. Substituting this into the ODE, $dy/dt = k_1 x - k_2 y$, and rearranging, $y' + k_2 y = Ak_1 e^{-k_1 t}$. Using the integrating factor $e^{k_2 t}$, $(ye^{k_2 t})' = Ak_1 e^{(k_2-k_1)t}$, giving $y = Ak_1 e^{-k_1 t}/(k_2 - k_1) + Ce^{-k_2 t}$, where C is constant. Using $y(0) = B$, we get $B = Ak_1/(k_2 - k_1) + C$, so $C = B - Ak_1/(k_2 - k_1)$. So, $x = Ae^{-k_1 t}$ and $y = Ak_1(e^{-k_1 t} - e^{-k_2 t})/(k_2 - k_1) + Be^{-k_2 t}$.

(b). See Fig. 3(b) for antihistamine levels. The maximum of about 1.75 units is reached in about four hours.

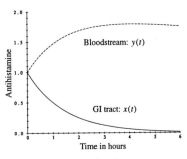

Problem 3(b).

4. From Example 1.8.1, we have that the amount of medication in the blood is given by $y(t) = (k_1 A/(k_1 - k_2))(\exp[-k_2 t] - \exp[-k_1 t])$. The maximum value of $y(t)$ occurs where $y'(t) = 0$. Setting $(k_1 A/(k_1 - k_2))(-k_2 \exp[-k_2 t] + k_1 \exp[-k_1 t])$ equal to 0, we obtain $k_2 \exp[-k_2 t] = k_1 \exp[-k_1 t]$. Taking the natural logarithm of both sides of this equation and simplifying, we find that $t = (\ln k_1 - \ln k_2)/(k_1 - k_2)$ is the time when $y(t)$ is largest.

5. **(a).** See Fig. 5(a) for antihistamine levels (the bottom curve corresponds to the smallest value of k_1, 0.0691, and the top curve to the largest, 1.5). The curves for the larger values of k_1 eventually cross the curves for the smaller values. This models the fact that effective clearance of the antihistamine from the blood occurs only after the medication level in the blood has peaked; if k_1 is small, it takes longer to peak.

(b). From a careful inspection of Fig. 5(a) we see that the solution curve corresponding to $k_1 = 0.11$ represents the lower limit of values of k_1 for which the medication level in the blood rises to 0.2 in exactly 2 hours. We also see that $k_1 = 0.3$ is the upper limit of values of k_1 for which the medication level does not exceed 0.8 over the given time span. So $0.11 \le k_1 \le 0.3$.

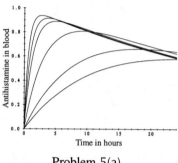

Problem 5(a).

6. **(a).** The boxes and arrows diagram is identical to the one given in Example 1.8.1, and the IVPs are identical to those given by (3). So, $x'(t) = -k_1 x$, $x(0) = A$, and $y'(t) = k_1 x - k_2 y$, $y(0) = 0$, where the rate constants k_1 and k_2 have different values for the decongestant than for the antihistamine. The boxes and arrows diagram is shown below:

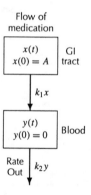

(b). The IVPs may be solved from the top down. Using the integrating factor $\exp[k_1 t]$ and applying the initial condition $x(0) = A$ for $x'(t) = -k_1 x$, we obtain $x(t) = A\exp[-k_1 t]$. Inserting this expression into the second rate equation $y'(t) = k_1 x - k_2 y = k_1 A \exp[-k_1 t] - k_2 y$, using the integrating factor technique again, and applying the initial condition $y(0) = 0$, we obtain (for $k_1 \ne k_2$), $y(t) = (k_1 A/(k_1 - k_2))(\exp[-k_2 t] - \exp[-k_1 t])$. See Fig. 6(b). Since the values of k_1 and k_2 for the decongestant are much larger than the corresponding values for the antihistamine, one expects the decongestant level in the GI tract to drop much faster than the antihistamine level. Similarly, the decongestant level in the blood should reach its peak more quickly and then decline more rapidly than is true for the antihistamine level. Comparison of Figure 6(b) with Figure 1.8.1 shows that this is indeed the case.

(c). See Fig. 6(c). The larger the value of k_2, the sooner the antihistamine level peaks, the smaller the peak level, and the more rapid the decline.

(d). The model IVPs are identical to those in Example 1.8.5, but with different values for the rate constants k_1 and k_2. That is, $x'(t) = 12\,\mathrm{sqw}(t, 25/3, 6) - k_1 x$, $x(0) = 0$, $y'(t) = k_1 x - k_2 y$, $y(0) = 0$, with $k_1 = 1.386$ and $k_2 = 0.1386$. See Fig. 6(d). For the same reasons as in part (b), the decongestant levels peak more rapidly and fall off farther and more rapidly than the antihistamine levels. The decongestant levels in the blood are essentially oscillating around an equilibrium level within 48 hours, but Figure 1.8.4 shows that that is not at all true for the antihistamine.

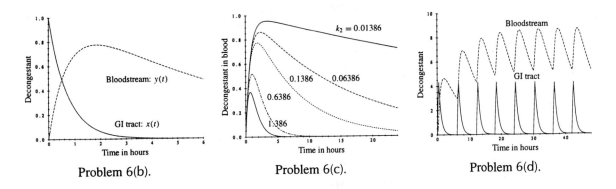

Problem 6(b). Problem 6(c). Problem 6(d).

(e). See Fig. 6(e), Graphs 1, 2, and 3, where the respective values of k_2 are 0.1386, 0.06386, and 0.01386. Assuming that the lowest value of k_2 is that of someone who is old, sick, or both, the rising decongestant level in Fig. 6(e), Graph 3 is alarming.

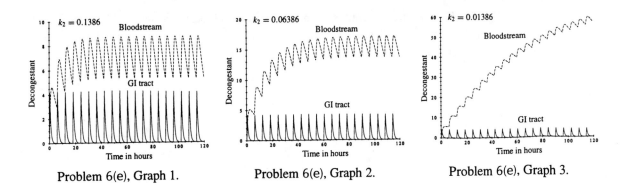

Problem 6(e), Graph 1. Problem 6(e), Graph 2. Problem 6(e), Graph 3.

7. **(a).** The model IVP for the amounts $x(t)$ and $y(t)$ of continuous-acting medication in the GI tract and bloodstream, respectively, is $x'(t) = I - k_1 x$, $x(0) = 0$, and $y'(t) = k_1 x - k_2 y$, $y(0) = 0$.

(b). The rate equations may be solved from the top down by using integrating factors. Using the integrating factor $e^{k_1 t}$ and the initial data $x(0) = 0$ to solve the equation $x'(t) =$

$I - k_1 x$ for $x(t)$, we see that $x(t) = I(1 - e^{-k_1 t})/k_1$. Inserting this expression for $x(t)$ into the second equation $y'(t) = k_1[I(1 - e^{-k_1 t})/k_1] - k_2 y$, using the integrating factor $e^{k_2 t}$ and the initial data $y(0) = 0$, we obtain a formula for $y(t)$. The solution is found to be $y(t) = (I/k_2)[1 + (k_2 e^{-k_1 t} - k_1 e^{-k_2 t})/(k_1 - k_2)]$.

(c). As $t \to \infty$, $x(t) \to I/k_1$ and $y(t) \to I/k_2$.

(d). See Fig. 7(d), Graphs 1 and 2. The antihistmine level in the GI-tract reaches equilibrium very quickly, but even after 200 hours the level in the blood is still rising although no longer very rapidly. It is a different story altogether with the decongestant where the levels are essentially at equilibrium in both compartments within 48 hours.

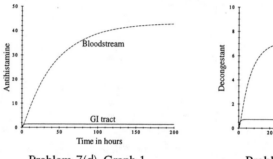

Problem 7(d), Graph 1.

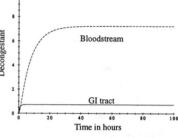

Problem 7(d), Graph 2.

(e). See Fig. 7(e), Graph 1 for decongestant levels and Fig. 7(e), Graph 2 for antihistamine levels. As expected, the levels of both medications in the blood continue to rise, particularly the antihistamine.

(f). See Fig. 7(f) for the antihistamine levels in the bloodstream over a 120-hour period. From the graph the antihistamine level exceeds the safe limit of 50 units after about 67 hours and is still rising.

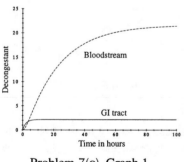

Problem 7(e), Graph 1.

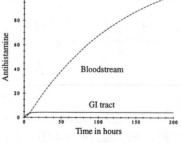

Problem 7(e), Graph 2.

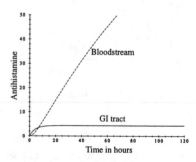

Problem 7(f).

1.9 **Change of Variables and Pursuit Models**

Suggestions for preparing a lecture

Topics: Homogeneous rate functions of order zero and the way to solve any ODE that involves such a rate function, the pursuit model, changing variables, scaling (if there is time). Sometimes we go on and give a second lecture on polar coordinates and scaling.

Remarks: Changing variables requires students to use the Chain Rule, and, although they have used it before in their studies (and maybe in this course), it doesn't hurt to remind them how it can be used. Of the various changes of variables mentioned in this section, there is usually lecture time for only two or three. We usually do the rescaling-the-variables example (because of its importance in preparing an IVP for computing) and the one involving functions that are homogeneous of order 0 (so that we can do the pursuit model at the end of the section).

Making up a problem set

Problems 1(**c**) and 2(**c**) (because of the neat graph!), one part of Problem 3, something from Problems 5 and 6 (so that students see how a change of variable can reduce a nonlinear to a linear ODE), Problem 7 (because of the importance of learning how to scale), and one of Problems 9–11 (depending on what was done in the lectures). **Warning:** The algebra in many of the problems can be daunting, and you may want to allow students to use a CAS.

Comments

A widely used method to deal with a difficult equation is to change the variables, hoping that the transformed equation will be simpler (and more easily solvable) in the new variables. There are hundreds of changes of variables for dealing with complicated (often nonlinear) ODEs; see, for example, *Handbook of Exact Solutions for Ordinary Differential Equations*, A.D. Polyanin, V.F. Zaitsev; CRC Press Inc., 1995, Boca Raton. See also D. Zwillinger's *Handbook of Differential Equations*, 2nd ed., 1992; Academic Press, San Diego. In this section, we consider a few variable changes that reduce some simple (but nonlinear) first-order ODEs to forms that can be solved by methods of this chapter. Bernoulli ODEs (Problem 9), Riccati ODEs (Problem 10), ODEs involving a function that is homogeneous of order 0 (Example 1.9.2)—these are all transformed to simpler ODEs by a variable change. Applications of variable changes are made to rescaling a model of vertical motion (Example 1.9.5, and Problem 8).

An interesting sidelight to the pursuit problem: from Example 1.9.4 and Problem 11 we see that when the goose reaches the nest it is flying directly into the wind. Is this always true?

As in Section 1.6, it often happens that solutions can only be defined by implicit formulas. So we get integral curves as graphs, rather than solution curves. A solution curve is an arc of an integral curve with no vertical tangents.

---

1. In the solutions below, each ODE is written in the form $y' = f(x, y)$, and it is shown that f is homogeneous of order 0 by verifying that $f(kx, ky) = f(x, y)$. The change of variable $y = xz$ converts the ODE into a separable ODE in x and z, and this new ODE is given, along with its solutions. Finally, the solutions of the original ODE are given, sometimes implicitly, by replacing z by y/x. **Caution:** There is a lot of algebraic manipulation here, so watch your steps!

(a). The ODE is $y' = (y + x)/x$. The rate function is homogeneous of order 0, because

$$f(kx, ky) = \frac{kx + ky}{kx} = \frac{x + y}{x} = f(x, y)$$

If $y = xz$, then $y' = xz' + z = (xz + x)/x = z + 1$, and $xz' = 1$. The solutions are $z = \ln|x| + C$. Using $y = xz$, $y = x \ln|x| + Cx$, where the domain is restricted to $x > 0$ or $x < 0$.

(b). The ODE may be written as $y' = (y - x)/(x - 4y)$, and we have that

$$f(kx, ky) = \frac{ky - kx}{kx - 4ky} = \frac{y - x}{x - 4y} = f(x, y)$$

and the rate function is homogeneous of order 0. After the variable change $y = xz$,

$$y' = xz' + z = (xz - x)/(x - 4xz) = (z - 1)/(1 - 4z)$$

$$xz' = (-1 + 4z^2)/(1 - 4z)$$

which separates to $[(1 - 4z)/(-1 + 4z^2)] \, dz = (1/x) \, dx$. Integral tables (or the partial fraction techniques) yield the (implicit) solutions

$$\frac{1}{4} \ln \left| \frac{2z - 1}{2z + 1} \right| - \frac{1}{2} \ln|4z^2 - 1| = \ln|x| + C$$

In x, y variables, we have $(1/4) \ln |(2y/x - 1)/(2y/x + 1)| - (1/2) \ln |4y^2/x^2 - 1| = \ln|x| + C$. After a certain amount of fiddling with the logarithm terms, this simplifies to $(2y - x)(2y + x)^3 = C$, where C is a constant.

(c). The ODE may be written as $y' = (x^2 - xy - y^2)/(xy)$, and we have that

$$f(kx, ky) = \frac{k^2 x^2 - k^2 xy - k^2 y^2}{k^2 xy} = f(x, y)$$

So the rate function is homogeneous of order 0. The variable change $y = xz$ leads to the ODE $xz' + z = (x^2 - x^2 z - x^2 z^2)/x^2 z = (1 - z - z^2)/z$, or $z'z/(1 - z - 2z^2) = 1/x$. This ODE integrates (after using partial fractions or integral tables) to $-\frac{1}{6} \ln|1 - 2z| - \frac{1}{3} \ln|1 + z| = \ln|x| + C$. Let $z = y/x$, rearrange, exponentiate, and obtain $x^3(x - 2y)(x + y)^2 = C$, where C is a constant.

(d). The ODE may be written as $y' = (2y^2 - x^2)/(xy)$, and we have that the rate function is homogeneous of order 0 because

$$f(kx, ky) = \frac{2k^2 y^2 - k^2 x^2}{k^2 xy} = \frac{2y^2 - x^2}{xy} = f(x, y)$$

Let $y = xz$ to obtain $xz' + z = (2x^2 z^2 - x^2)/(x^2 z) = (2z^2 - 1)/z$, or $xz' = (z^2 - 1)/z$. In separated form, we have $z \, dz/(z^2 - 1) = dx/x$, which integrates to $(\ln|z^2 - 1|)/2 = \ln|x| + K$, or replacing z by y/x, $y^2 - x^2 = Cx^4$. So $y = \pm x(1 + Cx^2)^{1/2}$, where C is a constant and x satisfies $1 + Cx^2 > 0$.

(e). The ODE is $y' = 4 + 7y/x + 2y^2/x^2$, and we have that f is homogeneous of order 0 because

$$f(kx, ky) = 4 + \frac{7ky}{kx} + \frac{2k^2 y^2}{k^2 x^2} = f(x, y)$$

Let $y = xz$; $xz' + z = 4 + 7z + 2z^2$ and $dz/(4 + 6z + 2z^2) = dx/x$, which integrates to $[\ln|(z+1)/(z+2)|]/2 = \ln|x| + K$, which in terms of x and y reduces to $\ln|(x + y)/(2x + y)|^{1/2} - \ln|x| = K$ or $\ln|(x + y)/(2x + y) \cdot (1/x^2)|^{1/2} = K$. Exponentiating, $x + y = Cx^2(2x + y)$, where C is any constant, or $y = (2Cx^3 - x)/(1 - Cx^2)$, where x is such that $1 - Cx^2 \neq 0$.

2. See Figs. 2(a)–2(e) for integral curves of the respective systems of Problems 1(a)–1(e). Solution curves are arcs without vertical tangents and lie on integral curves.

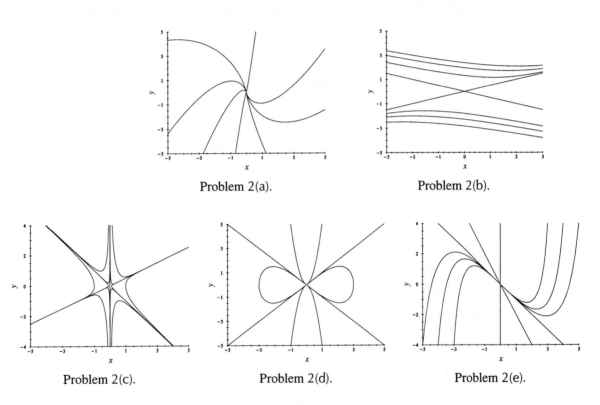

Problem 2(a).   Problem 2(b).

Problem 2(c).   Problem 2(d).   Problem 2(e).

3. In each case we find a "general" implicit solution in the form $H(x, y) = C$. We don't always attempt to solve for y in terms of x and C.

(a). Let $z = x + y$, in the ODE, $dy/dx = \cos(x + y)$. Then $dz/dx = 1 + dy/dx = 1 + \cos z$, so $(1 + \cos z)^{-1} dz = dx$. Using a table of integrals, we have $\tan(z/2) = x + C$, so $\tan[(x + y)/2] = x + C$. Then $x + y = 2\arctan(x + C) + n\pi$, so $y = -x + 2\arctan(x + C) + n\pi$, where C is any constant and n is any integer.

(b). Let $z = 2x + y$, so $dz/dx = 2 + dy/dx$. Then the ODE, $(2x + y + 1) + (4x + 2y + 3) dy/dx = 0$, becomes $z + 1 + (2z + 3)(dz/dx - 2) = 0$, which simplifies to $[(2z +$

$3)/(3z+5)]\,dz = dx$. Integrating, we have that $2z/3 - (\ln|3z+5|)/9 = x + C$, or $(4x + 2y)/3 - (1/9)\ln|6x + 3y + 5| = x + C$, where C is any constant and $|6x + 3y + 5| \neq 0$.

(c). Let $z = x + 2y$, so $dz/dx = 1 + 2dy/dx$. Then the ODE, $(x + 2y - 1)\,dx + 3(x + 2y)\,dy = 0$, becomes $(z - 1) + 3z(dz/dx - 1)/2 = 0$ or $3z\,dz/(z + 2) = dx$, which integrates to $3(z - 2\ln|z + 2|) = x + C$. In terms of x and y, $3(x + 2y - 2\ln|x + 2y + 2|) = x + C$, where C is any constant and $|x + 2y + 2| \neq 0$.

(d). The ODE, $e^{-y}(y' + 1) = xe^x$, may be written as $y' + 1 = xe^{x+y}$. Let $z = x + y$, so $dz/dx = 1 + dy/dx$. Then $dz/dx = 1 + y' = xe^z$, so $e^{-z}\,dz = x\,dx$. Integrating, $-e^{-z} = x^2/2 + C$, so $-z = \ln|-C - x^2/2|$. In terms of x and y, $x + y = -\ln|-C - x^2/2|$, $y = -x - \ln|C + x^2/2|$, where C is any constant and $C + x^2/2 \neq 0$. Alternatively, let $u = e^{-y}$. Then $u' - u = -xe^x$, which is linear, and can be solved by the integrating factor technique of Section 1.4.

4.   To solve $yy'' + (y')^2 = 1$ we first rewrite the ODE in terms of z, where $y = z^{1/2}$. We have $y' = z^{-1/2}z'/2$, and $y'' = -z^{-3/2}(z')^2/4 + z^{-1/2}z''/2$. So $yy'' + y'^2 = z^{1/2}[-z^{-3/2}(z')^2/4 + z^{-1/2}z''/2] + z^{-1}(z')^2/4$. The ODE, $y\,y'' + (y')^2 = 1$, becomes $z'' = 2$. Integrating twice, we have $z = x^2 + C_1 x + C_2$. The initial conditions $y(0) = 1$, $y'(0) = 0$ become $z(0) = 1$, $z'(0) = 0$. So, $C_2 = 1$, $C_1 = 0$. Since $y = z^{1/2}$, the solution is $y = (x^2 + 1)^{1/2}$.

5.   **(a).** Make the change of variable $z = 1/y$ in the logistic equation $y' = r(1 - y/K)y$. Since $y = 1/z$, we see that $y' = -z^{-2}z'$, so

$$-z^{-2}z' = r\left(1 - \frac{1}{Kz}\right)\left(\frac{1}{z}\right)$$

$$-z' = r\left(z - \frac{1}{K}\right)$$

$$z' = -rz + \frac{r}{K}$$

which is a linear ODE.

(b). Multiplying through by the integrating factor e^{rt}, we have

$$(ze^{rt})' = \frac{r}{K}e^{rt}$$

Integrating, we have

$$ze^{rt} = \left(\frac{r}{K}\right)\frac{e^{rt}}{r} + C$$

$$z = Ce^{-rt} + 1/K$$

Changing back to y, we have

$$y = \frac{1}{Ce^{-rt} + 1/K}$$

Using the initial condition $y(0) = y_0$ to evaluate C, we have that

$$y_0 = \frac{1}{C + 1/K}$$

$$C + 1/K = \frac{1}{y_0}$$

$$C = \frac{1}{y_0} - \frac{1}{K}$$

So

$$y = \frac{1}{(1/y_0 - 1/K)e^{-rt} + 1/K}$$

$$= \frac{Ky_0}{(K - y_0)e^{-rt} + y_0}$$

which is formula (15) of Section 1.6.

6. The ODE is $y' = (a + by)(c(t) + d(t)y)$.

(a). Introducing the new dependent variable by setting $y = [(1/z) - a]/b$ and using the Chain Rule, we have that

$$z' = \frac{-by'}{(a + by)^2} = \frac{-b(\alpha(t) + \beta(t)y)}{a + by} = -bz\left[\alpha(t) + \beta(t)\left(\frac{1}{bz} - \frac{a}{b}\right)\right]$$

since $z = [a + by]^{-1}$ and $y = [(1/z) - a]/b$. Note that we can rewrite the equation for z' in the form

$$\frac{dz(t)}{dt} = [a\beta(t) - b\alpha(t)]z(t) - \beta(t)$$

which is linear. The integration factor technique can be used to find a solution $z(t)$, and then $y(t) = [(1/z) - a]/b$ is a solution of the original ODE. We must remain within a t-interval on which $z(t)$ does not vanish.

(b). Comparing this ODE with the given standard form we see that

$$a = 3, \quad b = 1, \quad c(t) = 2t, \quad d(t) = t$$

and so the transformed ODE in z becomes

$$z' = tz - t$$

Using the integrating factor approach we have the general solution

$$z = 1 + Ce^{t^2/2}$$

where C is an arbitrary constant. So from **(a)**,

$$y = 1/(1 + Ce^{t^2/2}) - 3$$

for an arbitrary constant C is the general solution of the given ODE.

7. Set $x(t) = P(t)/K$ and $t = s/r$ in the logistic ODE $P'(t) = r(1 - P(t)/k)P(t)$. By the chain rule,

$$\frac{dx}{ds} = \frac{1}{K}\frac{dP}{dt}\frac{dt}{ds} = \frac{1}{K}r(1 - p/K)p/r$$

$$= \frac{1}{K}(1 - x)Kx = (1 - x)x$$

The advantage of using the scaled ODE is that there are no parameters in the rate function. This means that a solution of the scaled equation can be used to describe the behavior of solutions of the unscaled ODE regardless of the values of the positive constants r and K.

8. The ODE is $v' = -g - (k/m)v$. Let $t = aT$, $v = bV$, where a and b will be chosen such that the ODE for the motion of the whiffle ball in the rescaled time and velocity variables T and V is simpler than the original ODE $v' = -g - (k/m)v$. We have by the Chain Rule

$$\frac{dv}{dt} = \frac{dv}{dV}\frac{dV}{dT}\frac{dT}{dt} = (b/a)\,dV/dT = -g - (k/m)bV$$

Thus, $dV/dT = -ag/b - (ak/m)V$. If we let $a = m/k$, $b = mg/k$, then $dV/dT = -1 - V$. The solution of this ODE is $V(T) = Ce^{-T} - 1$. As $t \to \infty$, we have that $T \to \infty$, $V(T) \to -1$. Since $v = bV = mgV/k$, we have that the limiting velocity is $-mg/k$, as expected from Section 1.5.

9. The Bernoulli ODE is $dy/dt + p(t)y = q(t)y^b$.

   (a). Let $b \neq 0$, 1 and divide the ODE $dz/dt + (1 - b)p(t)z = (1 - b)q(t)$ through by y^b. Then multiply by $1 - b$ and the ODE can be written as

   $$\frac{d}{dt}(y^{1-b}) + (1 - b)p(t)y^{1-b} = (1 - b)q(t)$$

   Now make the change of variables $z = y^{1-b}$ and the ODE takes the form

   $$\frac{dz}{dt} + (1 - b)p(t)z = (1 - b)q(t) \tag{i}$$

   which is a linear ODE. If $y(t)$ solves the original ODE, then $z(t) = (y(t))^{1-b}$ solves the linear ODE (i). Conversely, if z is any solution to the last ODE, then $y = z^{1/(1-b)}$ is a solution of the original ODE, or perhaps several solutions after taking the root $1/(1 - b)$.

   (b). The logistic equation $P' = r(1 - P/K)P$ may be written as $P' - rP = -rP^2/K$ which has the Bernoulli form with $p(t) = -r$, $q(t) = -r/K$, $b = 2$. See Figure 9(b).

   (c). We may transform the Bernoulli ODE, $y' + t^{-1}y = y^{-4}$, to a linear ODE using the variable change $z = y^5$. We obtain $z' + 5t^{-1}z = 5$, where an integrating factor is $e^{5\ln t} = t^5$, and the solutions are given by $z = 5t/6 + C/t^5$. Then $y = z^{1/5} = (5t/6 + C/t^5)^{1/5}$, where $t > 0$ and C is an arbitrary constant. See Fig. 9(c) for solution curves.

   (d). $y' - t^{-1}y = -y^{-1}/2$ is a Bernoulli equation with $b = -1$. The change of variable $z = y^2$ transforms the y-equation into the linear ODE $z' - 2t^{-1}z = -1$, whose solutions are given by $z = t + Ct^2$. So, $y = \pm(t + Ct^2)^{1/2}$, where C is any constant and t lies in an interval where $t + Ct^2 > 0$. See Fig. 9(c).

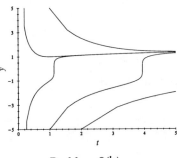

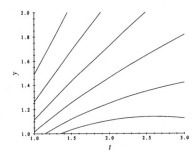

Problem 9(b).   Problem 9(c).

10. **(a).** Let the Riccati equation, $y' = ay + by^2 + F$, have a solution $y = g$. If $z = (y - g)^{-1}$, then $z' = -(y - g)^{-2}(y' - g') = -(y - g)^{-2}(ay + by^2 + F - ag - bg^2 - F) = -z^2(a/z + b(y^2 - g^2)) = -az - bz(y + g)$. Since $y - g = z^{-1}$, $y = g + z^{-1}$. So, $z' = -az - bz(2g + z^{-1})$, or $z' + (a + 2bg)z = -b$, which is linear in z. Once this has been solved by the integrating factor technique for $z(t)$, we obtain y from $y = g + z^{-1}$.

(b). First, verify that $y = g(t) = 1$ is a particular solution of the Riccati ODE $y' = (1 - 2t)y + ty^2 + t - 1$. Then, by the technique given in part **(a)**, the change of variable $z = (y - 1)^{-1}$ changes the ODE into the linear ODE, $z' + (1 - 2t + 2t)z = -t$, or $z' + z = -t$, whose solutions are given by $z = Ce^{-t} - t + 1$. So, solutions are given by $y = 1 + 1/z = 1 + [Ce^{-t} - t + 1]^{-1}$, where C is any constant and t is restricted to an interval for which $Ce^{-t} \neq -1 + t$.

(c). Direct calculation shows that the function $y = g(t) = e^t$ is a solution of the Riccati equation, $y' = e^{-t}y^2 + y - e^t$. So, the substitution $z = (y - e^t)^{-1}$ reduces the equation to the linear ODE, $z' + (1 + 2e^{-t}e^t)z = -e^{-t}$, or $z' + 3z = -e^{-t}$, whose solutions are given by $z = Ce^{-3t} - e^{-t}/2$. Solutions $y(t)$ are given by $y = e^t + 1/z = e^t + [Ce^{-3t} - e^{-t}/2]^{-1}$, where C is any constant and t is restricted to an interval for which $2Ce^{-3t} \neq e^{-t}$.

(d). Direct calculation shows that $y = t$ is a solution of the ODE rearranged into Riccati form, $y' = (-2t^4 + t^{-1})y + t^3y^2 + t^5$. The change of variable $z = (y - t)^{-1}$ reduces the equation to the linear ODE $z' + t^{-1}z = -t^3$, whose solutions are given by $z = Ct^{-1} - t^4/5$. So, solutions $y(t)$ are given by $y = t + 1/z = t + [Ct^{-1} - t^4/5]^{-1}$, where C is any constant and $t < (5C)^{1/5}$ or $t > (5C)^{1/5}$.

(e). The equation $P' = rP - (rP^2)/K - H$ is a Riccati equation with $a = r$, $b = -r/K$, $F = -H$. Let $K = 1600$, $r = 0.01$, $H = 3$. A direction calculation shows that $P(t) = g(t) = 1200$ is a solution. So, the change of variable $z = [P - 1200]^{-1}$ reduces the equation to the linear ODE, $z' + (0.01 - 0.015)z = 10^{-4}/16$, or $z' - 0.005z = 10^{-4}/16$, whose solutions are given by $z = Ce^{0.005t} - 1/800$. And so, $P = 1200 + 1/z = 1200 + 800(800Ce^{0.005t} - 1)^{-1}$, where C is any constant; t is restricted to an interval on which $800Ce^{0.005t} \neq 1$.

11. **(a).** The problem may be set up exactly as the pursuit problem in the text, except that the wind's velocity has component $-b/2$ along the x-axis and $b/2$ along the y-axis. So,

$x' = -b\cos\theta - b/2$ and $y' = -b\sin\theta + b/2$. Dividing, we have

$$\frac{y'}{x'} = \frac{dy}{dx} = \frac{2\sin\theta - 1}{2\cos\theta + 1} = \frac{2y - (x^2 + y^2)^{1/2}}{2x + (x^2 + y^2)^{1/2}}$$

Since this quotient is a homogeneous function of order 0 in x and y, we may introduce $y = xz$ to obtain the separable ODE $xdz/dx + z = [2z - (1+z^2)^{1/2}]/[2 + (1+z^2)^{1/2}]$, or

$$\frac{2 - (1+z^2)^{1/2}}{(z+1)(1+z^2)^{1/2}}\, dz = \frac{1}{x}\, dx$$

Let $z = u - 1$ to reduce the left side to a form found in integral tables. After solving and returning to x, y variables and simplifying, we obtain $[\sqrt{2}(x^2 + y^2)^{1/2} + x - y]^2 = a^{\sqrt{2}}(3 + 2\sqrt{2})|x + y|^{2-\sqrt{2}}$, where the initial data are used. The goose follows the path defined by this expression if it heads towards its nest while the wind blows from the southeast.

(b). If $b = 1$, then the equivalent IVP for the flight path is $x' = -\cos\theta - 1/2 = -x(x^2 + y^2)^{-1/2} - 1/2$, $y' = -\sin\theta + 1/2 = -y(x^2 + y^2)^{-1/2} + 1/2$, $x(0) = a$, $y(0) = 0$. See Fig. 11(b), Graph 1 for the flight paths if $a = 1, 2, \ldots 9$. The goose always reaches the nest, but overshoots the mark a bit and has to head into the wind at the end so that it comes into the nest tangent to the line $y = -x$. See Fig. 11(b), Graph 2 for a zoom on a portion of Fig. 11(b), Graph 1 that shows flight paths tangent to the line $y = -x$ (dashed) at the origin.

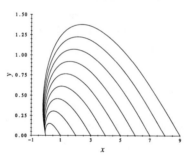

Problem 11(b), Graph 1.

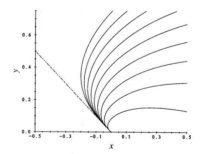

Problem 11(b), Graph 2.

12.   Group project.

Background Material: ODEs in Polar Coordinates

The ODE in rectangular coordinates x and y

$$dy/dx = f(x, y) \tag{i}$$

can be transformed into an equivalent ODE in polar coordinates r and θ. First recall the relations between polar and rectangular coordinates:

$$x = r\cos\theta, \qquad y = r\sin\theta, \qquad r = (x^2 + y^2)^{1/2}, \qquad \tan\theta = y/x$$

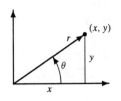

Now suppose that $y = g(x)$ defines a solution curve of the ODE (i). This solution curve can also be defined (perhaps only in part) by specifying r as a function of the polar angle θ, that is, $r = r(\theta)$. In polar coordinates the relation $y = g(x)$ becomes

$$r \sin \theta = g(r \cos \theta) \tag{ii}$$

Differentiating each side with respect to θ and using the Chain Rule, we have

$$\frac{dr}{d\theta} \sin \theta + r \cos \theta = g'(r \cos \theta) \left(\frac{dr}{d\theta} \cos \theta - r \sin \theta \right)$$

Since $y = g(x)$ solves $y' = f(x, y)$, then $g'(r \cos \theta) = f(r \cos \theta, r \sin \theta)$. So

$$\frac{dr}{d\theta} \sin \theta + r \cos \theta = f(r \cos \theta, r \sin \theta) \left(\frac{dr}{d\theta} \cos \theta - r \sin \theta \right) \tag{iii}$$

The ODE in (iii) is equivalent to $y' = f(x, y)$. Although ODE (iii) looks complicated, it is actually simpler than (i) for some functions $f(x, y)$. Let's look at an example.

Applying formula (iii) to the ODE

$$\frac{dy}{dx} = \frac{y - 2x}{2y + x} \tag{iv}$$

we have the equivalent ODE in polar coordinates:

$$\frac{dr}{d\theta} \sin \theta + r \cos \theta = \frac{r \sin \theta - 2r \cos \theta}{2r \sin \theta + r \cos \theta} \left(\frac{dr}{d\theta} \cos \theta - r \sin \theta \right) \tag{v}$$

After canceling the common factor r in the quotient, multiplying each side by $2 \sin \theta + \cos \theta$, simplifying, and again canceling a common factor, ODE (v) reduces, after some algebra, to the linear ODE in r:

$$\frac{dr}{d\theta} = -\frac{r}{2} \tag{vi}$$

The solutions of ODE (vi) are given by

$$r = Ce^{-\theta/2}$$

where C is any nonnegative constant.

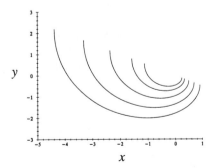

The corresponding curves in the xy-plane are spirals around the origin. The five solution curves shown are arcs of one of these spirals, arcs that have no vertical tangent line because only such arcs can be graphs of functions of x. The "top" arcs of the spiral are also solution curves, but they are not shown. Finally, note that taking logarithms of each side of the equation $r = Ce^{-\theta/2}$ and returning to the xy-variables, the solutions $y(x)$ of ODE (iv) satisfy the equation

$$\frac{1}{2}\ln(x^2 + y^2) = \ln C - \frac{1}{2}\arctan(y/x) \tag{vii}$$

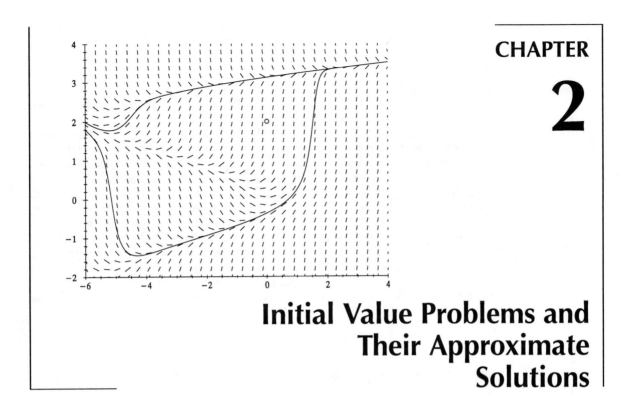

CHAPTER

2

Initial Value Problems and Their Approximate Solutions

The goal of this chapter is to deepen and extend some topics that were introduced in Chapter 1: existence and uniqueness of a solution for a first-order IVP, numerical approximation of a solution of an IVP, and a study of sensitivity via computer simulation. The early sections of the chapter have a bare-bones treatment of the topics, with many examples, and this should suffice for the most part. The last section of the chapter has a different character from the rest of the chapter in that the object of prime interest is a discrete dynamical system. Euler's Method applied to the logistic ODE is basically only a convenient artifice for initiating this study. We think that this section is a lot of fun, bringing in period doubling and apparently chaotic wandering.

We have tried to be mindful throughout that although numerical methods are important because they drive the engine that produces graphs of solutions, they are not the main object of interest in the course. Therefore we have tried to give numerical methods a featured spot as early as possible in the text, but we limit our coverage to concepts that would affect the ability to produce decent graphs. There are many numerical solvers available on the market today which can generate most of the graphs of solutions of ODEs which appear in this text. Most CAS packages contain numerical ODE solvers as well, but because of their symbolic manipulation facilities they are also capable of producing solution formulas for some classes of ODEs. A brief overview article on ODE solvers appeared in 1994 in *The College Mathematics Journal*'s Special Issue on Differential Equations.[1] Along with a short description of each solver, the article contains information about the academic cost of the package and a telephone number for obtaining more information (as of 1994).

The graphs your solver produces may look different from the graphs in the text. There are many reasons for this: sometimes the reasons are platform dependent, sometimes algorithm dependent,

---

[1] Andrew Flint and Ron Wood, "ODE Solvers in the Classroom", *College Math. Jour.*, *25*, No. 5 (1994), pp. 458–461.

and sometimes maybe you (or we) just hit the wrong keys on the keyboard. The way your machine does arithmetic and function evaluations is an example of the former. Memory may also be a problem (the data sets for some of our graphs are huge). If you have an adaptive solver (as we do), then your default settings may cause you to miss some important features in the graph. In any case, there are some things you can do to improve the appearance of your graphs. For Runge-Kutta solvers try changing the step size and shortening the solve-time interval. For adaptive solvers try changing the default settings for the absolute and relative error bounds, the number of plotted points, the maximum number of function evaluations, the maximum and minimum interval step size, etc. Please keep in mind that the graphs that appear in this text are not the result of a first pass. Often, we had to adjust the time span, the intervals for the state variables, the initial data, or the solver parameters before we could get a picture that was both informative and good-looking. Another reason why your graphs will often not look as good as those in the text: the printing technology that produced the graphs in the text is equivalent to a 1200 dpi laser printer. (We don't even see our graphs with this quality at home—1200 dpi laser printers are expensive!) But do the best you can with the equipment you have—a good picture may be worth more than many words and formulas. If you are asked to "sketch a curve," we mean "draw without a computer." Otherwise we use the word "plot."

2.1 Existence and Uniqueness

Suggestions for preparing a lecture

Topics: The Basic Questions, existence and uniqueness, and the Picard Iteration scheme.
Remarks: Sometimes we add a second lecture and do the details of Appendices A.1 and A.2.

Making up a problem set

One part each from Problems 1 and 2, 4 or one part of 5, 7, 8(a), 9 (if Picard Iterates are discussed).

Comments

The five Basic Questions posed in this section are at the heart of the theory of differential equations: Does an IVP have a solution? Is it unique? How can a solution be described? How sensitive is a solution to changes in parameters and data? How can a solution be extended; what happens to solutions as time advances? The focus in Chapter 1 has been on obtaining solution formulas. With the formulas in hand, it is often a simple matter to answer the questions. Sometimes solution formulas are too complicated to be of much help in understanding solution behavior. In any case, the questions have to be considered even in the absence of solution formulas. In this section we introduce examples and counterexamples needed to clarify the meaning of the questions, and we show approaches to answering them. When you read this section, you might tie the questions with a specific physical phenomenon that has already been discussed, for example, the "pollution in a tank" problem of Section 1.4. Existence: pollution enters the tank and the pollution level in the tank does build up (so the model IVP for tank pollution levels does have a solution). Uniqueness: if there are two identical tanks, identical amounts of pollutant flowing into the tanks, identical initial amounts of pollutant in the tanks, identical outflow rates, then pollution levels in the tanks are identical (uniqueness of the IVP's solution). Description of solution: words, formulas, pictures, graphs, computer output. Sensitivity/Continuity: change the pollutant levels a little in the inflow stream and the levels in the tank change a little (the solution of the IVP varies continuously with the data). Extension and Long-Term Behavior: after a long time one expects the concentration of pollutant in the tank to be pretty much the same as the concentration of pollutant in the inflow stream (and the formula for the solution of the IVP will mirror that expectation). So the Basic Questions are just the mathematicians' way of encoding in a few words and symbols the everyday experience of people who work with dynamical systems. In addition, the answers to the five Basic Questions are crucial to the use of numerical solvers. Without the Existence, Uniqueness, Extension, and Sensitivity properties, we would never be sure whether the approximation our numerical solver generates has anything to do with the IVP being addressed, nor would we have any idea about its "life-span."

This section focuses on the existence and the uniqueness of the solution of an IVP. The verifications are given in Appendices A.1 and A.2, although we describe in this section the Picard Iteration scheme which is the key idea of the proof of existence. The Picard idea is simple and we always like to illustrate how it works with a simple example where the iteration generates the Taylor series of the solution.

The question of uniqueness is also central, especially these days when the meaning of chaotic dynamics is a popular topic for scientific debate. If an IVP has an unique solution, then any dynamical system accurately modeled by the IVP is deterministic (not chaotic) in the sense that the present state of the system completely determines its future (and its past) states, at least over the time span where the model is valid. Nonuniqueness introduces an element of uncertainty; which path does a solution follow, and how can the system decide which path to follow? This is a good reason for reading some of the nonuniqueness examples: you might well contrast determinism and uncertainty in this connection. Even when the conditions of uniqueness are satisfied, some systems of ODEs or their numerical approximations appear to have chaotic solutions, but we postpone that discussion to Section 2.7 (numerical approximation) and Chapter 9 (ODEs).

Questions of continuity/sensitivity are left to Section 2.3 and to 2.4, while the use of numerical computer approximations is taken up in Sections 2.5 and 2.6.

---

1. **(a).** The IVP is $y' = e^t y - y^3$, $y(0) = 0$. The function $f = e^t y - y^3$ is continuous for all t and y since e^t, y, and y^3 are continuous. Moreover, $\partial f/\partial y = e^t - 3y^2$ is also continuous for all t and y. So, the hypotheses of the Existence and Uniqueness Theorem are satisfied in every rectangle, and the given initial value problem has exactly one solution.

   (b). The IVP is $y' = |t|y^2 - 1/(3y + t)$, $y(0) = 1$. The function $f = |t|y^2 - 1/(3y + t)$ and its partial derivative $\partial f/\partial y = 2|t|y + 3/(3y + t)^2$ are continuous in t and y throughout any region not intersected by the straight line $3y + t = 0$. Since the initial point $(0, 1)$ does not lie on that line, there is a rectangle containing the initial point in which the hypotheses of the Existence and Uniqueness Theorem hold (e.g., any rectangle in the region $3y + t > 0$); the initial value problem has exactly one solution in any of these rectangles that contains $(0, 1)$.

   (c). The IVP is $y' = |t||y|$, $y(0) = 1$. The function $f = |t||y|$ and its partial derivative $\partial f/\partial y = |t|$ if $y > 0$, $-|t|$ if $y < 0$ are continuous in any neighborhood of the point $(0,1)$ for which $y > 0$. So, by the Existence and Uniqueness Theorem the initial value problem has exactly one solution in any rectangle containing $(0, 1)$ and lying in the upper half plane $y > 0$.

2. **(a).** The IVP may be written in the form $y' = (t^2 + y^2)/2ty$, $y(0) = 0$. The rate function, $f(t) = (t^2 + y^2)/2ty$, is undefined on the coordinate axes $t = 0$, $y = 0$ of the ty-plane. Since the initial point $(0,1)$ is on the y-axis, the Existence Theorem does not apply.

(b). The function $y_1 = t^2$ is a solution of $ty' = 2y$, $y(0) = 0$, as we see by a direct substitution. The function $y_2(t) = t^2 \, \text{step} \, t$ is also a solution of the differential equation for $t < 0$ and for $t > 0$, and satisfies the initial condition $y(0) = 0$. To verify that it is differentiable at $t = 0$, we must show that $y_2'(0)$ exists. Specifically, we must show that $\lim_{t \to 0-} y_2'(t) = \lim_{t \to 0+} y_2'(t)$. We have that $\lim_{t \to 0-} (y_2(t) - y_2(0))/t = \lim_{t \to 0-} (0 - 0)/t = 0$, while $\lim_{t \to 0+} (y_2(t) - y_2(0))/t = \lim_{t \to 0+} (t^2 - 0)/t = \lim_{t \to 0+} t = 0$. Thus, $y_2'(0)$ exists, and so $y_2(t)$ is a second solution of the initial value problem. This does not contradict the Uniqueness Theorem since the rate function $f = 2y/t$ is not continuous at the initial point (0,0), and so the Uniqueness Theorem doesn't apply.

3.   Writing the ODE $ty' - y = t^{n+1}$ in the normal form $y' - y/t = t^n$ and using the integrating factor $1/t$, we have that $(y/t)' = t^{n-1}$. Integrating, $y/t = t^n/n + C$, so we find $y(t) = t^{n+1}/n + Ct$, for any positive n. This formula defines solutions of the original ODE for all t. Since each of these solutions satisfies $y(0) = 0$, the IVP has infinitely many solutions. Since the conditions of the Uniqueness Theorem don't apply at $t = 0$ where $f = y/t + t^n$ is discontinuous, there is no contradiction.

4.   For $y \geq 0$, any solution of $y' = |y|$, $y(0) = 0$ satisfies $y' = y$, $y(0) = 0$. Since for $y \geq 0$, every solution of $y' = y$ has the form $y = Ce^t$ for some constant C, the initial condition implies that $C = 0$ and $y = 0$ for $t \geq 0$. Similarly, for $y \leq 0$, we must have that $y' = -y$, $y(0) = 0$, and, hence, that $y = 0$. Thus, $y = 0$, all t, is the unique solution. Note, however, that $\partial|y|/\partial y = -1$ if $y < 0$, $+1$ if $y > 0$, and so $\partial f/\partial y$ is discontinuous on any region containing $y = 0$. In this case, we have shown that a unique solution exists even though the Existence and Uniqueness Theorem cannot be used.

5.   **(a).** To solve $y' = y \, \text{step}(t)$, $y(0) = 1$, first suppose $t \geq 0$ and solve $y' = y$, $y(0) = 1$ to obtain $y = e^t$, $t \geq 0$. For $t < 0$, solutions are $y = C$ since $y' = 0$, $t < 0$; set $C = 1$ to obtain the solution $y = 1$ with the property that $\lim_{t \to 0-} y(t) = \lim_{t \to 0+} y(t) = 1$. The function

$$ y = \begin{cases} e^t, & t \geq 0 \\ 1, & t < 0 \end{cases} $$

is continuous and satisfies the differential equation for $t \geq 0$ (with right-hand derivative $+1$ at $t = 0$) and for $t < 0$. However, this function is not differentiable at $t = 0$ since $\lim_{t \to 0-} y'(t) = \lim_{t \to 0-} (y(t) - y(0))/t = \lim_{t \to 0-} (1 - 1)/t = 0$, while $\lim_{t \to 0+} y'(t) = \lim_{t \to 0+} (y(t) - y(0))/t = \lim_{t \to 0+} (e^t - 1)/t = 1$. By our definition of solution, $y(t)$ is not a solution since it is not differentiable at $t = 0$. (Sometimes it is useful to say that a function such as $y(t)$ is a solution since its right- and left-hand derivatives are equal everywhere except at the point $t = 0$, where they exist but are not equal.) The graph of $y(t)$ has a kink at $t = 0$.

(b). The IVP is $2yy' = -1$, $y(1) = 0$. Antidifferentiating each side of $2yy' = -1$, we have $y^2 = -t + C$; $C = 1$ since we require that $y = 0$ when $t = 1$. So we obtain, apparently, *two* solutions, $y_1 = \sqrt{1-t}$ and $y_2 = -\sqrt{1-t}$ valid for $t \leq 1$. The problem is that the derivative of each solution at $t = 1$ is infinite. So, the derivative of neither "solution" exists at the initial point, and there is no solution at all unless we are willing to accept infinite slopes.

(c). The IVP is $y' = (1 - y^2)^{1/2}$, $y(0) = 1$. The rate function is $f(y) = (1 - y^2)^{1/2}$. Since the partial derivative of f with respect to y, $f_y(0, 1) = -y(1 - y^2)^{-1/2}\big|_{y=1}$ is undefined, the Existence and Uniqueness Theorem does not apply. The constant function $y_1(t) = 1$, all t, is a solution, but the problem may have additional solutions. In fact, if we separate variables, we have $(1 - y^2)^{-1/2} y' = (\arccos y)' = 1$. Integrating, $\arccos y = t + C$, so $y = \cos(t + C)$. Using the initial condition $y(0) = 1$, we can take $C = 0$. Since y' is nonnegative, we restrict t to the interval $-\pi \le t \le 0$ [or to any interval of the form $(2n - 1)\pi \le t \le 2n\pi$] on which $y = \cos t$ has a nonnegative slope. Notice that the constant function $y(t) = -1$ is also a solution to the ODE (but not the IVP). We can construct another solution of the initial value problem by setting $y = -1$ for $t < -\pi$, then $y = \cos t$ for $-\pi \le t \le 0$, and finally $y = +1$ for $t \ge 0$. Such a function is continuous for all t, differentiable for all t (since $y' = -\sin t$ for $-\pi \le t \le 0$ and $y'(t) = 0$ for $t \le -\pi$ and for $t \ge 0$), and satisfies the rate equation and initial condition. In fact, we see that we can build infinitely many solutions by linking the horizontal line $y(t) = -1$ for $t < (2n - 1)\pi$, to the horizontal line $y(t) = 1$ for $t > 2n\pi$, by means of the graph of $y(t) = \cos t$, $(2n - 1)\pi \le t \le 2n\pi$, n a nonpositive integer.

6. **(a).** Fig. 6(a) shows the direction field and the solution curves of $y' = 3y \sin y + t$ corresponding to the initial conditions $y(-6) = 0, 1, 1.5, 1.9, 1.95, 1.955, 1.956, 1.9565, 1.957,$ $1.96, 1.97, 2, 3, 4$. Wherever solutions curves spread apart, the solutions may be very sensitive to the initial conditions. When sketching solution curves, a small error in one region can therefore lead to large errors further along. Note the big blank region in Fig. 6(a). It seems that the only way to fill that region with solution curves is to put the initial point inside the region and solve forward and backward in time. You can be pretty certain that for points in this region as $t \to -6^+$, $y(t) \to 1.95\ldots$, and as t increases, $y(t)$ gets "trapped" on the slanted curve at the upper right. Remember that solution curves cannot cross each other, so the solution curve is trapped and must behave as indicated. You might zoom on the upper right and pull apart the apparently merging curves.

(b). The "expansion" of solution curves of $y' = 3y \sin y + t$ starting at $t_0 = -6$, $1.8 \le y_0 \le 2$ makes it very hard to hit the point $(0, 2)$. Note how these curves later seem to squeeze together and exit at the upper left of the rectangle. Figure 6(b), Graph 1 shows several of these solution curves which fail to hit $(0, 2)$. In Figure 6(b), Graph 2 we "cheated" in order to hit the point $(0, 2)$ from $(-6, y_0)$, where $1.8 \le y_0 \le 6$. We started at the target $(0, 2)$ and solved backward. The computer then gave us the number $y_0 = 1.956198501618254$, but this value can't be exactly right since data compression and expansion (not to mention the fact that numerical solvers only approximate values) always causes uncertainties.

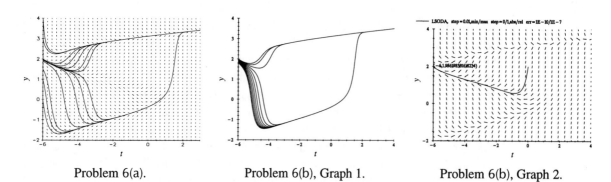

Problem 6(a).   Problem 6(b), Graph 1.   Problem 6(b), Graph 2.

7. **(a).** The equation $ty' + y = 2t$ can be written in normal form as $y' + y/t = 2$ on any t-interval not containing 0. An integrating factor is given by t. The general solution is $y = t + C/t$ for $t \neq 0$. Each value of C gives two distinct solutions, one defined on the t-interval $(-\infty, 0)$ and the other on $(0, \infty)$. Note that $y = t$ is a solution that is defined for all t.

(b). Applying the initial condition $y(0) = 0$ to $y(t) = t + C/t$, we find that $C = 0$. So, the solution of the IVP, $ty' + y = 2t$, $y(0) = 0$, is $y(t) = t$. If $y(0) = y_0 \neq 0$, then C is nonzero and there is no solution at all that is defined at $t = 0$. This does not contradict the Existence and Uniqueness Theorem, since $p(t) = 1/t$ is not defined at $t = 0$ and so the theorem doesn't apply.

(c). See Fig. 7(c). The solution curve of $y = t$ goes right through the origin. All other solution curves are arcs of hyperbolas and are asymptotic to one end or the other of the y-axis as $t \to 0^+$, or 0^-, and to one end or the other of the line $y = t$ as $t \to \pm\infty$. These properties are implied by Fig. 7(c), and also follow directly from the solution formula.

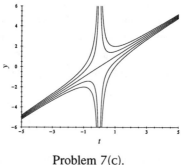

Problem 7(c).

8. **(a).** The IVP is $y' + 2y = \text{step}(1 - t)$, $y(0) = 0$. Using the integrating factor approach, we find a general solution to the differential equation $y' + 2y = q(t)$, with $q(t) = \text{step}(1 - t)$, to be $y(t) = Ce^{-2t} + e^{-2t} \int_0^t q(s)e^{2s}\, ds$. Applying the initial conditions $y(0) = 0$, we have

$$y(t) = e^{-2t} \int_0^t q(s)e^{2s}\, ds = \begin{cases} e^{-2t} \int_0^t e^{2s}\, ds, & t \leq 1 \\ e^{-2t} \int_0^1 e^{2s}\, ds, & t \geq 1 \end{cases}$$

$$= \begin{cases} (1 - e^{-2t})/2, & t \le 1 \\ e^{-2t}(e^2 - 1)/2, & t \ge 1 \end{cases}$$

[since the integrand is 0 for $t \ge 1$]. The solution graph rises as t increases toward 1, but falls for $t \ge 1$. Since $q(t)$ has a discontinuity at $t = 1$, the solution graph has a sharp corner at $t = 1$, and this is visible in the graph. See Fig. 8(a).

(b). Again using the integrating factor approach to solve the ODE $y' + p(t)y = 0$ with $p(t) = 1 + \text{step}(t - t)$, we have $y = y_0 e^{-\int_0^t p(r)\,dr}$. For $t \le 1$, $-\int_0^t p(r)\,dr = -2t$; for $t \ge 1$, $-\int_0^t p(r)\,dr = -\int_0^1 2\,ds - \int_1^t 1\,ds = -1 - t$. Since $y(0) = 2$, $y = 2$ and we have

$$y = \begin{cases} 2e^{-2t}, & t \le 1 \\ 2e^{-1-t}, & t \ge 1 \end{cases}$$

See Fig. 8(b), where the elbow at $t = 1$ is visible (barely).

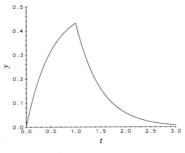

Problem 8(a).

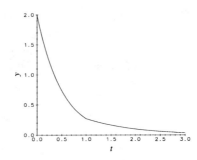

Problem 8(b).

9. The algorithm for producing the Picard iterates to approximate the solution of the IVP, $y' = y$, $y(0) = 1$ is

$$y_0(t) = 1, \quad y_1(t) = 1 + \int_0^t y_0(s)\,ds, \quad \dots, \quad y_{n+1}(t) = 1 + \int_0^t y_n(s)\,ds$$

So we have

$$y_1(t) = 1 + \int_0^t 1\,ds = 1 + t$$

$$y_2(t) = 1 + \int_0^t (1 + s)\,ds = 1 + t + t^2/2$$

$$y_3(t) = 1 + \int_0^t (1 + s + s^2/2)\,ds = 1 + t + t^2/2 + t^3/6$$

$$y_4(t) = 1 + \int_0^t (1 + s + s^2/2 + s^3/6)\,ds = 1 + t + t^2/2 + t^3/6 + t^4/24$$

Since the Taylor series for e^t is $1 + t + \cdots + t^n/n! + \cdots$ and we know that the solution of the IVP is e^t, we see that the Picard algorithm generates the Taylor series one term at a time.

10. Group project.

2.2 Extension and Long-Term Behavior

Suggestions for preparing a lecture

Topics: The Extension Principle and its consequences, periodic forced oscillations (via examples), a brief discussion of Sign Analysis and the time translation property for solutions and solution curves of autonomous ODE, state lines, and (very important) the fact that a positively bounded solution of a scalar autonomous ODE must approach an equilibrium solution as $t \to +\infty$.

Remarks: Long-term behavior is a topic that comes up often in this text. It lies at the heart of the modern approach to dynamical systems.

Making up a problem set

One part from 1 and 2, 4, one or two parts from 5 and 6 and 7, 10, one part from 11. Students also have fun doing the group project 12. If you assigned Problem 10(b) from Section 2.1, don't assign Problem 4 since these two problems overlap.

Comments

In this section we address a topic that is often left out in introductory courses: How far can a solution be extended backward and forward in time? We take up the example IVP, $y' = ky^2$, $y(0) = 1$, where the solution "reaches infinity" in finite time. This can be imagined in the form of a (highly unlikely) physical system modeled by the IVP, where the units of y are, say, miles, time is measured in hours, and the units of k are (miles $\cdot$ hours)$^{-1}$. If $k = 1$, then $y(t)$ is "beyond the end of the universe" within an hour. This also reminds us that a mathematical model has a limited region of usefulness in the space of time and state variables.

Solutions of $y' = f(t, y)$ and the corresponding solution curves cannot just "die" without some reason. A solution can only stop by time T if the solution curve is about to enter a region where the conditions on f given in the Extension Principle fail to hold, or if the solution blows up as time nears the value T (i.e., the solution has a finite escape time). The idea of a maximally extended solution and the existence of periodic forced oscillations are two important topics in this section. We like Example 2.2.7 on the transmission of a "message." We like to emphasize the long-term behavior of solutions of an autonomous ODE $y' = f(y)$: Every solution either tends to $\pm\infty$ or to an equilibrium solution as t increases to $+\infty$. This question of the long-term behavior of solutions of autonomous ODEs is a hot issue today, because it is directly related to chaos. We come back to the question in Chapter 9 for a system of autonomous ODEs. We explain in Section 9.2 just what the long-term behavior of a bounded solution of a planar autonomous system can be (chaos cannot occur in dimension 2, just as Theorem 2.2.2 says it can't occur in dimension 1). However, chaos can occur if an autonomous system has 3 state variables (see the Lorenz system of Section 9.4). What we just said applies to ODEs, but *not* to discrete dynamical systems. As we note in Section 2.7, chaotic wandering *is* possible with a single state variable if we are dealing with certain discrete one-dimensional transformations.

1.   **(a).**   Writing the ODE $yy' = -t$ as $(y^2)'/2 = -t$ and then integrating, we have $y^2/2 = -t^2/2 + C$, so $y = \pm\sqrt{2C - t^2}$. Using the initial condition $y(0) = 1$, we find $C = 1/2$

and we take the plus sign. So the maximally extended solution satisfying $y(0) = 1$ is $y = \sqrt{1 - t^2}$ and the maximal interval of existence of this solution is $-1 < t < 1$. At $t = \pm 1$, the solution's slope would be infinite. See Fig. 1(a) for the graph of the maximally extended solution. Note that if we were to try to extend the time interval to values of t for which $|t| > 1$, then y and y' would both be complex-valued.

(b). The separable ODE, $2y' = -y^3$, can be separated as $-2y^{-3}y' = 1$. We have $(y^{-2})' = 1$, which integrates to $y^{-2} = t + C$. So $y(t) = \pm(t + C)^{-1/2}$. Using $y(0) = -1$, we have $C = 1$ and we take the minus sign. This means that the maximally extended solution to this IVP is $y = -(t + 1)^{-1/2}$, and so the maximal interval of existence of this solution is $t > -1$. See Fig. 1(b).

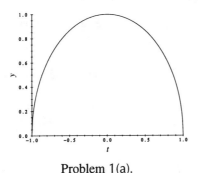

Problem 1(a).

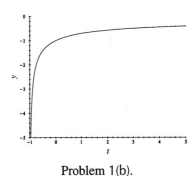

Problem 1(b).

2. **(a).** Separating variables in $y' = -t^2/[(y + 2)(y - 3)]$, we have $(y^2 - y - 6)\,dy = -t^2\,dt$, which integrates to $y^3/3 - y^2/2 - 6y = -t^3/3 + K$. Using the initial data $y(0) = 0$, we find $K = 0$. Multiplying through by 6, $2y^3 - 3y^2 - 36y = -2t^3$. If $y = -2$, then $-2t^3 = 44$ so $t = -(22)^{1/3}$. If $y = 3$, then $-2t^3 = -81$ so $t = (81/2)^{1/3}$. At $y = -2$ or $y = 3$, y' becomes infinite, so the solution can't be extended beyond the interval $-(22)^{1/3} < t < (81/2)^{1/3}$.

    (b). The variables in the ODE $(3y^2 - 4)y' = 3t^2$ are separated. Integrating yields $y^3 - 4y = t^3 + C$. Since $y(0) = 0$, $C = 0$ so $y^3 - 4y = t^3$. Since the rate function becomes infinite when $y = \mp\sqrt{4/3}$, no solution curve can cross either of these two horizontal lines in the ty-plane. Plugging $y = \mp\sqrt{4/3}$ into our solution formula yields $t = \pm(\frac{16}{3\sqrt{3}})^{1/3} \approx \pm 1.45$. So time t cannot be extended beyond the interval $|t| < (\frac{16}{3\sqrt{3}})^{1/3}$ because $y'(t)$ would become infinite at the points $\pm(\frac{16}{3\sqrt{3}})^{1/3}$. See Fig. 2(b).

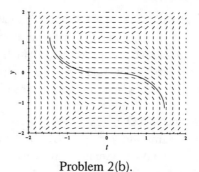

Problem 2(b).

3. Separating variables in the ODE, $y' = 1 - y^2$, and integrating, we obtain the general solution

$$\ln |(1 + y)/(1 - y)|^{1/2} = t + C, \quad \text{where } C \text{ is an arbitrary constant}$$

Exponentiating, squaring, dropping the absolute value signs, and redefining the constant of integration we have

$$\frac{1 + y}{1 - y} = ce^{2t}, \quad \text{where } c \text{ is an arbitrary constant} \tag{i}$$

Imposing the initial condition $y(0) = y_0$, we see that $c = (1 + y_0)/(1 - y_0)$. Solving (i) for y we obtain

$$y = \frac{c - e^{-2t}}{e^{-2t} + c}$$

Now if $y_0 < -1$, then we have that $-1 < c < 0$, and $y(t)$ escapes to $-\infty$ as $t \to t_e^-$, where $e^{-2t_e} + c = 0$. Solving $e^{-2t_e} + c = 0$ for the escape time t_e,

$$t_e = (-1/2) \ln(-c) = (-1/2) \ln \frac{1 + y_0}{y_0 - 1}$$

4. **(a).** The rate function $f(y) = 3y^{2/3}$ is continuous for all values of y, including $y = 0$. However, $df/dy = 2y^{-1/3}$ is discontinuous at $y = 0$ (it isn't even defined there). So the conditions of the Existence and Uniqueness Theorem are satisfied everywhere in the ty-plane except along the line $y = 0$ (i.e., on the t-axis). The conditions are not satisfied for any value of t_0 if we stipulate that $y(t_0) = 0$.

   (b). If we let $y_1(t) = 0$, all t, then $y_1'(t) = 0$, all t, and so $y_1(t) = 0$ is a solution of the IVP, $y' = 3y^{2/3}$, $y(0) = 0$. Now, suppose $y_2(t) = t^3$. Then $y_2' = 3t^2 = 3y_2^{2/3}$. Since $y_2(0) = 0$, $y_2(t)$ is another solution of the same IVP.

   (c). By the translation property since $y = t^3$ is a solution of the ODE, $y' = 3y^{2/3}$, so is $y = (t - t_0)^3$ for any t_0. So the IVP, $y' = 3y^{2/3}$, $y(t_0) = 0$, has solutions $y = 0$ and $y = (t - t_0)^3$. There are other solutions. It is not too hard to show that for any $a \le t_0$ and

$b \geq t_0$, the function

$$y(t) = \begin{cases} (t-a)^3 & t \leq a \\ 0 & a < t \leq b \\ (t-b)^3 & t > b \end{cases}$$

is a solution of this IVP. Note $y(t)$ is continuous for all t since $\lim_{t \to a^-} y(t) = \lim_{t \to b^+} y(t) = 0$. To show that each of these solutions is differentiable at the point $t = a$, proceed as follows:

$$\frac{y(t) - y(a)}{t - a} = \frac{(t-a)^3}{t-a} = (t-a)^2 \to 0 \text{ if } t \to a^-$$

so $\lim_{t \to a^-} y'(t) = 0$; the right-hand derivative $\lim_{t \to a^+} y'(t) = 0$ since $y(t) = 0$, $a < t \leq b$. So $y'(a)$ exists and is 0; similarly for $y'(b)$. We have constructed infinitely many solutions of the IVP.

5. Recall that $y = y_0$ is an equilibrium solution of $y' = f(y)$ if $f(y_0) = 0$.

    (a). The equilibrium solutions of $y' = (1-y)(y+1)^2$ are $y = \pm 1$. See Fig. 5(a). The solutions between $y = -1$ and $y = +1$ tend to $y = 1$ as $t \to +\infty$ and to $y = -1$ as $t \to -\infty$.

    (b). The equilibrium solutions of $y' = \sin(y/2)$ are $y = 2k\pi$, $k = 0, \pm 1, \pm 2, \ldots$. See Fig. 5(b). The solutions between $y = 0$ and $y = 2\pi$ tend to $y = 2\pi$ as $t \to +\infty$ because the rate function is positive in that band and to $y = -2\pi$ as $t \to -\infty$. The solutions between $y = -2\pi$ and $y = 0$ tend to $y = -2\pi$ as $t \to +\infty$ because the rate function is negative in the band and to $y = 2\pi$ as $t \to -\infty$. The behavior as $t \to \infty$ alternates in this fashion in adjacent bands between equilibrium solutions.

    (c). The equilibrium solutions of $y' = y(y-1)(y-2)$ are $y = 0$, $y = 1$, and $y = 2$. See Fig. 5(c). Solutions between $y = 1$ and $y = 2$ tend to $y = 1$ as $t \to +\infty$ and to $y = 2$ as $t \to -\infty$. Solutions between $y = 0$ and $y = 1$ tend to $y = 1$ as $t \to +\infty$ and to $y = 0$ as $t \to -\infty$.

    (d). The equilibrium solutions of $y' = 3y - ye^{y^2}$ are $y = 0$ and $y = \pm\sqrt{\ln 3}$. See Fig. 5(d). Solutions in the band between $y = 0$ and $y = \sqrt{\ln 3}$ tend to $y = \sqrt{\ln 3}$ as $t \to +\infty$ because $y' > 0$ in the band and to $y = 0$ as $t \to -\infty$. On the other hand, solutions in the band between $y = -\sqrt{\ln 3}$ and $y = 0$ tend to $y = -\sqrt{\ln 3}$ as $t \to +\infty$ and to $y = 0$ as $t \to -\infty$.

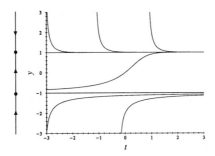

Problem 5(a).

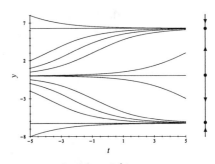

Problem 5(b).

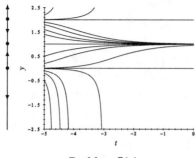

Problem 5(c).   Problem 5(d).

6.   **(a).** A solution curve of $y' = -y\cos t$ is rising whenever it lies in the region $-y\cos t > 0$, that is, whenever $-y$ and $\cos t$ are both positive or both negative. The regions in R where solution curves rise are the region $-8 < y < 0$, and either $-6 < t < -3\pi/2$ or $-\pi/2 < t < \pi/2$ or $3\pi/2 < t < 6$; and the region $0 < y < 8$, and either $-3\pi/2 < t < -\pi/2$ or $\pi/2 < t < 3\pi/2$. Curves fall elsewhere in the rectangle R.

   (b). See Fig. 6(b); the nullclines $y = 0$, $t = \pm\pi/2$, $t = \pm3\pi/2$ are the dashed lines.

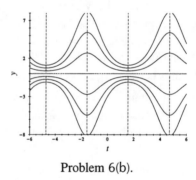

Problem 6(b).

7.   Solution curves rise on one side of a nullcline (and fall on the other) in all of these graphs. The nullclines in the graphs are dashed.

   (a). The ODE is $y' = (y + 3)(y - 2)$. See Fig 7(a). The nullclines are $y = -3$, $y = 2$. Notice that the nullclines are also solutions.

   (b). The ODE is $y' = 2t - y$. See Fig 7(b). The nullcline is $y = 2t$.

   (c). The ODE is $y' = ty - 1$. See Fig 7(c). The nullcline is $y = 1/t$.

   (d). The ODE is $y' = (1 - t)y$. See Fig 7(d). The nullclines are $y = 0$, $t = 1$. Notice that $y = 0$ is also a solution.

   (e). The ODE is $y' + (\sin t)y = t\cos t$. See Fig 7(e). The nullcline is $y\sin t = t\cos t$.

   (f). The ODE is $y' = y - t^2$. See Fig 7(f). The nullcline is $y = t^2$.

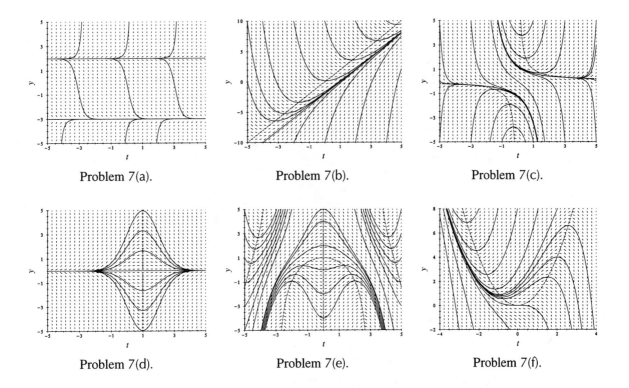

Problem 7(a). Problem 7(b). Problem 7(c).

Problem 7(d). Problem 7(e). Problem 7(f).

8. Suppose that the rate function $f(t, y)$ for the ODE $y' = f(t, y)$ always satisfies $f(t, y) \geq cy^2$ for all $t \geq 0$. So we have for $y > 0$:

$$y'(t) \geq cy^2(t), \quad y' - cy^2 \geq 0, \quad y^{-2}y' - c \geq 0, \quad (-(1/y) - ct)' \geq 0$$

But if the functions $g(t)$ and $h(t)$ satisfy $g(t) \geq h(t)$ on some interval $a \leq t \leq b$, then $\int_a^b g(t)\, dt \geq \int_a^b h(t)\, dt$. So we have on any interval $0 \leq t \leq T$ where $y(t)$ is defined that

$$\int_0^T (-(1/y) - ct)'\, dt \geq \int_0^T 0\, dt = 0$$

$$\left[-(1/y) - ct\right]_{t=0}^{t=T} \geq 0$$

$$-(1/y(T)) - cT + 1/y_0 \geq 0$$

where $y(0) = y_0 > 0$. So we have

$$-1/y(T) \geq cT - 1/y_0, \quad 1/y(T) \leq (1/y_0) - cT, \quad y(T) \geq \frac{1}{(1/y_0) - cT} \qquad \text{(i)}$$

where we have used several properties of inequalities. Now $T > 0$ has not been specified; we see from (i) that as $T \to (1/(cy_0))^-$, then the right side of (i) tends to $+\infty$, and that forces $y(t)$ to escape to $+\infty$ no later than at $t = T$.

9. **(a).** The ODE $y' = y^2$ separates to $y^{-2}dy = dt$. Integrating, $-1/y = t + C$ so if $y(0) = 5$, then $C = -1/5$ so $-1/y(t) + 1/5 = t$, and $y = 5/(1 - 5t)$, where we require that t be less than $1/5$. As $t \to (1/5)^-$, $y(t) \to +\infty$. As $t \to -\infty$, $y(t) \to 0^+$.

(b). The IVP is $y' = y^2$, $y(0) = 5$. See Fig. 9(b). The exact solution curve $y = 5/(1 - 5t)$, $t < 0.2$ is dashed. This solution tends to $+\infty$ as $t \to 0.2^-$, but the thirteen Picard iterates shown in Fig. 9(b) all sail right through the singularity at $t = 0.2$ as if it weren't even there. The graph of the Picard iterates $y_n(t)$ bends up toward the graph of the true solution as $t \to +\infty$. The larger n is, the more pronounced the bending and the closer the graph of the iterate is to the true solution.

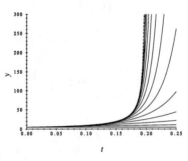

Problem 9(b).

10.   The IVP is $V_O' + V_O = \text{sqw}(t, 50, T)$, $V_O(0) = 0$. See Fig. 10, Graphs 1, 2, 3, in which T is, respectively, 0.1, 10, and 50 milliseconds. The message received improves as the period T increases, i.e., as the frequency decreases.

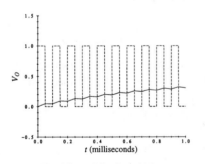
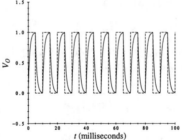
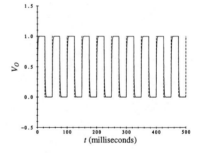

Problem 10, Graph 1.   Problem 10, Graph 2.   Problem 10, Graph 3.

11.   All of the solution curves [except for part **(c)**] tend to a periodic forced oscillation (Theorem 2.2.4) because the coefficient p_0 in every linear ODE, $y' + p_0 y = q(t)$, is positive [except for part **(c)**]. This means that the term $e^{-p_0 t}C$ in the general solution of the ODE tends to 0 as $t \to +\infty$, leaving the periodic forced oscillation as the only visible solution. The solutions of the ODE in part **(c)** are $y = C + \int_0^t 3\,\text{sqw}(s, 50, 2)\,ds$; since the integrand has value 3 over half of every time interval of length 2 (and is 0 otherwise), the value of the integral increases piecemeal. The constant C simply translates the solution curves up or down according to the sign of C. The corners on all of the solution curves except parts **(e)** and **(f)** reflect the discontinuities in the driving terms. The graphs for parts **(e)** and **(f)** are

smooth since the driving terms are continuous. The ODEs and some of their solution curves are shown below.

(a). $y' + y = 3\,\text{sqw}(t, 25, 3)$,

(b). $y' + 0.75y = 3\,\text{sqw}(t, 50, 1)$,

(c). $y' = 3\,\text{sqw}(t, 50, 2)$,

(d). $y' + 0.2y = \text{sqw}(t, 50, 1)$,

(e). $y' + y = \sin 3t$,

(f). $y' + 0.5y = 2\,\text{trw}(t, 50, 2)$.

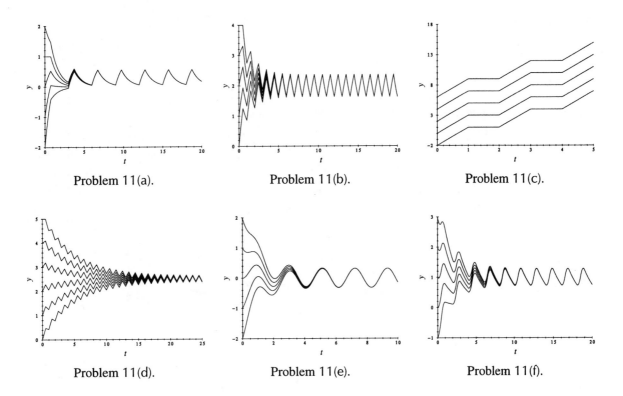

Problem 11(a).   Problem 11(b).   Problem 11(c).

Problem 11(d).   Problem 11(e).   Problem 11(f).

12. Although we rarely "solve" a group project, we give a partial solution here. Labeling the four bulleted items in the problem statement as **(a)–(d)**, here is what the "solution" looks like.

(a). From Sections 1.3 and 1.4, we see that the unique solution of the IVP $y' + p_0 y = q(t)$, $y(0) = y_0$ is given by $y(t) = e^{-p_0 t}y_0 + e^{-p_0 t}\int_0^t e^{p_0 s}q(s)ds$. This IVP has a periodic solution $y(t)$ with period T if and only if $y(T) = y(0) = y_0$. Applying this condition to the solution formula, we see that y_0 must satisfy the equation

$$y_0 = e^{-p_0 T}y_0 + e^{-p_0 T}\int_0^T e^{p_0 s}q(s)ds \tag{i}$$

which has a unique solution, $\bar{y}_0$, for y_0 because $p_0 \neq 0$ (hence, $e^{-p_0 T} \neq 1$). We shall show that the solution $\bar{y}(t)$ defined by

$$\bar{y}(t) = e^{-p_0 t}\bar{y}_0 + e^{-p_0 t}\int_0^t e^{p_0 s}q(s)ds$$

has period T by showing that for all t, $\bar{y}(t+T) = \bar{y}(t)$:

$$\bar{y}(t+T) = e^{-p_0(t+T)}\bar{y}_0 + e^{-p_0(t+T)}\int_0^{t+T} e^{p_0 s}q(s)\,ds$$

$$= e^{-p_0 t}\left[e^{-p_0 T}\bar{y}_0 + e^{-p_0 T}\int_0^T e^{p_0 s}q(s)\,ds\right] + e^{-p_0 t}\int_T^{t+T} e^{p_0(s-T)}q(s)\,ds$$

$$= e^{-p_0 t}\bar{y}_0 + e^{-p_0 t}\int_0^t e^{p_0 u}q(u)\,du \quad (\text{let } u = s - T \text{ and use } q(u+T) = q(u))$$

$$= \bar{y}(t)$$

where we have used the fact that $y_0 = \bar{y}_0$ solves equation (i). Also, the uniqueness of $\bar{y}_0$ implies that this IVP has the unique periodic solution $\bar{y}(t)$.

(b). Let $\bar{y}(t)$ denote the periodic solution in part **(a)**. The general solution of the ODE $y' + p_0 y = q(t)$, is $y = Ce^{-p_0 t} + \bar{y}(t)$, C an arbitrary constant. If $p_0 > 0$, then the first term is transient (i.e., dies out as $t \to \infty$) and so for any constant C, $y(t)$ rapidly begins to look like $\bar{y}(t)$. See Fig. 12(b) for the graph of $y(t)$ if $y(0) = 2$, $p_0 = 0.5$, $q(t) = \text{sww}(t, 50, 4)$.

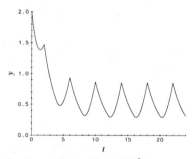

Section 2.2, Problem **12(b)**

(c). The unique periodic forced oscillation $\bar{y}(t)$ is the solution of the IVP $y' + p(t)y = q(t)$, $y(0) = \bar{y}_0$, where $\bar{y}_0$ is the unique solution of $y_0 = e^{-P_0(T)}y_0 + e^{-P_0(T)}\int_0^T e^{P_0(s)}q(s)ds$, where $P_0(T) = \int_0^T p(v)dv$.

(d). We know that in order for a continuous function $y(t)$ to be periodic with period T, it is necessary that $y(0) = y(T)$. As in part **(c)**, this implies that

$$y_0 = e^{-P_0(T)}y_0 + e^{-P_0(T)}\int_0^T e^{P_0(s)}q(s)ds$$

where $P_0(T) = \int_0^T p(v)dv$. Now if $P_0(T) = 0$, the only way this equation can hold is if the condition

$$\int_0^T e^{P_0(s)} q(s)\, ds = 0 \qquad\qquad (ii)$$

is satisfied, and in that case the equation above holds for all y_0. It follows that every solution

$$y = y_0 + e^{-P_0(t)} \int_0^t e^{P_0(s)} q(s)\,ds$$

of the ODE in this case is periodic with period T. If the integral condition (ii) is not satisfied, then no solution of the ODE is periodic with period T. An example of the first case is $y' + (\cos t)y = \cos t$, where $T = 2\pi$. An example of the second case is $y' + (\cos t)y = 2 + \cos t$, where again $T = 2\pi$. Note that

$$\int_0^{2\pi} e^{\sin s}(2 + \cos s)ds > 0$$

because the integrand is positive.

13. Group project.

14. Group project.

Background Material: Sign Analysis and Long-Term Behavior

Here is a verification of part of Theorem 2.2.3. Suppose that $y_1 < y_2$ are two consecutive zeros of $f(y)$; that is, $f(y_1) = f(y_2) = 0$, but $f(y)$ does not vanish in the interval $y_1 < y < y_2$. Then since $f(y)$ is continuous, $f(y)$ must be of one sign in the strip $y_1 < y < y_2$, say positive. Note that the lines $y = y_1$ and $y = y_2$ are solution curves for the ODE. Now, because f is positive, the solution that passes through the point (t_0, y_0), $y_1 < y_0 < y_2$, must be increasing as t increases, if the solution curve remains in the strip $y_1 < y < y_2$. To leave this strip as t increases, the solution curve would have to cross the equilibrium solution curve $y = y_2$, but the Uniqueness Theorem implies that this cannot happen. The Extension Theorem easily shows that our solution curve is defined for all $t \geq t_0$. Similarly, we see that for backward time our solution curve falls for decreasing t, but it never crosses the equilibrium solution $y = y_1$. The Extension Theorem shows again that the solution is defined for all $t \leq t_0$. If follows that the limits

$$\lim_{t\to+\infty} y(t) = b \leq y_2 \quad \text{and} \quad \lim_{t\to-\infty} y(t) = a \geq y_1$$

both exist. We now show that $a = y_1$ and $b = y_2$. By shifting this solution curve in t we can completely fill up the strip $a < y < b$ with the solution curves of the ODE. Now suppose that $b < y_2$, contrary to our assertion. Then, as noted above, the solution curve defined by the IVP $y' = f(y)$, $y(t_0) = b$, can be extended to the entire t-axis. Extending the solution forward, the solution must cross the line $y = b$ because $f(b) > 0$, and so must intersect a solution curve in the strip $a < y < b$, in contradiction to the Uniqueness Theorem. It must be that $b = y_2$. Similarly, it can be shown that $a = y_1$.

2.3   Sensitivity

Suggestions for preparing a lecture

Topics: Examples, examples, and more examples, with a general treatment of the imprecise use of the term "sensitivity" and the connection between continuity and sensitivity in terms of changing data parameters.

Making up a problem set

Problem 1, 2 or 3, one part of 5, 6. You might even ask some students to build and test an RC circuit for its ability to transmit low-frequency square waves.

Comments

Information about how the solution of an IVP changes as the data change is critical for a real understanding of the IVP, particularly if the IVP models some real phenomenon. The idea of sensitivity is fairly easy to grasp, even though we never give a precise definition. The informal explanation of the solution of an IVP being a continuous function of all the data is also easy to understand, but it must be emphasized that the time interval about t_0 must be small. Note that small changes in data may evolve over time into huge variations. This is good background for a later discussion on the Lorenz system (Section 9.4), where we speak of chaos. We only treat linear ODEs in this section. See Appendix A.4 for a discussion of the sensitivity question for the IVP, $y' = f(t, y)$, $y(t_0) = y_0$.

---

1.   The solution curves for the IVPs $y' = -y \sin t + ct \cos t$, $y(-4) = -6$, $c = 0.7, 0.8, \ldots,$ 1.3 in Fig. 2.3.1 in the text seem to be diverging as $t \to -1^-$, but the plot in Fig. 1 below for $-4 < t < 20$ shows that they alternately spread out and bunch together. It appears that as t increases, the curves oscillate with ever-increasing amplitude, after a while bunching up at the peaks and spreading out at the troughs, with successively larger spreads at each trough. This pattern is only our conjecture. Do you have a different conjecture?

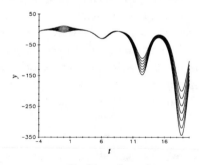

Problem 1.

2.   We apply the Bounded Input–Bounded Output Principle to the ODE, $P' + r(t)P = R(t)$, where $0 < r_0 \le r(t)$ and $|R(t)| \le R_0$. From the BIBO Principle, we have

$$P(t) \le e^{-r_0 t} P_0 + \frac{R_0}{r_0}(1 - e^{-r_0 t}) \le P_0 + R_0/r_0, \quad \text{for } t \ge 0$$

where we have dropped the absolute value signs on $P(t)$ since we are looking for an upper bound only.

3. We are given that the vat has a 100 gallon capacity, that r gallons of brine run into the vat per minute, and the salt concentration in the inflow stream is $c(t)$ lbs/gal. The brine exits the tank at the rate of r gal/min. Let $S(t)$ be the lbs of salt dissolved in the brine in the tank at time t. We are given that $S(0) = 5$. Then $S'(t) = $ Rate in $-$ Rate out $= rc(t) - rS/100$, $S(0) = 5$, where, say, $0 < r_0 \leq r \leq r_1$ and $0 \leq c(t) \leq c_0$. The Bounded Input–Bounded Output Principle applies to the ODE, $S' + rS/100 = rc$. Identifying $|y_0| = 5$, $p_0 = r_0/100$, and $M = r_1c_0$, we have that $|S(t)| \leq 5e^{-r_0t/100} + (100/r_0)r_1c_0(1 - e^{-r_0t/100}) \leq 5 + 100r_1c_0/r_0$. Since $c(t) = S(t)/100$, the safety rule that $c(t)$ must never exceed 0.1 lb/gal implies that we must be sure that $S(t) \leq 10$ at all times, which we can guarantee by setting $100r_1c_0/r_0 \leq 5$, that is, we require that $20r_1c_0 \leq r_0$ to ensure that the concentration of salt in the tank never exceeds 0.1 lbs/gal.

4. The solution curves of $y' = -cy/t$, $y(10) = 3$, $0 < t \leq 30$, $c = -1.5, -1, 0, \ldots, 1.5$ are shown in Graph 1. Graph 2 indicates that solutions are not very sensitive to changes in c if $|c + 1| \leq 0.5$ and $0 \leq t \leq 10$. Graphs 3 and 4 show that solutions are sensitive to changes in c if either $|c + 1| \leq 0.5$, $10 \leq t \leq 30$, or if $|c - 1| \leq 0.5$, $0 < t \leq 10$. In Graph 3, the solutions seem to become unbounded as $t \to +\infty$. Finally, in Graph 5, it appears that if $|c + 1| \leq 0.5$, $10 \leq t \leq 30$, then, although the solution curves seem to diverge for awhile, they don't continue to get farther apart. So we might say that solutions are sensitive to changes in c on the interval $10 \leq t \leq 30$, but perhaps not on a larger interval. The answers to sensitivity questions depend on the parameter ranges and the time spans under consideration.

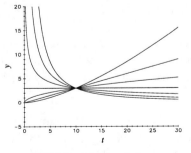

Problem 4, Graph 1.

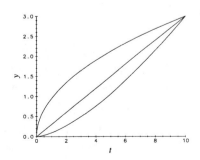

Problem 4, Graph 2.

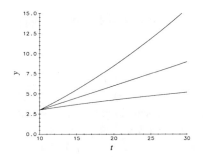

Problem 4, Graph 3.

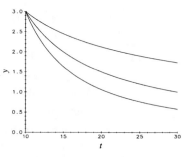

Problem 4, Graph 4.   Problem 4, Graph 5.

5.   **(a).** Using the integrating factor e^t to solve $y' = -y + e^{-t}$, we see that $(ye^t)' = 1$, so integrating and using $y(0) = a$, we have $y(t, a) = ae^{-t} + te^{-t}$; similarly for $y(0) = b$. We have

$$|y(t, a) - y(t, b)| = |a - b|e^{-t}, \quad t \geq 0$$

The solution is relatively insensitive to changes in initial data because as time goes on, the deviation between the solutions decays to zero.

(b). The solution of the IVP $y' = -y + c$, $y(0) = 1$ is $y_c(t) = e^{-t} + c(1 - e^{-t})$, so

$$|y_c(t) - y_5(t)| = |c - 5|(1 - e^{-t}) \leq 0.1(1 - e^{-t}), \quad t \geq 0$$

which increases from 0 at $t = 0$ to 0.1 as $t \to +\infty$. So, the solution is relatively insensitive to changes in y_0 because the deviation between y_c and y_5 never gets larger than the maximum deviation 0.1 of c from 5.

(c). The solution $y_c(t, a)$ of the IVP $y' = -y + ct$, $y(0) = a$ is obtained by using the integrating factor e^t and then integration by parts (or a table of integrals): $y_c(t, a) = ae^{-t} + c(e^{-t} + t - 1)$. Using this formula and the Triangle Inequality ($|A + B| \leq |A| + |B|$), we see that

$$|y_c(t, a) - y_1(t, b)| \leq |a - b|e^{-t} + |c - 1| \, |e^{-t} + t - 1|$$

As t increases from 0 to 10, the first term on the right decays to a small fraction of $|a - b|$, but the last term has increased quite a bit. So, on the interval $0 \leq t \leq 10$ the solution is insensitive to the changes in the initial data, but highly sensitive to the changes in the parameter c.

6.   The IVP is $V_O' + V_O = \text{trw}(t, 50, T)$, $V_O(0) = 0$. See Graphs 1, 2, 3, in which T is, respectively, 0.1, 10, and 50 milliseconds. The message received (solid) matches the message sent (dashed) more and more closely as the period T increases, i.e., as the frequency decreases.

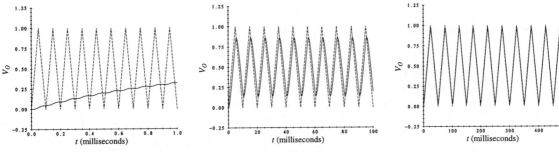

Problem 6, Graph 1.   Problem 6, Graph 2.   Problem 6, Graph 3.

7. The solution $y(t)$ of the IVP, $y' + p_0 y = M$, $y(0) = y_0$, is $y = y_0 e^{-p_0 t} + M(1 - e^{-p_0 t})/p_0$; that is, if y_0 and p_0 are positive, then y precisely equals the right-hand side of the inequality given in the Bounded Input–Bounded Output Principle. We have shown that for the specific case $p(t) = p_0$, $q(t) = M$, the inequality can't be improved. So, it is not in general possible to replace the right-hand side by a smaller quantity.

8. An integrating factor for the ODE, $y' + (t + 1)^{-2} y = e^{1/(t+1)}$, is the function $e^{-1/(t+1)}$. Solutions are given by $y = (C + t)e^{1/(t+1)}$, and so $y \to \infty$ as $t \to +\infty$. Note that $p(t) = (t + 1)^{-2} > 0$ but that there is no positive constant p_0 such that $p(t) \geq p_0$ for all $t \geq 0$. We have shown that for this specific ODE, $p(t) > 0$ and yet the BIBO estimate fails. So, the condition $p(t) \geq p_0 > 0$ cannot in general be weakened to $p(t) > 0$.

9. The ODE $y' + y = t$ has integrating factor e^t, and its solutions are given by $y = Ce^{-t} + t - 1$. So $y \to +\infty$ as $t \to +\infty$. Note that the input t is not bounded on the interval $t \geq 0$. We have shown that for this specific ODE, $q(t)$ is unbounded and BIBO fails. The condition $|q(t)| \leq M$ cannot in general be dropped.

10. **(a).** The variables in the ODE $y' = -y/(1 + y^2)$ may be separated: $-(y^{-1} + y)y' = 1$. Integration gives $t = C - \ln|y| - y^2/2$. This equation can't be solved for y in terms of elementary functions of t. To show that $y \to 0$ as $t \to \infty$, we must argue indirectly. We have that $t + \ln|y| + y^2/2 = C$, or $e^t|y|e^{y^2/2} = e^C$. As $t \to \infty$, $e^t \to \infty$ and since $e^{y^2/2} \geq e^0 = 1$ the only way the product on the left can remain constant is for $|y| \to 0$. Every solution $y = y(t)$ has the property that $y(t) \to 0$ as $t \to +\infty$.

    (b). To find the extreme values of the rate function $r(y) = -y(1 + y^2)^{-1} + 1$, differentiate and equate to 0 : $-(1 + y^2)^{-1} + 2y^2(1 + y^2)^{-2} = 0$. Solving for y, we obtain $y = \pm 1$. Since $r(1) = 1/2$ while $r(-1) = 3/2$ and $\lim_{y \to \pm\infty} r(y) = 1$, we see that the minimum value of the rate function is $1/2$, and $y'(t) \geq 1/2$.

    (c). Integrating each side of the inequality $y'(t) \geq 1/2$ from 0 to t, we have $y(t) - y_0 \geq t/2$ or $y(t) \geq y_0 + t/2$, and so $y(t) \to \infty$ as $t \to \infty$. We interpret this situation as one where there is bounded input [the function $+1$ on the right-hand side of $y' = -y(1 + y^2)^{-1} + 1$], while all solutions of the undriven equation $y' = -y(1 + y^2)^{-1}$ tend toward the equilibrium $y = 0$ [see part **(a)**]. This is the setting for the Bound Input–Bounded Output Theorem, except that the equation here is nonlinear and the Theorem need not apply. In this case, we see that the conclusion does not hold; the bounded input actually stimulates the production of an unbounded output.

(d). See Fig. 10(d), Graph 1 for solutions of $y' = -y/(1 + y^2)$ and Fig. 10(d), Graph 2 for solutions of the driven ODE $y' = -y/(1 + y^2) + 1$. The solutions of the first ODE $\to 0$ as $t \to +\infty$, while those of the second tend to ∞ as $t \to +\infty$.

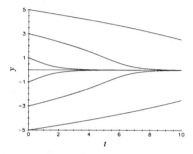
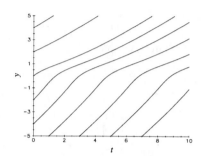

Problem 10(d), Graph 1.   Problem 10(d), Graph 2.

Background Material: Verification of Theorem 2.3.1

Here is the verification of Theorem 2.3.1 that was omitted from the text. Suppose that $p(t) \geq p_0$ where $p_0 \neq 0$. Then by properties of integrals and exponentials, we have for all positive values of t that

$$p_0 t \leq \int_0^t p(s)\, ds$$

$$e^{-\int_0^t p(s)\, ds} \leq e^{-p_0 t}$$

Since the solution of the IVP

$$y' + p(t) y = q(t), \quad y(0) = y_0$$

is given by

$$y(t) = e^{-\int_0^t p(s)\, ds} \left\{ y_0 + \int_0^t e^{\int_0^s p(r)\, dr} q(s)\, ds \right\}$$

$$= e^{-\int_0^t p(s)\, ds} y_0 + \int_0^t e^{-\int_s^t p(r)\, dr} q(s)\, ds$$

we have [using the triangle inequality and the fact that $|\int_s^t p(r)\, dr| \leq \int_s^t |p(r)|\, dr$ if $t \geq s$]:

$$|y(t)| \leq e^{-p_0 t} |y_0| + \int_0^t e^{-p_0(t-s)} M\, ds$$

$$\leq e^{-p_0 t} |y_0| + \frac{M}{p_0} e^{-p_0 t} \left[e^{p_0 t} - 1 \right]$$

$$\le e^{-p_0 t}|y_0| + \frac{M}{p_0}\left[1 - e^{-p_0 t}\right]$$

$$\le |y_0| + \frac{M}{p_0} \quad \text{(for } t \ge 0\text{)}$$

This verifies Theorem 2.3.1.

2.4 Introduction to Bifurcations

Suggestions for preparing a lecture

Topics: The saddle-node bifurcation for the scaled harvested/restocked logistic ODE, scaling of the general harvesting/restocking model, bifurcation diagrams.

Remarks: Students may find bifurcations a bit over their heads at first, but drawing lots of graphs helps them get the "big picture."

Making up a problem set

One part of Problem 1 and the corresponding part of Problem 2, one part of Problem 3 and the corresponding part of Problem 4 [tell students to do Problems 1 and 2 together and in either order; same instructions for Problems 3 and 4], one of Problems 5, 6, or 7.

Comments

Bifurcations are new to the introductory ODE syllabus, but they are an important part of any course that introduces the ideas of sensitivity to data changes and is serious about modeling physical systems. Although most changes in the data do not lead to drastic changes in solution behavior (at least not over a short span of time), there are those occasional data ranges where something completely new occurs, although the consequences may only become evident with time's advance. The term "bifurcation" is now applied to this kind of event, but it is not a precisely defined term. We introduce this idea via examples and suggestive (rather than exact) definitions. We introduce the saddle-node bifurcation in the context of a harvested/restocked and logistically changing population. We do the important task of rescaling so that the only remaining parameter in the model ODE is the harvesting/restocking rate. This is a good example to see just why rescaling is often done before computing. In reading this material, you might think of four stages after the scaling has been done. First, find the equation of the equilibrium solutions in terms of the parameter c and plot and discuss the meaning of their graphs in the cy-plane (the bifurcation diagram), clearly locating the bifurcation point. Next, show graphs of solutions before, at, and beyond bifurcation. Then replot the bifurcation diagram, but this time introducing solid or dashed arcs depending upon whether an equilibrium solution is an attractor or a repellor. Finally, plot the rate function $f = f(y, c)$ in a yf-plane for various values of c below, at, and above the bifurcation value, and show that at the bifurcation value c^* the graph of $f = f(y, c^*)$ is tangent to $f = 0$. You may want to complete these stages in a different order.

1. Find the equilibrium solutions y_i as functions of the parameter c. Then find the value of c where the complex equilibrium solutions merge into a single real equilibrium solution and then split into two real equilibrium solutions. The sign of the rate function near an equilibrium determines whether the equilibrium is an attractor or a repeller. For example,

if $f(y_1, c_1) = 0$ and if $f(y, c_1) > 0$ for y near y_1 but $y < y_1$, while $f(y, c_1) < 0$ for y near y_1 and $y > y_1$, then $y = y_1$ is an attracting equilibrium for $c = c_1$ because nearby solutions rise toward it or fall toward it as time increases. When drawing the bifurcation diagram in the cy-plane, use a solid arc for the attracting branch and a dashed arc for the repelling branch.

(a). The ODE is $y' = c - y^2$. Since $f(y, c) = c - y^2$, if $c > 0$ the equilibrium solutions are at $y = \pm\sqrt{c}$; if $c = 0$, $y = 0$ is the only equilibrium solution, while if $c < 0$, there are no real equilibrium solutions since $\sqrt{c}$ is complex. So $c = 0$ is the critical value. As c increases through 0, the complex conjugate numbers $\pm\sqrt{c}$ move toward each other in the complex plane, meet when $c = 0$, and split apart into a pair of real equilibrium solutions. The solution $y = \sqrt{c}$, $c > 0$, is an attractor since if $y < \sqrt{c}$ but near $\sqrt{c}$, then $f = c - y^2$ is positive and solutions of $y' = c - y^2$ rise up toward $y = \sqrt{c}$, while if $y > \sqrt{c}$, then $c - y^2$ is negative and solutions fall toward the equilibrium. A similar argument shows that $y = -\sqrt{c}$ is a repeller. See Fig. 1(a) for the bifurcation diagram.

(b). The ODE is $y' = c - 2y + y^2$. The zeros of $f(y, c) = c - 2y + y^2$ are $y_{1,2} = 1 \mp (1 - c)^{1/2}$, and these are real and distinct if $c < 1$, real and equal if $c = 1$, and complex conjugates if $c > 1$. For large values of y, the sign of $f(y, c)$ is positive because of the y^2 term. Then $f(y, c)$ changes sign and becomes negative as y decreases past y_2. So $y = y_2 = 1 + (1 - c)^{1/2}$ is a repeller. A similar argument shows that $y = y_1$ is an attractor. See Fig. 1(b) for the bifurcation diagram.

(c). The ODE is $y' = c + 2y + y^2$. The zeros of $f(y, c) = c + 2y + y^2$ are $y_{1,2} = -1 \mp \sqrt{1 - c}$, and are real and distinct for $c < 1$, real and identical for $c = 1$, and complex conjugates for $c > 1$. For large y, $f(y, c)$ is positive, but for $y < y_2$ and near y_2, $f(y, c)$ is negative. So $y = y_2$ is a repeller. A similar argument shows that $y = y_1$ is an attractor. See Fig. 1(c) for the bifurcation diagram.

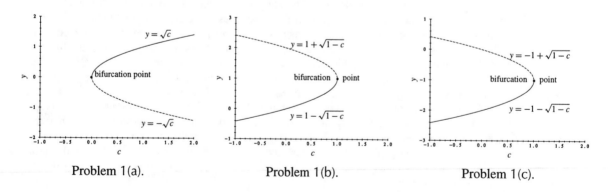

Problem 1(a).   Problem 1(b).   Problem 1(c).

2. **(a).** Figure 2(a), Graphs 1, 2, and 3 show, respectively, the falling solution curves where $c = -1$ and the rate function is always negative, solution curves that fall towards and away from the equilibrium $y = 0$ where $c = 0$, and solution curves falling and rising toward the attracting equilibrium $y = 1$ and rising and falling away from the repelling equilibrium $y = -1$ where $c = +1$.

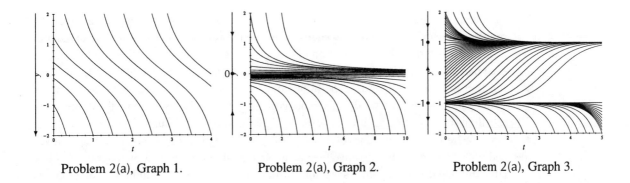

Problem 2(a), Graph 1.   Problem 2(a), Graph 2.   Problem 2(a), Graph 3.

(b). Figure 2(b), Graphs 1, 2, and 3 show, respectively, the rising and falling solution curves toward the attracting equilibrium $y = 0$ (where $c = 0$) and the falling and rising curves away from the repelling equilibrium $y = 2$, the curves rising toward and away from the single equilibrium at $y = 1$ ($c = 1$), and the rising curves where there is no real equilibrium at all ($c = 2$).

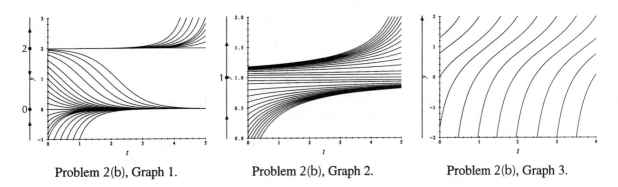

Problem 2(b), Graph 1.   Problem 2(b), Graph 2.   Problem 2(b), Graph 3.

(c). Figure 2(c), Graphs 1, 2, and 3 show, respectively, solution curves rising and falling toward the attracting equilibrium $y = -2$ (with $c = 0$) and falling and rising away from the repelling equilibrium $y = 0$, curves rising toward and away from the single equilibrium at $y = -1$ ($c = 1$), rising curves with no real equilibrium solutions ($c = 2$).

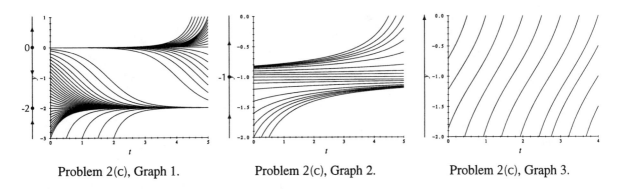

Problem 2(c), Graph 1.   Problem 2(c), Graph 2.   Problem 2(c), Graph 3.

3. First find the equilibrium solutions by finding the zeros of each rate function as functions of the parameter c. Then find the value of c where the zeros of the rate function merge into a single real zero of the rate function. For c on one side of this value the rate function in each part of this problem has a single real zero. For c on the other side of the value the rate function has three real zeros. The bifurcation diagram shows the real zeros as functions of the parameter c. The solid arcs denote attracting equilibrium solutions; the dashed arcs repelling solutions. For each value c_i of c each real equilibrium solution y_i is shown to be an attractor [repeller] if for values of y near y_i, $y > y_i$, the sign of $f(y, c_i)$ is negative [positive] and for $y < y_i$ the sign is positive [negative]. In the pitchfork bifurcation diagrams, arcs of attracting equilibrium solutions are solid, while arcs of repelling equilibrium solutions are dashed. See the text subsection on the pitchfork bifurcation for more details.

(a). The ODE is $y' = (c - 2y^2)y$. The zeros of $f = (c - 2y^2)y$ are $y_1 = 0$, $y_{2,3} = \mp\sqrt{c/2}$, and the latter two are real for $c > 0$, merge with y_1 for $c = 0$, and are complex for $c < 0$. For $c > 0$, y_2 and y_3 are attractors since $f(y, c)$ is negative for y near y_2 [y_3] if $y > y_2$ [$y > y_3$] and positive if $y < y_2$ [$y < y_3$]. For $c > 0$, y_1 is a repeller since $f(y, c)$ is positive for y near 0, $y > 0$, and negative for $y < 0$. For $c < 0$, y_1 is an attractor since $f(y, c)$ is negative for $y > 0$ and positive for $y < 0$. See Fig. 3(a).

(b). The ODE is $y' = -(c + y^2)y$. The zeros of $f = -(c + y^2)y$ are $y_1 = 0$, $y_{2,3} = \mp\sqrt{-c}$; the latter two are real for $c < 0$, merge with y_1 for $c = 0$, and are complex for $c > 0$. For $c < 0$, y_2 and y_3 are attractors since $f(y, c)$ is negative for y near y_2 [y_3] and $y > y_2$ [$y > y_3$] and positive for $y < y_2$ [$y < y_3$]. But y_1 is a repeller [attractor] for $c > 0$ [$c < 0$] since for y near 0 and $y > 0$ the sign of $f(y, c)$ is positive [negative] while for $y < 0$ the sign is negative [positive]. See Fig. 3(b).

(c). The ODE is $y' = (c - y^4)y$. The five zeros of $f = (c - y^4)y$ are $y_1 = 0$, $y_{2,3} = \mp c^{1/4}$, and $y_{4,5} = \mp i|c|^{1/4}$. The only zeros of interest here are y_1, y_2, y_3 because $y_{4,5}$ are complex, except if $c = 0$ in which case all five coincide at $y = 0$. The zeros $y_{2,3}$ are real for $c > 0$ where they are both attractors because $f(y, c)$ is negative for y near y_2 or y_3 with $y > y_2$ or $y > y_3$, but positive for $y < y_2$ or $y < y_3$. The zero y_1 is a repeller [attractor] for $c > 0$ [$c < 0$] because $f(y, c)$ is positive [negative] for y near 0 but $y > 0$ and negative [positive] for $y < 0$. See Fig. 3(c).

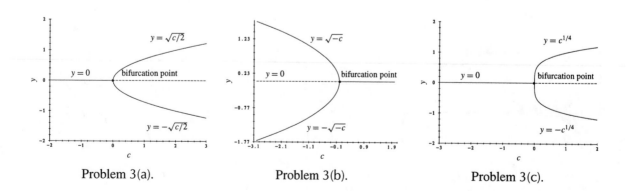

Problem 3(a).   Problem 3(b).   Problem 3(c).

4. In each case solution curves are drawn for three values of c, one below, one at, and the third above the value of c where the pitchfork bifurcation occurs. In each case there is a single, weakly attracting equilibrium solution at the value of c where equilibrium occurs.

(a). See Figure 4(a), Graphs 1, 2, and 3 for solution curves for $c = -2$, $c = 0$ (where the bifurcation occurs), and $c = 2$, respectively. For $c = -2$ we have a single attracting equilibrium $y = 0$, while for $c = 2$ there are three equilibrium solutions, $y_1 = 0$ (a repeller) and $y_{2,3} = \pm 1$ (both attractors).

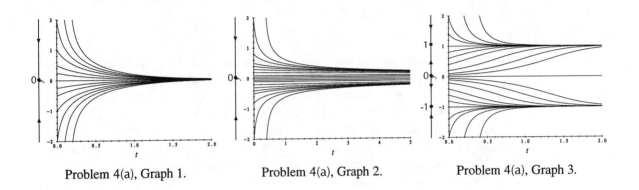

Problem 4(a), Graph 1.　　　Problem 4(a), Graph 2.　　　Problem 4(a), Graph 3.

(b). See Fig. 4(b), Graphs 1, 2, and 3 for solution curves for $c = -1$, $c = 0$ (the bifurcation value), and $c = 1$, respectively. At $c = -1$ there are three equilibrium solutions, $y_1 = 0$ (a repeller) and $y_{2,3} = \mp 1$ (attractor). At $c = 1$ there is a single attractor ($y = 0$).

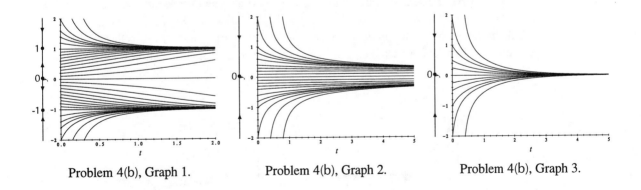

Problem 4(b), Graph 1.　　　Problem 4(b), Graph 2.　　　Problem 4(b), Graph 3.

(c). See Fig. 4(c), Graphs 1, 2, and 3 for solution curves for $c = -1$, $c = 0$ (the bifurcation value), $c = 1$, respectively. At $c = -1$ there is a single attracting equilibrium ($y = 0$) while at $c = 1$ there are three equilibrium solutions, $y = 0$ (a repeller), $y_{2,3} = \mp 1$ (attractors).

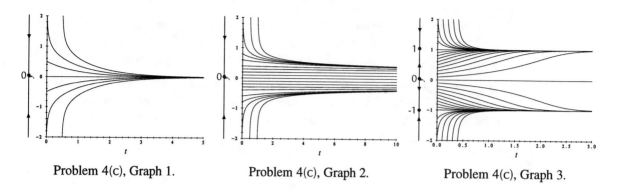

Problem 4(c), Graph 1.   Problem 4(c), Graph 2.   Problem 4(c), Graph 3.

5. In each case set the rate function $f(y, c)$ equal to zero to find the equilibrium solutions. One of these is $y_1 = 0$, the other, y_2, is a function of c and real for all c. As c crosses a critical value, y_2 moves toward y_1, merges with it at the critical value, and then moves away. Except at the critical value of c where the transcritical bifurcation occurs, one of y_1 and y_2 is an attractor and the other a repeller [because of the signs of $f(y, c)$] but as c crosses the bifurcation value the attractor becomes a repeller and vice-versa. The transcritical bifurcation diagram in the cy-plane shows the graphs of $y_1 = 0$ and y_2 as a function of c. Solid arcs denote an attractor, dashed arcs a repeller. As noted above, the signs of $f(y, c)$ near an equilibrium value of y are used to determine whether the equilibrium is an attractor or a repeller. At the bifurcation value of c, the equilibrium solution curve attracts on one side and repels on the other.

(a). The ODE is $y' = cy - y^2$. Here $f(y, c) = cy - y^2 = y(c - y)$, so $y_1 = 0$ and $y = c$ are the two equilibrium curves and the transcritical bifurcation occurs for $c = 0$. See Fig. 5(a), Graphs 1–3 for solution curves for $c = -1$ ($y_1 = 0$ is an attractor and $y_2 = -1$ is a repeller), $c = 0$ at the transcritical bifurcation value ($y = 0$ attracts on one side and repels on the other), $c = 1$ ($y_1 = 0$ is a repeller and $y_2 = 1$ is an attractor). Figure 5(a), Graph 4 shows the bifurcation diagram in the cy-plane: the two lines are $y = 0$, $y = c$. Observe how the attractor switches to a repeller, and vice versa as c crosses the bifurcation value 0.

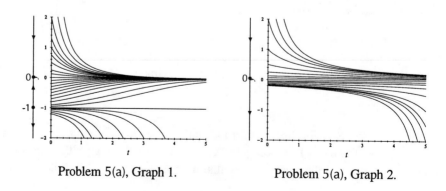

Problem 5(a), Graph 1.   Problem 5(a), Graph 2.

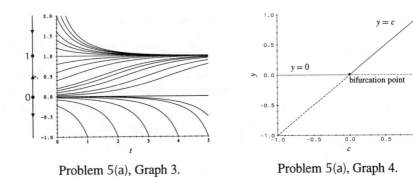

Problem 5(a), Graph 3. Problem 5(a), Graph 4.

(b). The ODE is $y' = cy + 10y^2$. In this part, $f = cy + 10y^2 = y(c + 10y)$, so the tran-scritical bifurcation occurs for $c = 0$ and $y_1 = 0$, $y_2 = -c/10$ are the two equilibrium curves. See Fig. 5(b), Graphs 1–3 for solution curves for $c = -10$ ($y_1 = 0$ is an attractor and $y_2 = 1$ is a repeller), $c = 0$ which is the bifurcation value ($y = 0$ attracts on one side and repels on the other), $c = 10$ ($y_1 = 0$ is a repeller and $y_2 = -1$ is an attractor). Figure 5(b), Graph 4 shows the transcritical bifurcation diagram in the cy-plane. The two lines $y_1 = 0$, $y_2 = c/10$ switch, respectively, from attractor to repeller and vice-versa as c increases through 0.

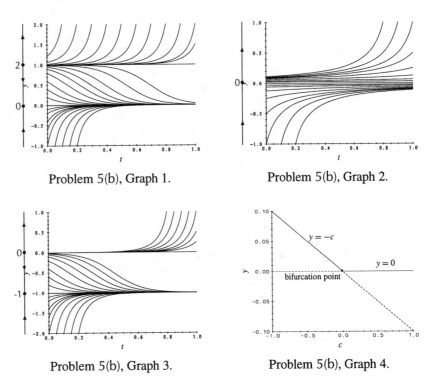

Problem 5(b), Graph 1. Problem 5(b), Graph 2.

Problem 5(b), Graph 3. Problem 5(b), Graph 4.

6. The ODE is $P' = r(1 - P/K)P - H$. If the harvesting/restocking rate is, say, Q_1, slightly below the tangent bifurcation value of $rK/4$, then the rate function has two zeros, $P_{1,2} =$

$K/2 \mp (K/2)(1 - 4Q_1/Kr)^{1/2}$, which are very close together. This means that in the tP-plane the two equilibrium solutions $y = P_1$, $y = P_2$, where $P_1 < P_2$, bound a very narrow band, with $y = P_1$ a repeller and $y = P_2$ an attractor. So if the population level is below the band, extinction is inevitable. Even if the level lies within the band, a small disturbance may send it below. Banning harvesting altogether would help in this situation since the model for subsequent times would be logistic and the population levels would then rise toward equilibrium. But any time population levels get too low, random events may have drastic effects and it is doubtful that even an outright ban on harvesting would have much effect.

7.　The ODE is $P' = (1 - P/1000)P - H$.

(a). Suppose that N hunting licenses are issued and (in the worst case scenario from the duck's viewpoint) each hunter bags the maximum of 20 ducks. Measuring time in years, the model ODE is $P' = (1 - P/1000)P - 20N$. The zeros of the rate function are given by using the quadratic formula

$$P_{1,2} = 500 \mp (10^6 - 8 \cdot 10^4 N)^{1/2}/2$$

So we must set N below the critical value which is where the quantity inside the square root moves from positive to negative (i.e., where it is zero). That value is $10^6 - 8 \cdot 10^4 N = 0$, or $N = 10^2/8 = 12.5$. So no more than 12 licenses should be issued.

(b). If N hunters bag 20 ducks each during the year, then $P = P_1$ and $P = P_2$ bound a narrow band centered at 500. The equilibrium point at $P = P_1$ is a repeller. If the duck population stays above P_1 (the lower edge of the band) then theoretically the duck population will tend to P_2. But if the duck population is below P_1, then continued harvesting by N hunters at the rate of 20 ducks per year per hunter will lead to duck extinction.

2.5　Approximate Solutions

Suggestions for preparing a lecture

　Topics:　Discussion of the general idea of numerically approximating a solution of an IVP, Euler's Method, Fourth-Order Runge Kutta, and (maybe) the method your students will be using.

　Remarks:　Our preference is to spend only a little time on numerical methods in the introductory course, more or less following the principle that it is better to "learn to drive before one studies auto mechanics."

Making up a problem set

Two parts from 1 and 2, 3.

Comments

How does a numerical solver approximate the solution of the first-order IVP $y' = f(t, y)$, $y(t_0) = y_0$? The underlying principle of all solvers is simple enough: approximate the rate function at a point, multiply by a time increment, and repeat. The differences in the various methods have to do with how they approximate the rate function and how the step size is adjusted. In addition, many solvers have adjustable controls on allowable absolute and relative errors, minimum and maximum internal step size (for adaptive methods), number of points plotted, and so on. This section is a brief introduction to a complex and fascinating area of numerical analysis—the search for the "best" ODE solver. Of course, no such thing exists since for every numerical solver yet invented there is some IVP that "breaks" it. Nevertheless, there are some basic notions that can be conveyed about solvers. We focus on one-step algorithms such as the Euler Method and Runge-Kutta Method, and use examples to illustrate what they can do. Actual numerical estimates are given below only for the Euler and RK4 methods. An excellent source for a more extensive treatment is L.F. Shampine's book, *Numerical Solutions of Ordinary Differential Equations*, Chapman & Hall, New York, 1994.

1. Approximate values for $y(1)$, where $y' = -y$, $y(0) = 1$, with step $h = 0.1$, will vary according to the calculator or numerical solver used. The exact answer is $1/e = 0.367879441...$. See Fig. 1 for Euler's Solution (dashed), RK4 (solid).

    (a). Euler's algorithm, $y_n = y_{n-1} - hy_{n-1}$, $1 \leq n \leq 10$, gives $y(1) \approx y_{10} = 0.348678$ (first six decimals).

    (b). Heun's algorithm becomes $y_n = y_{n-1} + h(-y_{n-1} - (y_{n-1} - hy_{n-1}))/2$, $1 \leq n \leq 10$.

    (c). Fourth-order Runge-Kutta, $y_n = y_{n-1} + h(k_1 + 2k_2 + 3k_3 + k_4)/6$, where $1 \leq n \leq 10$, $k_1 = -y_{n-1}$, $k_2 = -(y_{n-1} + hk_1/2)$, $k_3 = -(y_{n-1} + hk_2/2)$, $k_4 = -(y_{n-1} + hk_3)$, gives $y(1) \approx y_{10} = 0.367879$ (first six decimals).

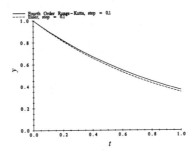

Problem 1.

2. If $y' = -y$, $y(0) = 1$, then $y(t) = e^{-t}$, and $y(1) = 1/e$. The first nine decimals of $1/e$ are 0.367879441. The first six decimals of the Euler approximations to $y(1)$ for $h = 0.01$, 0.001, 0.0001 are given in part **(a)**, and the first twelve decimals of the RK4 approximations in part **(b)**.

    (a). The three Euler approximations to $1/e$ for the three values of $h = 0.01$, 0.001, 0.0001 are 0.366032, 0.367695, 0.387861. The three Euler polygons are nearly coincident [Fig. 2(a), Graph 1]. Zooming in on the right end of the polygons separates the arcs [Fig. 2(a), Graph 2].

(b). The three RK4 approximations to $1/e$ for $h = 0.01, 0.001, 0.0001$ are 0.367879441203, 0.367879441171, and 0.367879441171. The three RK4 polygons are so close [Fig. 2(b)] that our solver graphics couldn't separate them even by zooming in.

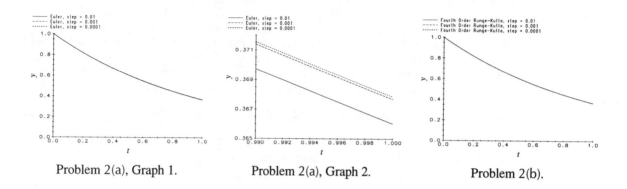

Problem 2(a), Graph 1.   Problem 2(a), Graph 2.   Problem 2(b).

3.   Euler's method [part **(a)**] and RK4 [part **(b)**] are used to approximate $y(1)$ if $y' = -y^3 + t^2$, $y(0) = 0$, and the step size h is $0.1, 0.01, 0.001$.

(a). The first six decimals of the respective Euler approximations to $y(1)$ for the three values of h are 0.283786, 0.325058, and 0.329236. Figure 3(a) shows the three Euler polygons.

(b). The first six decimals of the respective RK4 approximations to $y(1)$ are 0.329699, 0.329700, and 0.329700. The RK4 polygons are very close together [Fig. 3(b), Graph 1], but zooming in on the end segment of the polygons separates the $h = 0.01$ polygon from the other two [Fig. 3(b), Graph 2].

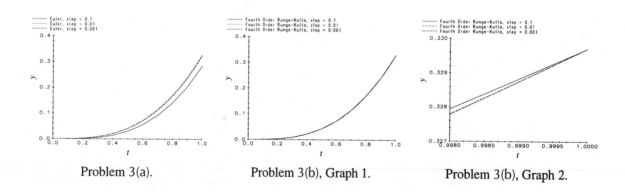

Problem 3(a).   Problem 3(b), Graph 1.   Problem 3(b), Graph 2.

4.   **(a).** The exact solution of $y' = 2y$, $y(0) = 1$ is $y = e^{2t}$, and so at T we have $y(T) = e^{2T}$. Euler's formula gives $y(T) \approx y_N = y_{N-1} + 2hy_{N-1} = (1 + 2h)y_{N-1} = (1 + 2h)^2 y_{N-2} = \cdots = (1 + 2h)^N y_0 = (1 + 2h)^N = (1 + 2T/N)^N$ since $h = T/N$. Thus, $y_N \to e^{2T}$ as $N \to \infty$ since $e^x = \lim_{N \to \infty}(1 + x/N)^N$.

(b). This is exactly as in **(a)**, except $y' = -y$, $y(0) = 1$ and so $y(T) = e^{-T}$. Euler's method gives $y_N = y_{N-1} - hy_{N-1} = (1-h)y_{N-1} = (1-h)^2 y_{N-2} = \cdots = (1-h)^N = (1-T/N)^N \to e^{-T}$ as $N \to \infty$, since $e^x = \lim_{N\to\infty}(1+x/N)^N$.

5.   We shall interpret a as t_j and b as t_{j+1} in an RK4 process for solving $y' = f(t)$, $y(t_0) = y_0$ over an interval containing a, b, $a < b$. Suppose y_j has been calculated as an estimate for $y(t_0 + jh)$, where $y'(t) = f(t)$, $y(t_0) = y_0$, $t_0 \le t \le T$. Then $y_{j+1} = y_j + \int_{t_j}^{t_{j+1}} f(t)dt \approx y_j + (h/6)[f(t_j) + 4f(t_j + h/2) + f(t_{j+1})]$ by Simpson's formula, where $h = t_{j+1} - t_j$. Fourth-order Runge-Kutta gives $y_{j+1} = y_j + (h/6)[k_1 + 2k_2 + 2k_3 + k_4]$, where $k_1 = f(t_j)$, $k_2 = f(t_j + h/2) = k_3$, $k_4 = f(t_j + h) = f(t_{j+1})$. So, Simpson's method and fourth-order Runge-Kutta yield identical values for y_{j+1}, once y_j has been calculated.

2.6   Computer Implementation

Suggestions for preparing a lecture

Topics: A general discussion of computer "lies" and some specific examples of good solvers doing bad things. You might want to do some of Problems 3, 4 on reversibility of time (or rather the failure of reversibility) when applying RK4 to solve an IVP which has a strongly attracting solution.

Remarks: We recommend assigning only two or three problems here because the computing may take some time, and it is important that students explain their computational results.

Making up a problem set

Several parts of Problems 2, 3, 4, 5, 8. The problems we like best are 2, 3, and 4. The qualitative behavior of Euler Solutions for the logistic equation (addressed in Problem 5) is explored in depth in Section 2.7.

Comments

Given any numerical algorithm for solving ODEs, there will be some IVP for which the algorithm works poorly, if at all. This is a fact of numerical algorithms, so we address here a few of the things that can go wrong when using a solver. Problems with numerical algorithms show how important it is not to place unquestioning faith in their reliability. By this point in the text we hope that the reader will have begun to see that each of the three approaches to ODEs (finding solution formulas, using theory, computing numerical approximations) has both advantages and drawbacks. Understanding the pluses and minuses of each approach contributes to a deeper understanding of the behavior of the ODEs themselves.

Tips on using numerical solvers

In Section 2.2 we talked about extending a solution by taking the final values of the state variables and using them as initial data for the extended solution. There are at least two reasons why this procedure is not recommended when using numerical solvers:

- The solver's user interface may report numerical data with less precision than the solver itself uses for calculations (i.e., the solver may report and accept data with 4 digits of precision but use 15 digits of precision internally. Look at the ODE $y' = 3y \sin y + t$ some of whose solutions appear in the chapter cover figure. Imagine that your solver produces an approximate value for $y(t)$ at $t_{final} = -6$ and that you wish to extend your solution forward in time. Can you see the trouble you would have if y_{final} is very near 0?

- Some solvers are adaptive in that they make decisions that affect the solver engine at every time step. These decisions are based on the recent history of the solution process. This is especially true of solvers based on multistep methods like the one used to produce the figures in this text. If the solution process is restarted at one point, then the history is lost and the solver is left to its own devices on how to start the solution process. Should this happen at a sensitive solution point like the one mentioned above, then some surprises may be in store for the user.

---

1. **(a).** Separating variables in the ODE, $y' = y^3$, we have $(y^{-3})' = dt$. Integrating we get $-y^{-2}/2 = t + C$ Using the initial condition $y(0) = 1$ we get, $y^{-2} = 1 - 2t$, that is, $y = (1 - 2t)^{-1/2}$. Therefore, the t-interval on which this maximally extended solution is defined is $(-\infty, 1/2)$.

   (b). The Euler formula in this case with $h = 0.05$ is $y_n = y_{n-1} + 0.05 y_{n-1}^3$, $n = 1, 2, \ldots, 20$, with $y_0 = 1$. The Euler Solution (solid curve in Fig. 1) actually exits the rectangle before n reaches 20.

   (c). See Fig. 1 for the graphs of $y(t)$ (dashed) (actually Runge-Kutta with step 0.0001) and the Euler Solution (solid curve). The Euler Solution stepped across the singularity at $t = 1/2$ and started tracking a neighboring solution curve.

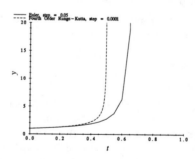

Problem 1.

2. **(a).** Since the solutions $z(t)$ and $y(t)$ coincide at $t = t_1$, the Existence and Uniqueness Theorem implies that $z(t) = y(t)$ for all t where these solutions are defined and, subsequently, $z(t_0) = y(t_0) = y_0$.

   (b). The IVP is $y' = 3y \sin y - t$, $y(0) = 0.4$. Use RK4 to obtain

   $$y_n = y_{n-1} + (k_1 + 2k_2 + 2k_3 + k_4)/60$$

   where $k_1 = 3y_{n-1} \sin y_{n-1} - t_{n-1}$, $k_2 = 3(y_{n-1} + 0.05k_1) \sin(y_{n-1} + 0.05k_1) - (t_{n-1} + 0.05)$, $k_3 = 3(y_{n-1} + 0.05k_2) \sin(y_{n-1} + 0.05k_2) - (t_{n-1} + 0.05)$, and $k_4 = 3(y_{n-1} + $

$0.1k_3) \sin(y_{n-1} + 0.1k_3) - (t_n)$, $n = 1, 2, \ldots, 80$, with $y_0 = 0.4$. So, we find that $y(8) \approx y_{80} = -6.2830725$. See Fig. 2(b), where 15 decimals of the computed value of y_{80} are printed out.

(c). The IVP is $z' = 3z \sin z - t$, $z(8) = y(8)$. Use RK4 to obtain

$$z_n = z_{n-1} + (k_1 + 2k_2 + 2k_3 + k_4)/60$$

where $k_1 = 3z_{n-1} \sin z_{n-1} - t_{n-1}$, $k_2 = 3(z_{n-1} + 0.05k_1) \sin(z_{n-1} + 0.05k_1) - (t_{n-1} + 0.05)$, $k_3 = 3(z_{n-1} + 0.05k_2) \sin(z_{n-1} + 0.05k_2) - (t_{n-1} + 0.05)$, and finally $k_4 = 3(z_{n-1} + 0.1k_3) \sin(z_{n-1} + 0.1k_3) - t_n$, $n = 1, 2, \ldots, 80$, with $z_{80} = y_{80} = -6.2830725$. The upper curve in Fig. 2(c) shows that $z(0) \approx 6.3$, vastly different from the value of $y(0) = 0.4$. Solution curves merge and diverge in complex ways. Solution curves with very different initial values for $t = 0$ may appear to merge as t approaches 8. The effect of the numerical approximation is that the computed solution at $t = 8$ may no longer correspond to the solution beginning at the given initial point. When the IVP is then solved backward from $t = 8$, the solver follows some other one of the diverging solution curves.

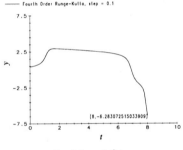

Problem 2(b).

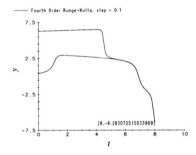

Problem 2(c).

3. **(a).** To solve $y' + 2y = \cos t$, we use the integrating factor e^{2t} and a table of integrals (or integration by parts). We see that $(ye^{2t})' = e^{2t} \cos t$. Integrating, $ye^{2t} = \int e^{2t} \cos t \, dt + C$, so $y = Ce^{-2t} + 0.4 \cos t + 0.2 \sin t$, where C is an arbitrary constant. As $t \to +\infty$, $y(t) \to y_p(t) = 0.4 \cos t + 0.2 \sin t$.

(b). See Fig. 3(b) for solution curves with initial points $(0, \pm 1.5)$, $(0, 0.4)$, $(2.5, 1.5)$, $(5, -1.5)$, $(7.5, 1.5)$, $(10, -1.5)$, $(12.5, 1.5)$, $(15, -1.5)$, and $(17.5, 1.5)$. As t increases, all of these solution curves converge to the graph of the particular solution y_p found in part **(a)**.

(c). Fig. 3(c) is the forward solution curve of $y' + 2y = \cos t$, $y(0) = 0.4$ as approximated by RK4 with step size $h = 0.1$; the computed approximation to $y(20)$ is $0.34582020\ldots$.

(d). The lower curve in Fig. 3(d) shows what happens when the solver tries to go backward from the computed approximation to $y(20)$ with a step size $h = 0.1$. The approximation of $y(20)$ is of course not exact, so when we solve backwards from this approximation we trace out a very different solution curve.

(e). See Fig. 3(e), Graphs 1 and 2, corresponding to $h = 0.01, 0.001$. The problem with going backward in time is that the forward solutions converge very rapidly to the particular solution $y_p = 0.4 \cos t + 0.2 \sin t$, but numerical solvers can only give approximate answers. Thus, $y(20) \approx y_p(20)$ can only be computed approximately, no matter how small a step size is used. In effect, when the backward solutions are attempted, they start at the different computed approximations corresponding to 0.01 and 0.001. While solving backwards, small errors occur. Due to sensitivity, these cause the failure to accurately solve backwards. The computed values of the approximations at $t = 20$ are printed on the graphs.

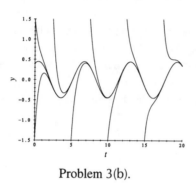

Problem 3(b).

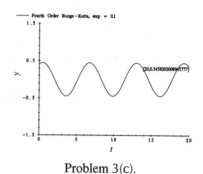

Problem 3(c).

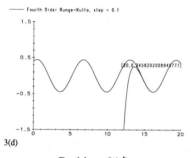

3(d)

Problem 3(d)

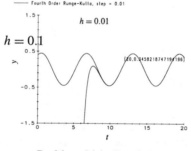

Problem 3(e), Graph 1.

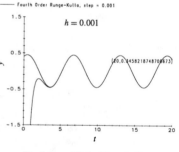

Problem 3(e), Graph 2.

4. **(a).** Using Euler's Method for the IVP, $y' = 1 + ty \cos y$, $y(0) = 1$, we have that $y_n = y_{n-1} + 0.1(1 + t_{n-1} y_{n-1} \cos y_{n-1})$, $n = 1, 2, \ldots, 280$, with $y_0 = 1$. See Fig. 4(a). The Euler Solution (solid) for $h = 0.1$ becomes unstable at $t \approx 20$, but seems to be all right if h is shortened by an order of magnitude to $h = 0.01$ (dashed).

    (b). Use RK4 to obtain that $y_n = y_{n-1} + (k_1 + 2k_2 + 2k_3 + k_4)/60$ where $k_1 = 1 + t_{n-1} y_{n-1} \cos y_{n-1}$, $k_2 = 1 + (t_{n-1} + 0.05)(y_{n-1} + 0.05k_1) \cos(y_{n-1} + 0.05k_1)$, $k_3 = 1 + (t_{n-1} + 0.05)(y_{n-1} + 0.05k_2) \cos(y_{n-1} + 0.05k_2)$ and $k_4 = 1 + t_n (y_{n-1} + 0.1k_3) \cos(y_{n-1} + 0.1k_3)$. See Fig. 4(b). The RK4 solution goes wild if $h = 0.1$ (solid), but seems to behave nicely if $h = 0.01$ (dashed). We see the hazards of using one-step methods with a large step size over too long a span of time.

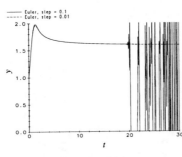

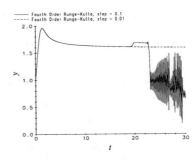

Problem 4(a).

Problem 4(b).

5. **(a).** The IVP is $y' = (1 - y)y$, $y(0) = y_0$. There are two equilibrium solutions, $y(t) = 1$ and $y(t) = 0$, corresponding to initial conditions $y_0 = 1$ and $y_0 = 0$, respectively. Separating variables in the ODE, $y' = y(1 - y)$, we have $[1/y + 1/(1 - y)]y' = 1$, so $\ln|y/(1 - y)| = t + C$. Since $y(0) = y_0$, $C = \ln|y_0/(1 - y_0)|$. So

$$\ln|y/(1 - y)| = \ln|y_0/(1 - y_0)| + t$$

$$\ln\left|\frac{y(1 - y_0)}{(1 - y)y_0}\right| = t$$

$$\frac{y(1 - y_0)}{(1 - y)y_0} = e^t$$

Solving for y,

$$y = \frac{y_0 e^t}{1 - y_0 + y_0 e^t} = \frac{y_0}{y_0 + (1 - y_0)e^{-t}}, \qquad t_0 > 0$$

As $t \to +\infty$, $y \to 1$.

(b). Use Euler's Method to obtain $y_n = y_{n-1} + 0.75 y_{n-1}(1 - y_{n-1})$, $n = 1, 2, \ldots, 11$, with $h = 0.75$ and $y_0 = 2$. See Fig. 5(b), where the Euler Solution is solid. The first segment of the Euler Solution falls steeply. The step size of $h = 0.75$ is long enough that the right endpoint of the segment lies far below the equilibrium level of $y = 1$ that the true solution approaches. From that point on the Euler Solution moves upward toward the equilibrium solution, as do all true solution curves in the region $0 < y < 1$.

(c). If $h = 1.5$, then $y_n = y_{n-1} + 1.5 y_{n-1}(1 - y_{n-1})$, $n = 1, 2, \ldots, 6$, with $y_0 = 1.4$. See Fig. 5(c), where the Euler Solution is the solid curve. The first segment of the Euler Solution again goes way below the equilibrium solution $y = 1$, but, because the step size $h = 1.5$ is so large, some of the later segments also overshoot. It appears that the Euler Solution oscillates about the equilibrium, $y = 1$, but approaches equilibrium as n increases.

(d). If $h = 2.5$, then $y_n = y_{n-1} + 2.5 y_{n-1}(1 - y_{n-1})$, $n = 1, 2, 3, 5$, with $y_0 = 1.3$. See Fig. 5(d); the Euler Solution is the solid curve. With step size $h = 2.5$, the Euler Solution overshoots badly at each step, sometimes too high, other times too low.

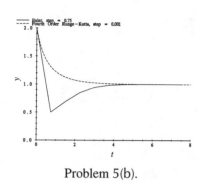

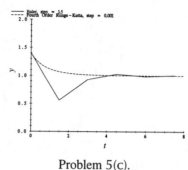

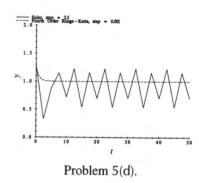

Problem 5(b).   Problem 5(c).   Problem 5(d).

6. The formula for the exact solution of $y' = ry(1 - y)$, $y(0) = y_0 > 0$ is $y = y_0[y_0 + (1 - y_0)e^{rt}]^{-1}$. This formula is used in parts **(a)** and **(b)** to plot the exact solution. The Euler Solution is defined by the algorithm $y_{n+1} = y_n + rh(1 - y_n)$.

(a). See Fig. 6(a), Graphs 1 and 2 for the Euler Solutions of $y' = 10y(1 - y)$, $y_0 = 0.3, 2.0$. In each case $h = 0.05$, and so $a = rh = 0.5$. See Fig. 6(a), Graph 1 for the Euler Solution (solid) corresponding to $y_0 = 0.3$. The Euler Solution follows the true solution (dashed) quite closely. The Euler Solution (solid) of the IVP with $y_0 = 2$ reaches the equilibrium state $y = 1$ in one step and remains there forever [Fig. 6(a), Graph 2].

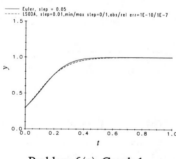

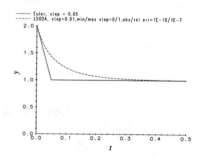

Problem 6(a), Graph 1.   Problem 6(a), Graph 2.

(b). See Fig. 6(b), Graphs 1 and 2 for the Euler Solutions with $r = 100$, $h = 0.015$ (so $a = rh = 1.5$) corresponding to initial values $y_0 = 0.5, 1.5$ at $t = 0$. The second segment of the Euler Solution through the initial point $t_0 = 0$, $y_0 = 0.5$ [Fig 6(b), Graph 1] drops below the line $y = 0$, and from then on the Euler Solution continues to fall because $y' < 0$ below $y = 0$. With a shorter step size the segments of the Euler Solution through the point $t_0 = 0$, $y_0 = 1.5$ never drop below $y = 0$; the resulting Euler Solution [see solid curve in Fig. 6(b), Graph 2] seems to oscillate about and converge to the true solution (dashed curve).

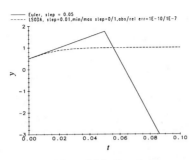

Problem 6(b), Graph 1.   Problem 6(b), Graph 2.

7. **(a).** The rate function is $-y(1-y)^2$. Since $(1-y)^2$ is positive for $y \neq 1$, y' is negative for $y > 0$, so for $y_0 > 1$, solution curves of the ODE approach the equilibrium solution $y = 1$ from above, whereas for $y_0 < 1$ solutions diverge from the equilibrium $y = 1$ and fall toward the equilibrium $y = 0$. So, $y = 1$ is semistable, attracting solutions from above and repelling solutions from below.

(b). The Euler Solutions in Fig. 7(b) correspond to $h = 0.1$, 0.5, 1, 1.5, and 2 from top to bottom. The $h = 0.1$ curve is almost identical to the actual solution curve. For $h = 0.1$, 0.5, and 1, the Euler Solution approaches the correct equilibrium solution $y = 1$ in the limit because no segment of the Euler Solution ever dips below $y = 1$. But for $h = 1.5$ and 2, the Euler Solution overshoots the semistable equilibrium $y = 1$, drops into the region $0 < y < 1$, where y' is negative, and heads for the stable equilibrium $y = 0$. Note how the Euler Solution for $h = 2$ oscillates about $y = 0$. In a situation like this, a good adaptive solver would reduce the step size as the solution curve approaches $y = 1$.

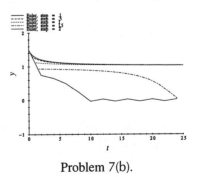

Problem 7(b).

8. **(a).** Figure 8(a), Graphs 1, 2, and 3 show, respectively, the tx- and ty-component plots, and the xy orbits for the two solutions of the system $x' = y$, $y' = -4x$ through the given initial points: $(x_0, y_0) = (0, 1)$ yielding $A = 1$, $\phi = 0$, and $(x_0, y_0) = (0, 2)$ yielding $A = 2$, $\phi = 0$. The graphs reveal periodic oscillations of period π of the state variables in time.

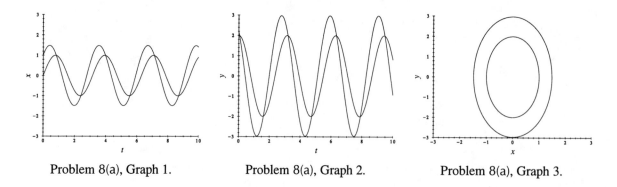

Problem 8(a), Graph 1.   Problem 8(a), Graph 2.   Problem 8(a), Graph 3.

(b). Use RK4 with $h = 0.1$ for ordinary differential systems and obtain that $x_n = x_{n-1} + (k_1 + 2k_2 + 2k_3 + k_4)/60$, $y_n = y_{n-1} + (p_1 + 2p_2 + 2p_3 + p_4)/60$, where $k_1 = y_{n-1}$, $k_2 = y_{n-1} + 0.05p_1$, $k_3 = y_{n-1} + 0.05p_2$ and $k_4 = y_{n-1} + 0.1p_3$. In addition, we have $p_1 = -4x_{n-1}$, $p_2 = -4(x_{n-1} + 0.05k_1)$, $p_3 = -4(x_{n-1} + 0.05k_2)$ and $p_4 = -4(x_{n-1} + 0.1k_3)$. See Fig. 8(b) for the graph of $x = x(t)$, $0 \le t \le 100$, using RK4 with $h = 0.1$.

(c). When fewer solution points are plotted, the amplitude of the sine wave appears to undergo a periodic modulation. This occurs because the larger interval between plotted points causes us to miss some of the peaks. Figure 8(c), Graphs 1, 2, and 3 show the solutions, plotting only every fifth, tenth, and twentieth point, respectively. We see that the amplitude modulation becomes quite noticeable. Note that in Fig. 8(c), Graph 3 not only does the amplitude of oscillation appear to be modulated, but the frequency appears to have decreased. Apparently, the sampling rate is so low that a number of the oscillations are completely missed. This second effect is known as aliasing. These figures illustrate the importance of plotting solution curves at adequate resolution.

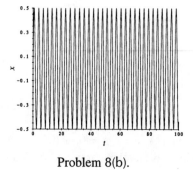

Problem 8(b).

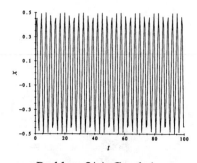

Problem 8(c), Graph 1.

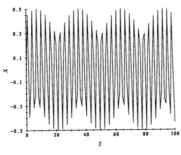

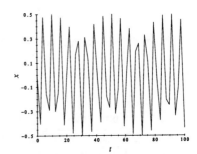

Problem 8(c), Graph 2.   Problem 8(c), Graph 3.

Background Material: Effect of Round-off Error

Suppose Euler's Method is used to estimate $y(3)$, where $y(t)$ solves the IVP,

$$y' = y, \qquad y(0) = 0$$

The plot below shows the error in approximating $y(3)$ as a function of the total number of steps.

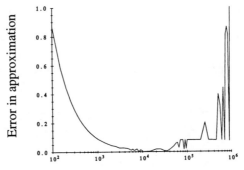

Total number of steps.

2.7 Euler's Method, the Logistic ODE, and Chaos

Suggestions for preparing a lecture

Topics: This is a tough subject to cover in one lecture, but if that's all the time you have, then you might do what we do: do several examples of period doubling in the Euler algorithm for the logistic IVP, $y' = (1 - y)y$, $y(0) = y_0$, where the parameter to be varied is the step size h. End with a seemingly chaotic plot like Figure 2.7.6.

Remarks: Only if there is time do we talk about the material at the end of the section. One possibility: make the whole section a group project.

Making up a problem set

Problems 1, 2, 3, or 4.

Comments

Section 2.7 takes up the application of a single numerical algorithm (Euler's Method with step size h) and what happens when it is applied to approximate solutions of the logistic ODE, $y' = (1 - y)y$. The results are startling and show how sensitive the Euler algorithm, $y_n = y_{n-1} + h(1 - y_{n-1})y_{n-1}$, is to changes in the step size. You may want to ignore ODEs entirely and approach the Euler Method as a discrete dynamical system that depends on the parameter h. From this viewpoint, we see period doubling sequences of values of h, sequences that are now known to lead to chaotic wandering. However, a single section on the topic of chaotic wandering and discrete dynamical systems can hardly do justice to this fascinating topic. For this reason we have listed in the text (and also in the Background Material below) references for further reading in the area. **Warning:** You can easily get hooked on the graphical images of chaotic wandering, to the detriment of all other topics in the syllabus. We shall return to chaos in Section 9.4 when we consider the Lorenz system. We have avoided giving a mathematical definition of chaos and have opted instead to use the descriptive term, chaotic wandering. There is a good reason for avoiding the technical definition—there is as yet no agreement on just what the definition should be! See Devaney [*An Introduction to Chaotic Dynamical Systems*, Addison-Wesley, New York, 1989] for one definition and M. Vellekoop and R. Berglund [*Amer. Math. Monthly*, April 1994, pp. 353–355] for comments on Devaney's definition. We describe the bifurcation diagrams (Figs. 2.7.7, 2.7.8) for the Euler algorithm as h increases.

The book by James Gleick, *Chaos*, Viking Penguin Inc., Harrisonburg, VA, 1987, tells the story of how the modern developments in chaos came about.

1. Let u be a value of y_0 giving rise to a 2-cycle in the recursion relation $y_{n+1} = y_n + hy_n(1 - y_n)$ and put $v = u + hu(1 - u)$. Then for u to give rise to a 2-cycle, we must have $u = v + hv(1 - v)$. That is, $y_k = u$ implies $y_{k+1} = v$ implies $y_{k+2} = u$ since u gives rise to a 2-cycle. We can plug the expression for u in terms of v into the expression for v in terms of u. So, $v = v + hv(1 - v) + h[v + hv(1 - v)][1 - v - hv(1 - v)]$, or (after some algebra), $hv(1 - v)[h^2v^2 - (2h + h^2)v + h + 2] = 0$. If $v = 0$ or $v = 1$, then $y_n = 0$ for all n, or $y_n = 1$ for all n in which case we have a constant solution rather than a 2-cycle. Otherwise, v must be a root of the quadratic, $h^2v^2 - (2h + h^2)v + h + 2 = 0$, which has only the two roots $v_{1,2} = (2 + h \pm \sqrt{h^2 - 4})/(2h)$. There are only these 2 values of y_0 in $0 < y_0 < \frac{(1+h)}{h}$ that give rise to 2-cycles.

2. See Graph 1 ($r = 100$, $h = 0.023$) and Graph 2 ($r = 100$, $h = 0.025$). The exact solution curve of the IVP $y' = 100y(1 - y)$, $y(0) = 1.3$ is the dashed curve in Graphs 1, 2. The Euler Solution (solid) in Fig. 2, Graph 1 eventually seems to tend to a 2-cycle that oscillates about the equilibrium solution $y = 1$. In Fig. 2, Graph 2 the Euler Solution (solid) appears to approach a 4-cycle oscillating about the equilibrium solution $y = 1$.

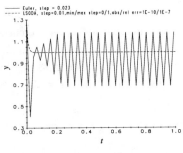

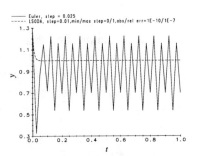

Problem 2, Graph 1

Problem 2, Graph 2

3. **(a).** Figure 3(a), Graphs 1–4 show the segments of the Euler Solution for the IVP, $y' = y(1 - y)$, $y(0) = 1.2$, $0 \le t \le 400$ with $r = 1$, $h = 2.65$. The segments are plotted over the subintervals $[0, 100]$, $[100, 200]$, $[200, 300]$, $[300, 400]$. There appear to be no cycles and no particular pattern—and so chaotic wandering about the equilibrium $y = 1$.

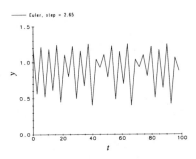

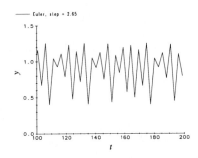

Problem 3(a), Graph 1

Problem 3(a), Graph 2

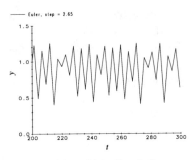

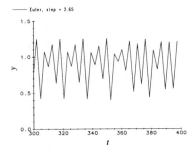

Problem 3(a), Graph 3

Problem 3(a), Graph 4

(b). Figure 3(b), Graphs 1–4 show segments of the Euler Solution for the IVP, $y' = 2.65y(1 - y)$, $y(0) = 1.2$, $0 \le t \le 400/2.65$ with $r = 2.65$. $h = 1$. Segments are plotted over the subintervals $[0, 100/2.65]$, $[100/2.65, 200/2.65]$, $[200/2.65, 300/2.65]$, $[300/2.65, 400/2.65]$. The graphs look identical to the corresponding graphs in part **(a)**

because $rh = 2.65$ in both parts. By scaling the timeline by the value of r in part **(b)**, this means in effect that the step size remains constant in the two parts. Since $rh = 2.65$ is fixed, the graphs look identical. If $r = r_0 > 0$, $h = h_0 > 0$, and $r_0 h_0 = 2.65$, then we will always get graphs of chaotically wandering Euler Solutions that look identical on t-intervals $[A/r_0, B/r_0]$, regardless of the value of r_0, if we set $h = 2.65/r_0$.

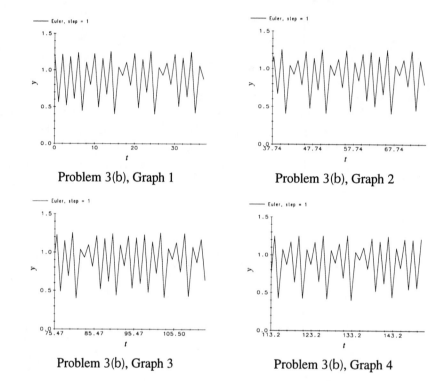

Problem 3(b), Graph 1   Problem 3(b), Graph 2

Problem 3(b), Graph 3   Problem 3(b), Graph 4

4.   See Figure 4, Graphs 1–4 for the respective plots of the Euler values for the IVP $y' = y(1 - y)$, $y(0) = y_0$, with $y_0 = 0.25$, 0.5, 0.75, 1.2. Although the values seem to be wandering chaotically over the interval $0.3 \le y \le 1.28$ regardless of the value of y_0, there seem to be bands of higher density and other bands of low density and distinct high-density lines that mark the boundaries of the bands.

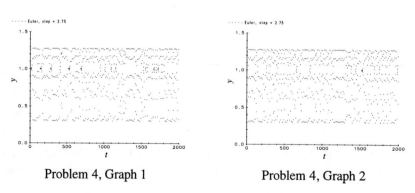

Problem 4, Graph 1   Problem 4, Graph 2

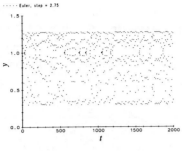

Problem 4, Graph 3   Problem 4, Graph 4

5. Group project. The plot below illustrates the geometry of the parabola given by plotting y_n versus y_{n-1}.

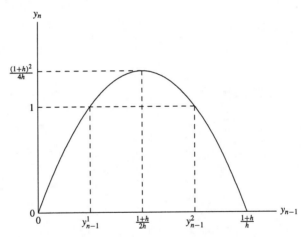

6. Group project.

Background Material: A 32-cycle of the logistic equation

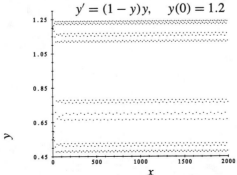

$$y' = (1 - y)y, \quad y(0) = 1.2$$

Approach to the Euler points of a 32-cycle Euler Solution: $h = 2.569$.

The figure above shows that the Euler Solution with $h = 2.569$ and $y_0 = 1.2$ tends to a 32-cycle Euler Solution, but the vertical separation of some of the lines of Euler points is so small that distinct levels are hard to see. In fact, each of the sixteen visible levels actually has two lines, but only zooming will split these into visibly different lines.

Second-Order
Differential Equations

The up-and-down motion of a body suspended by a spring may be modeled by a second-order ODE (Section 3.1). Why second-order? Like most dynamical systems where forces are involved, Newton's Second Law comes into play:

mass × acceleration of body = sum of forces acting on body

Acceleration is the second derivative with respect to time of the state variable representing the body's position. The proper use of Newton's Laws involves vectors, but the motion here is only one-dimensional (along the local vertical under the point of suspension of the spring), and we can get away with real-valued state variables. The vector treatment of Newton's Second Law is given in Section 4.1, where we model pendulum motion.

All the basic facts about constant-coefficient second-order linear ODEs appear in the pages of this chapter—and more besides. Our approach, for the most part, is fairly standard, but here and there we have added some material that gives some deeper insight into some corners that are often overlooked. The only model in the chapter appears in the first section [we like to start off our chapters with a model whenever possible]. In addition to the standard linear spring, we also develop hard and soft springs, and aging springs. It is no more trouble to model all these types of springs in one fell swoop, so that is what we do. The hard and soft spring ODEs are nonlinear and provide us with a nice opportunity to compare solutions of nonlinear ODEs and their linearizations (see especially Fig. 3.1.3). We take up the aging spring model again in Chapter 11, where we show how to solve it using series methods.

The geometry of solutions of second-order ODEs is a bit more complicated than for first-order ODEs. Orbits and time state curves are introduced in Section 3.2 along with several examples. Section 3.2 also extends the Fundamental Theorem to a second-order initial value problem. This theorem gives conditions that ensure that an IVP has exactly one solution and also gives the extension and sensitivity properties of that solution. Sections 3.3–3.6 introduce constant-coefficient linear second-order ODEs with and without initial data. The basic tool for finding solution formulas is the use of the operator $P(D) = D^2 + aD + b$ where a and b are real numbers. So we spend some time talking about this operator. The emphasis of this chapter is on linear ODEs, because the solution set of a second-order linear ODE can be characterized in simple terms. In Sections 3.3 and 3.4 we show how to find all solutions if the second-order linear ODE is undriven and has constant coefficients. In Section 3.5 we look at undriven linear ODEs with periodic solutions and the difficulty in tracking such solutions numerically.

The Method of Undetermined Coefficients for linear homogeneous second-order ODEs with constant coefficients and polynomial-exponential driving terms appears in examples in Section 3.3 and is extensively developed in Section 3.6.

In Section 3.7 we present the general theory of linear ODEs using Wronskians and basic solution sets. Variation of parameters "works" for every second-order linear ODE, $y'' + a(t)y' + b(t)y = f(t)$, and we present the method in a lengthy problem at the end of Section 3.7.

3.1   Springs: Linear and Nonlinear Models

Suggestions for preparing a lecture

Topics: The models for the motion of a spring, linearization of a nonlinear spring ODE, solution curves and orbits for the modeling ODEs.

Remarks: Before linearizing the ODE of a nonlinear spring about an equilibrium point, it might be helpful to review Taylor series of a function of two variables (see Appendix B.2, item 8).

Making up a problem set

Parts of Problems 1 and 3, Problem 6. Problem 7 is a good "solver tip" problem you may want to assign.

Comments

In this section we model various kinds of springs: Hooke's Law springs where the spring force is proportional to displacement, hard and soft springs where the force has extra terms in the cube of the displacement, and the aging spring, which like the Hooke's Law case leads to a linear ODE but with a time-varying coefficient of proportionality. The regions of validity of the model ODEs are discussed. Of particular importance, we show how the nonlinear spring model ODEs can be linearized near an equilibrium point.

This section is mostly introductory, presenting not much theory, but lots of examples and comments about the spring model. Note that force models (typically) lead to second-order ODEs in a position variable and that second-order ODEs require two state variables, usually position and velocity. Second-order ODEs play a starring role in an ODE course because of the force laws. We often refer to the spring models later in this chapter and in subsequent chapters.

---

1.   **(a).**   The motion satisfies equation (8) of the text, $mz'' + cz' + kz = f(t)$. Since there is no damping or driving force, $c = 0$ and $f = 0$, and the equation becomes $z'' + (k/m)z = 0$; so we need to find k/m. Since $kh = mg$, $k/m = g/h$, where $g = 384$ in/sec^2 [see

Example 3.1.1], $k/m = g/h = (384 \text{ in}/\text{sec}^2)/(24 \text{ in}) = 16 \text{ sec}^{-2}$. So the equation of motion is $z'' + 16z = 0$.

(b). See Fig. 1(b) for the four orbits of the ODE, $z'' + 16z = 0$, where $z(0) = 15, 10, 5, 2$ and $z'(0) = 0$.

(c). See Fig. 1(c) for the plot in tz-space (top) and in tz'-space (bottom). The motion is periodic, and all nonconstant solutions seem to have period $T \approx 1.6$ sec (reading from the tz-graph at top right).

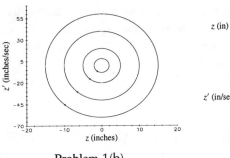

Problem 1(b).

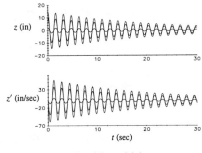

Problem 1(c).

2. **(a).** As in Problem 1, we have $k/m = g/h = 16 \text{ sec}^{-2}$. This time $c/m = cg/mg = [(2.60 \times 10^{-4} \text{ lb} \cdot \text{sec}/\text{in})(384 \text{ in}/\text{sec}^2)]/(1 \text{ lb}) = 0.1 \text{ sec}^{-1}$. So the ODE is $z'' + 0.1z' + 16z = 0$.

(b). See Fig. 2(b) for the three orbits; the initial data is $z(0) = 15, 10, 2$ and $z'(0) = 0$. Adjacent loops of the innermost spiral are so close together that they seem to be on top of one another, but that only shows the limitations of computer graphics.

(c). See Fig. 2(c) for the plot in tz-space (top) and in tz'-space (bottom). The amplitude of the motion seems to be decaying to zero as the damping dissipates energy and oscillations have smaller and smaller amplitudes. Note, however, that the "period" of the oscillations never changes. The period seems to be $T \approx 1.6$ seconds. If you compare this period with that of the motion in Problem 1, it seems to be the same. The damping term doesn't seem to affect the period, although it does make the amplitudes of the oscillations die down as time advances.

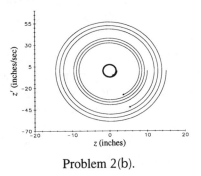

Problem 2(b).

Problem 2(c).

3. In part **(b)**, the unbounded orbits appear outside the region where the soft spring model is valid. See Example 3.1.3, Example 3.1.4, and the discussion after equation (5) in the text. Whenever we interpret the results of our model in terms of the system being modeled, we must be careful to restrict our attention to the regions where our model is valid. For springs, the model is valid only when the displacement is within the elastic limits of the spring. That is, the model is only valid for displacements yielding a restoring spring force, positive for $y < 0$ and negative for $y > 0$. For the soft springs of **(b)** and **(c)**, the elastic limit is no greater than $(10)^{1/2} \approx 3.16$, since for $y > (10)^{1/2}$, $S(y)$ will have the wrong sign for a restoring force. A cyclical orbit [as in **(a)** and **(b)**] indicates that after a certain period of time the system returns to its original configuration, and since time does not appear explicitly in the equations of motion (i.e., the ODEs are autonomous) the system repeats its motion periodically. Generally, the period will be different for different orbits, as a component plot will show. In **(c)** the motion is damped, so energy is dissipated and the amplitude decays with time. In every case $m = 1$ and the corresponding system is $y' = v$, $v' = S(y) - cy'$.

(a). Here we have $y' = v$, $v' = -0.2y - 0.02y^3$, $0 \le t \le 25$. See Figs. 3(a). The equilibrium point at the origin of Fig. 3(a), Graph 1 and the corresponding straight line solution curve in Fig. 3(a), Graph 2 correspond to the spring at rest at $y = 0$ in its position of static equilibrium. The other three solutions are periodic and correspond to the periodic oscillations of the undamped spring. The periods of the three solutions corresponding to $y_0 = 0$, $v_0 = 1, 3, 9$ are approximately 12.3, 8.7, and 5.5, respectively; the period appears to *decrease* as the amplitude increases. Apparently the stiffness of the spring is such that the spring oscillates faster and with greater amplitude if it is given a greater initial velocity.

(b). In this case the system is $y' = v$, $v' = -0.2y + 0.02y^3$, $0 \le t \le 30$. See Figs. 3(b) for orbits and solution curves with the given initial points. The point at $(0, 0)$ in Fig. 3(b), Graph 1 and the straight line $y = 0$ in Fig. 3(b), Graph 2 correspond to the spring at rest in static equilibrium. The initial points $y_0 = 0$, $v_0 = 0.4$, 0.9 give periodic solutions with respective periods of approximately 14.5, 19 and amplitudes of approximately 0.9, 2.4. Here the periods seem to *increase* with increasing amplitude and initial velocity. The remaining solutions, plotted along with their orbits, correspond to the spring stretching or compressing far beyond the limits of validity of the model. See the curves in Figs. 3(c) that "escape" as t increases.

(c). Here, $y' = v$, $v' = -y + 0.1y^3 - 0.1v$, $0 \le t \le 40$. The solution with initial point $y_0 = 0$, $v_0 = 2.44$ is just close enough to the rest point of static equilibrium $y = 0$, $v = 0$ [see Fig. 3(c), Graphs 1, 2], that the damping forces the motion of the spring into decaying oscillations around the equilibrium. The solution with initial data $y_0 = 0$, $v_0 = 2.46$ has just that extra amount of initial velocity (and so, energy) that it appears to move beyond the limits of validity of the model. See the orbit in Fig. 3(c), Graph 1 and the solution curve in Fig. 3(c), Graph 2 that "escape" to infinity as t increases.

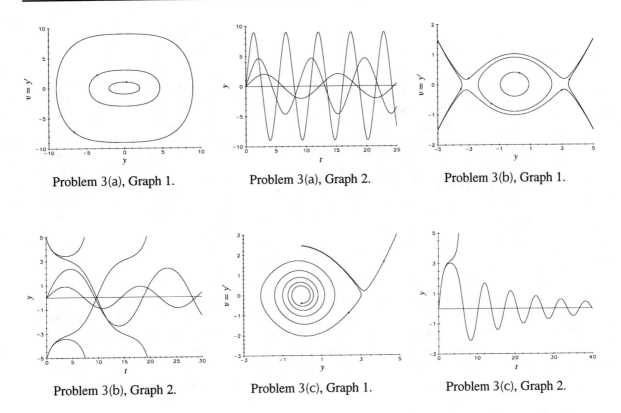

Problem 3(a), Graph 1.   Problem 3(a), Graph 2.   Problem 3(b), Graph 1.

Problem 3(b), Graph 2.   Problem 3(c), Graph 1.   Problem 3(c), Graph 2.

(d). Following Example 3.1.5, we rewrite $y'' = -y + 0.1y^3 - 0.1y'$ as $y'' = F(y, y')$. Notice first that $F(0, 0) = 0$. Taking partial derivatives, $F_y = -1 + 0.3y^2$ and $F_{y'} = -0.1$. So $F_y(0, 0) = -1$ and $F_{y'}(0, 0) = -0.1$. So the linear approximation of F at $(0, 0$ is $G(y, y') = -y - 0.1y'$. Thus, the linear approximation to the original ODE is $y'' = -y - 0.1y'$. See the curves in Figs. 4, Graph 1 and Graph 2 for the nonlinear ODE and the curves in Figs. 4, Graph 3 and Graph 4 for the linear approximation. There are no observable differences between the nonlinear curves and the linear approximation curves.

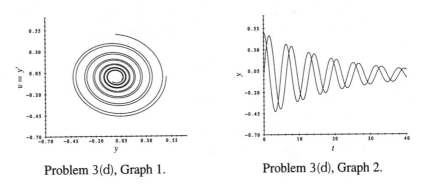

Problem 3(d), Graph 1.   Problem 3(d), Graph 2.

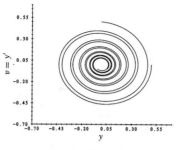

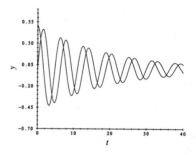

Problem 3(d), Graph 3.   Problem 3(d), Graph 4.

4. Since z measures the displacement from the rest position of the spring, the gravitational force is exactly balanced by the spring force, and the equation of motion of the mass will not involve the term mg (see the argument in the text). Let z be positive above the rest position and negative below. By Newton's Second Law, Hooke's Law, and the given law of magnetic attraction, we have that $mz'' = -kz - \alpha/(b+z)^2$, where α is a positive constant. The ODE is valid only for $z > -b$ (i.e., the magnet is above the plate). The ODE is nonlinear, and solutions z cannot be expressed in terms of elementary functions of t.

5. We will try to account for gravity by simply changing variables to shift the equilibrium position of the spring from h in the old variable y to 0 in the new variable z as done in equation (8) in the text. Substituting $y = z - h$ in the hard/soft spring models, $ay'' + cy' + ky \pm jy^3 = -mg + f(t)$, with $S(y)$ given by (3) or (4) yields $mz'' + cz' + kz \pm j(z^3 - 3z^2h + 3zh^2) = f(t)$, where we have used the fact that the equilibrium displacement h is the value of y at which the gravitational force is exactly balanced by the spring force, so h satisfies $k(-h) \pm j(-h)^3 = -mg$. Since our transformed ODE for motion about the equilibrium changes by a nonzero term $\pm j(-3z^2h + 3zh^2) = \pm 3zhj(-z + h)$ when gravity is introduced, gravity *does* affect the motion of hard and soft springs more than just changing the position of static equilibrium, and it does so by changing the form of the ODE of the motion near the point of static equilibrium. In the Hooke's Law spring, this substitution does not change the equation of motion since $j = 0$ under Hooke's Law, so the Hooke's Law spring's motion is not affected by gravity.

6. We have $k = 1.01$ N/m, mass $= 1$ kg, $c = 0.2$ N· sec /m and $f(t) = $ sqw$(t, 50, 2\pi)$ N. By equation (8) in the text, $mz'' + cz' + kz = f(t)$, and the fact that the spring is at equilibrium at $t = 0$, the IVP is $z'' + 0.2z' + 1.01z = $ sqw$(t, 50, 2\pi)$, $z(0) = 0$, $z'(0) = 0$, where time and distance are measured in seconds and meters (denoted by m). See Fig. 6 for the plot of z vs. t for this IVP; the driving force $f = $ sqw$(t, 50, 2\pi)$ is plotted in the upper graph. The response appears to become periodic with the same frequency as the periodic driving term, but slightly out of phase.

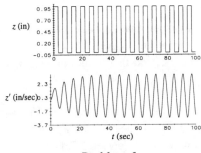

z (in)

z' (in/sec)

t (sec)

Problem 6.

7.   Group project.

3.2   Second Order ODEs and Their Properties

Suggestions for preparing a lecture

Topics: The Fundamental Theorem (Theorem 3.2.1), properties of solutions of a second-order autonomous ODE, examples, graphs of orbits, component curves, time-state curves.

Remarks: There is a lot of material here so don't feel it has to be covered in detail the first time through. The properties are reviewed now and again throughout the chapter.

Making up a problem set

Problems 1**(b)**, **(e)**, 2**(b)**, 4, 10**(b)**, 11**(a)**. You may also want to assign (or else discuss in the lecture) group problem 12 (because of the unusual way a new state variable is introduced to accomplish a particular graphical task).

Comments

Whenever we introduce a new class of ODEs (second-order ODEs in this chapter), we need to assure ourselves that IVPs have unique solutions. As usual, this requires that the functions in the ODE satisfy certain continuity and smoothness conditions. That is the reason Theorem 3.2.1 is given here. In addition, we want to know whether the solution of an IVP responds in a continuous fashion to changes in the initial data—and this is where the Continuity [or Sensitivity] part of Theorem 3.2.1 comes in. We also define and name the various graphs corresponding to a solution of an IVP. If the ODE has the form $y'' = F(t, y, y')$ and $y = y(t)$ is a solution, we discuss and illustrate solution curves [y vs. t], velocity curves [y' vs. t], orbits [y' vs. y], and time-state curves [the parametrically-defined curve $t = t$, $y = y(t)$, $y' = y'(t)$ in tyy'-space]. It takes a while to absorb all of this, and so we have given lots of examples with pictures. Each graph gives its own kind of information about the behavior of a solution; if we don't have a solution formula, we need all the help we can get to determine the behavior of a solution. Group problem 12 should be noted because it gives a way to trick your solver so that it will plot a time-state curve and its orbit on the same set of axes. As in the group project in Section 3.1, the method involves adding extra ODEs.

1. To verify that each IVP satisfies Theorem 3.2.1, we must write the ODE in the form $y'' = F(t, y, y')$ and show that F, $\partial F/\partial y$, and $\partial F/\partial y'$ are continuous throughout tyy'-space. The ty-solution curves and ty'-velocity curves are plotted in Graph 1, then the orbits in the yy'-plane in Graph 2, and finally the time-state curves (t is the vertical axis) in tyy'-space in Graph 3.

   (a). The ODE can be rewritten as $y'' = F(t, y, y') = -y$. The function $F = -y$ is continuous throughout tyy'-space as are $\partial F/\partial y = -1$, and $\partial F/\partial y' = 0$, so Theorem 3.2.1 is satisfied. The solutions $y = y(t)$ and velocities $v = y'(t)$ of the IVPs $y'' + y = 0$, $y(0) = 0.5$, 1.0, 1.5, $y'(0) = 0$ seem to be periodic with a common period T a little larger than 3. There is a single equilibrium solution, $y = 0$ for all t, whose graph is a straight line. The nonconstant orbits seem to be simple closed curves (which again suggests that the solutions are periodic). The equilibrium solution corresponds to a point orbit at the origin. The time-state curves are helices, while the equilibrium time-state curve is a vertical line. See Figs. 1(a).

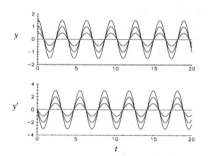

Problem 1(a), Graph 1.

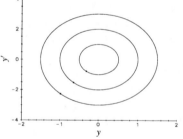

Problem 1(a), Graph 2.

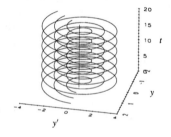

Problem 1(a), Graph 3.

   (b). The ODE can be rewritten as $y'' = F(t, y, y') = -0.1y' - 4y$. Since $F = -0.1y' - 4y$, $\partial F/\partial y = -4$, and $\partial F/\partial y' = -0.1$. are all continuous, Theorem 3.2.1 is satisfied. The solutions of the IVPs $y'' + 0.1y' + 4y = 0$, $y(0) = 0.5$, 1.0, 1.5, $y'(0) = 0$ seem to be oscillating with a decaying amplitude that seems to tend to 0 as $t \to +\infty$ [Fig. 1(b), Graph 1]. The single equilibrium solution, $y = 0$, all t, corresponds to a straight line. The orbits in Fig. 1(b), Graph 2 spiral toward the origin as time increases. The point at the origin corresponds to the equilibrium. The time-state curves of Fig. 1(b), Graph 3 spiral inward and upward and approach the vertical line $y = y' = 0$ corresponding to the equilibrium state as $t \to \infty$.

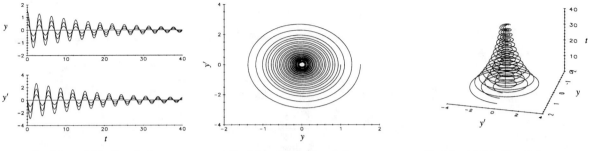

Problem 1(b), Graph 1.   Problem 1(b), Graph 2.   Problem 1(b), Graph 3.

(c). The ODE can be rewritten as $y'' = F(t, y, y') = 0.1y' - 4y$. The function $F = 0.1y' - 4y$ is continuous throughout tyy'-space as are $\partial F/\partial y = -4$ and $\partial F/\partial y' = 0.1$, so Theorem 3.2.1 is satisfied. Here the IVPs are $y'' - 0.1y' + 4y = 0$, $y(0) = 0, 1, 2$, $y'(0) = 0$, but we solve *backward* from $t = 0$ to $t = -100$. There is a single equilibrium solution, $y = 0$, all t. See Figs. 1(c).

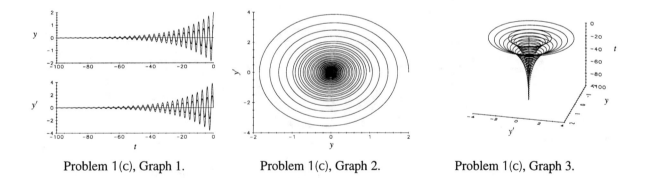

Problem 1(c), Graph 1.   Problem 1(c), Graph 2.   Problem 1(c), Graph 3.

(d). The ODE can be rewritten as $y'' = F(t, y, y') = -0.2y' - 10y - 0.2y^3$. The function $F = -0.2y' - 10y - 0.2y^3$ is continous throughout tyy'-space, as are $\partial F/\partial y = -10 - 0.6y^2$ and $\partial F/\partial y' = -0.2$, so Theorem 3.2.1 is satisfied. Here the IVP is $y'' + 0.2y' + 10y + 0.2y^3 = -9.8$, $y(0) = -4$, $y'(0) = 0$. The equilibrium solution, $y = 0$, all t, satisfies $y(0) = 0$, $y'(0) = 0$. Although the ODE is nonlinear, the graphs resemble those of part **(b)**, and the nonconstant solution curve shown seems to approach the equilibrium solution as $t \to \infty$. See Figs. 1(d).

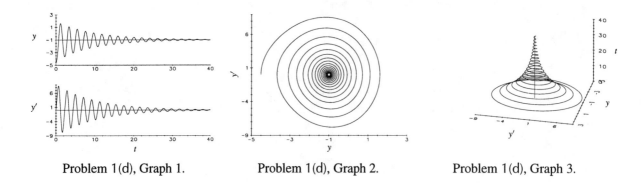

Problem 1(d), Graph 1.   Problem 1(d), Graph 2.   Problem 1(d), Graph 3.

(e). The ODE can be rewritten as $y'' = F(t, y, y') = -(0.1y^2 - 0.08)y' - y^3$. The function $F = -(0.1y^2 - 0.08)y' - y^3$ is continuous, as are $\partial F/\partial y = -0.2yy' - 3y^2$ and $\partial F/\partial y' = -0.08$, so Theorem 3.2.1 is satisfied. The solution of the nonlinear IVP $y'' + (0.1y^2 - 0.08)y' + y^3 = 0$, $y(0) = 0$, $y'(0) = 0.5$ seems to oscillate with a slowly growing amplitude [Fig. 1(e), Graph 1]. The orbit [Fig. 1(e), Graph 2] suggests that as $t \to +\infty$ the orbit spirals to the orbit of a slightly tilted, periodic "racetrack" orbit. Turning back to the solution curve in Fig. 1(e), Graph 1, we see from the solution curve for large t that the period of the racetrack orbit is about 4.4. The time-state curve is shown in Fig. 1(e), Graph 3.

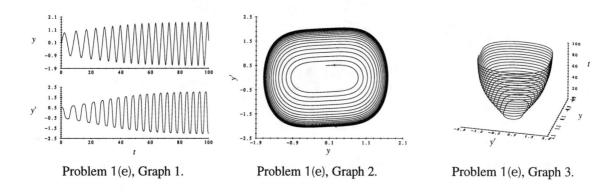

Problem 1(e), Graph 1.   Problem 1(e), Graph 2.   Problem 1(e), Graph 3.

(f). The ODE can be rewritten as $y'' = -0.05y' - 25y + \sin(5.5t)$. The function $F = -0.05y' - 25y + \sin(5.5t)$ is continuous throughout tyy'-space, as are $\partial F/\partial y = -25$ and $\partial F/\partial y' = -0.05$, so Theorem 3.2.1 is satisfied. The solution of the linear nonautonomous IVP $y'' + 0.05y' + 25y = \sin(5.5t)$, $y(0) = y'(0) = 0$ seems to have a diminishing amplitude, but it is not clear whether or not the amplitude tends to 0 as $t \to \infty$. See Fig. 1(f), Graph 1. The orbit is self-intersecting, but not periodic [Fig. 1(f), Graph 2]. This is no contradiction since the ODE is nonautonomous. The time-state curve in Fig. 1(f), Graph 3 does *not* self-intersect.

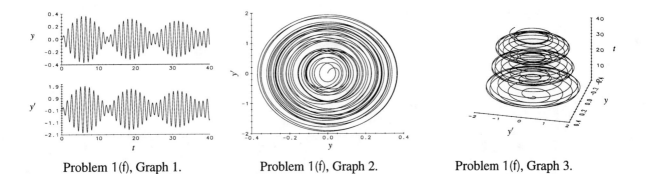

Problem 1(f), Graph 1.   Problem 1(f), Graph 2.   Problem 1(f), Graph 3.

2. **(a).** All solution curves are concave down because $y'' = -1$ is everywhere negative. See Fig. 2(a).

(b). Solution curves are concave down when y is positive because then $y'' = -y$ is negative, but concave up when y is negative. A solution curve has an inflection point whenever $y = 0$. See Fig. 2(b) for some ty-curves. Note how the arcs of solution curves above the t-axis are concave down, while those below the t-axis are concave up.

(c). The inflection points of solution curves of $y'' = y^2 - t$ lie on the graph of $y = \pm\sqrt{t}$. Above and below this parabola (dashed), solution curves are concave up, but inside the parabola they are concave down. See Fig. 2(c) for the solution curves starting at $(0, 0.75)$ and $(0, 0.925)$.

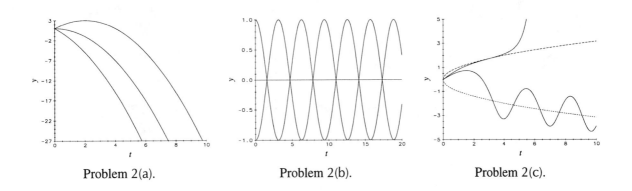

Problem 2(a).   Problem 2(b).   Problem 2(c).

3. See Fig. 3 for graphs of several solution curves of $t(y''y + (y')^2) + y'y = 1$ with initial points $y(1) = 1$, $|y'(1)| = |v_0| \le 10$.

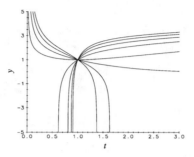

Problem 3.

4.   **(a).** The periods of the three solutions in Fig. 3.2.6 appear to be approximately 5, 7.5, 9.5, corresponding to the solutions of decreasing amplitude. Increasing the value of v_0 increases the amplitude of the first oscillation and decreases the period of the motion.

(b). See Figs. 4(b) for the graphs of the orbit in the yy'-plane, the ty- and ty'-component curves, and the time-state curve for the IVP, $y'' + y = \cos(1.1t)$, $y(0) = 5$, $y'(0) = 0$. From the component curves it appears that the period is about 62.

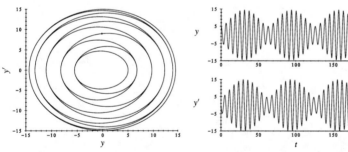

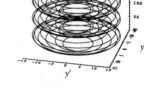

Problem 4(b), Graph 1.   Problem 4(b), Graph 2.   Problem 4(b), Graph 3.

5.   To solve the IVP $y'' = 2y'y$, $y(0) = 1$, $y'(0) = 1$ we use the hint to see that $y'' = (y^2)'$. Integrating and applying the initial conditions, we see that $y' = y^2$, so $y'/y^2 = 1$. Integrating again and applying the initial conditions, we obtain $-1/y + 1 = t$. Solving for y, we have $y = 1/(1 - t)$. The largest t-interval for this IVP is $t < 1$. See Figs. 5, Graph 1 and Graph 2 for the solution graph and orbit, respectively.

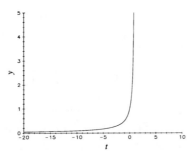
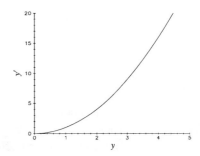

Problem 5, Graph 1.   Problem 5, Graph 2.

6. Multiply the ODE $y'' = F(y)$ by y' to obtain $y'y'' = y'F(y)$. Integrating, we have $\int y'y''dt = \int y'F(y)dt = \int F(y)(dy/dt)\,dt = \int F(y)dy$. So, $y'^2(t)/2 = C + \int^y F(s)\,ds$, where C is a constant, is an equation for the orbits. The value of C could be determined from initial data. For the ODE $y'' = y'F(y) = F(y)dy/dt$ we can integrate each side to obtain $y' = C + \int^y F(s)\,ds$, which defines an orbit for each value of C.

7. Look for solutions to the IVP $t^2y'' - 2ty' + 2y = 0$, $y(0) = 0$, $y'(0) = 0$ in the form $y = Ct^\alpha$. Substituting this expression into the system, we obtain $C\alpha(\alpha - 1)t^\alpha - 2C\alpha t^\alpha + 2Ct^\alpha = Ct^\alpha[\alpha(\alpha - 1) - 2\alpha + 2] = 0$. For a nontrivial solution, $C \neq 0$, and so $\alpha(\alpha - 1) - 2\alpha + 2 = \alpha^2 - 3\alpha + 2$ must be zero. So, $\alpha = 1$ or $\alpha = 2$. This means that the original ODE has solutions $y = C_1t + C_2t^2$ where C_1 and C_2 are arbitrary constants. In order to satisfy the initial conditions, $y(0) = 0$, $y'(0) = 0$, we must have $C_1 = 0$. The IVP has infinitely many solutions $y = C_2t^2$, where C_2 is an arbitrary constant. This conclusion does not contradict Theorem 3.2.1 because $F(t, y, y') = (2ty' - 2y)/t^2$ is not continuous with respect to t at $t = 0$.

8. Multiply the ODE $y'' + \omega^2 y = 0$ by y' to obtain $y'y'' + \omega^2 yy' = 0$. This can be written as $((y')^2)'/2 + (\omega^2 y^2)'/2 = 0$, which integrates to $(y')^2/2 + (\omega^2 y^2)/2 = C$, where C is any non-negative constant. The graph in the yy'-plane is an ellipse if $C > 0$ since ω^2 is positive.

9. If we set $v = y'(t)$, $v\,dv/dy = y''(t)$ [see Section 1.7], then the ODE, $y'' = -ky - jy^3$ becomes the first-order ODE

$$v\frac{dv}{dy} = -ky - jy^3$$

which is separable. Separating the variables,

$$v\,dv = (-ky - jy^3)\,dy$$

Integrating each side, we have

$$\frac{1}{2}v^2 = -\frac{1}{2}ky^2 - \frac{j}{4}y^4 + C_1$$

$$v^2 = (y')^2 = C - ky^2 - jy^4/2$$

where $C = 2C_1$ is a positive constant. For each positive value of C, the graph is an oval-shaped curve in the yv-plane which cuts the y-axis at the real roots of $C - ky^2 - jy^4/2$.

These real roots can be found by using the quadratic formula for y^2, and then taking the square root of the positive roots. In fact, we have that the real roots are given by

$$y = \pm \left[-\frac{k}{j} + \left(\frac{k^2}{j^2} + \frac{2C}{j} \right)^{1/2} \right]^{1/2}$$

10. **(a).** The component curves of the Painlevé transcendant $y'' = y^2 - t$ correspond to initial values $y(0) = 0$, $-5 \leq y'(0) \leq 2$. Close inspection suggests that if $y'(0)$ is close to -5, then the corresponding solution curve becomes unbounded in finite time. As solution curves cross the parabola $t = y^2$, their concavity switches.

(b). The three orbits of $y'' = y^2 - 0.1y^3 - t$ shown in the figure correspond to the initial conditions $y(0) = 0$, $y'(0) = 0, -2.5, -5.0$. See Figs. 10(b) for the time-state and the component curves.

(c). See Fig. 10(c) for the solution curve of the transcendent $y'' = y^2 - t$ (solid curve) and of the perturbed transcendent $y'' = y^2 - 0.1y^3 - t$ (dashed curve). The magnitude of $0.1y^3$ for the range of y-values shown is small compared to the magnitude of $y^2 - t$, so the perturbation term $-0.1y^3$ has only a small effect on the solution curve.

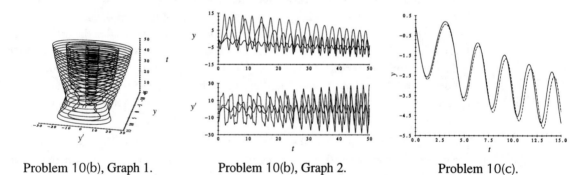

Problem 10(b), Graph 1.   Problem 10(b), Graph 2.   Problem 10(c).

11. According to Theorem 3.2.1, if F, $\partial F/\partial y$, and $\partial F/\partial y'$ are continuous for all t, y, y', then the IVP, $y'' = F(t, y, y')$, $y(t_0) = y_0$, $y'(t_0) = v_0$ has exactly one solution.

(a). The functions $y_1 = \sin t$, and $y_2 = t$ both have value 0 at $t_0 = 0$, and the derivatives at $t_0 = 0$ have common value 1. So if these two functions were both solutions of $y'' = F(t, y, y')$, then there would be *two* solutions satisfying the conditions $y(0) = 0$, $y'(0) = 1$, and this would contradict Theorem 3.2.1. So $y_1(t)$ and $y_2(t)$ cannot both be solutions of $y'' = F(t, y, y')$.

(b). Note that $y_1 = e^t$ and $y_2 = 1 + t + t^3/3$ both satisfy the conditions $y(0) = 0$, $y'(0) = 1$. So they can't both solve the ODE $y'' = F(t, y, y')$ since that would violate the uniqueness part of Theorem 3.2.1.

12. Group project.

3.3 Undriven Constant Coefficient Linear ODEs, I

Suggestions for preparing a lecture

Topics: Differentiation operator, polynomial operators, calculations with polynomial operators, solution formula techniques, general solution.

Remarks: In the background material at the end of this manual section, we approach the topic from the perspective of basic solution sets and the Wronskian. These topics come up in Section 3.7, but in a more general setting. Because we cannot assume that the reader is familiar with the notion of a linear space and linear independence, we avoid a direct approach using these concepts.

Making up a problem set

Four or five parts from Problems 1–3, Problem 4.

Comments

Section 3.3 lays the foundation for a study of undriven second-order linear ODEs with constant coefficients and the corresponding IVPs. Only the case where the characteristic roots are real is discussed. We take the linear operator approach, but gently. The next section treats undriven ODEs with constant coefficients whose characteristic roots are complex numbers.

1. Although it is not necessary to include the step using operator notation, we put it in to help you grow accustomed to its use and to help you see how the characteristic polynomial $P(r)$ relates to the polynomial operator $P(D)$.

   (a). The ODE $y'' = D^2[y] = 0$ has the characteristic polynomial r^2 which has a double root, $r = 0$. Using Theorem 3.3.1, the general solution is $y(t) = C_1 e^{0t} + C_2 t e^{0t} = C_1 + C_2 t$, where C_1 and C_2 are arbitrary reals. Notice that you could also integrate both sides twice to get the same result.

   (b). The ODE $y'' + y' - 2y = (D^2 + D - 2)[y] = 0$ has the characteristic polynomial $r^2 + r - 2$ which has roots $r_1 = -2$ and $r_2 = 1$. Using Theorem 3.3.1, the general solution is $y(t) = C_1 e^{r_1 t} + C_2 e^{r_2 t} = C_1 e^{-2t} + C_2 e^t$, where C_1 and C_2 are arbitrary reals.

   (c). The ODE $y'' - 4y' + 4y = (D^2 - 4D + 4)[y] = 0$ has the characteristic polynomial $r^2 - 4r + 4$ which has a double root, $r = 2$. Using Theorem 3.3.1, the general solution is $y(t) = C_1 e^{2t} + C_2 t e^{2t}$, where C_1 and C_2 are arbitrary reals.

   (d). The ODE $y'' - 4y = (D^2 - 4)[y] = 0$ has the characteristic polynomial $r^2 - 4$ which has roots $r_1 = 2$ and $r_2 = -2$. So the general solution is $y(t) = C_1 e^{2t} + C_2 e^{-2t}$.

   (e). The ODE $5y'' - 10y' = (5D^2 - 10D)[y] = 0$ has the characteristic polynomial $5r^2 - 10r$ which has roots $r_1 = 0$ and $r_2 = 2$. So the general solution is $y(t) = C_1 + C_2 e^{2t}$.

   (f). The ODE $2y'' + 12y' + 18y = (2D^2 + 12D + 18)[y] = 0$ has the characteristic polynomial $2r^2 + 12r + 18$ which has a double root, $r = -3$. So the general solution is $y(t) = C_1 e^{-3t} + C_2 t e^{-3t}$.

   (g). The ODE $y'' - 6y' + 9y = (D^2 - 6D + 9)[y] = 0$ has the characteristic polynomial $r^2 - 6r + 9$ which has a double root, $r = 3$. So the general solution is $y(t) = C_1 e^{3t} + C_2 t e^{3t}$.

(h). The ODE $y'' + 4y' - y = (D^2 + 4D - 1)[y] = 0$ has the characteristic polynomial $r^2 + 4r - 1$ which has roots $-2 + \sqrt{5}$ and $-2 - \sqrt{5}$. So the general solution is $y(t) = C_1 e^{(-2+\sqrt{5})t} + C_2 e^{(-2-\sqrt{5})t}$.

(i). The ODE $y'' + 2y' + y = (D^2 + 2D + 1)[y] = 0$ has the characteristic polynomial $r^2 + 2r + 1$ which has a double root, $r = -1$. So the general solution is $y(t) = C_1 e^{-t} + C_2 t e^{-t}$.

(j). The ODE $4y'' - 4y' + y = (4D^2 - 4D + 1) = 0$ has the characteristic polynomial $4r^2 - 4r + 1$ which has a double root, $r = -1/2$. So the general solution is $y(t) = C_1 e^{-t/2} + C_2 t e^{-t/2}$.

(k). The ODE $y'' - 2y' + y = (D^2 - 2D + 1)[y] = 0$ has the characteristic polynomial $r^2 - 2r + 1$ which has a double root, $r = 1$. So the general solution is $y(t) = C_1 e^t + C_2 t e^t$.

(l). The ODE $y'' - 10y' + 25y = (D^2 - 10D + 25)[y] = 0$ has the characteristic polynomial $r^2 - 10r + 25$ which has a double root, $r = 5$. So the general solution is $y(t) = C_1 e^{5t} + C_2 t e^{5t}$.

2. **(a).** The ODE $y'' + y' = (D^2 + D)[y] = 0$ has characteristic polynomial $r^2 + r$ which has roots $r_1 = 0$ and $r_2 = -1$. So the general solution is $y(t) = C_1 + C_2 e^{-t}$. Using $y'(0) = 2$ gives us $C_2 = -2$ which along with $y(0) = 1$ gives us $C_1 = 3$. So our final solution is $y(t) = 3 - 2e^{-t}$. See Fig. 2(a).

(b). The ODE $y'' + 3y' + 2y = (D^2 + 3D + 2)[y] = 0$ has characteristic polynomial $r^2 + 3r + 2$ which has roots -1 and -2. The general solution is $y(t) = C_1 e^{-t} + C_2 e^{-2t}$. Applying initial conditions $y(0) = 0$ and $y'(0) = 1$ gives $C_1 + C_2 = 0$ and $-C_1 - 2C_2 = 1$ so $C_1 = 1$ and $C_2 = -1$. The final solution is $y(t) = e^{-t} - e^{-2t}$. See Fig. 2(b).

(c). The characteristic polynomial of $y'' - 9y = (D^2 - 9)[y] = 0$ has roots ± 3. The general solution is $y(t) = C_1 e^{3t} + C_2 e^{-3t}$. Applying initial conditions $y(0) = 2$ and $y'(0) = -1$ gives $C_1 = 5/6$ and $C_2 = 7/6$. The final solution is $y(t) = 5/6 e^{3t} + 7/6 e^{-3t}$. See Fig. 2(c).

(d). The characteristic polynomial of $y'' - 4y' + 4y = (D^2 - 4D + 4)[y] = 0$ has a double root, 2. The general solution is $y(t) = C_1 e^{2t} + C_2 t e^{2t}$. Applying initial conditions $y(0) = 1$ and $y'(0) = 1$ gives $C_1 = 1$ and $C_2 = -1$. The final solution is $y(t) = e^{2t} - t e^{2t}$. See Fig. 2(d).

(e). The characteristic polynomial of $y'' - 25y = (D^2 - 25)[D] = 0$ has roots ± 5. The general solution is $y(t) = C_1 e^{5t} + C_2 e^{-5t}$. Applying initial conditions $y(1) = 0$ and $y'(1) = 1$ gives $C_1 = e^{-5}/10$ and $C_2 = -e^5/10$. The final solution is $y(t) = \left(e^{5(t-1)} - e^{5(1-t)} \right)/10$. See Fig. 2(e).

(f). The characteristic polynomial of $y'' + y' - 6y = 0$ has roots -3 and 2. The general solution is $y(t) = C_1 e^{-3t} + C_2 e^{2t}$. Applying initial conditions $y(0) = 1$ and $y'(0) = -1$ gives $C_1 = 3/5$ and $C_2 = 2/5$. The final solution is $y(t) = \left(3e^{-3t} + 2e^{2t} \right)/5$. See Fig. 2(f).

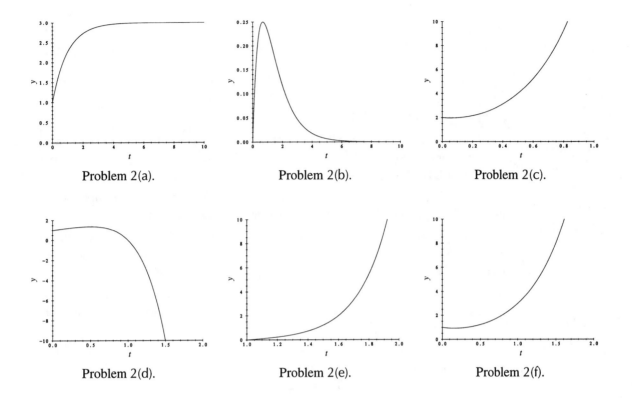

Problem 2(a).   Problem 2(b).   Problem 2(c).

Problem 2(d).   Problem 2(e).   Problem 2(f).

3. Theorem 3.3.1 shows us that all possible solutions of $y'' + ay' + by = (D^2 + aD + b)[y] = P(D)[y] = 0$ are of the form $C_1 e^{r_1 t} + C_2 e^{r_2 t}$ or $C_1 e^{r_1 t} + C_2 t e^{r_1 t}$ where $P(r) = r^2 + ar + b = (r - r_1)(r - r_2)$. Solving for these cases is a matter of matching terms in the given solution with the appropriate roots. The values of C_1 and C_2 don't affect our choice of differential equation (these values are fixed when initial conditions are chosen).

(a). In the case of $e^t - e^{-t}$ we see that the roots are $r_1 = 1$ and $r_2 = -1$ (we don't really care about C_1 or C_2). So the characteristic polynomial $P(r)$ must have as roots 1 and -1. So $P(r) = (r - r_1)(r - r_2) = (r - 1)(r + 1) = r^2 - 1$. So we know that $P(D) = D^2 - 1$ and $P(D)[y] = (D^2 - 1)[y] = y'' - y$. Our original ODE is $y'' - y = 0$.

(b). In the case of $e^t - te^t$ we see that both roots are 1 (This is why we can have t as a coefficient of $-te^t$). So $P(r) = (r - 1)^2 = r^2 - 2r + 1$. So $P(D) = D^2 - 2D + 1$ and $P(D)[y] = (D^2 - 2D + 1)[y] = y'' - 2y' + y$. So the original ODE is $y'' - 2y' + y = 0$.

(c). In the case of $e^t + e^{-2t}$ we see that the roots are $r_1 = 1$ and $r_2 = -2$. So $P(r) = (r - 1)(r + 2) = r^2 + r - 2$. So $P(D) = D^2 + D - 2$ and $P(D)[y] = (D^2 + D - 2)[y] = y'' + y' - 2y$. So the original ODE is $y'' + y' - 2y = 0$.

(d). In the case of $1 + e^{-3t}$ we see that the roots are $r_1 = 0$ and $r_2 = -3$. (Any time there is a constant in the solution, one of the roots must be zero. Note that $e^{0 \cdot t} = 1$.) So $P(r) = r(r + 3) = r^2 + 3r$. So $P(D) = D^2 + 3D$ and $P(D)[y] = (D^2 + 3D)[y] = y'' + 3y'$. So the original ODE is $y'' + 3y' = 0$.

(e). In the case of $e^{2t} + 10000e^{3t}$ we see that the roots are $r_1 = 2$ and $r_2 = 3$. So $P(r) = (r-2)(r-3) = r^2 - 5r + 6$. So $P(D) = D^2 - 5D + 6$ and $P(D)[y] = (D^2 - 5D + 6)[y] = y'' - 5y' + 6y$. So the original ODE is $y'' - 5y' + 6y = 0$.

(f). In the case of $e^{\sqrt{2}t} + e^{-\sqrt{2}t}$ we see that the roots are $r_1 = \sqrt{2}$ and $r_2 = -\sqrt{2}$. So $P(r) = (r - \sqrt{2})(r + \sqrt{2}) = r^2 - 2$. So $P(D) = D^2 - 2$ and $P(D)[y] = (D^2 - 2)[y] = y'' - 2y$. So the original ODE is $y'' - 2y = 0$.

(g). In the case of $e^{\pi t} - 3$ we see that the roots are $r_1 = \pi$ and $r_2 = 0$. So $P(r) = (r - \pi)r = r^2 - r\pi$ giving us $P(D) = D^2 - D\pi$ and $P(D)[y] = (D^2 - D\pi)[y] = y'' - \pi y'$. So the original ODE is $y'' - \pi y' = 0$.

(h). In the case of e^{t^2} we see that this cannot be written in an acceptable form because of the t^2 in the exponent. This cannot result from an undriven linear ODE.

(i). In the case of $t + 2$ we see that it can be written in the form $2e^{0t} + te^{0t}$ where both roots are zero. So $P(r) = r^2$ giving $P(D) = D^2$ and $P(D)[y] = D^2[y] = y''$. So the original ODE must be $y'' = 0$.

(j). In the case of $t^2 e^{-t}$ we see that the t^2 in the coefficient cannot occur in an undriven linear *second order* ODE (It could occur in a third or higher order ODE, but we aren't concerned with that here).

4. **(a).** Applying the initial conditions $y(-1) = 2$, $y'(-1) = 0$ to the general solution $y(t) = C_1 + C_2 t$ found in Problem 1**(a)** we have $C_1 - C_2 = 2$, $C_2 = 0$. So $C_1 = 2$, $C_2 = 0$ and $y(t) = 2$, for all t, solves the IVP. See Fig. 4(a).

(b). Applying the initial conditions $y(-1) = 2$, $y'(-1) = 0$ to the general solution $y(t) = C_1 e^{-2t} + C_2 e^t$ found in Problem 1**(b)**, we have $C_1 e^2 + C_2 e^{-1} = 2$ and $-2C_1 e^2 + C_2 e^{-1} = 0$. Solving we obtain $C_1 = (2/3)e^{-2}$, $C_2 = (4/3)e$. So $y = (2/3)e^{-2}e^{-2t} + (4/3)e^t$ is the solution of the IVP. See Fig. 4(b).

(c). Applying the initial conditions $y(-1) = 2$, $y'(-1) = 0$ to the general solution $y(t) = C_1 e^{2t} + C_2 t e^{2t}$ found in Problem 1**(c)** we have $(C_1 - C_2)e^{-1} = 2$ and $2C_1 e^{-2} + C_2(-2e^{-2} + e^{-2}) = 0$. Solving, we obtain $C_1 = -2e^2$, $C_2 = -4e^2$ and so $y = (-2e^2 - 4e^2 t)e^{2t}$ solves the IVP. See Fig. 4(c).

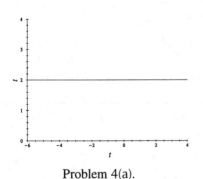

Problem 4(a).

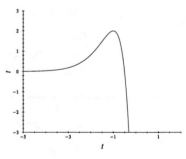

Problem 4(b).

Problem 4(c).

5. **(a).** $(D - 2)[e^{-t}] = D[e^{-t}] - 2e^{-t} = -e^{-t} - 2e^{-t} = -3e^{-t}$.

(b). $(D+3)[e^{-t}] = D[e^{-t}] + 3e^{-t} = -e^{-t} + 3e^{-t} = 2e^{-t}$.

(c). $(D^2 + D - 6)[e^{-t}] = D^2[e^{-t}] + D[e^{-t}] - 6e^{-t} = e^{-t} - e^{-t} - 6e^{-t} = -6e^{-t}$.

(d). Using the fact from part **(b)** that $(D+3)[e^{-t}] = 2e^{-t}$,

$$(D-2)(D+3)[e^{-t}] = (D-2)[2e^{-t}]$$
$$= -2e^{-t} - 4e^{-t}$$
$$= -6e^{-t}$$

which equals the result in part **(c)** since $D^2 + D - 6 = (D-2)(D+3)$.

(e). Using the fact from part **(a)** that $(D-2)[e^{-t}] = -3e^{-t}$,

$$(D+3)(D-2)[e^{-t}] = (D+3)[-3e^{-t}]$$
$$= 3e^{-t} - 9e^{-t}$$
$$= -6e^{-t}$$

which equals the results in parts **(c)** and **(d)** since $D^2 + D - 6 = (D+3)(D-2) = (D-2)(D+3)$.

(f). $(D+1)^2[e^{-t}] = (D^2 + 2D + 1)[e^{-t}] = e^{-t} - 2e^{-t} + e^{-t} = 0$.

6. **(a).** For $y(t) = 2e^{5t}$, $D[y(t)] = 10e^{5t} = 5y(t)$. So, $(D-5)[2e^{5t}] = 0$. Any $P(D) = (D-k)(D-5)$ will work.

(b). For $y(t) = e^{2t} + e^{-t}$, $(D-2)[e^{2t}] = 0$ and $(D+1)[e^{-t}] = 0$. So,

$$(D-2)(D+1)[e^{2t} + e^{-t}] = (D+1)(D-2)[e^{2t}] + (D-2)(D+1)[e^{-t}]$$
$$= (D+1)[0] + (D-2)[0]$$
$$= 0$$

So the polynomial $(D-2)(D+1) = D^2 - D - 2$ satisfies $P(D)[e^{2t} + e^{-t}] = 0$.

(c). For $y(t) = te^t$, $D[y(t)] = te^t + e^t = y(t) + e^t$. So, $(D-1)[y(t)] = e^t$. Note that $(D-1)[e^t] = 0$. Therefore,

$$(D-1)\big[(D-1)[y(t)]\big] = (D-1)[e^t]$$
$$= 0$$

So the polynomial $(D-1)^2 = D^2 - 2D + 1$ satisfies $P(D)[te^t] = 0$.

(d). For $y(t) = \sin t$, $D^2[y(t)] = D^2[\sin t] = -\sin t = -y(t)$. So, $(D^2 + 1)[\sin t] = 0$ and we choose $P(D) = D^2 + 1$.

7. **(a).** We will guess that $y(t) = Ae^{-t}$ for some value of A. Then

$$(D-2)[Ae^{-t}] = -Ae^{-t} - 2Ae^{-t}$$
$$= -3Ae^{-t}$$
$$= e^{-t}$$

is accomplished by setting $A = -1/3$, so $y(t) = (-1/3)e^{-t}$.

(b). We will guess that $y(t) = At + B$ for some choice of A, B. Then

$$(D - 2)[y(t)] = A - 2At - 2B$$

$$= t - 1$$

So we need $-2A = 1$ and $A - 2B = -1$. So, $y(t) = -1/2t + 1/4$.

(c). We will guess that $y(t) = A \cos t + B \sin t + C$ for some choice of A, B, C. Then

$$(D - 2)[y(t)] = -A \sin t + B \cos t - 2A \cos t - 2B \sin t - 2C$$

$$= (-A - 2B) \sin t + (B - 2A) \cos t - 2C$$

$$= \sin t + 4$$

So $A = -1/5$, $B = -2/5$, $C = -2$, and $y(t) = -1/5 \cos t - 2/5 \sin t - 2$.

8. **(a).** Writing the ODE $y'' - y = \sin 2t$ as $(D^2 - 1)[y] = \sin 2t$ and factoring the operator we obtain $(D + 1)(D - 1)[y] = \sin 2t$. If we put $v = (D - 1)[y]$, then v solves the ODE $(D + 1)[v] = \sin 2t$, or written differently: $v' + v = \sin 2t$. Multiplying through by the integrating factor e^t, the ODE becomes $(ve^t)' = e^t \sin 2t$. Using Table 1.3.1 and integrating we obtain

$$ve^t = (1/5)e^t(\sin 2t - 2\cos 2t) + C_1$$

where C_1 is an arbitrary real, and so

$$v = (1/5)(\sin 2t - 2\cos 2t) + C_1 e^{-t}$$

Now we must solve $(D - 1)[y] = v$. That is,

$$y' - y = (1/5)(\sin 2t - 2\cos 2t) + C_1 e^{-t}$$

Multiplying through by the integrating factor e^{-t}, this ODE becomes

$$(ye^{-t})' = (1/5)(e^{-t} \sin 2t - 2e^{-t} \cos 2t) + C_1 e^{-2t}$$

Integrating and using Table 1.3.1 again we obtain

$$ye^{-t} = (1/25)e^{-t}(-\sin 2t - 2\cos 2t) - (2/25)e^{-t}(-\cos 2t + 2\sin 2t) - (C_1/2)e^{-2t} + C_2$$

where C_1 and C_2 are arbitrary constants. Solving for y we have that

$$y = -(1/5)\sin 2t + C_1 e^{-t} + C_2 e^t$$

(b). Writing the ODE $y'' - y' - 2y = \cos t$ in operator form $(D^2 - D - 2)[y] = \cos t$, we notice that the two characteristic roots are $r_1 = 2$, $r_2 = -1$. Factoring the operator, the ODE becomes $(D - 2)(D + 1)[y] = \cos t$. If we put $v = (D + 1)[y]$, then v satisfies the first-order ODE $(D - 2)[v] = \cos t$. Solving for v we multiply the ODE $v' - 2v = \cos t$ by the integrating factor e^{-2t} to obtain $(ve^{-2t})' = e^{-2t} \cos t$. Using Table 1.3.1 (A Short Table of Antiderivatives) we have, after integrating, that $ve^{-2t} = (1/5)e^{-2t}(-2\cos t + \sin t) + C_1$, where C_1 is an arbitrary real. Solving for v, we next have to solve the ODE $(D + 1)[y] = v$, that is, $y' + y = (1/5)(-2\cos t + \sin t) + C_1 e^{2t}$. Multiplying through by the integrating factor e^t, the ODE becomes $(ye^t)' = -(2/5)e^t \cos t + (1/5)e^t \sin t + C_1 e^{3t}$. Using Table 1.3.1 again and integrating we have

$$ye^t = -(2/5)(1/2)e^t(\cos t + \sin t) + (1/5)(1/2)e^t(\sin t - \cos t) + (C_1/3)e^{3t} + C_2$$

where C_1 and C_2 are arbitrary reals. Solving for y, and replacing $C_1/3$ by just C_1 (which is just as arbitrary as $C_1/3$), we have

$$y = -(1/10)\sin t - (3/10)\cos t + C_1 e^{2t} + C_2 e^{-t}$$

where C_1 and C_2 are arbitrary reals.

(c). Writing the ODE $y'' + 2y' + y = e^{-t}$ as $(D^2 + 2D + 1)[y] = e^{-t}$ and factoring the operator we have $(D+1)^2[y] = e^{-t}$. So any solution v of $(D+1)[v] = e^{-t}$ produces a solution $y(t)$ of the original ODE if $(D+1)[y] = v$. First let's find all functions $v(t)$ which solve $v' + v = e^{-t}$. Multiplying through by the integrating factor e^t, this ODE becomes $(ve^t)' = 1$. Integrating we have $ve^t = t + C_1$, so $v = C_1 e^{-t} + te^{-t}$. Now we look at the ODE $(D+1)[y] = C_1 e^{-t} + te^{-t}$ or $y' + y = C_1 e^{-t} + te^{-t}$. Multiplying through by the integrating factor e^t, this ODE becomes $(ye^t)' = C_1 + t$. Integrating, we obtain $ye^t = C_1 t + t^2/2 + C_2$. Solving for y we have

$$y = (C_1 t + C_2)e^{-t} + (t^2/2)e^{-t}$$

where C_1 and C_2 are arbitrary reals is the general solution of the original ODE.

9. If y solves the ODE $(D-1)(D+2)(D-3)[y] = 0$, then $v = (D+2)(D-3)[y]$ solves the first-order ODE $(D-1)[v] = 0$. Solving this ODE we obtain $v = C_1 e^t$, where C_1 is an arbitrary constant. Now if y solves $(D+2)(D-3)[y] = v$, then $w = (D-3)[y]$ solves the ODE $(D+2)[w] = v = C_1 e^t$. Now let's solve $w' + 2w = C_1 e^t$. Multiplying through by the integrating factor e^{2t}, this ODE becomes $(we^{2t})' = C_1 e^{3t}$. Integrating, we have $we^{2t} = (C_1/3)e^{3t} + C_2$, where C_2 is an arbitrary real. So we may just as well write $w = C_1 e^t + C_2 e^{-2t}$ where we have replaced $C_1/3$ by C_1 (which is just as arbitrary as $C_1/3$). Finally, we solve the ODE $(D-3)[y] = w = C_1 e^t + C_2 e^{-2t}$ for y. Write the ODE as $y' - 3y = C_1 e^t + C_2 e^{-2t}$ and multiply through by the integrating factor e^{-3t} to get $(ye^{-3t})' = C_1 e^{-2t} + C_2 e^{-5t}$. Integration gives us that $ye^{-3t} = -(C_1/2)e^{-2t} - (C_2/5)e^{-5t} + C_3$, where C_3 is an arbitrary constant. Solving for y we get the general solution of the original ODE,

$$y = C_1 e^t + C_2 e^{-2t} + C_3 e^{3t}$$

where we have replaced $-C_1/2$ by C_1 and $-C_2/5$ by C_2 (because C_1 and C_2 are just as arbitrary as $-C_1/2$ and $C_2/5$).

   Notice that this result suggests that Theorem 3.3.1 can be extended to linear ODEs of any order.

10. Let $u(t)$ and $v(t)$ be two solutions of the IVP, $y'' + ay' + by = 0$, $y(t_0) = 0$, $y'(t_0) = 0$, on some t-interval I. From the Closure Property the function $w = u - v$ is also a solution of $y'' + ay' + by = 0$. Now since $w(t_0) = 0$, and $w'(t_0) = 0$ we see that $w(t)$ solves the IVP (10) in the text and so $w = 0$, for all t in I. It follows that $0 = u(t) - v(t)$, so $u(t) = v(t)$ for all t in I and the claim is established.

There is another approach to finding a general solution formula for the constant-coefficient ODE $P(D)[y] = 0$, which will be especially useful when the coefficients of $P(D)$ are not constants. Let's see how this approach goes.

Background Material: Another General Solution Formula Approach

From the General Solution Theorem 3.3.1 we saw that if the solution pair $\{y_1, y_2\}$ is cleverly chosen, then all solutions of the ODE $P(D)[y] = 0$ are given by taking all linear combinations of these solutions. It turns out that there are infinitely many solution pairs $\{y_1, y_2\}$ that will also do the job. To see this, suppose that $\{y_1(t), y_2(t)\}$ is a pair of solutions of the ODE $P(D)[y] = 0$. If c_1 and c_2 are any constants, then from the Closure Property we know that the linear combination $c_1 y_1 + c_2 y_2$ is also a solution of that ODE.

Now if $z(t)$ is any solution at all, then we claim that there are unique constants $\tilde{c}_1$ and $\tilde{c}_2$ such that $z = \tilde{c}_1 y_1 + \tilde{c}_2 y_2$, provided that the determinant

$$W(t) = \det \begin{bmatrix} y_1(t) & y_2(t) \\ y_1'(t) & y_2'(t) \end{bmatrix} = y_1(t)y_2'(t) - y_1'(t)y_2(t) \tag{i}$$

is nonzero at a point t_0 (any point t_0 will do). The function $W(t)$ is called the *Wronskian* of the solution pair $\{y_1, y_2\}$ and is sometimes written as $W[y_1, y_2](t)$ to emphasize the dependence on the solution pair.

Suppose now that t_0 is a point where $W(t_0) \neq 0$. Let's look at the solvability of the algebraic equations

$$\begin{align} c_1 y_1(t_0) + c_2 y_2(t_0) &= z(t_0) \\ c_1 y_1'(t_0) + c_2 y_2'(t_0) &= z'(t_0) \end{align} \tag{ii}$$

for the constants c_1 and c_2. The determinant of this algebraic system is $W[y_1, y_2](t_0)$, the Wronskian of the solution pair $\{y_1, y_2\}$ evaluated at t_0. Since by assumption $W[y_1, y_2](t_0) \neq 0$, the algebraic system (ii) has a unique solution $\tilde{c}_1, \tilde{c}_2$. Now put

$$w(t) = \tilde{c}_1 y_1(t) + \tilde{c}_2 y_2(t) - z(t)$$

and notice that $w(t)$ solves IVP (10), and so $w(t)$ is the zero function. This implies that $z(t) = \tilde{c}_1 y_1(t) + \tilde{c}_2 y_2(t)$, and so $z(t)$ is indeed a linear combination of $y_1(t)$ and $y_2(t)$. It follows that

$$y = c_1 y_1 + c_2 y_2, \quad c_1, c_2 \text{ arbitrary constants}$$

is the general solution of $P(D)[y] = 0$.

So apparently any solution pair whose Wronskian is nonzero at some (any) point is very important. We need a name for such solution pairs:

❖ **Basic Solution Set**. The pair of solutions $\{y_1, y_2\}$ of the ODE $y'' + ay' + by = 0$ is called a *basic solution set* if there is a point t_0 such that the Wronskian $W[y_1, y_2](t_0) \neq 0$.

Actually, the Wronskian of a solution pair is either always zero or never zero.

So the general solution of $P(D)[y] = 0$ is given by $y = c_1 y_1 + c_2 y_2$, for arbitrary constants c_1 and c_2, if $\{y_1, y_2\}$ is a basic solution set. As we show next, basic solution sets are easy to construct if the roots of the characteristic polynomial $r^2 + ar + b$ are real.

Let's examine the solution pairs that appear in the general solution to the ODE $y'' + ay' + by = 0$. If the roots r_1 and r_2 of the characteristic polynomial satisfy $r_1 \neq r_2$, we have the solution pair $y_1 = e^{r_1 t}$ and $y_2 = e^{r_2 t}$. The Wronskian

$$W[y_1, y_2] = e^{r_1 t}(e^{r_2 t})' - (e^{r_1 t})'(e^{r_2 t})$$

$$= (r_2 - r_1)e^{(r_1 + r_2)t}$$

is always nonzero, and so this solution pair is basic.

If $r_1 = r_2$, we have the solution pair $y_1 = e^{r_1 t}$ and $y_2 = te^{r_1 t}$. The Wronskian

$$W[y_1, y_2] = e^{r_1 t}(te^{r_1 t})' - (e^{r_1 t})'(te^{r_1 t})$$

$$= e^{r_1 t}(e^{r_1 t} + r_1 te^{r_1 t}) - (r_1 e^{r_1 t})(te^{r_1 t})$$

$$= e^{2r_1 t}$$

is never zero, and so this solution pair is also basic.

3.4 Undriven Constant Coefficient Linear ODEs, II

Suggestions for preparing a lecture

Topics: Finding all solutions of the constant-coefficient ODE $y'' + ay' + by = 0$, characteristic polynomial, getting real-valued solutions from complex-valued solutions.

Remarks: It's important for students to master the technique of extracting real-valued solutions from complex-valued ones. This concept comes up again in chapters 4 and 7 in a significant way.

Making up a problem set

Problem 1, one or two parts of 2 (part **(e)** may be a bit difficult for most students), one or two parts of 3 or 4, 5 or 6, one part of 7 or 8.

Comments

We characterize the solution set of the second-order, constant coefficient, undriven, linear ODE, no matter whether the characteristic roots are real or complex. The differentiation operator is generalized to act on complex-valued functions of a real variable. Example 3.4.3 shows how to get real solutions of a real ODE from complex solutions. We like to do a lot of graphing of solution curves, especially with $y(t_0)$ fixed, but $y'(t_0)$ varying, as initial data. This highlights once more that solution curves of normalized second-order ODEs require two items of initial data to determine a unique solution, that solution curves of a second-order ODE can intersect with impunity, and that one can imagine each point (t_0, y_0) in the ty-plane as a source of solution curves emerging from the point in all directions (except vertical). Problems 5 and 6 are particularly important in the later applications because they give the conditions for stability or asymptotic stability (or as engineers might say in the latter case, all solutions are "transients").

---

1. The general solution of $P(D)[y] = 0$ is given in each case.

(a). Since the characteristic polynomial of $P(D) = D^2 - D - 2$ is $r^2 - r - 2$ whose roots are -1 and 2, we have $y = c_1 e^{-t} + c_2 e^{2t}$, where c_1 and c_2 are arbitrary real constants.

(b). The characteristic polynomial of $D^2 - 4D + 5$ is $r^2 - 4r + 5$ whose roots are $2 \pm i$, so we take as a basic solution set $\{\text{Re}[e^{(2+i)t}], \text{Im}[e^{(2-i)t}]\}$. So, $y = e^{2t}(c_1 \cos t + c_2 \sin t)$ gives the real-valued solutions, where c_1 and c_2 are arbitrary real constants.

(c). The characteristic polynomial of $D^2 - 4D + 4$ is $r^2 - 4r + 4$ which has a double root $r = 2$. The general solution is $y = (c_1 + c_2 t)e^{2t}$, where c_1 and c_2 are arbitrary real constants.

2. In each case apply Theorem 3.4.2 or Theorem 3.4.1 in reverse; that is, if $y = e^{r_1 t}$ is the solution of the constant-coefficient ODE $P(D)[y] = 0$, then r_1 is a root of the characteristic polynomial of $P(D)$. So, if r_1 and r_2, $r_1 \neq r_2$ are both roots, then $(r - r_1)(r - r_2)$ is the characteristic polynomial and $P(D)$ is $(D - r_1)(D - r_2) = D^2 - (r_1 + r_2)D + r_1 r_2$. Proceed similarly with double or complex roots.

(a). Let $r_1 = -5$, $r_2 = -2$ so the characteristic polynomial is $(r+5)(r+2) = r^2 + 7r + 10$. The ODE is $y'' + 7y' + 10y = 0$.

(b). Here $r = 4$ must be a double root, the characteristic polynomial is $(r - 4)^2 = r^2 - 8r + 16$, and the ODE is $y'' - 8y + 16 = 0$.

(c). In this case a pair of complex roots $-4 \pm i$ must be involved. The characteristic polynomial is $(r - r_1)(r - r_2) = (r + 4 - i)(r + 4 + i) = r^2 + 8r + 17$. The ODE is $y'' + 8y' + 17y = 0$.

(d). Here the roots are $3 \pm 5i$ and the characteristic polynomial is $(r - (3 + 5i))(r - (3 - 5i)) = r^2 - 6r + 34$. The ODE is $y'' - 6y' + 34y = 0$.

(e). Here we extend the ideas of this section to what must be a fourth-order constant coefficient, linear ODE. The polynomial has roots i (because of the term $\cos t$) and $2 - i$ (because of the term $e^{(2-i)t}$). Since the characteristic polynomial has real coefficients, the complex conjugates of these roots must also be roots, so we need a polynomial with roots $\pm i$ and $2 \pm i$. So, the polynomial is $(r + i)(r - i)(r - 2 - i)(r - 2 + i) = (r^2 + 1)(r^2 - 4r + 5)$. The ODE is $(D^2 + 1)(D^2 - 4D + 5)[y] = y'''' - 4y''' + 6y'' - 4y' + 5y = 0$. In terms of the hint given in the statement of the problem, $P(D) = D^2 + 1$ and $Q(D) = D^2 - 4D + 5$.

3. In each part, solution curves are plotted in Graph 1, with the corresponding orbits plotted in Graph 2; c_1 and c_2 are arbitrary real constants.

(a). The characteristic polynomial for $y'' + y' = (D^2 + D)[y] = 0$ is $r^2 + r$, which factors to $(r + 1)r$. Thus, solutions of the ODE are given by e^{-t}, $e^{0 \cdot t}$, and the linear combinations of these functions. That is, the real-valued solutions are given by $y = c_1 e^{-t} + c_2$. See Figs. 3(a). From the figures and the solution formulas we see that as $t \to +\infty$, $y(t) \to c_2$, while (if $c_1 \neq 0$), as $t \to -\infty$, $y(t) \to \pm\infty$ depending on the sign of c_1. The orbits are slanted rays that approach the y-axis in the yy'-plane as $t \to +\infty$.

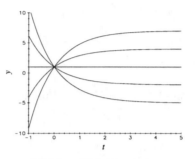

Problem 3(a), Graph 1.

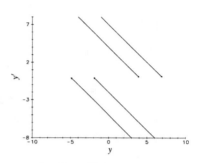

Problem 3(a), Graph 2.

(b). The characteristic polynomial for $y'' + 2y' + 65y = (D^2 + 2D + 65)[y] = 0$ is $r^2 + 2r + 65$ whose roots are are $-1 \pm 8i$. Solutions of the ODE are given by $y_1 = e^{(-1+8i)t} = e^{-t}(\cos 8t + i\sin 8t)$, $y_2 = e^{(-1-8i)t} = e^{-t}(\cos 8t - i\sin 8t)$, and the linear combinations of these functions. So the real-valued solutions are given by $y = e^{-t}(c_1 \cos 8t + c_2 \sin 8t)$. See Fig. 3(b), Graph 1. From the formula and the graphs we see that solutions are decaying oscillations with sinusoids of period $\pi/4$ that tend to 0 as $t \to +\infty$, but become unbounded as $t \to -\infty$. The orbits in the yy'-plane [Fig. 3(b), Graph 2] are spirals that wind clockwise toward the point $y = y' = 0$ as $t \to +\infty$.

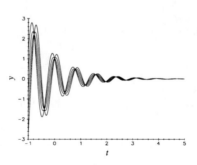

Problem 3(b), Graph 1.

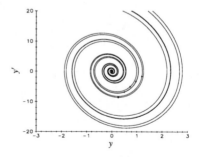

Problem 3(b), Graph 2.

(c). The characteristic polynomial of the ODE $y'' + 3y' + 2y = (D^2 + 3D + 2)[y] = 0$ is $r^2 + 3r + 2$ which factors to $(r + 1)(r + 2)$. Solutions of the ODE are given by e^{-t}, e^{-2t}, and the linear combinations of these functions. So the real-valued solutions are given by $y = c_1 e^{-t} + c_2 e^{-2t}$. See Fig. 3(c), Graph 1. From the formula and the pictures as $t \to +\infty$ solutions decay to 0, but as $t \to -\infty$, solutions become unbounded. The orbits tend to the origin of the yy'-plane as $t \to +\infty$ [Fig. 3(c), Graph 2].

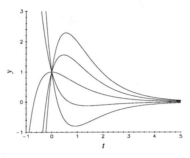

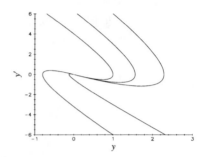

Problem 3(c), Graph 1.   Problem 3(c), Graph 2.

(d). The characteristic polynomial of the ODE $y'' + 10y = (D^2 + 10)[y] = 0$ is $r^2 + 10$ whose roots are $\pm i\sqrt{10}$. Solutions of the ODE are given by $e^{\pm i\sqrt{10}t} = \cos\sqrt{10}t \pm i\sin\sqrt{10}t$, and the linear combinations of these functions. So the real-valued solutions are given by $y = c_1\cos\sqrt{10}t + c_2\sin\sqrt{10}t$. See Fig. 3(d), Graph 1. Solutions are periodic oscillations of period $2\pi/\sqrt{10}$, and the orbits in the yy'-plane are ovals (i.e., cycles) [Fig. 3(d), Graph 2].

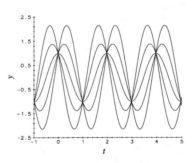

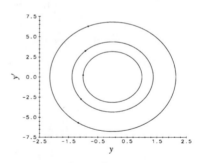

Problem 3(d), Graph 1.   Problem 3(d), Graph 2.

(e). The characteristic polynomial of the ODE $y'' - y/9 = (D^2 - 1/9)[y] = 0$ is $r^2 - 1/9$ which factors to $(r + 1/3)(r - 1/3)$. Solutions of the ODE are given by $e^{-t/3}$, $e^{t/3}$, and the linear combinations of these functions. The real-valued solutions are given by $y = c_1e^{-t/3} + c_2e^{t/3}$. See Fig. 3(e), Graph 1. The solution curves become unbounded as $t \to \pm\infty$ if c_1 and c_2 are nonzero. The time span used here is fairly short, so we only see arcs of orbits (in some cases) [Fig. 3(e), Graph 2] that would move out of the rectangle if the time span, both forward and backward, were longer. In Fig. 3(e), Graph 3 we have enlarged the xy-rectangle and lengthened the time span, so we can really see what is going on.

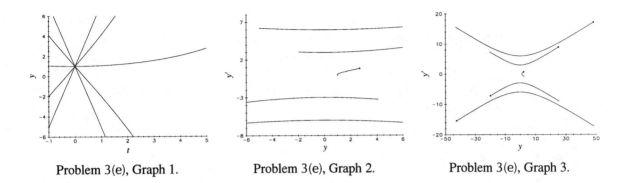

Problem 3(e), Graph 1.　　　　　Problem 3(e), Graph 2.　　　　　Problem 3(e), Graph 3.

(f). The characteristic polynomial of the ODE $y'' - 3y'/4 + y/8 = (D^2 - 3D/4 + 1/8)[y] = 0$ is $r^2 - 3r/4 + 1/8$ which factors to $(r - 1/2)(r - 1/4)$. Solutions of the ODE are given by $e^{t/2}$, $e^{t/4}$, and linear combinations of these functions. That is, the real-valued solutions are given by $y = c_1 e^{t/2} + c_2 e^{t/4}$, where c_1 and c_2 are any real constants. See Fig. 3(f), Graph 1. As t increases to infinity the solutions become unbounded, but as $t \to -\infty$ both $y(t)$ and $y'(t) \to 0$. Figure 3(f), Graph 2 isn't very revealing, so we extend the time span to $-20 \le t \le 5$ and enlarge the xy-screen in Fig. 3(f), Graph 3 and we get a pretty good picture of orbital behavior.

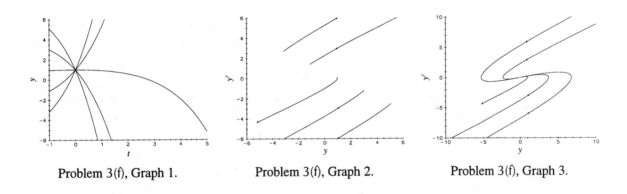

Problem 3(f), Graph 1.　　　　　Problem 3(f), Graph 2.　　　　　Problem 3(f), Graph 3.

4.　(a). Since the characteristic polynomial for $y'' + 2y' + 2y = (D^2 + 2D + 2)[y] = 0$ is $r^2 + 2r + 2$ whose roots are $-1 \pm i$, the general real-valued solution is $y = e^{-t}(c_1 \cos t + c_2 \sin t)$. The initial data $y(0) = 1$, $y'(0) = 0$ are used to find c_1 and c_2: $c_1 = 1$ and $c_2 = 1$. The solution is $y = e^{-t}(\cos t + \sin t)$. See Fig. 4(a). As $t \to +\infty$, we have that $y(t) \to 0$ in the fashion of a decaying sinusoid, but the amplitudes are so small it is hard to see the oscillations.

(b). Since the characteristic polynomial for $y'' + 4y' + 4y = (D^2 + 4D + 4)[y] = 0$ is $r^2 + 4r + 4$ which has the double root -2, the general real-valued solution is $y = e^{-2t}(c_1 + c_2 t)$. The initial data $y(1) = 2$, $y'(1) = 0$ are used to find c_1 and c_2: $c_1 = -2e^2$ and $c_2 = 4e^2$. The solution is $y = e^{2(1-t)}(4t - 2)$. See Fig. 4(b). As $t \to +\infty$, the solution decays to zero.

(c). The characteristic polynomial for $y'' + y' - 6y = (D^2 + D - 6)[y] = 0$ is $r^2 + r - 6$ which has the roots -3 and 2. The general real-valued solution of the ODE is $y = c_1 e^{-3t} + c_2 e^{2t}$. The initial data $y(-1) = 1$, $y'(-1) = -1$ imply that $c_1 e^3 + c_2 e^{-2} = 1$, and $-3c_1 e^3 + 2c_2 e^{-2} = -1$. So $c_1 = 3e^{-3}/5$, $c_2 = 2e^2/5$ and $y = 3e^{-3(t+1)}/5 + 2e^{2(t+1)}/5$. See Fig. 4(c). As $t \to +\infty$ the second term in the above solution formula tends to ∞ and the first tends to 0.

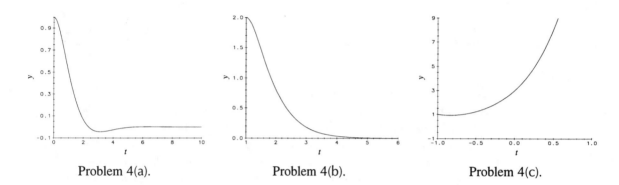

Problem 4(a).   Problem 4(b).   Problem 4(c).

5.   **(a).** If $a > 0$ and $b > 0$, then the roots of the characteristic polynomial $r^2 + ar + b$ are $-a/2 \pm (a^2/4 - b)^{1/2}$. If $(a/2)^2 > b$, then both roots r_1 and r_2 are negative since $((a/2)^2 - b)^{1/2} < a/2$. In this case, the general solution $y = c_1 e^{r_1 t} + c_2 e^{r_2 t} \to 0$ as $t \to +\infty$. If $b = (a/2)^2$, then there is a double root $-a/2$ and the general solution is $y = c_1 e^{r_1 t} + c_2 t e^{r_1 t}$ and again $y \to 0$ as $t \to +\infty$ since r_1 is negative and $e^{r_1 t} \to 0$ faster than $t \to +\infty$ (apply L'Hopital's Rule to $t/e^{-r_1 t}$). If $b > (a/2)^2$, then the roots $\alpha \pm i\beta$ are imaginary with real part $-a/2 = \alpha$, which is negative. The general solution $y = e^{\alpha t}(c_1 \cos \beta t + c_2 \sin \beta t) \to 0$ as $t \to +\infty$ because $e^{\alpha t} \to 0$, while $c_1 \cos \beta t + c_2 \sin \beta t$ oscillates but remains bounded.

(b). Suppose all solutions $y(t) \to 0$ as $t \to +\infty$. Then the real parts of r_1 and r_2 must both be negative as can be seen by inspecting the three formulas in (14). Let $r^2 + ar + b = (r - r_1)(r - r_2) = r^2 - (r_1 + r_2)r + r_1 r_2$. Now either r_1 and r_2 are both real, or else $r_1 = \alpha + i\beta$, $r_2 = \alpha - i\beta$, α and $\beta \neq 0$ real. In either case $a = -(r_1 + r_2)$, $b = r_1 r_2$. In the first case, $a > 0$ and $b > 0$ since r_1 and r_2 are negative. In the second case $a > 0$ and $b > 0$ since $a = -(r_1 + r_2) = -2\alpha$, where α is negative and $b = r_1 r_2 = \alpha^2 + \beta^2 > 0$.

6.   **(a).** The roots of the characteristic polynomial $r^2 + ar + b$ are r_1 and r_2 where $r_1 + r_2 = -a$ and $r_1 r_2 = b$ since $(r - r_1)(r - r_2) = r^2 - (r_1 + r_2)r + r_1 r_2$. Suppose first that $a > 0$ and $b > 0$. Then the roots have negative real parts so that the product of the roots is negative and real but the sum is positive and real. So solutions are linear combinations of functions of the form $e^{r_i t}$, $e^{\alpha t} \cos \beta t$, $e^{\alpha t} \sin \beta t$, or $t e^{rt}$, where r_i, α, r are negative real numbers and $\beta \neq 0$. But if $t \geq 0$, then $|e^{r_i t}| \leq 1$, $|e^{\alpha t} \cos \beta t| \leq 1$, $|e^{\alpha t} \sin \beta t| \leq 1$, and $|t e^{rt}| \leq e^{-r}$. [The function $|t e^{rt}|$ achieves its maximum at $t = -1$ with value e^{-r}.] All these possible solutions are positively bounded. Since any linear combination of positively bounded functions is also positively bounded, every solution is positively bounded. [This follows from the Triangle Inequality, $|y(t)| = |c_1 y_1(t) + c_2 y_2(t)| \leq |c_1||y_1(t)| + |c_2||y_2(t)|$.] Now suppose $a = 0$,

$b > 0$. Then, the roots of $r^2 + b$ are $\pm i\sqrt{-b}$, and solutions are given by $y = c_1 e^{i\sqrt{b}t} + c_2 e^{-i\sqrt{b}t}$ and are positively bounded since $|y| \le |c_1||e^{i\sqrt{b}t}| + |c_2||e^{-i\sqrt{b}t}| = |c_1| + |c_2|$. Finally, consider the case $a > 0$, $b = 0$. The roots of $r^2 + ar$ are 0 and $-a$. The solutions are given by $y = c_1 + c_2 e^{-at}$ and $|y| \le |c_1| + |c_2|$, so y is again positively bounded.

(b). Suppose now that all solutions of $y'' + ay' + by = 0$ are positively bounded. Suppose $a < 0$. Then the root $r_1 = -a/2 + ((a/2)^2 - b)^{1/2}$ is positive or has a positive real part since $-a/2 > 0$ and the square root term is nonnegative or imaginary. The corresponding solution contains a term $e^{\alpha t}$ where $\alpha > 0$, and this term $\to \infty$ as $t \to +\infty$. The contradiction implies that a must be ≥ 0. Suppose $b < 0$. Since $r_1 r_2 = b$, where r_1 and r_2 are the roots of $r^2 + ar + b$, we see that r_1 and r_2 must be real and of opposite sign, say $r_1 > 0$, $r_2 < 0$. The solution $e^{r_1 t} \to \infty$ as $t \to \infty$. This contradiction shows that $b \ge 0$. Now suppose a and b are both zero. The ODE is $y'' = 0$. The solution $y = t$ is unbounded. This contradiction shows that not both a and b can be 0.

7. **(a).** Using the quadratic formula, we find that the roots of the characteristic polynomial $r^2 + ir + 2$ are $-2i$ and i. The general solution of the ODE $y'' + iy' + 2y = 0$ is $y = k_1 e^{-2it} + k_2 e^{it}$, where k_1 and k_2 are any complex constants.

   (b). The function $k_1 e^{-2it} + k_2 e^{it}$ is real if and only if one summand is the conjugate of the other for all t. But if $k_1 e^{-2it} = \overline{k_2 e^{it}} = \bar{k}_2 e^{-it}$, then $k_1/\bar{k}_2 = e^{it}$ for all t. This is impossible, since the value of e^{it} is not constant. So there are no nontrivial real solutions.

   (c). Let z be a root of $r^2 + i$. Then $z = r_0 e^{i\theta}$ and $z^2 = r_0^2 e^{2i\theta} = -i = e^{(3\pi/2 + 2n\pi)i}$. Matching magnitudes and angles, we have that $r_0 = 1$ and $2\theta = 3\pi/2 + 2n\pi$. The only two distinct angles correspond to $n = 0, 1$. So, $r_0 = 1$, $\theta_0 = 3\pi/4$, $\theta_1 = 7\pi/4$, and the two roots of $r^2 + i$ are $e^{3\pi i/4} = (-1 + i)/\sqrt{2}$ and $e^{7\pi i/4} = (1 - i)/\sqrt{2}$. The general solution is $y = k_1 e^{(-1+i)t/\sqrt{2}} + k_2 e^{(1-i)t/\sqrt{2}}$, where k_1 and k_2 are any complex constants.

8. **(a).** The characteristic polynomial is $r^2 + (1 + i)r + i$, which factors to $(r + 1)(r + i)$. Solutions of the ODE are given by e^{-t}, $e^{-it} = \cos t - i\sin t$, and the linear combinations of these functions. The real-valued solutions are given by $y = ce^{-t}$, where c is any real constant.

   (b). The characteristic polynomial is $r^2 + (-1 + 2i)r - 1 - i$, which factors to $(r + i)(r - 1 + i)$. Solutions are e^{-it} and $e^{(1-i)t}$, and solutions are all imaginary, except for the trivial solution $y = 0$.

Background Material: Another Approach to Finding Solution Formulas

In the background material at the end of Section 3.3 of this Student Resource Manual, we showed that if we could find a basic solution pair $\{y_1, y_2\}$ for the ODE $y'' + ay' + by = 0$, then the general solution could be written as

$$y = c_1 y_1 + c_2 y_2$$

where c_1 and c_2 were arbitrary constants. In that section we assumed that the solutions y_1 and y_2 were real-valued, so the arbitrary constants were reals. It will come as no surprise that the notion of a basic solution pair extends to complex-valued solutions, but with the difference that the constants c_1 and c_2 this time are arbitrary complex numbers. For example, in Theorem 3.4.2 note that if r_1 and r_2 are the characteristic roots, the ODE becomes

$$(D - r_1)(D - r_2)[y] = 0$$

so $y_1 = e^{r_1 t}$ and $y_2 = e^{r_2 t}$ are solutions. If $r_1 \neq r_2$ (r_1 and r_2 need not be real), then the Wronskian $W[e^{r_1 t}, e^{r_2 t}] = (r_2 - r_1)e^{(r_1 + r_2)t} \neq 0$, so $\{e^{r_1 t}, e^{r_2 t}\}$ is a basic solution pair. If $r_1 = r_2$, notice that $(D - r_1)^2[te^{r_1 t}] = 0$, so we have the solution pair $y_1 = e^{r_1 t}$, $y_2 = te^{r_1 t}$. Since the Wronskian $W[e^{r_1 t}, te^{r_1 t}] = e^{2r_1 t} \neq 0$, it follows that $\{e^{r_1 t}, te^{r_1 t}\}$ is a basic solution pair in this case.

The case $r_1 = \bar{r}_2 = \alpha + i\beta$ in Theorem 3.4.1 merits a special comment. In this case we have the two complex-valued solutions $e^{(\alpha + i\beta)t}$ and $e^{(\alpha - i\beta)t}$. Now from text formula (8),

$$e^{(\alpha + i\beta)t} = e^{\alpha t} \cos \beta t + i e^{\alpha t} \sin \beta t$$

Since $P(D) = D^2 + aD + b$ is a linear operator we see that

$$0 = P(D)[e^{\alpha t} \cos \beta t + i e^{\alpha t} \sin \beta t]$$
$$= P(D)[e^{\alpha t} \cos \beta t] + i P(D)[e^{\alpha t} \sin \beta t]$$

So *if the coefficients a and b are real* we conclude that

$$P(D)[e^{\alpha t} \cos \beta t] = 0, \quad \text{and} \quad P(D)[e^{\alpha t} \sin \beta t] = 0$$

so $y_1 = e^{\alpha t} \cos \beta t$ and $y_2 = e^{\alpha t} \sin \beta t$ comprise a real-valued solution pair. Since the Wronskian

$$W[e^{\alpha t} \cos \beta t, e^{\alpha t} \sin \beta t] = \beta e^{2\alpha t} \neq 0$$

it follows that $\{e^{\alpha t} \cos \beta t, e^{\alpha t} \sin \beta t\}$ is a basic solution pair for real-valued solutions.

In practice we prefer to use the basic solution pair approach, mostly because it generalizes to linear undriven ODEs with *nonconstant* coefficients (we show this in Section 3.7), and to higher order linear ODEs as well.

From Theorem 3.4.1, solutions of the constant-coefficient ODE $P(D)[y] = 0$ have oscillatory long-term behavior only if the roots of the characteristic equation $P(r) = 0$ are not real.

3.5   Periodic Solutions and Simple Harmonic Motion

Suggestions for preparing a lecture

Topics: A review of periodic functions with all the terminology associated with them, discussion of when the sum of two periodic functions is periodic (and how to find the period of the sum), the simple harmonic oscillator and how to write all solutions as a single sinusoid, a little on problems in tracking periodic functions numerically.

Remarks: All the properties of periodic functions as solution of ODEs are gathered together here in one place.

Making up a problem set

Problems 1, 3, 4. Problem 5 is a favorite one of ours, but it would be a challenging one for most students.

Comments

Periodic solutions play an important role in the study of ODEs, so we thought we should give them a section of their own. Readers probably have seen periodic functions before, but they may have forgotten the terminology associated with them. It is especially important to read the discussion of when the sum of two periodic functions is periodic (and how to find the period of the sum). The simple harmonic motion examples (Examples 3.5.1 and 3.5.3) should be read carefully. Example 3.5.2 brings up the important point that when tracking a periodic solution (or an oscillating solution for that matter) with a numerical solver, decreasing the number of plotted points per unit time causes an interesting modulation effect on the output (aliasing). Not much is made of this phenomenon other than to say that when using a numerical solver it's important to try various numbers of plotted points before accepting the results. Incidentally, the aliasing effect is the reason stage-coach wheels in the movies often seem to rotate slowly or even rotate backwards. It all depends on the frame rate of the movie cameras.

---

1. **(a).** If $f(t)$ has (fundamental) period T, then T is the smallest positive number such that for any time t, $f(t + T) = f(t)$. Let $g(t) = f(\omega t)$ and suppose that $g(t)$ has (fundamental) period T_1; that is, T_1 is the smallest positive number such that $g(t + T_1) = g(t)$. Then $f(\omega(t + T_1)) = f(\omega t)$ by the definition of $g(t)$, so $f(\omega t + \omega T_1) = f(\omega t)$. But the smallest positive value of T_1 for which this holds satisfies $\omega T_1 = T$, since T is the period of f. So, $T_1 = T/\omega$ is the period of $g(t) = f(\omega t)$.

   (b). If $f(t)$ has period T and $g(t)$ has period S then we know that for any time t, $f(t + T) = f(t)$ and $g(t + S) = g(t)$. If $h(t) = f(t) + g(t)$ is periodic then there is a k such that $h(t + k) = h(t)$, that is, such that $f(t + k) + g(t + k) = f(t) + g(t)$ where we have used the definition of $h(t)$. The only way this equality can hold for *all* t is if $f(t + k) = f(t)$ and $g(t + k) = g(t)$. If we choose m and n to be the smallest positive integers such that $k = mT = nS$ (which is always possible if T/S is rational since we can write $T/S = m/n$ as a reduced fraction), then $f(t + k) = f(t + mT) = f(t)$ and $g(t + k) = g(t + nS) = g(t)$ since we are simply evaluating the functions over several periods. This means that $f(t + k) + g(t + k) = f(t) + g(t)$, so $h(t + k) = h(t)$. Since we chose m and n to be as small as possible, k is as small as possible as well. So the (fundamental) period of $h(t)$ is k.

   (c). See Fig. 1(c) for the graphs of $\sin 2t$ (long dashes) and $\sin 5t$ (short dashes), with periods $T = \pi$ and $S = 2\pi/5$, respectively. Using the notation of part **(b)**, we have $m = 2$ and $n = 5$ so that $k = 2(\pi) = 5(2\pi/5) = 2\pi$. The solid curve is the graph of $\sin 2t + \sin 5t$, which does indeed have period 2π.

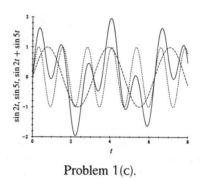

Problem 1(c).

(d). Suppose the ratio T/S of the periods of f and g is irrational. Consider the functions as starting out at $t = 0$ "in phase" with each other. They will never again be exactly in phase, since no integer multiple of T will exactly equal an integer multiple of S. That is, there is no value k such that $f(t + k) = f(t)$ for all t and $g(t + k) = g(t)$ for all t simultaneously. We will never have that $f(t + k) + g(t + k) = f(t) + g(t)$ for all t, so $h(t) = f(t) + g(t)$ is not periodic.

2. Suppose that $(D^2 + aD + b)[y] = 0$ has a periodic solution. We will look for restrictions of a and b. The characteristic polynomial of $P(D) = D^2 + aD + b$ is $r^2 + ar + b$. The roots are $(-a \pm \sqrt{a^2 - 4b})/2$. Using Theorem 3.4.1 we see that if the roots are real $(a^2 - 4b \geq 0)$, then we get $C_1 e^{r_1 t} + C_2 e^{r_2 t}$ or $C_1 e^{r_1 t} + C_2 t e^{r_1 t}$, both of which are clearly not periodic. If the roots are complex the general solution is $C_1 e^{at/2} \cos(\sqrt{4b - a}/2) + C_2 e^{at/2} \sin(\sqrt{4b - a}/2)$. If $a \neq 0$ then $e^{at/2}$ increases or decreases with time, so the solution cannot be periodic. We must have $a = 0$, in which case we get $C_1 \cos \sqrt{b} + C_2 \sin \sqrt{b}$ which, for $b > 0$, is the sum of two periodic functions with the same period. Therefore, $a = 0$ and $b > 0$.

    Now suppose that $P(D) = D^2 + b$, $b > 0$. The roots are $\pm ib$, so the general solution is $y(t) = C_1 \cos bt + C_2 \sin bt$, which is periodic.

3. To solve $(D^2 + 9)[y] = -3 \cos 2t$, first note that the undriven solution is $C_1 \cos 3t + C_2 \sin 3t$. We need only find any particular solution to the driven equation. Guess a solution of the form $y_d = A \cos 2t + B \sin 2t$. Plugging this into the ODE, $(D^2 + 9)[A \cos 2t + B \sin 2t] = -4A \cos 2t + 9A \cos 2t - 4B \sin 2t + 9 \sin 2t = -3 \cos 2t$. We get $5A \cos 2t = -3 \cos 2t$, $5B \sin 2t = 0$. So we must set $A = -3/5$, $B = 0$. The general solution is $y(t) = C_1 \cos 3t + C_2 \sin 3t - (3/5) \cos 2t$, which is periodic.

4. To determine whether the ODE $(D^2 + 4)[y] = \sin 2t$ has any periodic solutions, we will find and examine the general solution. Note that the undriven solution is $C_1 \sin 2t + C_2 \cos 2t$. Now assume a particular solution of the form $y_d = At \sin 2t + Bt \cos 2t$. Plugging this into the ODE, $(D^2 + 4)[At \sin 2t + Bt \cos 2t] = 4A \cos 2t - 4At \sin 2t - 4B \sin 2t - 4Bt \sin 2t + 4A \sin 2t + 4B \cos 2t = \sin 2t$. Equating the coefficients like terms on either side of the equation, we need $4A + 4B = 0$, $-4A - 4B = 0$, and $-4B + 4A = 1$. The first two equations give us the same information, $B = -A$. Plugging this into the third equation, $8A = 1$, so $A = 1/8$ and $B = -1/8$. The general solution of the driven equation is $y(t) =$

$C_1 \sin 2t + C_2 \cos 2t + (1/8)t \sin 2t - (1/8)t \cos 2t$. Resonance occurs no matter what we choose for C_1 and C_2, so there are no periodic solutions.

5. Group Project

3.6 Driven Constant Coefficient Linear ODEs

Suggestions for preparing a lecture

Topics: Polynomial-exponential functions; driven, constant coefficient linear ODEs; their general solutions (use examples); the Method of Undetermined Coefficients.

Remarks: Before assigning a specific problem, look at the solutions and decide whether you want your students to get involved in the particular algebraic intricacies of that problem.

Making up a problem set

We don't have any favorite problems here—they all are straightforward, but involve a lot of algebraic manipulation. The graphs in Problem 9 are visually attractive. We also like Problem 11 because it shows that undetermined coefficients can also be used to find periodic forced oscillations of a first-order linear ODE driven by a sinusoid.

Comments

We try to do two things in this section: first, to get across the idea of the Method of Undetermined Coefficients, and second, to lead the reader through the algebraic intricacies of actually constructing a particular solution. A CAS would be helpful for the algebra. We have included many examples, because that is probably the only way to see how to carry out the undetermined coefficients approach. Although we do some numerical solving/plotting, the geometry of solution graphs is *not* the point of this section. The point is the construction of solution formulas. **Warning:** some problems involve a lot of crank work.

1. Use the Euler formulas, $\cos\theta = (e^{i\theta} + e^{-i\theta})/2$, $\sin\theta = i(e^{-i\theta} - e^{i\theta})/2$.

    (a). $\cos 2t - \sin t = (e^{2it} + e^{-2it})/2 + (e^{it} - e^{-it})i/2 = [i(e^{it} - e^{-it}) + (e^{2it} + e^{-2it})]/2$.

    (b). $t\sin^2 t = t[(e^{-it} - e^{it})i/2]^2 = t[2 - (e^{2it} + e^{-2it})]/4$.

    (c). $t^2 \sin 2t - (1+t)\cos^2 t = t^2 \sin 2t - (1+t)(1 + \cos 2t)/2 = t^2(-e^{2it} + e^{-2it})i/2 - (1+t)[2 + (e^{2it} + e^{-2it})]/4 = -(1+t)/2 - (1 + t + 2it^2)e^{2it}/4 + (2it^2 - t - 1)e^{-2it}/4$.

    (d). $\sin^3 t = (-e^{it} + e^{-it})^3 i^3/8 = i(e^{3it} - 3e^{it} + 3e^{-it} - e^{-3it})/8$.

    (e). $(1-t)e^{it}\cos 3t = (1-t)e^{it}(e^{3it} + e^{-3it})/2 = (1-t)(e^{4it} + e^{-2it})/2$.

    (f). Since $\sin^2 3t = (1 - \cos 6t)/2 = (2 - e^{6it} - e^{-6it})/4$, we have $(i + t - t^2)e^{(3+i)t}\sin^2 3t = (i + t - t^2)e^{(3+i)t}/2 - (i + t - t^2)(e^{(3+7i)t} + e^{(3-5i)t})/4$.

2. **(a).** $\cos 2t - \sin t = \text{Re}[e^{2it} + ie^{it}]$ because $e^{2it} = \cos 2t + i\sin 2t$ and $ie^{it} = i\cos t - \sin t$.

    (b). $t\sin^2 t = \text{Re}[t(1 - e^{2it})/2]$ since $\sin^2 t = (1 - \cos t)/2$ and $\cos t = \text{Re}[e^{2it}]$.

(c). $t^2 \sin 2t - (1+t)\cos^2 t = \text{Re}[it^2 e^{-2it} - (1+t)(1+e^{2it})/2]$ since $\sin 2t = \text{Re}[ie^{-2it}]$ and $\cos^2 t = (1+\cos 2t)/2$, while $\cos 2t = \text{Re}[e^{2it}]$.

3. In finding an annihilator a factor t^k suggests a root of multiplicity $k+1$ for the characteristic polynomial of the annihilator, while a complex exponent suggests a pair of complex conjugate roots.

(a). We want to annihilate $t^2 e^{-it}$. The factor t^2 suggests a root of multiplicity 3; since $(D+i)[e^{-it}] = 0$, we guess and then verify that $(D+i)^3[t^2 e^{-it}] = 0$. Any polynomial operator with $(D+i)^3$ as a factor is an annihilator. So, an annihilator with real coefficients is $[(D+i)(D-i)]^3 = (D^2+1)^3$.

(b). We have to annihilate $te^{-t}\cos 2t$. Note that $te^{-t}\cos 2t = \text{Re}[te^{(-1+2i)t}]$. We need only find an annihilator with real coefficients for $te^{(-1+2i)t}$. Now from the polynomial operator identity (5),

$$(D - (-1+2i))^2[te^{(-1+2i)t}] = e^{(-1+2i)t}D^2[t] = 0$$

So, an annihilator with real coefficients is

$$[(D - (-1+2i))(D - (-1-2i))]^2 = (D^2+2D+5)^2$$

(c). The function to be annihilated is $t + \sin t$. An annihilator with real coefficients for $t + \sin t$ is $D^2(D^2+1)$, since $D^2[t] = 0$ and $(D^2+1)[\sin t] = 0$.

(d). We need to annihilate $\sin t + \cos 2t + t^2$. An annihilator with real coefficients for $\sin t + \cos 2t + t^2$ is $D^3(D^2+1)(D^2+4)$; D^3 annihilates t^2, D^2+1 takes care of $\sin t$, and D^2+4 annihilates $\cos 2t$.

4. If e^t is a solution of the undriven ODE $P_1(D)[y] = 0$ and te^{-3t} is a solution of the undriven ODE $P_2(D)[y] = 0$, then both are solutions of the undriven ODE $P(D)[y] = 0$ where $P(D) = P_1(D)P_2(D)$. This can be easily shown by using the fact that $P(D)[e^t] = P_2(D)P_1(D)[e^t] = P_2(D)[0] = 0$ and similarly $P(D)[te^{-3t}] = P_1(D)P_2(D)[te^{-3t}] = P_1(D)[0] = 0$. So all we need to do is find $P_1(D)$ and $P_2(D)$. We know that $(D-1)[e^t] = 0$. So let $P_1(D) = D - 1$. Because of the t as a coefficient of te^{-3t} we need $P_2(D)$ to have the root -3 with multiplicity 2. So $P_2(D) = (D+3)^2$. So

$$P(D)[y] = (D-1)(D+3)^2[y] = (D^3+5D^2+3D-9)[y] = 0$$

is an appropriate ODE.

5. **(a).** The IVP is $y'' - 4y = 2 - 8t$, $y(0) = 0$, $y'(0) = 5$. Note that $P(D) = (D-2)(D+2)$. A particular solution y_d of $(D-2)(D+2)[y] = 2 - 8t$ has the form

$$y_d = A + Bt$$

Since $(D-2)(D+2)[A + Bt] = -4(A + Bt) = 2 - 8t$, it follows that $A = -1/2$ and $B = 2$. So, a particular solution is $y_d = 2t - 1/2$. The characteristic polynomial $r^2 - 4$ has roots $r = \pm 2$, so the general solution of the undriven equation is $c_1 e^{-2t} + c_2 e^{2t}$. The general solution of the driven ODE is

$$y = c_1 e^{-2t} + c_2 e^{2t} + 2t - 1/2$$

Since $y(0) = 0$, $y'(0) = 5$, the solution of this IVP is

$$y = -e^{-2t}/2 + e^{2t} + 2t - 1/2$$

(b). The IVP is $y'' + 9y = 81t^2 + 14\cos 4t$, $y(0) = 0$, $y'(0) = 3$. The ODE, $y'' + 9y = 81t^2 + 14\cos 4t$, has $P(D) = D^2 + 9$. First we find a particular solution for $(D^2 + 9)[y] = 81t^2$. Let's guess $y_d = At^2 + Bt + C$. Now $(D^2 + 9)[y_d] = (D^2 + 9)[At^2 + Bt + C] = 2A + 9(At^2 + Bt + C) = 9At^2 + 9Bt + 2A + 9C$. Matching coefficients we have $9A = 81$, $9B = 0$, $2A + 9C = 0$. So $A = 9$, $B = 0$, $C = -2$, and $y_d = 9t^2 - 2$. Next, let's find a particular solution for $(D^2 + 9)[y] = 14\cos 4t$. Note that $\mathrm{Re}[14e^{i4t}] = 14\cos 4t$, so we will first look for a particular solution of the ODE $(D^2 + 9)[z] = 14e^{i4t}$ and then take the real part of that solution. Guessing $z_d = Ae^{i4t}$, we see that $(D^2 + 9)[Ae^{i4t}] = A((4i)^2 + 9)e^{i4t} = -7Ae^{4it}$. Matching coefficients we have $-7A = 14$, so $z_d = -2e^{i4t}$ is a particular solution of the z-ODE. Now $y_d = \mathrm{Re}[z_d]$ solves the y-ODE, so $y_d = -2\cos 4t$. So a particular solution of the original ODE is the sum of the two particular solutions just computed: $9t^2 - 2 - 2\cos 4t$. The characteristic polynomial $r^2 + 9$ has roots $r = \pm 3i$, so the general solution of the undriven equation is $C_1 \cos 3t + C_2 \sin 3t$. The general solution of the driven ODE is

$$y = C_1 \cos 3t + C_2 \sin 3t - 2 + 9t^2 - 2\cos 4t$$

Imposing the initial conditions, $y(0) = 0$, $y'(0) = 3$, in order to determine the values of C_1 and C_2, we have $0 = C_1 - 4$, $3C_2 = 3$ so the solution of the given IVP is $y = 4\cos 3t + \sin 3t - 2\cos 4t + 9t^2 - 2$.

(c). The IVP is $y'' + y = 10e^{2t}$, $y(0) = 0$, $y'(0) = 0$. The ODE, $y'' + y = 10e^{2t}$, has $P(D) = D^2 + 1$. A particular solution has the form

$$y_d = Ae^{2t}$$

Insert y_d into the ODE to obtain

$$5Ae^{2t} = 10e^{2t}$$

Matching coefficients, we see that $A = 2$, so $y_d = 2e^{2t}$ is a particular solution. The characteristic polynomial $r^2 + 1$ has roots $\pm i$, so the general solution of the undriven equation is $C_1 \cos t + C_2 \sin t$. The general solution of the driven ODE is

$$y = C_1 \cos t + C_2 \sin t + 2e^{2t}$$

Imposing the initial conditions, $y(0) = 0$, $y'(0) = 0$, we see that $0 = C_1 + 2$, $0 = C_2 + 4$, so the solution of the IVP is

$$y = -2\cos t - 4\sin t + 2e^{2t}$$

(d). The IVP is $y'' - y = e^{-t}(2\sin t + 4\cos t)$, $y(0) = y'(0) = 1$. The ODE, $y'' - y = e^{-t}(\sin t + 4\cos t)$ has $P(D) = D^2 - 1$. A particular solution has the form

$$y_d = Ae^{-t}\sin t + Be^{-t}\cos t$$

Inserting y_d into the ODE, we have

$$(2B - A)e^{-t}\sin t - (2A + B)e^{-t}\cos t = e^{-t}\sin t + 4e^{-t}\cos t$$

Matching coefficients of like terms

$$2B - A = 2, \quad -2A - B = 4$$

we have $A = -2$, $B = 0$, so $y_d = -2e^{-t}\sin t$ is a particular solution. The characteristic polynomial $r^2 - 1$ has roots ± 1, so the general solution of the undriven equation is $C_1 e^t + C_2 e^{-t}$. The general solution of the driven ODE is

$$y = C_1 e^t + C_2 e^{-t} - 2e^{-t}\sin t$$

The initial conditions, $y(0) = 1$, $y'(0) = 1$ imply that

$$1 = C_1 + C_2, \quad 1 = C_1 - C_2 - 2$$

so $C_1 = 2$, $C_2 = -1$. The solution of the IVP is

$$y = 2e^t - e^{-t} - 2e^{-t}\sin t$$

(e). The IVP is $y'' - 3y' + 2y = 8t^2 + 12e^{-t}$, $y(0) = 0$, $y'(0) = 2$. The ODE, $y'' - 3y' + 2y = 8t^2 + 12e^{-t}$, has $P(D) = (D - 1)(D - 2)$. A particular solution has the form

$$y_d = A + Bt + Ct^2 + De^{-t}$$

Inserting y_d into the ODE, we have

$$2A - 3B + 2C + (2B - 6C)t + 2Ct^2 + 6De^{-t} = 8t^2 + 12e^{-t}$$

Matching coefficients of like terms, we have

$$2A - 3B + 2C = 0, \quad 2B - 6C = 0, \quad 2C = 8, \quad 6D = 12$$

so $A = 14$, $B = 12$, $C = 4$, $D = 2$, and $y_d = 14 + 12t + 4t^2 + 2e^{-t}$ is a particular solution. The characteristic polynomial $r^2 - 3r + 2$ has roots 1, 2, so the general solution of the undriven equation is $C_1 e^t + C_2 e^{2t}$. The general solution of the driven ODE is

$$y = C_1 e^t + C_2 e^{2t} + 14 + 12t + 4t^2 + 2e^{-t}$$

Imposing the initial conditions, $y(0) = 0$, $y'(0) = 2$, we have

$$0 = C_1 + C_2 + 16, \quad 2 = C_1 + 2C_2 + 10$$

so $C_1 = -24$, $C_2 = 8$. The IVP has the solution

$$y = 8e^{2t} - 24e^t + 2e^{-t} + 4t^2 + 12t + 14$$

6. In each case y_u is the general real-valued solution of the undriven ODE which we learned how to find in Section 3.3, while y_d is a particular solution of the driven ODE; c_1 and c_2 are arbitrary real constants [complex constants in part **(h)**]. The general solution is $y(t) = y_u + y_d$.

(a). For the ODE $y'' - y' - 2y = (D^2 - D - 2)[y] = 2\sin 2t$, the roots of the characteristic polynomial $r^2 - r - 2$ are $-1, 2$. We have $y_u = c_1 e^{-t} + c_2 e^{2t}$. Let's guess $y_d = A\cos 2t +$

$B \sin 2t$ and so $y_d'' - y_d' - 2y_d = -6A \cos 2t - 6B \sin 2t + 2A \sin 2t - 2B \cos 2t = 2 \sin 2t$. So, $-6A - 2B = 0$ and $2A - 6B = 2$. Solving for A and B, we obtain $A = 1/10$ and $B = -3/10$. Hence, $y = y_u + y_d = c_1 e^{-t} + c_2 e^{2t} + (\cos 2t - 3 \sin 2t)/10$. See Fig. 6(a).

(b). From (a), $y_u = c_1 e^{-t} + c_2 e^{2t}$ for the ODE $y'' - y' - 2y = (D^2 - D - 2)[y] = t^2 + 4t$, since the roots of the characteristic polynomial $r^2 - r - 2$ are $-1, 2$. Let's guess that $y_d = A + Bt + Ct^2$ and so $y_d'' - y_d' - 2y_d = 2C - B - 2A - 2Bt - 2Ct - 2Ct^2 = t^2 + 4t$. So $-2A - B + 2C = 0$, $-2B - 2C = 4$, $-2C = 1$. So $A = 1/4$, $B = -3/2$ and $C = -1/2$, and $y = y_u + y_d = c_1 e^{-t} + c_2 e^{2t} + (1 - 6t - 2t^2)/4$. See Fig. 6(b).

(c). The ODE $y'' - 2y' + y = (D^2 - 2D + 1)[y] = -te^{-t}$ has characteristic polynomial $r^2 - 2r + 1$ with a double root 1, so $y_u = (c_1 + c_2 t)e^t$ and we look for y_d in the form $y_d = t^2(A + Bt)e^t$. Since $y_d'' - 2y_d' + y_d = 2Ae^t + 6Bte^t$, which equals $-te^t$ if $A = 0$ and $B = -1/6$, then $y = (c_1 + c_2 t)e^t - t^3 e^t/6$. See Fig. 6(c).

(d). The ODE $y'' - 2y' + y = (D^2 - 2D + 1)[y] = 2e^t$ has characteristic polynomial $r^2 - 2r + 1$ with a double root 1, so $y_u = (c_1 + c_2 t)e^t$ and we look for y_d in the form $y_d = At^2 e^t$. Then $y_d'' - 2y_d' + y_d = 2Ae^t$, which equals $2e^t$ if $A = 1$. So $y = (c_1 + c_2 t + t^2)e^t$. See Fig. 6(d).

(e). The ODE $y'' + 2y' + y = (D^2 + 2D + 1)[y] = e^t \cos t$ has characteristic polynomial $r^2 + 2r + 1$ with double root -1, so $y_u = (c_1 + c_2 t)e^{-t}$. Since $e^t \cos t = \text{Re}[e^{(1+i)t}]$, we will first seek a particular solution $\tilde{y}_d$ of the ODE $\tilde{y}_d'' + 2\tilde{y}_d' + \tilde{y}_d = e^{(1+i)t}$ and then take the real part of this solution. Let $\tilde{y}_d = Ae^{(1+i)t}$. Substituting in this value for $\tilde{y}$ we find that $A = 1/(3 + 4i) = (3 - 4i)/25$. Then, $y_d = \text{Re}[(3 - 4i)e^{(1+i)t}/25] = e^t(3 \cos t + 4 \sin t)/25$. Hence, $y = (c_1 + c_2 t)e^{-t} + e^t(3 \cos t + 4 \sin t)/25$. See Fig. 6(e).

(f). For the ODE $y'' + y' + y = (D^2 + D + 1)[y] = \sin^2 t$ we have the characteristic polynomial $r^2 + r + 1$ with complex roots $-1/2 \pm \sqrt{3}/2$, so $y_u = e^{-t/2}(c_1 \cos(\sqrt{3}t/2) + c_2 \sin(\sqrt{3}t/2))$. Since $\sin^2 t = (1 - \cos 2t)/2 = 1/2 - \text{Re}[e^{2it}]/2$, we will first seek a solution $\tilde{y}_d$ of the ODE $\tilde{y}'' + \tilde{y}' + y = 1/2 - e^{2it}/2$ and then take the real part of this solution. If we let $\tilde{y}_d = A + Be^{2it}$, we have that $\tilde{y}_d'' + \tilde{y}_d' + \tilde{y}_d = A + B[(2i)^2 + 2i + 1]e^{2it} = A + B(-3 + 2i)e^{2it} = 1/2 - e^{2it}/2$ if $A = 1/2$, $B = -1/[2(-3 + 2i)] = (3 + 2i)/26$. Hence, $y_d = \text{Re}[1/2 + (3 + 2i)e^{2it}/26] = 1/2 + (3 \cos 2t - 2 \sin 2t)/26$. Then $y = y_u + y_d$. See Fig. 6(f).

(g). The ODE $y'' + 4y' + 5y = (D^2 + 4D + 5)[y] = e^{-t} + 15t$ has characteristic polynomial $r^2 + 4r + 5$ with complex roots $-2 \pm i$, so $y_u = e^{-2t}(c_1 \cos t + c_2 \sin t)$. We can guess y_d to have the form $y_d = Ae^{-t} + B + Ct$. We have $y_d'' + 4y_d' + 5y_d = 2Ae^{-t} + (5B + 4C) + 5Ct$, which equals $e^{-t} + 15t$ if $A = 1/2$, $B = -12/5$, $C = 3$. Then $y = e^{-2t}(c_1 \cos t + c_2 \sin t) + e^{-t}/2 - 12/5 + 3t$. See Fig. 6(g).

(h). The ODE $y'' + 4y = (D^2 + 4)[y] = e^{2it}$ has characteristic polynomial $r^2 + 4$ with imaginary roots $\pm 2i$, so $y_u = k_1 e^{2it} + k_2 e^{-2it}$ and we can guess that y_d has the form $y_d = Ate^{2it}$. We have that $y_d'' + 4y_d = 4iAe^{2it}$, which equals e^{2it} if $A = -i/4$. Thus, $y = k_1 e^{2it} + k_2 e^{-2it} - ite^{2it}/4$.

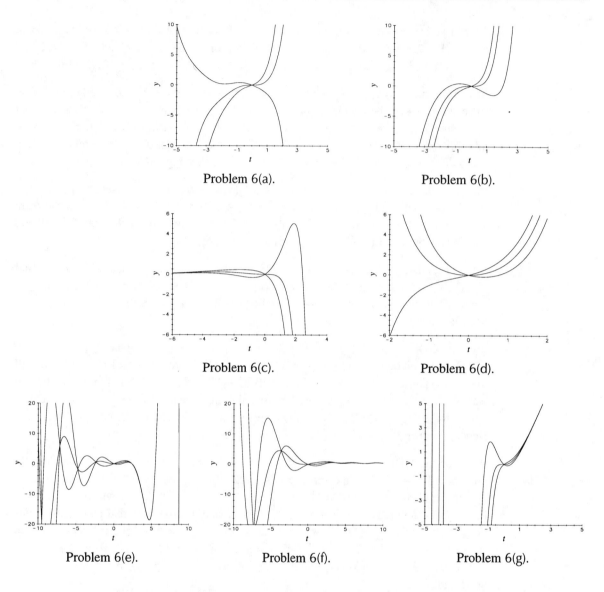

Problem 6(a).   Problem 6(b).

Problem 6(c).   Problem 6(d).

Problem 6(e).   Problem 6(f).   Problem 6(g).

7. **(a).** First we find a particular solution of the ODE $(D^2 - D - 2)[y] = 2e^t$. Writing $P(D) = D^2 - D - 2$, we guess a particular solution of the form $y_d = Ae^t$. $P(D)[Ae^t] = -2Ae^t$, so we set $A = -1$. Since the roots of $P(r)$ are $r_1 = 2$, $r_2 = -1$ it follows from Theorem 3.4.1 that $y = C_1 e^{2t} + C_2 e^{-t}$, where C_1 and C_2 are arbitrary constants, is the general solution of $P(D)[y] = 0$. So $y = C_1 e^{2t} + C_2 e^{-t} - e^t$ is the general solution of $P(D)[y] = f(t)$. The arbitrary constants C_1 and C_2 in the general solution $y = C_1 e^{2t} + C_2 e^{-t} - e^t$ are determined by the initial data; $y(0) = 0 = C_1 + C_2 - 1$, $y'(0) = 1 = 2C_1 - C_2 - 1$. So, $C_1 = 1$, $C_2 = 0$, and the solution of the IVP is $y = e^{2t} - e^t$.

(b). Guess a particular solution of the ODE $y'' + 2y' + 2y = (D^2 + 2D + 2)[y] = 2t$ of the form $y_d = At + B$. Since $y_d'' + 2y_d' + 2y_d = (At + B)'' + 2(At + B)' + 2(At + B) =$

$2A + 2At + 2B$. Matching coefficients we have $2A = 2$, $2A + 2B = 0$, and solving we see that $A = 1$, $B = -1$. So $y_d = t - 1$. The characteristic roots of $r^2 + 2r + 1$ are $r_1 = -1 + i$ and $r_2 = -1 - i$. So by Theorem 3.4.1 we see that $y = C_1 e^{-t} \cos t + C_2 e^{-t} \sin t + t - 1$ is the general solution of the driven ODE $y'' + 2y' + 2y = t - 1$. The constants of the general solution $y = C_1 e^{-t} \cos t + C_2 e^{-t} \sin t + t - 1$ are set by the data: $y(1) = 1 = C_1 e^{-1} \cos 1 + C_2 e^{-1} \sin 1$, and $y'(1) = 0 = e^{-1}[(-\cos 1 - \sin 1)C_1 + (-\sin 1 + \cos 1)C_2] + 1$. Solving for C_1 and C_2, we have $C_1 = e \cos 1$, $C_2 = e \sin 1$. So, $y = (e \cos 1)e^{-t} \cos t + (e \sin 1)e^{-t} \sin t + t - 1 = e^{1-t} \cos(t - 1) + t - 1$, where we have used a trigonometric identity.

8. **(a).** For $n = 1$, we have that $(D - r_0)[he^{r_0 t}] = (he^{r_0 t})' - r_0(he^{r_0 t}) = (h'e^{r_0 t} + r_0 he^{r_0 t}) - r_0 he^{r_0 t} = e^{r_0 t}h' = e^{r_0 t}D[h]$. For $n = 2$, we have that $(D - r_0)^2[he^{r_0 t}] = (D - r_0)[(D - r_0)[he^{r_0 t}]] = (D - r_0)[D[h]e^{r_0 t}] = e^{r_0 t}D[Dh] = e^{r_0 t}D^2[h]$, where we have used the result for $n = 1$ twice. Similarly, if $(D - r_0)^{n-1}[he^{r_0 t}] = e^{r_0 t}D^{n-1}[h]$, then $(D - r_0)^n[he^{r_0 t}] = (D - r_0)[(D - r_0)^{n-1}[he^{r_0 t}]] = (D - r_0)[e^{r_0 t}D^{n-1}[h]] = e^{r_0 t}D[D^{n-1}[h]] = e^{r_0 t}D^n[h]$ as claimed. The assertion has been proved by induction on n.

   (b). By part (a), $(D - r_0)^n[pe^{r_0 t}] = e^{r_0 t}p^{(n)}(t) = 0$, since the nth derivative $p^{(n)}(t)$ of $p(t)$ is $D^n[a_0 + a_1 t + \cdots + a_{n-1}t^{n-1}]$ which is zero for all t. So $y = p(t)e^{r_0 t}$ is a solution of $(D - r_0)^n[y] = 0$ for all polynomials $p(t)$ of degree $n - 1$ or less.

   (c). The operator $P(D)$ is $D^2 + aD + b$. We want to show that $P(D)[h(t)e^{st}] = e^{st}P(D + s)[h]$. Notice that

   $$D[he^{st}] = h'e^{st} + hse^{st} = e^{st}(D + s)[h]$$

   Applying D again to the above identity we have that

   $$D^2[he^{st}] = D[e^{st}(D + s)[h]] = e^{st}(D + s)^2[h]$$

   Since

   $$(D^2 + aD + b)[he^{st}] = D^2[he^{st}] + aD[he^{st}] + bhe^{st}$$
   $$= e^{st}\left((D + s)^2 + a(D + s) + b\right)[h]$$

   the validity of the claim is established.

9. To find a solution for the equation $y'' + 25y = \sin \omega t$ where $\omega \neq 5$, first we find the undriven solution and then we find a particular solution. The general real-valued solution of the undriven equation $y'' + 25y = 0$ is $y = C_1 \cos 5t + C_2 \sin 5t$. To find a particular solution, first note that $\sin \omega t = \text{Im}[e^{i\omega t}]$. So we first solve $(D^2 + 25)[z] = e^{i\omega t}$ where $y = \text{Im}[z]$. Substituting $z = Ae^{i\omega t}$ into the ODE yields $A = 1/(25 - \omega^2)$. Setting $y = \text{Im}[z]$ gives $y = \sin \omega t/(25 - \omega^2)$. The general solution of the driven ODE $y'' + 25y = \sin \omega t$ is, for $\omega \neq 5$,

   $$y = C_1 \cos 5t + C_2 \sin 5t + \frac{1}{25 - \omega^2} \sin \omega t$$

   See Figs. 9(a), Graphs 1 and 2 for the orbits of the solutions where $y(0) = y'(0) = 0$, $\omega = 4$ and then $\omega = 1$. In Fig. 9(a), Graph 3, we have superimposed two orbits by writing a new system

$$x' = y$$

$$y' = -25x + \sin(zt)$$

$$z' = 0$$

and using initial data $x = 0$, $y = 0$, $z = 4$, 1. You see why we love to call this the "heart and eyes problem."

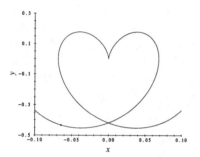

Problem 9, Graph 1.

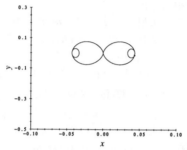

Problem 9, Graph 2.

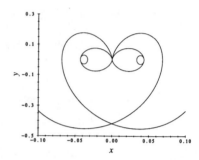

Problem 9, Graph 3.

10.   Group project.

11.   **(a).** If $z_d(t)$ is a solution of the ODE $z' + az = Ae^{i\omega t}$, then, since the coefficients in the operator $P(D) = D + a$ are real numbers, it follows from Theorem 3.6.3 that $y_d = \text{Re}[z_d]$ solves the same ODE but with the driving term replaced by $\text{Re}[Ae^{i\omega t}] = A\cos\omega t$.

(b). Comparing the formula $(D + a)[Ce^{i\omega t}] = C(i\omega + a)e^{i\omega t}$ with the ODE $(D + a)[z] = Ae^{i\omega t}$, we see that we can take $z_d = Ce^{i\omega t}$ if $C(i\omega + a) = A$. So, $z_d = (A/(i\omega + a))e^{i\omega t}$. After multiplying numerator and denominator of the coefficient fraction by $a - i\omega$, we obtain the desired formula for z_d.

(c). Writing

$$y_d = \text{Re}[z_d] = \frac{Aa}{a^2 + \omega^2}\cos\omega t + \frac{A\omega}{a^2 + \omega^2}\sin\omega t$$

and choosing an angle φ as indicated we have

$$y_d = \frac{A}{\sqrt{a^2 + \omega^2}}\left\{\frac{a}{\sqrt{a^2 + \omega^2}}\cos\omega t + \frac{\omega}{\sqrt{a^2 + \omega^2}}\sin\omega t\right\}$$

$$= \frac{A}{\sqrt{a^2 + \omega^2}}\{\cos\varphi\cos\omega t + \sin\varphi\sin\omega t\}$$

$$= \frac{A}{\sqrt{a^2 + \omega^2}}\cos(\omega t - \varphi)$$

where we have used the formula $\cos(\alpha - \beta) = \cos\alpha\cos\beta + \sin\alpha\sin\beta$.

(d). The general solution of the undriven ODE $y' + ay = 0$ is $y_u = Ce^{-at}$, where C is an arbitrary real number. So the general solution of $y' + ay = A\cos\omega t$ is $y = Ce^{-at} + y_d$. If $C \neq 0$, then the solution is not periodic. There is only one periodic solution of the given ODE (namely, when $C = 0$).

(e). We leave this to the group.

3.7 The General Theory of Linear ODEs

Suggestions for preparing a lecture

Topics: The Fundamental Theorem for Linear ODEs, Linearity Property and Superposition Principle for $P(D)$, basic solution sets, Wronskians, Abel's Theorem, General Solution Theorem 3.7.5.

Remarks: Do an Euler equation example.

Making up a problem set

Problems 3, 4, and 7 (because it is kind of curious). The group problem 8 on the Method of Variation of Parameters is important, but you probably shouldn't assign it unless you intend to use it later.

Comments

The two main results in this section, the General Solution Theorem and the Fundamental Theorem for Linear ODEs, are mostly of theoretical value, but form a nice way to pull things together as the chapter closes. If you intend to read about series methods in Chapter 11, then we suggest that you at least cover the highlights of this section. We made the Method of Variation of Parameters into a long group problem which outlines the development of the method. Although it's nice to know the statement of the Method, it's a lot of work to do examples where the driving term is not a polynomial-exponential.

So we concentrate on constant-coefficient operators $P(D)$ and polynomial-exponential driving terms. This covers most of the important applications in a first ODE course.

---

1. **(a).** Since $P(D)[y] = (tD^2 + D)[y] = ty'' + y' = (ty')'$, the null space of $tD^2 + D$ is found by integrating $(ty')' = 0$ to obtain $ty' = c_1$, so $y' = c_1 t^{-1}$. Integrating again, $y = c_1 \ln t + c_2$, $0 < t < \infty$, c_1 and c_2 arbitrary constants, which gives all the functions $y(t)$ in the null space.

(b). Substituting $y = t^2/2$ into $P(D)[y] = (tD^2 + D)[y]$ we get $2t$, so $y = t^2/2$ is a particular solution of $P(D)[y] = 2t$. By Theorem 3.7.5 the general solution may be written as the sum of $t^2/2$ and the functions in the null space, $y = c_1 \ln t + c_2 + t^2/2$, $0 < t < \infty$. If the conditions $y(0) = a$ and $y'(0) = b$ are imposed, we must set $c_1 = 0$ since $\ln 0$ is not defined. We have $y = c_2 + t^2/2$, $c_2 = a$, and $b = 0$. The initial value problem $ty'' + y' = 2t$, $y(0) = a$, $y'(0) = 0$, has the unique solution $y = a + t^2/2$, $-\infty < t < \infty$.

(c). In normal linear form we have that $y'' + y'/t = 2$, and the coefficient $1/t$ of y' is discontinuous on every interval containing the origin. There is no way to define the coefficient at $t = 0$ to make it continuous. So Theorem 3.7.1 does not apply, and there is no contradiction.

2.   By definition, the null space of $P(D)$ determines the general solution of $P(D)[y] = 0$. $P(D)y = (1-t^2)y'' - 2ty' = [(1-t^2)y']' = 0$ can be solved by integration: $(1-t^2)y' = c_1^*$, so $y' = c_1^*(1-t^2)^{-1}$. Integrating again,

$$y = \frac{c_1^*}{2} \ln \frac{1+t}{1-t} + c_2 = c_1 \ln \frac{1+t}{1-t} + c_2, \quad |t| < 1$$

where c_1 and c_2 are any constants. The null space consists of all of these functions $y(t)$.

3.   The functions y_1 and y_2 are shown to be solutions by direct substitution in the given ODE. The solution satisfying the given initial data is found by imposing the initial conditions on $y = c_1 y_1(t) + c_2 y_2(t)$ to determine c_1 and c_2. Since the normalized ODEs have discontinuous coefficients at $t = 0$, Theorem 3.7.1 applies only to intervals not containing 0.

(a). The IVP is $t^2 y'' + ty' - y = 0$, $y(1) = 0$, $y'(1) = 0$. We have $y = c_1 t^{-1} + c_2 t$, with $y(1) = 0 = c_1 + c_2$ and $y'(1) = -1 = -c_1 + c_2$. So $c_1 = 1/2 = -c_2$ and $y = (t^{-1} - t)/2$, $t > 0$. See Fig. 3(a).

(b). The IVP is $t^2 y'' - ty' + y = 0$, $y(1) = 1$, $y'(1) = 0$. Here, $y = c_1 t + c_2 t \ln t$, with $y(1) = 1 = c_1$ and $y'(1) = 0 = c_1 + c_2$. So $c_1 = 1$, $c_2 = -1$ and $y = t - t \ln t$, $t > 0$. See Fig. 3(b).

(c). The IVP is $t^2 y'' + ty' - 4y = 0$, $y(0) = 0$, $y'(0) = 0$. In this case, $y = c_1 t^2 + c_2 t^{-2}$, but the initial data are given only at $t = 0$ where Theorem 3.7.1 is not applicable. If $c_2 = 0$, then the solution $y = c_1 t^2$ satisfies both initial conditions for all constants c_1. See Fig. 3(c) for the graphs of several solutions.

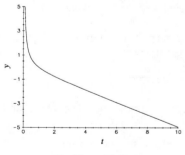

Problem 3(a).   Problem 3(b).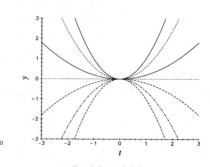

Problem 3(c).

4.   The Euler ODE is $t^2 y'' + pty' + qy = 0$, $t > 0$.

(a). Insert $y = t^r$ into the ODE, $t^2 y'' + pty' + qy = 0$, to obtain

$$t^r r(r-1) + t^r pr + qt^r = t^r(r(r-1) + pr + q)$$

$$= t^r(r^2 + (p-1)r + q) = 0$$

Since t^r is only 0 at $t = 0$, we must have that $y = t^r$ solves the Euler ODE if and only if r is a root of the quadratic $Q = r^2 + (p - 1)r + q$. If r_1 and r_2 are distinct real roots of Q, then $y_1 = t^{r_1}$ and $y_2 = t^{r_2}$ solve the Euler ODE for $t > 0$. Since the Wronskian

$$W[y_1, y_2](1) = \left[y_1(t)y_2'(t) - y_1'(t)y_2(t) \right]_{t=1} = \left[r_2 t^{r_1} t^{r_2-1} - r_1 t^{r_1-1} t^{r_2} \right]_{t=1} = r_2 - r_1 \neq 0,$$

y_1 and y_2 form a basic solution set.

If $r = \alpha + i\beta$, $\beta \neq 0$, is a complex root, then

$$t^{\alpha+i\beta} = t^\alpha t^{i\beta} = t^\alpha e^{i\beta \ln t} \quad \text{[for } t > 0\text{]}$$

But by the Euler formula,

$$e^{i\beta \ln t} = \cos(\beta \ln t) + i \sin(\beta \ln t)$$

So we have in this case that

$$t^r = t^\alpha \cos(\beta \ln t) + it^\alpha \sin(\beta \ln t)$$

Taking the real and the complex parts, we have the two real solutions

$$y_1 = t^\alpha \cos(\beta \ln t), \quad y_2 = t^\alpha \sin(\beta \ln t)$$

if $r = \alpha + i\beta$, $\beta \neq 0$, is a root of $r^2 + (p - 1)r + q$. Here $\{y_1, y_2\}$ is a basic solution set because the Wronskian of y_1 and y_2 is not 0 for $t = 1$. We show this directly by calculating

$$W[y_1, y_2]_{t=1} = \left[y_1(t)y_2'(t) - y_1'(t)y_2(t) \right]_{t=1}$$

$$= 1 \cdot \left[\alpha t^{\alpha-1} \sin(\beta \ln t) + \beta t^{\alpha-1} \cdot \frac{1}{t} \cos(\beta \ln t) \right]_{t=1} - 0 = \beta$$

since $y_1(1) = 1$ and $y_2(1) = 0$.

(b). We will change the independent variable t in the Euler ODE, $t^2 y'' + pty' + qy = 0$, to s, where $t = e^s$, $t > 0$, so $s = \ln t$. Using the Chain Rule, we have

$$\frac{dy}{dt} = \frac{dy}{ds}\frac{ds}{dt} = \frac{dy}{ds}\frac{1}{t} = \frac{dy}{ds}e^{-s}$$

$$\frac{d^2 y}{dt^2} = \frac{d}{ds}\left(\frac{dy}{dt}\right)\frac{ds}{dt} = \frac{d}{ds}\left(\frac{dy}{ds}e^{-s}\right)\frac{1}{t}$$

$$= \frac{d^2 y}{ds^2} \cdot \frac{e^{-s}}{t} + \frac{dy}{ds}\left(-e^{-s}\right)\frac{1}{t}$$

$$= \frac{d^2 y}{ds^2}\frac{1}{t^2} - \frac{dy}{ds}\frac{1}{t^2}$$

So,

$$0 = t^2 y'' + pty' + qy = \frac{d^2 y}{ds^2} - \frac{dy}{ds} + p\frac{dy}{ds} + qy$$

$$= \frac{d^2 y}{ds^2} + (p-1)\frac{dy}{ds} + qy$$

(c). The polynomial $Q(r)$ for the Euler ODE, $t^2 y'' - 2ty' + 2y = 0$ is $Q = r^2 - 3r + 2 = (r-1)(r-2)$. The roots are 1 and 2, so a basic solution set is $y_1 = t$, $y_2 = t^2$. The general solution of the ODE is $y = C_1 t + C_2 t^2$. Imposing the initial conditions $y(1) = 0$ and $y'(1) = -1$, we see that $0 = C_1 + C_2$ and $-1 = C_1 + 2C_2$. So, $C_1 = 1$ and $C_2 = -1$; the solution is $y = t - t^2$.

(d). Following the procedure suggested in part **(b)**, set $t = e^s$. The IVP, $t^2 y'' + 2ty' + 2y = 0$, $y(1) = 0$, $y'(1) = 1$, becomes

$$\frac{d^2 y}{ds^2} + \frac{dy}{ds} + 2y = 0, \quad y(0) = 0, \quad y'(0) = 1$$

Since the characteristic polynomial of the new ODE with constant coefficients is $r^2 + r + 2$ with roots $r_{1,2} = -1/2 \pm \sqrt{7}i/2$, we have

$$y = e^{-s/2} \left[C_1 \cos(\sqrt{7}s/2) + C_2 \sin(\sqrt{7}s/2) \right]$$

Imposing the initial conditions we have

$$0 = C_1, \quad 1 = \sqrt{7}C_2$$

and the solution in terms of the s-variable is

$$y = \frac{1}{\sqrt{7}} e^{-s/2} \sin(\sqrt{7}s/2)$$

In terms of the t-variable (where $s = \ln t$), the solution is

$$y = \frac{1}{\sqrt{7}} e^{-(\ln t)/2} \sin(\sqrt{7}(\ln t)/2)$$

$$= \frac{1}{\sqrt{7}} t^{-1/2} \sin(\sqrt{7}(\ln t)/2)$$

5. Suppose $y = u(t)$ is a solution of $y'' + a(t)y' + b(t)y = 0$.

(a). Let $z(t)$ satisfy $uz'' + (2u' + au)z' = 0$. Then if $y = uz$ is substituted into $y'' + ay' + by = 0$, we have $(u''z + 2u'z' + uz'') + a(u'z + uz') + buz = uz'' + z'(2u' + au) + z(u'' + au' + bu) = 0$ since $u'' + au' + bu = 0$ and $uz'' + (2u' + au)z' = 0$. So $y = u(t)z(t)$ is a solution of $y'' + a(t)y' + b(t)y = 0$.

(b). Let $v = z'$. Then the ODE $uz'' + (2u' + a(t)u)z' = 0$ from part **(a)** may be written in terms of v and normalized as $v' + (2u'/u + a)v = 0$, which is linear. Its integrating factor is $u^2(t) \exp \int^t a(r)\, dr$. So $v(t) = [u(t)]^{-2} \exp[-\int^t a(r)\, dr]$ is one solution of the ODE for v. We find z by an integration: $z(t) = \int_{t_0}^t v(s)\, ds$.

(c). We have that $W[u, uz] = u(uz)' - (uz)u' = u^2 z' = u^2 v = \exp[-\int^t a(r)\, dr] > 0$ where we have used $z' = v$ and v from part **(b)**, $v(t) = [u(t)]^{-2} \exp[-\int^t a(r)\, dr]$. Since the Wronskian is not 0 and $\{u, uz\}$ is a set of solutions of the normal ODE $y'' + ay' + by = 0$, the set $\{u, uz\}$ is a basic solution set.

(d). Using the preceding results, we see that if $u = e^t$ is one solution of $ty'' - (t+2)y' + 2y = 0$ then a second solution is given by $y = uz = e^t z$, where $z = \int^t e^{-2s} \exp(\int^s (1 +$

$2/r) \, dr) \, ds$ and we have normalized the ODE to $y'' - (1 + 2/t)y' + 2t^{-1}y = 0$. Carrying out the integration, we have that $z = \int^t e^{-2s}(s^2 e^s) \, ds = \int^t s^2 e^{-s} \, ds = -t^2 e^{-t} - 2e^{-t}(t + 1) = -e^{-t}(t^2 + 2t + 2)$. So, $uz = -(t^2 + 2t + 2)$. Since by part **(c)** $\{e^t, -(t^2 + 2t + 2)\}$ forms a basic solution set, we obtain all solutions from u and uz in the form $y = c_1 e^t + c_2(2 + 2t + t^2)$, where c_1 and c_2 are arbitrary constants.

6. Suppose that $W[y_1, y_2]$ is the Wronskian of solutions y_1 and y_2 of $P(D)[y] = 0$, where $P(D) = D^2 + a(t)D + b(t)$.

(a). Since $W = y_1 y_2' - y_1' y_2$, $W' = y_1' y_2' + y_1 y_2'' - y_1' y_2' - y_1'' y_2 = y_1 y_2'' - y_1'' y_2 = y_1(-by_2 - ay_2') - y_2(-by_1 - ay_1') = -a(y_1 y_2' - y_1' y_2) = -aW$, using the fact that y_1 and y_2 are solutions of $y'' + ay' + by = 0$. W satisfies the linear ODE, $W' + a(t)W = 0$, so $W(t) = W(t_0) \exp[-\int_{t_0}^t a(s) \, ds]$.

(b). Set $y_1 = u(t)$, $y_2 = v(t)$ in Abel's formula to obtain the ODE for v, $uv' - u'v = \exp(-\int^t a(s) \, ds)$, where we have set $W(t_0) = 1$ and dropped the lower limit (permissible since we are only after some second solution, and any multiple of a solution is again a solution). Then $\{u, v\}$ is a basic solution set since $W[u, v](t) = \exp\{-\int^t a(s) \, ds\} \neq 0$.

(c). Normalizing the ODE, we see that $a(t) = -(1 + 2/t)$. Given one solution $u = e^t$ of the ODE (normalized), $y'' - (1 + 2/t)y' + 2y/t = 0$ and then plugging into the formula in part **(b)** yields the ODE for v, $e^t v' - e^t v = \exp(\int^t(1 + 2/s) \, ds) = \exp(t + 2\ln t) = t^2 e^t$. Normalizing this ODE for v, $v' - v = t^2$. The integrating factor is e^{-t}, so $(e^{-t}v)' = t^2 e^{-t}$. Using a table of integrals, we have that $v = Ce^t - t^2 - 2t - 2$. We may as well set $C = 0$ and let $v = -t^2 - 2t - 2$; thus, for $t > 0$, $\{u, v\}$ is fundamental since $W[u, v] = e^t v' - e^t v = t^2 e^t > 0$.

7. Let $\{u(t), v(t)\}$ be a basic solution set for the ODE, $y'' + a(t)y' + b(t)y = 0$. Let c and d be consecutive zeros of $u(t)$. We will show first that $v(t)$ has at least one zero between c and d. Let $W(t) = W[u, v](t)$ be the Wronskian of the basic solution pair $\{u, v\}$. Since $W(t) \neq 0$ for all t we may as well assume that $W(t) > 0$ [if $W(t) < 0$, then an analogous argument works]. So, $0 < W(c) = -v(c)u'(c)$, and $0 < W(d) = -v(d)u'(d)$, where we have used $u(c) = u(d) = 0$. Now from the Vanishing Data Theorem 3.7.2 we see that $u'(c) \neq 0$ and $u'(d) \neq 0$ since $u(t)$ and $v(t)$ are not identically 0. Since $0 < -v(c)u'(c)$ and $0 < -v(d)u'(d)$ and since c and d are consecutive zeros of u, it must be that $u'(c)$ and $u'(d)$ are of opposite sign. It follows that $v(c)$ and $v(d)$ are of opposite sign, so v must have a zero at some point between c and d. A similar argument shows that $u(t)$ must have at least one zero between any two consecutive zeros of $v(t)$. Therefore, between any two consecutive zeros of one solution there is precisely one zero of the other solution. That's why the two curves sketched in the margin *cannot* form a basic solution set for the ODE, $y'' + a(t)y' + b(t)y = 0$.

8. Group project.

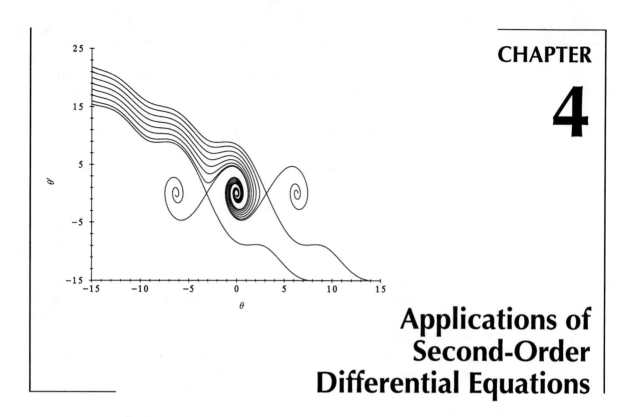

Applications of Second-Order Differential Equations

There are two main applications in this chapter: the pendulum and electrical circuits. The pendulum models reappear numerous times in later chapters. In Section 4.1, a derivation is given for the model ODE of the simple pendulum starting from Newton's Laws of Motion. In Section 4.4 Kirchhoff's Voltage and Current Laws are used to derive ODEs that model current and charge in electrical circuits.

The material in Section 4.2 is a general treatment of beats and resonance for linear second-order ODEs with constant coefficients. We have written this section so that it can be covered independently of the applications sections. Engineering students may already have seen frequency response modeling in a systems or lumped parameters course, so the material of Section 4.3 will not come as a big surprise to them. Our approach builds on the properties of linear ODEs which were developed in Chapter 3.

Periodic forced oscillations have been touched upon at various places in this text other than in Sections 4.2 and 4.3. Problem 12 in Section 2.2 outlines a proof of the existence of a unique periodic forced oscillation for a first-order, periodically driven and linear ODE, and (for good measure) we give a simpler version in Group Problem 5 in Section 4.2. Group Problem 9 in Section 4.3 does the same for second-order linear ODEs, as does the group problem in Section 7.9 for linear systems.

4.1 Newton's Laws: The Pendulum

Suggestions for preparing a lecture

Topics: Newton's Second Law, enough of vectors to understand the law, derivation of the pendulum ODE.

Remarks: You may want to lecture on some of the material in Problems 5, 7, and 8.

Making up a problem set

Two or three parts of 1, 3 or 6, 5. Problem 9 is a golden oldie but hard; if you assign it be prepared to give hints.

Comments

The proper language for Newtonian mechanics is the language of vectors. We have avoided vectors up to this point by considering motion only in one dimension—for example, a falling body, a vertically oscillating spring. But at some point or other vectors should be introduced so that Newton's three laws of motion can be fully understood. The approach to vectors in this section is traditional, and nonaxiomatic. Vectors are thought of as arrows, but denoted by boldface type. The dot product is introduced and some of its properties are listed (Problem 1). Unit coordinate vectors $\mathbf{i}$, $\mathbf{j}$, $\mathbf{k}$, $\hat{\mathbf{x}}$, $\hat{\mathbf{y}}$, $\hat{\mathbf{z}}$, $\hat{\mathbf{r}}$, $\hat{\boldsymbol{\theta}}$ are used. The main application of the vector approach is the derivation of the equation of motion of the simple pendulum. We have opted to leave the derivation of various properties of the solutions and orbits of the simple undamped pendulum (both nonlinear and linearized) to the problem set (Problems 5–8).

1. **(a).** Let $\mathbf{u} = (u_1, u_2, u_3)$. The value of $\|\mathbf{u}\|^2$ is the square of the distance from the origin $(0, 0, 0)$ to the point (u_1, u_2, u_3). So, $\|\mathbf{u}\|^2 = \left(\sqrt{(u_1 - 0)^2 + (u_2 - 0)^2 + (u_3 - 0)^2}\right)^2 = u_1^2 + u_2^2 + u_3^2$ for any vector $\mathbf{u}$.

    (b). Let $\mathbf{u} = (u_1, u_2, u_3)$ and $\mathbf{v} = (v_1, v_2, v_3)$ each have its tail at the origin. Suppose first that neither $\mathbf{u}$ nor $\mathbf{v}$ is the zero vector. By definition, $\mathbf{u} \cdot \mathbf{v} = \|\mathbf{u}\| \|\mathbf{v}\| \cos \theta$, where θ is the angle between $\mathbf{u}$ and $\mathbf{v}$. Using part **(a)** and the Cosine Law, we have that

    $$\mathbf{u} \cdot \mathbf{v} = \|\mathbf{u}\| \|\mathbf{v}\| \cos \theta$$
    $$= \sqrt{u_1^2 + u_2^2 + u_3^2} \sqrt{v_1^2 + v_2^2 + v_3^2} \cos \theta$$
    $$= \frac{1}{2} \left[(u_1^2 + u_2^2 + u_3^2) + (v_1^2 + v_2^2 + v_3^2) - \left((u_1 - v_1)^2 + (u_2 - v_2)^2 + (u_3 - v_3)^2\right) \right]$$

    since the Cosine Law states that for any triangle with sides of respective lengths a, b, and c and angle θ between the sides of length a and b, $c^2 = a^2 + b^2 - 2ab\cos\theta$ or $ab\cos\theta = \frac{1}{2}\left(a^2 + b^2 - c^2\right)$. The triangle in this case has sides $\mathbf{u}$, $\mathbf{v}$, and $\mathbf{u} - \mathbf{v}$. Squaring and then cancelling terms inside the brackets for the expression for $\|\mathbf{u}\| \|\mathbf{v}\| \cos \theta$, we obtain

    $$\mathbf{u} \cdot \mathbf{v} = u_1 v_1 + u_2 v_2 + u_3 v_3$$

Finally, suppose **u** (or **v**) is the zero vector. Then **u** · **v** is 0 because $\|\mathbf{u}\| = 0$ (or $\|\mathbf{v}\| = 0$). We also have $u_1 v_1 + u_2 v_2 + u_3 v_3 = 0$ in this case because $\mathbf{u} = (0, 0, 0)$ (or $\mathbf{v} = (0, 0, 0)$). So, for any **u** and **v**, $\mathbf{u} \cdot \mathbf{v} = u_1 v_1 + u_2 v_2 + u_3 v_3$.

(c). Let $\mathbf{u} = (u_1, u_2, u_3)$ and $\mathbf{v} = (v_1, v_2, v_3)$. Then $\mathbf{u} \cdot \mathbf{v} = u_1 v_1 + u_2 v_2 + u_3 v_3 = v_1 u_1 + v_2 u_2 + v_3 u_3 = \mathbf{v} \cdot \mathbf{u}$.

(d). Let $\mathbf{u} = (u_1, u_2, u_3)$, $\mathbf{v} = (v_1, v_2, v_3)$, and $\mathbf{w} = (w_1, w_2, w_3)$. Then

$$(\alpha\mathbf{u} + \beta\mathbf{w}) \cdot \mathbf{v} = (\alpha u_1 + \beta w_1)v_1 + (\alpha u_2 + \beta w_2)v_2 + (\alpha u_3 + \beta w_3)v_3$$

$$= \alpha(u_1 v_1 + u_2 v_2 + u_3 v_3) + \beta(w_1 v_1 + w_2 v_2 + w_3 v_3)$$

$$= \alpha\mathbf{u} \cdot \mathbf{v} + \beta\mathbf{w} \cdot \mathbf{v}$$

(e). Let $\mathbf{u} = (u_1, u_2, u_3)$. Then $\mathbf{u} \cdot \mathbf{u} = u_1^2 + u_2^2 + u_3^2 \geq 0$ for all **u**. Since a sum of squares of real numbers is 0 if and only if each number is 0, then $\mathbf{u} \cdot \mathbf{u} = u_1^2 + u_2^2 + u_3^2 = 0$ if and only if $\mathbf{u} = \mathbf{0}$.

(f). Let **u** and **v** be nonzero vectors, since if either is zero then we immediately have $0 \leq 0$, and the Cauchy-Schwarz inequality holds. If $\mathbf{u} \neq \mathbf{0}$, $\mathbf{v} \neq \mathbf{0}$, then $|\mathbf{u} \cdot \mathbf{v}| = |\|\mathbf{u}\| \|\mathbf{v}\| \cos\theta| = \|\mathbf{u}\| \|\mathbf{v}\| |\cos\theta| \leq \|\mathbf{u}\| \|\mathbf{v}\|$, since $|\cos\theta| \leq 1$ for all θ.

2. Since we are looking for the position relative to P, let $P = (0, 0, 0)$. Then the final position is given by the sum of the vectors $\mathbf{u}_1 = -20\mathbf{i}$, $\mathbf{u}_2 = -8\cos(10°)\mathbf{j} + 8\sin(10°)\mathbf{k} = -7.88\mathbf{j} + 1.39\mathbf{k}$, and $\mathbf{u}_3 = 42\mathbf{i}$, where the axis **i** is positive to the north, **j** is positive to the west, and **k** is positive upwards. The final position is approximately $22\mathbf{i} - 7.88\mathbf{j} + 1.39\mathbf{k}$, which translates to 22 miles North, 7.88 miles East, and 1.39 miles above P.

3. Applying Newton's Second Law (in the vertical direction) to the piston, we have $mx'' =$ force of gravity $+$ force due to gas pressure. The magnitude of the force due to gas pressure is $P \cdot A$ where A is the area of the piston. But $P = nRT/V = nRT/(x \cdot A)$, so the force due to gas pressure is $P \cdot A = nRT/x$. Since the force of gravity is directed downward while the force due to gas pressure is directed upward, $mx'' = -mg + nRT/x$.

4. **(a).** The equation of motion is $\theta'' + g\theta/L = 0$. Solving for the real solution, we see that $\theta(t) = C_1 \cos(t\sqrt{g/L}) + C_2 \sin(t\sqrt{g/L})$. The period both of $C_1 \cos(t\sqrt{g/L})$ and of $C_2 \sin(t\sqrt{g/L})$ is $2\pi/\sqrt{g/L} = 2\pi\sqrt{L/g}$. Since the sum of two periodic functions with period T is also a periodic function of period T, we know that the period of the pendulum is $T = 2\pi\sqrt{L/g}$.

(b). Set $2\pi\sqrt{L/g} = 1$ and obtain $L = g/(4\pi^2)$. In mks units where $g = 9.8 \ m/s^2$, we have that $L \approx \frac{1}{4}$ meter.

(c). From part **(a)** we have $\theta = C_1 \cos(t\sqrt{g/L}) + C_2 \sin(t\sqrt{g/L})$. Let's let the pendulum be at its lowest point $\theta = 0$ at $t = 0$. But $\theta(0) = 0$ forces $C_1 = 0$, so $\theta = C_2 \sin(t\sqrt{g/L})$. Since the pendulum swings with amplitude 1 radian, $C_2 = 1$, so $\theta = \sin(t\sqrt{g/L})$. Then $\theta' = \sqrt{g/L} \cos(t\sqrt{g/L})$ and at $t = 0$, $\theta' = \sqrt{g/L} = \sqrt{9.8/1} \approx 3.13$ rad/sec. Every time the pendulum passes through the lowest point, $\theta' = \sqrt{g/L} \approx 3.13$ rad/sec or $-\sqrt{g/L} \approx -3.13$ rad/sec.

Since $\theta = 0$ at $t = 0$, $|\theta| = 1$ first at $t = (\pi/2)\sqrt{g/L}$. Now $\theta'' = -g/L\sin(t\sqrt{g/L}) = -g/L = -9.8$ rad/(sec)^2 when $t = (\pi/2)\sqrt{g/L}$. Depending on which end of the swing the pendulum is at, $\theta'' = -9.8$ rad/sec^2 or $\theta'' = 9.8$ rad/sec^2.

5. **(a).** See Fig. 5(a) for some of the orbits of the ODE $mL\theta'' + cL\theta' + mg\sin\theta = 0$ with $m = 1$, $c = 0$, and $g/l = 10$, including the over-the-top orbits (wavy lines at top and bottom) and the back-and-forth orbits seen here as ovals. Can you plot the separatrices (we didn't here, but did in text Figure 4.1.1)?

(b). See Fig. 5(b). In part **(a)** there is no damping, and so the pendulum either swings back and forth forever (closed orbits), or makes full revolutions forever (wavy orbits at top and bottom of plots), or else falls from a position infinitesimally close to the vertically upright position and then rises infinitesimally close to that position but with θ increased or diminished by 2π (a separatrix). But in the system for the chapter cover figure there is damping, and so most orbits tend toward the equilibrium positions $\theta = 2n\pi$, $\theta' = 0$, corresponding to the pendulum hanging straight down. The points $\theta = (2n + 1)\pi$, $\theta' = 0$ are also equilibria, corresponding to the pendulum balancing straight up. Only two orbits (each a separatrix) tend to each of these points as t increases, one from the upper half plane $v = \theta' > 0$ with $\theta \to (2n + 1)\pi^-$, the other from the lower half plane $v = \theta' < 0$ with $\theta \to (2n + 1)\pi^+$.

(c). See Fig. 5(c), Graph 1 for orbits of the linearized, undamped and undriven pendulum ODE, $\theta'' + 10\theta = 0$. The orbits are the ellipses $(\theta')^2 + 10\theta^2 = $ constant. Close to the equilibrium point $\theta = 0$, $\theta' = 0$, these ellipses resemble the oval orbits of the nonlinear ODE, $\theta'' + 10\sin\theta = 0$. But at a distance from the point $\theta = 0$, $\theta' = 0$ the orbits of the linearized ODE don't look like the orbits of the nonlinear ODE at all. The linear ODE has only the single equilibrium point at the origin; the nonlinear ODE has infinitely many equilibrium points. All nonconstant orbits of the linear ODE are closed curves that surround the origin, and that is certainly not true for the nonlinear ODE. So we see that the linearized ODE, $\theta'' + 10\theta = 0$, may be a pretty good approximate model of the simple undamped pendulum dynamics near the origin, but is a total failure elsewhere, failing completely to model the other equilibrium points, over-the-top motions, and so on.

See Fig. 5(c), Graph 2 for orbits of the linearized, damped and undriven pendulum ODE, $\theta'' + \theta' + 10\theta = 0$. The inward spiraling orbits resemble the central inward spiraling orbits of the chapter cover figure, but at a distance from the equilibrium point $\theta = 0$, $\theta' = 0$, this linearized ODE is a terrible approximation to the nonlinear ODE, $\theta'' + \theta' + 10\sin\theta = 0$ for many of the same reasons mentioned in the preceding paragraphs.

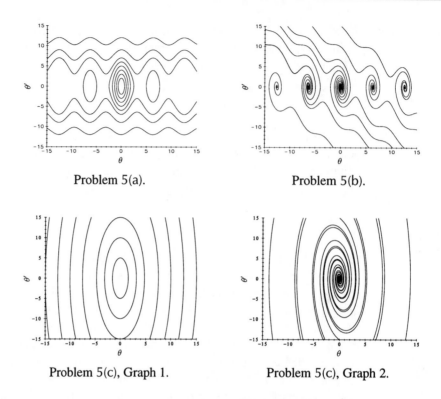

Problem 5(a).   Problem 5(b).

Problem 5(c), Graph 1.   Problem 5(c), Graph 2.

6. If $L = L(t)$, then equations (8) in the text must be altered to take that fact into account. We have that $\mathbf{v} = \mathbf{R}' = (L\hat{\mathbf{r}})' = L'\hat{\mathbf{r}} + L\hat{\mathbf{r}}' = L'\hat{\mathbf{r}} + L\hat{\boldsymbol{\theta}}\theta'$ and $\mathbf{a} = \mathbf{R}'' = (L'\hat{\mathbf{r}} + L\theta'\hat{\boldsymbol{\theta}})' = L''\hat{\mathbf{r}} + L'\hat{\mathbf{r}}' + L'\theta'\hat{\boldsymbol{\theta}} + L\theta''\hat{\boldsymbol{\theta}} + L\theta'\hat{\boldsymbol{\theta}}' = L''\hat{\mathbf{r}} + 2L'\theta'\hat{\boldsymbol{\theta}} + L\theta''\hat{\boldsymbol{\theta}} - L'(\theta')^2\hat{\mathbf{r}}$. Then from Newton's Second Law in the $\hat{\boldsymbol{\theta}}$ direction, we have $(m\mathbf{a})_{\hat{\boldsymbol{\theta}}} = \mathbf{F}_{\hat{\boldsymbol{\theta}}}$ or $m(2L'\theta' + L\theta'') = -cL\theta' - mg\sin\theta + F(t)$, or $mL\theta'' + (2mL' + cL)\theta' + mg\sin\theta = F(t)$, where $F(t)$ is the component of the external force in the $\hat{\boldsymbol{\theta}}$ direction.

7. **(a).** Multiplying the ODE $mL\theta'' + mg\sin\theta = 0$ by θ', we have $mL\theta'\theta'' + mg\theta'\sin\theta = 0$. Now $mL\theta'\theta'' = [mL(\theta')^2/2]'$ and $mg\theta'\sin\theta = [-mg\cos\theta]'$, so the ODE becomes $mL\theta'\theta'' + mg\theta'\sin\theta = [mL(\theta')^2/2 - mg\cos\theta]' = 0$, and we have that $mL(\theta')^2/2 - mg\cos\theta = c$, where c is a constant. See the plots in Fig. 5**(a)** for some of these orbits.

(b). The total energy at any time t is given by $\frac{1}{2}m(L\theta')^2 + mgL(1 - \cos\theta) = L(c + mg\cos\theta) + mgL(1 - \cos\theta) = cL + mgL$, where we have used the result of part **(a)**. Since $cL + mgL$ is constant for all t, we see that the total energy is the same at time t as it is at time 0. That is, energy is conserved.

8. Group Project 8 can be approached as follows. The solutions of the linearized pendulum ODE, $mL\theta'' + mg\theta = 0$, have the form $\theta = A\cos(t\sqrt{g/L} - \delta)$, where A and δ are arbitrary constants; the period is $2\pi\sqrt{L/g}$. Since $\theta' = -A\sqrt{g/L}\sin(t\sqrt{g/L} - \delta)$, we have that $\theta^2 + L(\theta')^2/g = A^2$, which for each $A > 0$ is the equation of an ellipse in the $\theta\theta'$-plane. The major axis lies along the θ-axis if $L/g > 1$, along the θ'-axis if $L/g < 1$, while the orbit is a circle if $L = g$.

If $\theta(0) = \theta_0$ and $\theta'(0) = 0$, then the orbit of the nonlinear pendulum ODE, $mL\theta'' + mg\sin\theta = 0$, is defined by

$$\frac{1}{2}mL(\theta')^2 - mg\cos\theta = -mg\cos\theta_0$$

as can be seen from Problem 7. The motion is periodic if the graph of the orbit in the $\theta\theta'$-plane is a simple closed curve not passing through any equilibrium points. The equation of the orbit is $L(\theta')^2 = 2g(\cos\theta - \cos\theta_0)$ and has the property that if (θ, θ') is a point on the orbit, then the points $(-\theta, \theta')$, $(\theta, -\theta')$, and $(-\theta, -\theta')$ also are on the orbit. We only need to show that the orbit in the first quadrant is an arc reaching from the positive θ'-axis to the positive θ-axis, then reflect this arc through the axes to obtain the simple closed curve which is the full orbit. None of the equilibrium points $\theta = n\pi$, $\theta' = 0$ lies on the given orbit since $0 < \theta_0 < \pi$. The arc of the orbit in the first quadrant falls steadily from the point $\theta = 0$, $\theta' = [2g(1 - \cos\theta_0)/L]^{1/2}$ to the point $\theta = \theta_0$, $\theta' = 0$, as we see directly from the equation of the orbit. After the reflections, we obtain the simple, closed orbit corresponding to a periodic solution.

In the first quadrant, we have that $\theta' = \sqrt{2g/L}\sqrt{\cos\theta - \cos\theta_0}$. Separating variables and integrating from $t = 0$ to $t = T/4$, where T is the period once around the full orbit, we have that $4\int_0^{\theta_0}(\cos\theta - \cos\theta_0)^{-1/2}\,d\theta = T(2g/L)^{1/2}$, where we have used the fact that the arc in the first quadrant is traversed in one fourth of the period. Note that here we have initialized time at $t_0 = 0$, then set $\theta = 0$ at $t_0 = 0$ and $\theta = \theta_0$ at $t = T/4$. The integral may be transformed into the elliptic integral given in the problem by the changes in constants and variables, $k = \sin\theta_0/2$, $\sin\phi = (1/k)\sin\theta/2$. The function $\sin\phi$ is defined by the latter equation since $\sin\theta_0/2 \geq \sin\theta/2 > 0$ if $0 \leq \theta \leq \theta_0 < \pi$. To change from the dummy variable of integration θ to the variable ϕ, note that $\cos\theta - \cos\theta_0 = 2(\sin^2\theta_0/2 - \sin^2\theta/2) = 2k^2(1 - \sin^2\phi) = 2k^2\cos^2\phi$ by a double angle formula of trigonometry. Next we have that $\cos\phi\,d\phi = [(2k)^{-1}\cos(\theta/2)]\,d\theta = (2k)^{-1}[1 - \sin^2(\theta/2)]^{1/2}\,d\theta = (2k)^{-1}(1 - k^2\sin^2\phi)^{1/2}\,d\theta$, or $d\theta = 2k\cos\phi\,d\phi(1 - k^2\sin^2\phi)^{-1/2}$. So $(\cos\theta - \cos\theta_0)^{-1/2}\,d\theta = (2k^2\cos^2\phi)^{-1/2}\,d\theta = 2^{1/2}(1 - k^2\sin^2\phi)^{-1/2}\,d\phi$. Since $\phi = 0$ when $\theta = 0$ and $\phi = \pi/2$ when $\theta = \theta_0$, the limits on the transformed integral are as given and the θ integral transforms to the ϕ integral. We expect the nonlinear pendulum motion modeled by $mL\theta'' = -mg\sin\theta$, $\theta(0) = \theta_0$, $\theta'(0) = 0$ to have a longer period than the motion of the linearized pendulum modeled by $mL\theta'' = -mg\theta$, $\theta(0) = \theta_0$, $\theta'(0) = 0$, since the magnitude of the restoring force of the nonlinear pendulum is less than that of the linearized pendulum, $|-mg\sin\theta| < |-mg\theta|$ with equality only at $\theta = 0$. So the nonlinear pendulum moves more slowly and has a longer period than the linearized pendulum, given identical initial conditions.

As $\theta_0 \to 0$, $k = \sin\theta_0/2 \to 0$ and the period T given by the integral tends to the value $4\sqrt{L/g}\int_0^{\pi/2}d\phi = 2\pi\sqrt{L/g}$. This is expected since for small $|\theta|$, $\sin\theta \approx \theta$, and the period of the nonlinear pendulum motion for small $|\theta_0|$, $\theta_0' = 0$, should be close to $2\pi\sqrt{L/g}$, which is the period of the linear pendulum motion regardless of initial data. At the other extreme as $\theta_0 \to \pi$, one expects the period to tend to ∞ since $\theta_0 = \pi$ corresponds to the pendulum standing at rest vertically upwards. With a slight displacement to $\pi - \varepsilon$, the pendulum would slowly move downward, gathering speed as it goes, then slow down as it climbs back up towards the top to come to rest at $-\pi + \varepsilon$ before it reverses direction. The smaller the value of ε, the longer the period.

9. Group project. Here is a sketch of the solution of the bug problem (Group Problem 9). Orient the table so that the center is at the origin and the x- and y-axes lie along the diagonals of the table. As the bugs move, these positions form the corners of a rotating and shrinking square. Since the problem is symmetric with respect to the paths of the four bugs, we will only find the path of one of the bugs, bug A say. Let $\mathbf{R}(t) = r(t)\hat{\mathbf{r}}$ be the position vector of bug A, and let $\hat{\boldsymbol{\theta}}$ be the unit vector orthogonal to $\hat{\mathbf{r}}$ and pointing toward bug B, the bug adjacent to A in the counterclockwise sense. Because of the symmetries of the problem, the velocity vector $\mathbf{R}'(t)$ always makes an angle of $135°$ with $\hat{\mathbf{r}}$ and an angle of $45°$ with $\hat{\boldsymbol{\theta}}$. But this implies that the component of $\mathbf{R}'(t)$ along $\hat{\mathbf{r}}$ is equal in magnitude but opposite in sign to the component along $\hat{\boldsymbol{\theta}}$. Since $\mathbf{R}' = r'\hat{\mathbf{r}} + r\hat{\mathbf{r}}' = r'\hat{\mathbf{r}} + r\theta'\hat{\boldsymbol{\theta}}$, the components referred to are r' and $r\theta'$, respectively. We must have $r' = -r\theta'$, that is, $dr/d\theta = -r$, and $r = Ce^{-\theta}$. Since $r = a/\sqrt{2}$ at $\theta = 0$, it follows that $C = a/\sqrt{2}$, and so we have $r = ae^{-\theta}/\sqrt{2}$. Considering the bugs as dimensionless points, we see that the path of each bug involves an infinitely increasing θ; $r \to 0$ as $\theta \to +\infty$.

And now for the kicker: although the angle θ goes to ∞, it takes only a finite time for the four bugs to meet at the center of the table! This follows from the fact that the speed, $\|\mathbf{R}'(t)\| = ((r')^2 + (r\theta')^2)^{1/2}$ is a positive constant, say b. Since $r' = -r\theta'$ (as above), we see that $b^2 = (r')^2 + (r\theta')^2 = 2(r')^2$. So, $r' = -b/\sqrt{2}$ (remember that $r' < 0$). Using the initial data $r(0) = a/\sqrt{2}$, we have that $r(t) = (-bt + a)/\sqrt{2}$. When $t = a/b$, $r(t) = 0$ and the bugs meet, probably pretty dizzy after all that whirling around the origin. Since $\theta' = -r'/r = -(\ln r)' = [\ln(1/r)]'$, we have that

$$\theta = \ln\left(\frac{1}{1 - (b/a)t}\right)$$

and we see that $\theta \to +\infty$ as $t \to (b/a)^-$.

4.2   Beats and Resonance

Suggestions for preparing a lecture

Topics: Free oscillations, forced oscillations, beats and resonance.

Making up a problem set

Parts of 1, 2, 4. Group problem 6 is fun. In fact, it is fun to do the experiment in class with a beaker of water and a ping pong ball (or a child's hollow play ball).

Comments

In this section we see how second-order constant coefficient linear ODEs respond to a sinusoidal driving force. ODEs of this kind often appear in the sciences and engineering. The notions of beats, resonance, free and forced oscillations (both in the undamped and damped cases) all arise in this setting. Take a pair of tuning forks tuned to slightly different frequencies and listen to the beats.

1. In each part, the equation has the form $y'' + \omega_0^2 y = A \cos \omega t$.

(a). For the ODE $y'' + 9y = 5 \cos 2t$, $A = 5$, $\omega_0 = 3$, and $\omega = 2$, so the solution exhibits beats. By (11) of Case 2, the general solution is $y(t) = C_1 \cos 3t + C_2 \sin 3t + \cos 2t$. Since $y(0) = y'(0) = 0$, $C_1 = -1$ and $C_2 = 0$, so the solution is $y(t) = \cos 2t - \cos 3t$.

(b). For the ODE $y'' + 4y = \cos 3t$, $A = 1$, $\omega_0 = 4$, and $\omega = 3$, so the solution exhibits beats. By (11) of Case 2, the general solution is $y(t) = C_1 \cos 4t + C_2 \sin 4t + (1/7) \cos 3t$. Since $y(0) = 1$ and $y'(0) = -1$, $C_1 = 6/7$ and $C_2 = -1/4$, so the solution is $y(t) = (6/7) \cos 4t - (1/4) \sin 4t + (1/7) \cos 3t$.

(c). For the ODE $y'' + y = \cos t$, $A = 1$ and $\omega_0 = \omega = 1$, so resonance occurs. By Case 1 in the text, $y = C_1 \cos t + C_2 \sin t + (1/2)t \sin t$. Since $y(0) = 1$ and $y'(0) = 0$, $C_1 = 1$ and $C_2 = 0$, so the solution is $y = \cos t + (1/2)t \sin t$.

2. The ODE is $y'' + 25y = \sin \omega t$. In each part the initial data $(0, 0)$, $(0, 10)$, and $(4, 0)$ have been used for $(y(0), y'(0))$. Overall we see that the solution curve with the initial values $y(0) = 0$, $y'(0) = 0$ shows most clearly the features of beats, resonance, or irrationally related frequencies connected with the particular value of ω.

(a). Fig. 2(a), Graph 1 shows beats, but the beats are somewhat obscured for the other two data sets [Figs. 2(a), Graphs 2 and 3].

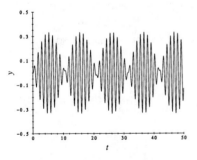
Problem 2(a), Graph 1.

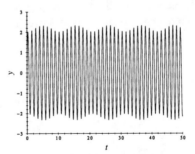
Problem 2(a), Graph 2.

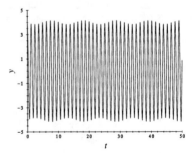
Problem 2(a), Graph 3.

(b). Now the input frequency 5.2 is close to the natural frequency 5.0, and the beats are pronounced, but the beat frequency of $(5.2 - 5.0)/2 = 0.1$ is very small [Fig. 2(b), Graph 1]. The beats are distinct, but not as sharp for the other two data sets [Figs. 2(b), Graphs 2 and 3].

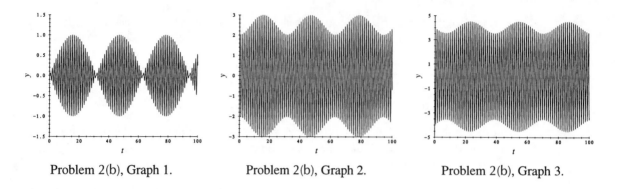

Problem 2(b), Graph 1.   Problem 2(b), Graph 2.   Problem 2(b), Graph 3.

(c). Here the input frequency 5 matches the natural frequency, and we have pure resonance since there is no damping. Figs. 2(c), Graphs 1–3 show three aspects of resonance.

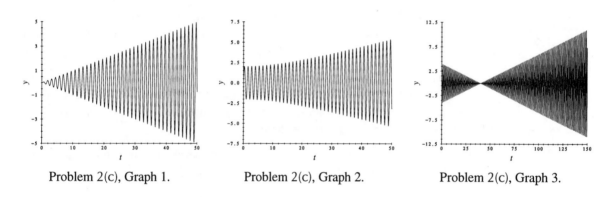

Problem 2(c), Graph 1.   Problem 2(c), Graph 2.   Problem 2(c), Graph 3.

(d). The natural frequency 5 and the input frequency π are not rationally related. Fig. 2(d), Graph 1 shows what the superposition of the two frequencies looks like with the first data set. For the other two data sets, the two frequencies are still there, but not nearly so obvious.

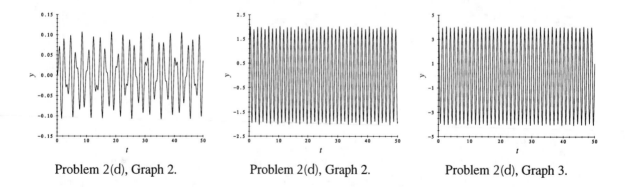

Problem 2(d), Graph 2.   Problem 2(d), Graph 2.   Problem 2(d), Graph 3.

3. Solving the IVP $y'' + \omega_0^2 y = A \sin \omega t$, $y(0) = 0$, $y'(0) = -A/(\omega_0 + \omega)$ as in Case 2 in the text, we obtain the general solution

$$y(t) = C_1 \cos \omega_0 t + C_2 \sin \omega_0 t + \frac{A}{\omega_0^2 - \omega^2} \sin \omega t$$

Since $y(0) = 0$, we have that $C_1 = 0$. Using $y'(0) = -A/(\omega_0 + \omega)$, we have $C_2 \omega_0 + A\omega/(\omega_0^2 - \omega^2) = -A/(\omega_0 + \omega)$, so $C_2 = -A/(\omega_0^2 - \omega^2)$. So

$$y(t) = A/(\omega_0^2 - \omega^2)[\sin \omega t - \sin \omega_0 t]$$

Using the hint,

$$y(t) = [2A/(\omega_0{}^2 - \omega^2)] \sin((\omega - \omega_0)t/2) \cos((\omega + \omega_0)t/2)$$

which exhibits the phenomenon of beats [see the explanation following (12)].

4. **(a).** Suppose that $x(t)$ is the distance from the midpoint of the block at time t to the surface of the water. Since the block is floating in equilibrium half-submerged, the mass of the block must equal the mass, $(hL^2/2)\rho$, of the displaced water. Now suppose the block is pushed down a distance $x < h/2$ into the water. Then the buoyant force acting on it must be $L^2 x \rho g$, the weight of the extra water displaced. By Newton's Second Law, $(hL^2/2)\rho x'' = -L^2 x \rho g$, which may be written as $x'' + 2gx/h = 0$. The solutions are $x = C_1 \cos t\sqrt{2g/h} + C_2 \sin t\sqrt{2g/h}$ of period $T = 2\pi\sqrt{h/2g}$, and so the block oscillates vertically with period $2\pi\sqrt{h/2g}$.

   (b). Repeat part **(a)**, but with a sphere of radius R floating half-submerged in equilibrium. The mass of the sphere must be the mass of the displaced water, which is $(2\pi R^3/3)\rho$, where ρ is the density of water. Suppose that the sphere is pushed into the water a distance x, $0 < x < R$. The extra volume of water displaced can be found by calculus (or by tables of geometric formulas) to be $\pi x(6R^2 - 2x^2)/6$. Newton's Second Law gives the equation of motion, $2\pi R^3 \rho x''/3 = -\pi x(6R^2 - 2x^2)\rho g/6 \approx -\pi R^2 \rho g x$, if x is very small compared to R. Rearranging, we have $x'' + 3gx/2R = 0$, and the solutions have the form $x = C_1 \cos(t\sqrt{3g/2R}) + C_2 \sin(t\sqrt{3g/2R})$, that is, a harmonic oscillator with period $T = 2\pi\sqrt{2R/3g}$. If x is not small compared to R, however, we cannot drop the term $-2x^2$, and the equation of motion is nonlinear: $x'' + (3g/2R^3)(R^2 x - x^3/3) = 0$. This ODE has the same form as the undamped soft spring ODE given in Section 3.1.

5. Here is the basic approach to the solution of the group problem. See Figs. 5, Graphs 1 and 2 for the solution curves of the two IVPs, $y'' + 4y = \cos \omega t$, $y(0) = y'(0) = 0$, $\omega = 3$, π. Figure 5, Graph 1 corresponds to $\omega = 3$, Fig. 5, Graph 2 to $\omega = \pi$. Only the upper part of the solution curve is plotted for $\omega = \pi$ so that the non-periodicity of the solution is clearly visible. The natural frequency for both solutions is $\omega_0 = 2$. The first solution curve is periodic of period 2π, but you can't easily detect the forced oscillation of period $2\pi/\omega = 2\pi/3$. The second curve is not periodic at all. In general, the solutions are given by formula (11) with $\omega_0 = 2$, $\omega = 3$, π: $y(t) = C_1 \cos 2t + C_2 \sin 2t + A/(4 - \omega^2) \cos \omega t$. If $\omega = 3$, then all solutions have period 2π as noted. If $\omega = \pi$, then the ratio of ω_0 and ω is irrational, and there are no periodic solutions unless $C_1 = C_2 = 0$. What do these results imply about the problem of detecting a periodic forced oscillation by examining the graph

of one solution of $y'' + \omega_0 y = A \cos \omega t$?

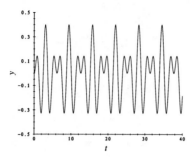

Problem 5, Graph 1.

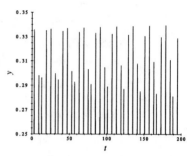

Problem 5, Graph 2.

6. Group project.

7. Group project.

4.3 Frequency Response Modeling

Suggestions for preparing a lecture

Topics: Forced damped oscillations, periodic forced oscillations, Bodé plots, parameter identification.

Remarks: The material of this section is standard fare for all engineering students.

Making up a problem set

Problem 1, one of 2–5, 6. You may want to assign group problem 7 or 8. For the more mathematically inclined students who know something about matrices, group problem 9 is intriguing.

Comments

Frequency response modeling is a meat-and-potatoes item on the engineering menu. Mathematicians are less accustomed to seeing the damped resonance phenomena couched in the engineering language of gain, phase shift, and Bodé plots used here, but once we get used to the new terms they seem natural and down-to-earth. The only real problem that we see is not the new terminology, but the need for care in the algebraic manipulation. We also touch on one of the most important engineering problems, the problem of parameter identification: How can you determine the coefficients of an ODE from a knowledge of the input and output?

1.   The text finds the unique periodic solution (11) of $y'' + 2cy' + k^2 y = F_0 \sin \omega t$. Proceeding in an analogous manner, replacing $\sin \omega t$ with $\cos \omega t$, but this time extracting the real instead of the imaginary part of $F_0 M(\omega) \exp[i(\omega t + \varphi(\omega))]$, we obtain

$$y_p = F_0 M(\omega) \cos(\omega t + \varphi(\omega))$$

where $M(\omega) = [(k^2 - \omega^2)^2 + 4c^2\omega^2]^{-1/2}$ and $\varphi(\omega) = \cot^{-1}\left((\omega^2 - k^2)/2c\omega\right)$ with $-\pi \leq \varphi(\omega) \leq 0$. See the discussion in the text following the second equation in (10) for the reason why $-\pi \leq \varphi(\omega) \leq 0$.

2. Figure 2 shows the input $3\sin 2t$ (dashed) of circular frequency $\omega = 2$ and the output $y(t)$ (solid) of the IVP, $y'' + 2y' + 4y = 3\sin 2t$, $y(0) = 0.5$, $y'(0) = 0.5$. $M(2)$ may be estimated from the plot by dividing the amplitude of the steady-state response by 3, the input amplitude. The phase shift may be estimated by finding how far to the right the output is shifted. The magnitude of this time shift is $|\varphi(\omega)/\omega| = |\varphi(2)/2|$ here. Multiplying this value by $\omega = 2$ gives us $-\varphi(2)$ since we're looking at how far to the *right* the output is shifted (a "lag time"). Using formula (10) with $2c = 2$ and $k^2 = 4$, we obtain $M(2) = 0.25$ and $\varphi(2) = -\pi/2$. The transfer function is $H(r) = 1/P(r) = 1/(r^2 + 2r + 4)$.

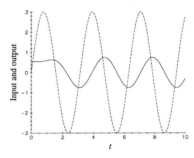

Problem 2.

3. Here, using the notation from ODE (1), $y'' + 2cy' + k^2 y = F_0 \sin \omega t$, we have $2c = 1$ and $k^2 = 2$ for the ODE, $y'' + y' + 2y = F_0 \sin \omega t$. So, from (10) we have that

$$M(\omega) = \frac{1}{\sqrt{(2 - \omega^2)^2 + \omega^2}} = \frac{1}{\sqrt{4 - 3\omega^2 + \omega^4}}, \qquad \varphi(\omega) = \cot^{-1}\left(\frac{\omega^2 - 2}{\omega}\right)$$

(a). $M(\omega)$ and $\varphi(\omega)$ can be found graphically by solving the ODE for various values of ω. $M(\omega)$ and $\varphi(\omega)/\omega$ can then be read off the plots. Plots should be similar to the plot given in Fig. 4.3.4.

(b). The formulas given above may be used to plot $M(\omega)$ and $\varphi(\omega)$, using the units of Fig. 4.3.3. See Figs. 3(b).

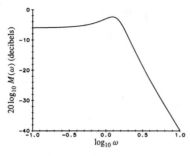

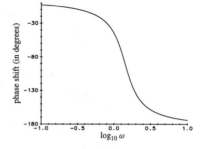

Problem 3(b), Graph 1.   Problem 3(b), Graph 2.

4. To find c and k from Fig. 4.3.3, pick a particular frequency ω_0 and find the corresponding values of $M(\omega_0)$ and $\varphi(\omega_0)$ from the Bodé plot. Formulas (10) then provide us with two equations in the unknowns $2c$ and k^2, allowing us to solve for approximate values for $2c$ and k^2. Answers may vary significantly, due to the exponential frequency scale which amplifies small errors in reading values off the plot.

5. **(a).** The Hooke's Law spring constant k satisfies the relationship $k \times$ deflection $= mg$. Using $g = 9.8$ m/s^2, $m = 1$ kg, deflection $= 0.0127$ m, we find $k \approx 772 N/m$. Solutions of $y'' + 2cy' + ky = 0$, where $c^2 < k$ so that the roots of the characteristic polynomial have non-zero imaginary part, have the form $y = Ae^{-ct}\cos\sqrt{k - c^2}t + Be^{-ct}\sin\sqrt{k - c^2}t$. Amplitudes decay by a factor of $0.01016/0.00254 = 4$ in 20 cycles here, so we need $e^{-c(20T)} = 1/4$, that is, $T = (1/20c)\ln 4$. Now $T = 2\pi/\sqrt{k - c^2}$, so we obtain $c = (1/40\pi)(\ln 4)\sqrt{k - c^2}$. Solving for c with $k = 772 N/m$, we obtain $2c = 0.613/s$.

(b). Resonance occurs at the value $\omega = \omega_r$ for which $M(\omega)$ is maximized, or (equivalently) where $(k - \omega^2)^2 + 4c^2\omega^2$ is minimized, where we have used (14) after replacing k^2 in the formula for $M(\omega)$ by k (since our ODE is $y'' + 2cy' + ky = 0$). Differentiating this expression with respect to ω and setting the derivative equal to 0, we find $\omega_r = \sqrt{k - 2c^2} \approx 27.8$/sec, where we have used the values of k and c given in **(a)**, $k = 772$ N/m and $2c = 0.613$/s. Then the maximal amplitude is $A \cdot M(\omega_r) = A/(2c\sqrt{k - c^2}) \approx 0.06A$.

6. Guessing a particular solution of the form $A\sin t + B\cos t$ for the ODE $y'' + 0.5y' + 16y = 100\sin t$, we see that $A = 60/9.01$, $B = -2/9.01$, so $y_p = (60\sin t - 2\cos t)/9.01$ is the unique periodic response. The roots of the characteristic polynomial $r^2 + 0.5r + 16$ are $-1/4 \pm i\sqrt{255/16}$, so the transients are given by $y_{tr} = C_1 e^{-t/4}\cos\beta t + C_2 e^{-t/4}\sin\beta t$, where $\beta = \sqrt{255/16} \approx 3.992$. Applying the initial conditions $y(0) = y'(0) = 0$ to $y = y_p + y_{tr}$, we obtain $C_1 \approx 0.222$ and $C_2 \approx -1.654$. So,

$$y_{tr} = 0.222e^{-t/4}\cos(3.992t) - 1.654e^{-t/4}\sin(3.992t)$$

See Fig. 6 for the solution curve (solid) of the IVP, along with the periodic component y_p (long dashes) and transient component y_{tr} (short dashes) of the solution.

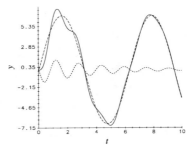

Problem 6.

7. Group project.

8. Group project.

9. Group project.

4.4 Electrical Circuits

Suggestions for preparing a lecture

Topics: A quick run through on voltage and current (the water analogy can help in this regard) and resistors, inductors, and capacitors. Then create a model of the simple circuit, and do the model for tuning a radio.

Remarks: Most students find it difficult to model circuits, so a little bit goes a long way.

Making up a problem set

Any one of 1–6, 7, 8. Group problem 9 matches up with Problem 7.

Comments

Electrical circuits provide a very nice application of second-order constant coefficient linear ODEs, an application of immense importance in today's technological world of communication channels. We start from scratch here and talk about charge carriers, voltage, currents, the whole bit. The going is heavy at times, but no one says that electricity is easy to understand. When all of the characters in the electrical circuit story have been introduced, we turn to the simple RLC circuit and solve the corresponding model ODE for the current in the circuit—no new solution techniques here, just the new language of circuits. The solution calculations are neither easier nor harder than before.

---

1. **(a).** From IVP (5), $LI'' + RI' + I/C = E'(t)$, we have, assuming that $R = 20$, $L = 10$, $C = 0.05$, $E(t) = 12$, and $10I'' + 20I' + I/0.05 = E' = 0$, or $I'' + 2I' + 2I = 0$. In addition, inserting $q_0 = 0.6$, $I_0 = 1$ into the boundary data formulas given in (5), we have $I(0) = 1$, $I'(0) = -2$. Since the roots of the characteristic polynomial are $-1 \pm i$, the general solution is $I(t) = k_1 e^{-t} \cos t + k_2 e^{-t} \sin t$. Applying the initial conditions, $k_1 = 1$

and $k_2 = -1$. So, $I(t) = e^{-t}[\cos t - \sin t]$, which is oscillatory with circular frequency 1 and decaying amplitude.

(b). Since $e^{-t} \to 0$ as $t \to \infty$, a damped oscillation in the current occurs. See Fig. 1(b); since e^{-t} decreases to zero very rapidly, the oscillations of $\cos t - \sin t$ aren't visible in the figure, although they are there.

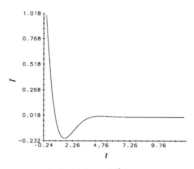

Problem 1(b).

(c). Overdamping occurs when the characteristic polynomial $Lr^2 + Rr + 1/C$ has distinct real roots, that is, if $R^2 > 4L/C$. In this case, that means when $R^2 > 800$, that is, when $R > 20\sqrt{2}$ ohms.

2. In every case we have $E(t) = E_0 = 0$, $q_0 = 10^{-3}$, $I_0 = 0$. IVP (5) becomes

$$LI'' + RI' + I/C = 0, \quad I_0 = 0, \quad I'_0 = -10^{-3}/LC$$

The characteristic polynomial of the ODE is $Lr^2 + Rr + C^{-1} = 0$, whose roots are $r_1 = -R/2L + \sqrt{R^2 - 4LC^{-1}}/2L, r_2 = -R/2L - \sqrt{R^2 - 4LC^{-1}}/2L$. If the roots are distinct,

$$I(t) = k_1 e^{r_1 t} + k_2 e^{r_2 t}$$

The initial data imply that $k_1 = -k_2 = [(r_2 - r_1)10^3 LC]^{-1}$. Plugging in values for L, R, and C will give us the current $I(t)$. The charge $q(t)$ is obtained from the formula

$$q(t) = q_0 + \int_0^t I(s)\,ds = 10^{-3} + k_1 \left(e^{r_1 t}/r_1 - e^{r_2 t}/r_2\right)\Big|_0^t$$

$$= 10^{-3} + k_1 \left[(e^{r_1 t} - 1)/r_1 - (e^{r_2 t} - 1)/r_2\right]$$

In each case we give only the values of r_1, r_2, k_1, and k_2.

(a). If $L = 0.3$ H, $R = 15$ Ω, $C = 3 \times 10^{-2}$ F, then $r_1 \approx -2.33$, $r_2 \approx -47.7$, $k_1 = -k_2 \approx -0.0024$.

(b). If $L = 1$ H, $R = 1000$ Ω, $C = 4 \times 10^{-4}$ F, then $r_1 \approx -2.51$, $r_2 \approx -997$, $k_1 = -k_2 \approx -0.0025$.

(c). If $L = 2.5$ H, $R = 500$ Ω, $C = 10^{-6}$ F, then $r_1 \approx -100 + 624i$, $r_2 \approx -100 - 624i$, $k_1 = -k_2 \approx 0.32i$.

3. Let q_1 denote the charge on the capacitor C_1, q_2 the charge on the capacitor C_2. Then by Kirchhoff's Voltage Law for the left and the right loops, respectively, $q_1/C_1 + q_2/C_2 = 0$ and $Rq_2' + q_2/C_2 = 0$, where $q_1(0) = C_1 E_0$.

   The IVP for $q_2(t)$ is $q_2' = -q_2/RC_2$, $q_2(0) = -C_2 E_0$, and so $q_2(t) = -C_2 E_0 e^{-t/RC_2}$. So $q_1 = -(C_1/C_2)q_2 = C_1 E_0 e^{-t/RC_2}$ and $I_1(t) = q_1'(t) = -C_1 E_0 e^{-t/RC_2}/C_2 R = (-1/6) \times 10^{-6} E_0 e^{-t/6}$ since $C_1 = 10^{-6}$, $C_2 = 2 \times 10^{-6}$, $R = 3 \times 10^6$.

4. The IVP to be solved is $10q'' + 20q' + 100q = 30\cos 2t$, where $q(0) = 0$ and $q'(0) = I(0) = 0$. We first find a particular solution of the form $q_p = A\sin 2t + B\cos 2t$. Inserting q_p into the ODE and matching coefficients of like terms to find A and B, we see that $6A - 4B = 0$, $4A + 6B = 3$, and so $A = 3/13$, $B = 9/26$ and $q_p = (3/13)\sin 2t + (9/26)\cos 2t$. Since the roots of the characteristic polynomial $10r^2 + 20r + 100$ are $-1 \pm 3i$, the general solution of the charge ODE is $q(t) = e^{-t}[k_1\cos 3t + k_2\sin 3t] + q_p(t)$. The constants k_1 and k_2 may be found from the initial data, $q(0) = q'(0) = 0$: $0 = k_1 + 9/26$, $0 = -k_1 + 3k_2 + 6/13$, and so $k_1 = -9/26$, $k_2 = -7/26$. The solution is $q(t) = (1/26)[-e^{-t}(9\cos 3t + 7\sin 3t) + 6\sin 2t + 9\cos 2t]$.

5. According to IVP (5), we have $2I'' + 7 \times 10^4 I' + 4I = E' = (60)' = 0$, $I(0) = 0$, $I'(0) = 60/2 = 30$. The solution is $I(t) = k_1 e^{r_1 t} + k_2 e^{r_2 t}$, where r_1 and r_2 are the roots of $r^2 + 3.5 \times 10^4 r + 2 = 0$ and $0 = k_1 + k_2$, $30 = r_1 k_1 + r_2 k_2$. Carrying out the calculations, we have $k_1 = 30/(r_1 - r_2) = -k_2$, and $r_1 = -34999.99994$, while $r_2 = -0.00006$. So, $k_1 \approx -0.000857 = -k_2$. The charge $q(t) = q_0 + \int_0^t I(s)\,ds = (k_1)/(r_1)[e^{r_1 t} - 1] - (k_1)/(r_2)[e^{r_2 t} - 1]$ coulombs. The dominant term is $-(k_1)/(r_2)[e^{r_2 t} - 1] \approx -14.29[e^{-0.00006t} - 1]$. If $t = 0.1$ sec, $q(0.1) \approx 0.000086$ coulombs.

6. **(a).** The ODE is $Lq'' + q/C = E_0\cos\omega t$, or $q'' + q/LC = (E_0/L)\cos\omega t$, where $\omega = 1/\sqrt{LC}$. The general solution of the undriven ODE is $k_1\cos\omega t + k_2\sin\omega t$. The frequency ω of the impressed voltage matches the natural frequency $1/\sqrt{LC}$ of the oscillations of the undriven system. There is a particular solution $At\cos\omega t + Bt\sin\omega t$. Matching coefficients, we have $A = 0$, $B = E_0/(2\omega L)$. So, $q = k_1\cos\omega t + (k_2 + E_0 t/2\omega L)\sin\omega t$ and $q(t)$ undergoes unbounded oscillations as $t \to \infty$ because of the term $t\sin\omega t$.

(b). If there is a positive resistance R, then the ODE is $Lq'' + Rq' + q/C = E_0\cos\omega t$. The roots of the characteristic polynomial have negative real parts: $r_1 = -R/2L \pm (R^2/4L^2 - 1/LC)^{1/2}$, so there is no frequency ω term in the solution set of the undriven system. A particular solution of the driven system may be found of the form $q_p(t) = A\cos\omega t + B\sin\omega t$. There are no terms of the form $At\cos\omega t$ or $Bt\sin\omega t$ in the solution $q(t)$. In this case $q(t) = k_1 e^{r_1 t} + k_2 e^{r_2 t} + A\cos\omega t + B\sin\omega t$, for arbitrary k_1 and k_2 and certain A and B. For $t \geq 0$, we have that $|k_1 e^{r_1 t}| \leq |k_1|$ and $|k_2 e^{r_2 t}| \leq |k_2|$, since r_1 and r_2 have negative real parts. Also, $|A\cos\omega t| \leq |A|$ and $|B\sin\omega t| \leq |B|$. So, $|q(t)|$ is bounded for $t \geq 0$.

7. With $L = 1$, $C = 1/25$ we know that the tuned circuit will home in on the frequency $\omega = 5$ input signal since $\omega^2 = 25 = 1/LC$ [see (16)]. The question is what will the various values of the resistance do to the amplitude of the response? See Fig. 7, Graph 1 for the graphs of the current vs. time in the three cases $R = 1.0$ (solid), $R = 0.1$ (short dashes), and $R = 0.01$ (long dashes). Apparently the lower the resistance, the larger the amplitude of the response; in the language of Section 4.3, the greater the gain. We have tuned the circuit to the input frequency ω by choosing C so that $1/LC = \omega^2$, and the resulting steady-state current is $I_p = (E_0/R)\cos(\omega t + \varphi)$ where $\omega E_0 = 4$ so $E_0 = 4/\omega = 4/5$. Since the amplitude of the input for the current ODE is ωE_0 (*not* E_0, which is the amplitude of the charge ODE), the gain is $(E_0/R)/E_0\omega = 1/R\omega$, i.e., $1/(5R)$ since $\omega = 5$. The amplitudes of the response currents seen in Fig. 7, Graph 1 approach $E_0/R = 4/5$ when $R = 1$ (short dashes), $(4/5)/(0.1) = 8$ when $R = 0.1$ (long dashes), and $(4/5)/(0.01) = 80$ when $R = 0.01$ (solid). The graphs are not of I_p, but include some transients as well. As t increases the graphs approach the graph of I_p. This is particularly evident if $R = 1.0$ and 0.1, where over the interval $90 \leq t \leq 100$, we are essentially looking at $I_p(t)$ with respective amplitudes $4/5$ ($R = 1.0$), 8 ($R = 0.1$). But, the graph corresponding to $R = 0.01$ has a long way to go before its amplitude approaches 80. Fig. 7, Graph 2 shows the current if $R = 0.01$; the ODE is solved over $0 \leq t \leq 1000$, but plotted only for $975 \leq t \leq 1000$, by which point the amplitude has approached 80.

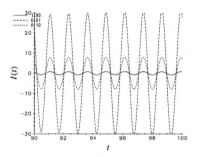

Problem 7, Graph 1.

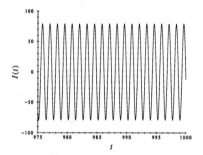

Problem 7, Graph 2.

8.　Kirchhoff's Current Law applied to the points a and b in the circuit implies that $I_2 = I_4 + I_5$ and $I_3 = I_1 + I_5$. So, $I_5 = I_3 - I_1$ and $I_4 = I_2 - I_3 + I_1$, and the currents through the various circuit elements may all be expressed in terms of the state variables I_1, I_2, and I_3. Apply Kirchhoff's Voltage Law to the outer loop of the circuit to obtain $E(t) = LI_1' + R_1 I_1 + R_2 I_3$. Apply the derivative of Kirchhoff's Voltage Law to the lower left and lower right loops, respectively, to obtain $E'(t) = I_2/C + R_1 I_4' = I_2/C + R_1(I_2' - I_3' + I_1')$ and $I_5/C + R_2 I_3' = R_1 I_4'$ or $(I_3 - I_1)/C + R_2 I_3' = R_1(I_2' - I_3' + I_1')$. This gives us three first-order, coupled rate equations for the currents I_1, I_2, and I_3:

$$LI_1' + R_1 I_1 + R_2 I_3 = E(t)$$

$$R_1 I_1' + R_1 I_2' - R_1 I_3' + I_2/C = E'(t)$$

$$R_1 I_1' + R_1 I_2' - (R_1 + R_2)I_3' + I_1/C - I_3/C = 0$$

9.　Group project.

10.　Group project.

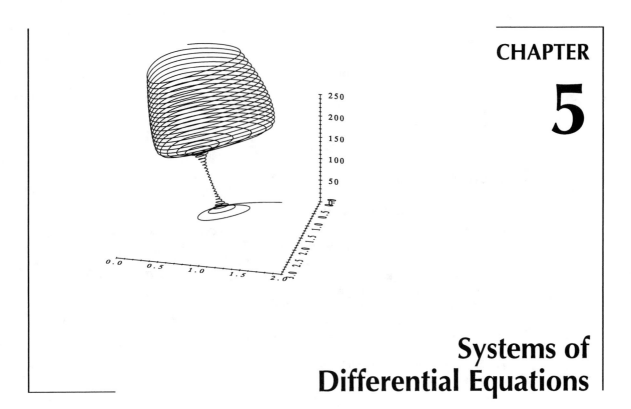

CHAPTER

5

Systems of
Differential Equations

Systems of ODEs have been used in earlier chapters, but informally. In Chapters 5, 7, 8, and 9 we describe the theory and practice of systems. Since solution formulas for systems of ODEs (particularly when they are not linear) are even rarer than they are for first- or second-order ODEs, we start out with computer simulations. That way, you can see what component curves, time-state curves, orbits, and so on look like. Now for some general observations about the contents of Chapter 5. Many models are presented in this chapter: coupled springs and masses, chemical reactions, models of predator-prey dynamics, competing or cooperating species, and models of the spread of a disease. The aim of this profusion of models is to show the naturalness of systems of ODEs as models of natural processes. Our experience is that people take to systems right away if there are models around to give some physical reality to the mathematical constructs. The chapter cover figure shows a time-state orbit of a model system for an autocatalytic reaction (see Example 5.1.5).

As we have done in Chapters 2 and 3, we introduce a version of the Fundamental Theorem (Theorem 5.2.1) for the existence, uniqueness, extension, and continuity/sensitivity properties of the solutions of initial value problems for systems of ODEs. Since this is its third appearance, readers will know what to expect, and have an idea of why we keep talking about these basic ideas. The sensitivity of solutions to changes in data again comes up in the text and in the problems. Our goal is to hit this idea so often that everyone will begin to understand its importance in the applications of ODEs. This is an area where computer simulations are a critical factor in conveying the effects of parameter changes.

Population models (like the Lotka-Volterra models of Section 5.4) are controversial because they are often viewed as reductionist; that is, they vastly oversimplify the reality of population dynamics. And that is certainly true in many cases. However, these models do provide a way to think about interacting species, and they are the first step in most attempts to gain an understanding of how populations interact. The possum disease model of Section 5.5 is due to Professor Graeme Wake and his colleagues in New Zealand. We have included it here because it is still in the development stage and models an important problem for the New Zealand economy. We think that every ODE course should have at least one open question, model, or theorem—something where the final "answer" isn't yet known. The possum model is one of these.

5.1   First-Order Systems

Suggestions for preparing a lecture

Topics: From a n-th order scalar ODE to a system, an example of converting a second-order ODE to a system, a model leading naturally to a system (e.g., the spring-mass system or the auto-catalator), the vocabulary of systems, solutions, orbits, and so on. If you have time for a second lecture and are heavily into modeling, you might do the springs and masses in one lecture and the chemical kinetics and the autocatalator in another.

Remarks: When lecturing on the contents of this section, we like to show how to convert a scalar ODE to a system and illustrate the process with a rehash of one of the falling body models from Section 1.5 (e.g., the whiffle ball model). Then we do the chemical kinetics modeling and the autocatalator, or the springs and masses, sometimes both. Whichever models or examples are used, graphs of component curves, orbits, and time-state curves should be shown to give visual reality to the ideas.

Making up a problem set

Although the text here emphasizes the graphical aspects of solutions of systems, students are often more comfortable with solution formulas, especially at the start of a serious study of systems. For this reason Problems 1, 2, 3, 5, 6 all ask for explicit solutions. We usually select three or four parts of Problems 1–3, 5, 6 [we like Problem 2(**d**)]. In addition we assign Problem 4 (if we have done springs and masses) and parts of Problems 7 and 8 (if we have done chemical kinetics). Assigning a group problem (like 9 and 10) is a good idea because it gets the students deeply involved with systems and graphics. **Warning:** The solution process for Problem 6 makes use of 2×2 determinants.

Comments

Systems of first-order ODEs have appeared several times in models in earlier chapters. In this section we show that a system of springs and masses may by modeled by a system of four linear first-order ODEs; the state variables in the model are the positions and the velocities of the masses suspended by the springs. The basic modeling elements for Hooke's Law springs appear in Section 3.1 and are used in constructing the model. We will come back to this model in Chapter 7, where we introduce the terminology of normal modes and natural frequencies to describe the periodic oscillations. We also introduce a new kind of model in the autocatalator system—a model of chemical kinetics. The basic principle of Chemical Mass Action is given and discussed; from the principle we can write out the rate equations for each of the species involved in the chemical reaction. The rate equations for autocatalytic reactions are nonlinear, and their solutions behave quite strangely.

     The examples in this section are discussed entirely in the light of their computer simulations. The aim of the examples is to show the naturalness of systems in modeling phenomena and to give the user of this book experience with computer simulation of systems. We also introduce vector notation for modeling systems, but we do not use the boldface notation of Section 4.1. Component curves, orbits, and time-state curves are defined and graphed. Linear systems are defined, but their full treatment is delayed to Chapter 7.

Experiments

Both of the processes described in this section can be illustrated with experiments. Physicists can give you springs and masses that can slide back and forth horizontally along a smooth surface. Then you can experiment with initial displacements from equilibrium that will give in-phase, out-of-phase oscillations, and non-periodic motions.

You can also illustrate the autocatalytic chemical process for yourself if you can line up some chemical equipment. If you want to do this experiment, take a look at *Chemical Demonstrations* by Lee R. Summerlin and James L. Ealy; Americal Chemical Society, Washington, 1985. Summerlin and Ealy give you all the information you need to do three autocatalytic experiments. In each experiment the autocatalytic oscillations are seen as color changes in a chemical solution. The actual reactions taking place are quite complicated, so the model given in this section can be viewed as just a rough "caricature" of what is going on in an actual reaction.

---

1. **(a).** The characteristic polynomial $r^2 - 4$ for $y'' - 4y = 0$ has roots ± 2, so the general solution is $y = C_1 e^{-2t} + C_2 e^{2t}$. If $x_1 = y$ and $x_2 = y'$, the equivalent system is $x_1' = x_2$, $x_2' = 4x_1$ with general solution $x_1 = y = C_1 e^{-2t} + C_2 e^{2t}$, $x_2 = y' = -2C_1 e^{-2t} + 2C_2 e^{2t}$.

(b). The characteristic polynomial $r^2 + 9$ for $y'' + 9y = 0$ has roots $\pm 3i$, so $y = C_1 \cos 3t + C_2 \sin 3t$ is the general solution. The equivalent system is $x_1' = x_2$, $x_2' = -9x_1$ with general solution $x_1 = y = C_1 \cos 3t + C_2 \sin 3t$, $x_2 = y' = -3C_1 \sin 3t + 3C_2 \cos 3t$.

(c). The characteristic polynomial $r^2 + 5r + 4$ for $y'' + 5y' + 4y = 0$ has roots -4, -1, so the general solution is $y = C_1 e^{-4t} + C_2 e^{-t}$. The equivalent system is $x_1' = x_2$, $x_2' = -4x_1 - 5x_2$ with general solution $x_1 = y = C_1 e^{-4t} + C_2 e^{-t}$, $x_2 = y' = -4C_1 e^{-4t} - C_2 e^{-t}$.

(d). Let $x_1 = y$, so $x_2 = x_1' = y'$ and $y'' = x_2' = -2x_1 - 2x_2 = -2y - 2y'$. The scalar ODE that is equivalent to the system $x_1' = x_2$, $x_2' = -2x_1 - 2x_2$ is $y'' + 2y' + 2y = 0$. The characteristic polynomial is $r^2 + 2r + 2$, which has roots $-1 \pm i$. The general solution is $y = e^{-t}(C_1 \cos t + C_2 \sin t)$. The general solution of the system is $x_1 = y = e^{-t}(C_1 \cos t + C_2 \sin t)$, $x_2 = y' = e^{-t}[(C_2 - C_1) \cos t - (C_1 + C_2) \sin t]$.

(e). Let $x_1 = y$, so $x_2 = x_1' = y'$ and $y'' = x_2' = -16x_1 = -16y$. The scalar ODE that is equivalent to the system, $x_1' = x_2$, $x_2' = -16x_1$, is $y'' + 16y = 0$. The characteristic polynomial is $r^2 + 16$, which has roots $\pm 4i$. The general solution is $y = C_1 \cos 4t + C_2 \sin 4t$. The general solution of the system is $x_1 = y = C_1 \cos 4t + C_2 \sin 4t$, $x_2 = y' = -4C_1 \sin 4t + 4C_2 \cos 4t$.

(f). The ODE, $y''' + 6y'' + 11y' + 6y = 0$, has characteristic polynomial $r^3 + 6r^2 + 11r + 6 = (r + 3)(r + 2)(r + 1)$. The general solution is $y = C_1 e^{-3t} + C_2 e^{-2t} + C_3 e^{-t}$. The equivalent system is obtained by setting $x_1 = y$, $x_2 = y'$, $x_3 = y''$; the system is $x_1' = x_2$, $x_2' = x_3$, $x_3' = -6x_1 - 11x_2 - 6x_3$, whose general solution is $x_1 = C_1 e^{-3t} + C_2 e^{-2t} + C_3 e^{-t}$, $x_2 = -3C_1 e^{-3t} - 2C_2 e^{-2t} - C_3 e^{-t}$, $x_3 = 9C_1 e^{-3t} + 4C_2 e^{-2t} + C_3 e^{-t}$.

2. The system $x' = y$, $y' = -bx - ay + A \cos \omega t$ is equivalent to the scalar ODE $x'' + ax' + bx = A \cos \omega t$. The general solution of the corresponding undriven linear second-order ODE, $x'' + ax' + b = 0$, is given by $x = C_1 e^{r_1 t} + C_2 e^{r_2 t}$ if $r_1 \neq r_2$, and $x = C_1 e^{r_1 t} + t C_2 e^{r_1 t}$ if $r_1 = r_2$, where $r_{1,2} = (-a \pm \sqrt{a^2 - 4b})/2$ are the roots of $r^2 + ar + b = 0$, and C_1, C_2 are arbitrary constants. In each case the first graph shows x- and y-component graphs, the second graph is the orbit in the xy-plane, and the third graph shows the time-state curve.

(a). The system $x' = y$, $y' = -x - y$ is equivalent to the scalar ODE $x'' + x' + x = 0$. Because the roots of $r^2 + r + 1$ are $r_{1,2} = (-1 \pm \sqrt{3}i)/2$, the general solution of the scalar ODE is

$$x = e^{-t/2}\left(C_1 \cos(\sqrt{3}t/2) + C_2 \sin(\sqrt{3}t/2)\right)$$

Using the initial data, $x(0) = 1$, $y(0) = x'(0) = 0$ to determine $C_1 = 1$ and $C_2 = 1/\sqrt{3}$, we have

$$x(t) = e^{-t/2}\left[\cos(\sqrt{3}t/2) + \frac{1}{\sqrt{3}}\sin(\sqrt{3}t/2)\right]$$

$$y(t) = x'(t) = -\frac{2}{\sqrt{3}}e^{-t/2}\sin(\sqrt{3}t/2)$$

See Figs. 2(a). The sinusoidal oscillations are there, but barely visible in the graphs since the decaying exponential $e^{-t/2}$ quickly damps the amplitudes of the oscillations.

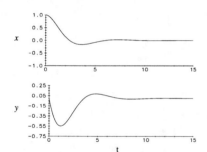

Problem 2(a), Graph 1.

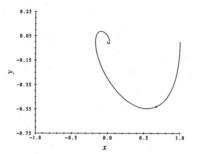

Problem 2(a), Graph 2.

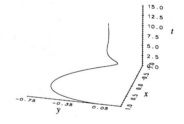

Problem 2(a), Graph 3.

(b). The system $x' = y$, $y' = -x - 3y$ is equivalent to the scalar ODE $x'' + 3x' + x = 0$. Because the roots of $r^2 + 3r + 1$ are $r_{1,2} = (-3 \pm \sqrt{5})/2$, the general solution of the scalar ODE is $x = C_1 e^{(-3+\sqrt{5})t/2} + C_2 e^{(-3-\sqrt{5})t/2}$. Using the initial data $x(0) = 1$, $y(0) = 0$ to determine $C_1 = (\sqrt{5} + 3)/2\sqrt{5}$ and $C_2 = (\sqrt{5} - 3)/2\sqrt{5}$, we have

$$x(t) = \frac{\sqrt{5} + 3}{2\sqrt{5}}e^{(-3+\sqrt{5})t/2} + \frac{\sqrt{5} - 3}{2\sqrt{5}}e^{(-3-\sqrt{5})t/2}$$

$$y(t) = x'(t) = -\frac{\sqrt{5}}{5}e^{(-3+\sqrt{5})t/2} + \frac{\sqrt{5}}{5}e^{(-3-\sqrt{5})t/2}$$

See Figs. 2(b). The damping constant is large enough here that the solutions do not involve sinusoids at all, but only exponentials which decay to 0 as $t \to \infty$. The graphs show this behavior.

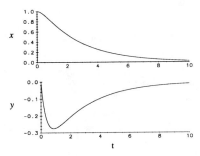

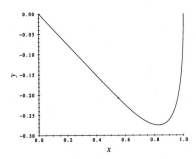

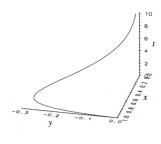

Problem 2(b), Graph 1.   Problem 2(b), Graph 2.   Problem 2(b), Graph 3.

(c). The system $x' = y$, $y' = -25x + \cos 5t$ is equivalent to the scalar ODE $x'' + 25x = \cos 5t$. The roots of $r^2 + 25$ are $r_{1,2} = \pm 5i$, and the general solution of the undriven scalar ODE is $x = C_1 \cos 5t + C_2 \sin 5t$. A particular solution of the driven ODE is $(t \sin 5t)/10$. Thus, the general solution is $x = C_1 \cos 5t + C_2 \sin 5t + (t \sin 5t)/10$. Using the initial data $x(0) = 0$, $y(0) = 0$, we obtain $x(t) = (t \sin 5t)/10$ and $y(t) = x'(t) = (\sin 5t + 5t \cos 5t)/10$. See Figs. 2(c). Note that this is an example of resonance (see Section 4.2), and the graphs show oscillations with increasing amplitudes.

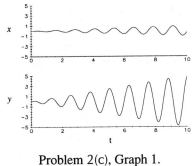

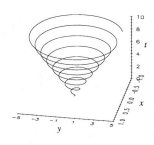

Problem 2(c), Graph 1.   Problem 2(c), Graph 2.   Problem 2(c), Graph 3.

(d). The system $x' = y$, $y' = -25x + \cos(5.5t)$ is equivalent to the scalar ODE $x'' + 25x = \cos(5.5t)$. The roots of $r^2 + 25$ are $r_{1,2} = \pm 5i$, and the general solution of the undriven scalar ODE is $x = C_1 \cos 5t + C_2 \sin 5t$. A particular solution of the driven ODE is $-(4 \cos 5.5t)/21$. Thus, the general solution is $x = C_1 \cos 5t + C_2 \sin 5t - (4 \cos 5.5t)/21$. Using the initial conditions $x(0) = 0$, $y(0) = 0$, we have $x(t) = (4/21)(\cos 5t - \cos 5.5t)$ and $y(t) = x'(t) = (4/21)(5.5 \sin 5.5t - 5 \sin 5t)$. See Figs. 2(d). Note the beats (see Section 4.2). The graphs show the characteristic pulsations of the beats when the natural frequency (5 in this case) and the driving frequency (5.5 in this case) are close in value.

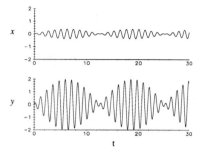

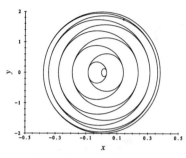

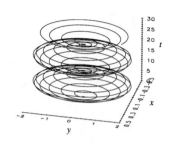

Problem 2(d), Graph 1.   Problem 2(d), Graph 2.   Problem 2(d), Graph 3.

3. These linear cascade problems are all solved by first solving the "source ODEs" and insert-
    ing the solutions in the ODEs at the next cascade level, solving those ODEs and repeating
    the process until the "final products" have been found. In every case we use the integrating
    factor technique of Sections 1.3 and 1.4.

    (a). The solution of the first IVP in the system $x_1' = -3x_1$, $x_2' = x_1$, $x_3' = 2x_1$, $x_1(0) = 0$, $x_2(0) = x_3(0) = 0$, is $x_1 = 10e^{-3t}$. Inserting this solution for x_1 into the second and
    third equations, $x_2' = 10e^{-3t}$ and $x_3' = 20e^{-3t}$. Integrating these and applying the initial
    conditions, we obtain the solution $x_1 = 10e^{-3t}$, $x_2 = 10(1 - e^{-3t})/3$, $x_3 = 20(1 - e^{-3t})/3$.
    The limiting values as $t \to +\infty$ are $x_1 = 0$, $x_2 = 10/3$ and $x_3 = 20/3$. The source variable
    is x_1; x_2 and x_3 are final products.

    (b). The solution of the first ODE in the cascade system $x_1' = -x_1$, $x_2' = x_1 - 3x_2$, $x_1(0) = 10$, $x_2(0) = 20$ is $x_1 = 10e^{-t}$. Inserting this solution for x_1 into the second equation,
    $x_2' = 10e^{-t} - 3x_2$. Using the integrating factor e^{3t} and applying the initial condition, we
    obtain the solution $x_1 = 10e^{-t}$, $x_2 = 5e^{-t} + 15e^{-3t}$. The limiting values as $t \to +\infty$ are
    $x_1 = x_2 = 0$; x_1 is the source variable, but there is no final product because both x_1 and x_2
    die out.

    (c). The solutions of the first and second IVPs in the system $x_1' = -2x_1$, $x_2' = -3x_2$,
    $x_3' = 2x_1 + 3x_2$, $x_1(0) = 1$, $x_2(0) = 2$, $x_3(0) = 0$ are $x_1 = e^{-2t}$, $x_2 = 2e^{-3t}$. Inserting these
    into the third equation, $x_3' = 2e^{-2t} + 6e^{-3t}$. Integrating and applying the initial conditions,
    we obtain the solution $x_1 = e^{-2t}$, $x_2 = 2e^{-3t}$, $x_3 = 3 - e^{-2t} - 2e^{-3t}$. The limiting values
    as $t \to +\infty$ are $x_1 = x_2 = 0$, $x_3 = 3$; x_1 and x_2 are source variables, and x_3 is the final
    product variable.

4. The system that models the positions (relative to equilibrium) and velocities of the coupled
    springs and masses is $x_1' = x_2$, $x_2' = -x_1 + \alpha x_3$, $x_3' = x_4$, $x_4' = x_1 - x_3$, where x_1 and x_3
    are the respective positions of the upper and lower masses, x_2 and x_4 are the corresponding
    velocities, and α is the parameter k_2/m_1.

        Graphs 1 and 2 in each of parts **(a)**, **(b)**, **(c)** show the x_1 (solid) and the x_3 (dashed)
    component graphs with initial data $(-\sqrt{\alpha}, 0, 1, 0)$ and $(\sqrt{\alpha}, 0, 1, 0)$, respectively. These
    graphs show that if the top mass is placed so that $x_1 = \pm\sqrt{\alpha}$ and the lower mass so that
    $x_3 = 1$, and then both released, the subsequent motion of both springs is periodic with a
    common period. The period is longer if $x_1 = \sqrt{\alpha}$ than if $x_1 = -\sqrt{\alpha}$. Graph 3 shows the

$x_1x_3x_2$-projection of the solution with initial data $(2\sqrt{\alpha}, 0, 0, 0)$ as well as two periodic solution projections corresponding to initial data $(\pm\sqrt{\alpha}, 0, 1, 0)$. The two periodic solutions correspond to the oval cycles, while the solution with initial data $(2\sqrt{\alpha}, 0, 0, 0)$ generates the curve that appears to move on the surface of a doughnut (torus). Graph 4 is the projection of these three solutions into the x_1x_3-plane. We see the Lissajous figure corresponding to the initial data $(2\sqrt{\alpha}, 0, 0, 0)$, while the two periodic solutions project onto the crossed line segments. The Lissajous curve is not periodic.

(a). See Figs. 4(a) where $\alpha = 0.05$.

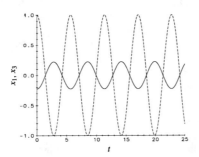

Problem 4(a), Graph 1.

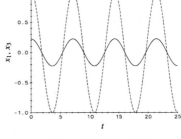

Problem 4(a), Graph 2.

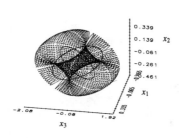

Problem 4(a), Graph 3.

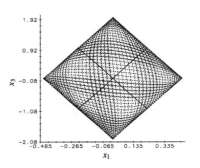

Problem 4(a), Graph 4.

(b). See Figs. 4(b) where $\alpha = 0.5$.

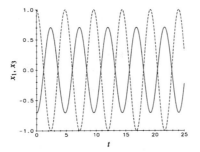

Problem 4(b), Graph 1.

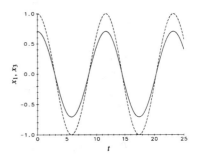

Problem 4(b), Graph 2.

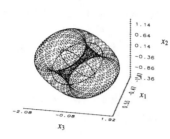

Problem 4(b), Graph 3.

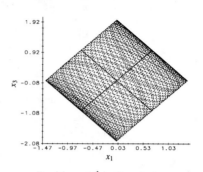

Problem 4(b), Graph 4.

(c). See Figs. 4(c) where $\alpha = 0.95$.

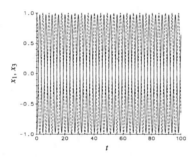

Problem 4(c), Graph 1.

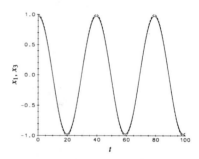

Problem 4(c), Graph 2.

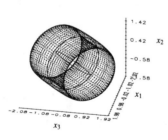

Problem 4(c), Graph 3.

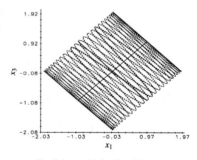

Problem 4(c), Graph 4.

5. **(a).** For the system $x_1' = x_2$, $x_2' = -3x_2$, $f_1 = x_2$ and $f_2 = -3x_2$. Separating variables
    in the ODE, $dx_2/dx_1 = f_2/f_1 = -3x_2/x_1$, we have $dx_2/x_2 = -3dx_1/x_1$, implying that
    $\ln|x_2| = -3\ln|x_1| + K$. So $x_2 = Cx_1^{-3}$ is a family of solutions of $dx_2/dx_1 = -3x_2/x_1$ and
    gives a family of orbits of the original system. Note that $x_1 = 0$ is excluded because of the
    division by x_1 in the ODE, although $x_1(t) = 0$, for all t, is an orbit of the original system.
    This shows that some orbits may be lost by this solution process. See Fig. 5(a).

    (b). For the system $x_1' = x_1x_2$, $x_2' = x_1^2 + x_2^2$, $f_1 = x_1x_2$ and $f_2 = x_1^2 + x_2^2$. The method of
    homogeneous functions (Section 1.9) leads to a family of solutions $x_2^2 = 2x_1^2\ln|x_1| + Cx_1^2$

of the ODE, $dx_2/dx_1 = (x_1^2 + x_2^2)/(x_1 x_2)$ [set $x_2 = zx_1$, rewrite the ODE in terms of z and x_1, separate variables, solve and then replace z by x_2/x_1.] As in (a), an orbit is lost in the process, namely, $x_1(t) = 0$. See Fig. 5(b).

(c). For the system $x_1' = x_2$, $x_2' = -e^{-x_1}$, $f_1 = x_2$ and $f_2 = -e^{-x_1}$. We have that $dx_2/dx_1 = -e^{-x_1}/x_2$. Separating variables and solving, we obtain $x_2^2 = 2e^{-x_1} + C$; note that for each negative C, x_1 must be restricted to an interval for which $2e^{-x_1} + C \geq 0$, that is, $x_1 \leq \ln(-2/C)$. Note also that $x_2(t)$ is always decreasing because x_2' is negative. Since $x_2' = -e^{-x_1}$, the magnitude of x_2' is huge if x_1 is negative. See Fig. 5(c).

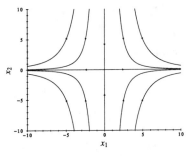

Problem 5(a).

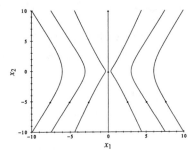

Problem 5(b).

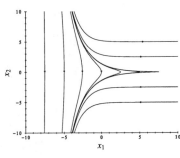

Problem 5(c).

6. **(a).** The system is $x_1' = x_1 + 3x_2$, $x_2' = x_1 - x_2$. Following the suggestion, the constants must be chosen so that for all t, $x_1' = r\alpha e^{rt} + s\beta e^{st} = x_1 + 3x_2 = (\alpha + 3\gamma)e^{rt} + (\beta + 3\delta)e^{st}$ and $x_2' = r\gamma e^{rt} + s\delta e^{st} = x_1 - x_2 = (\alpha - \gamma)e^{rt} + (\beta - \delta)e^{st}$. Matching coefficients of e^{rt} on each side, we have that $r\alpha = \alpha + 3\gamma$ and $r\gamma = \alpha - \gamma$, or $(1 - r)\alpha + 3\gamma = 0$ and $\alpha + (-1 - r)\gamma = 0$. Aside from the solution $\alpha = \gamma = 0$, these equations have nontrivial solutions α, γ if the determinant of the coefficient matrix $\begin{bmatrix} 1 - r & 3 \\ 1 & -1 - r \end{bmatrix}$ vanishes, that is, if $(1 - r)(-1 - r) - 3 = r^2 - 4 = 0$, or $r = \pm 2$. Similarly, matching coefficients of e^{st} leads to equations of exactly the same form, s, γ, δ replacing r, α, β, respectively. So we may as well set $r = -2$. Solving the resulting equations for s, α, β, γ, and δ, $\gamma = -\alpha$ and $s = 2$ with $\beta = 3\delta$. Renaming α and δ as C_1 and C_2, arbitrary constants, we have the solutions $x_1 = C_1 e^{-2t} + 3C_2 e^{2t}$, $x_2 = -C_1 e^{-2t} + C_2 e^{2t}$. Note that we have not shown that these are *all* solutions of the system.

(b). The system is $x_1' = 2x_1 - x_2$, $x_2' = 3x_1 - 2x_2$. Constants must be chosen so that for all t, $x_1' = r\alpha e^{rt} + s\beta e^{st} = 2x_1 - x_2 = (2\alpha - \gamma)e^{rt} + (2\beta - \gamma)e^{st}$ and $x_2' = r\gamma e^{rt} + s\delta e^{st} = 3x_1 - 2x_2 = (3\alpha - 2\gamma)e^{rt} + (3\beta - 2\delta)e^{st}$. Equating the coefficients of e^{rt} on each side of each equation, we have $r\alpha = 2\alpha - \gamma$ and $r\gamma = 3\alpha - 2\gamma$, or $(2 - r)\alpha - \gamma = 0$ and $3\alpha + (-2 - r)\gamma = 0$. We have nontrivial solutions α, γ if $\det \begin{bmatrix} 2 - r & -1 \\ 3 & -2 - r \end{bmatrix} = 0$, i.e., if $(2 - r)(-2 - r) + 3 = r^2 - 1 = 0$, or $r = \pm 1$. The equations relating s, β, δ are exactly the same, s, β, δ replacing r, α, γ, respectively. We may set $r = -1$ with $\gamma = 3\alpha$ and $s = 1$ with $\delta = \beta$. Renaming α and β as C_1 and C_2, we have the solutions $x_1 = C_1 e^{-t} + C_2 e^t$,

$x_2 = 3C_1 e^{-t} + C_2 e^t$, where C_1 and C_2 are arbitrary constants. As in part **(a)**, we have not shown that these are *all* solutions of the system.

7.　Use the discussion above text system (17) and also the Chemical Law of Mass Action as guides. Lowercase letters w, x, y, z denote concentrations.

(a). The speed of the first step of the reaction schematic, $X + Y \overset{k_1}{\to} Z \overset{k_2}{\to} W$, is $k_1 xy$, while the speed of the second step is $k_2 z$. So the ODEs are $x' = -k_1 xy$, $y' = -k_1 xy$, $z' = k_1 xy - k_2 z$, $w' = k_2 z$.

(b). The speed of the reaction with the schematic, $X + 2Y \overset{k}{\to} Z$, is kxy^2. Notice that Y is consumed at twice the rate of X. The three ODEs are $x' = -kxy^2$, $y' = -2kxy^2$, $z' = kxy^2$.

(c). The speed of the reaction with the schematic, $X + 2Y \overset{k}{\to} 6Y + W$, is kxy^2. Since 2 units of Y are consumed to generate 6 units of Y, the net gain is 4 units of Y. So the ODEs are $x' = -kxy^2$, $y' = 4kxy^2$, $w' = kxy^2$.

8.　**(a).** The unusual feature of the graphs of $x(t)$ and $y(t)$ where $w(0) = 50$ [Fig. 8(a), Graph 1], 100 [Fig. 8(b), Graph 2] is that there is nothing unusual. At least not in comparison with the oscillations in Figure 5.1.5, where $w(0) = 500$. Apparently, the oscillations are triggered by a large initial concentration of the reactant. Otherwise, the intermediates X and Y rise and then fall in the expected way.

(b). Computer simulations suggest that the oscillations in the concentrations of the intermediates are first seen if $w(0)$ is about 120. See Fig. 8(b).

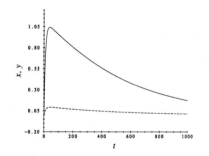

Problem 8(a), Graph 1.

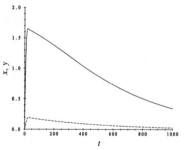

Problem 8(a), Graph 2.

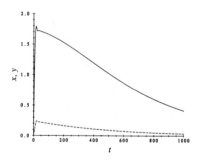

Problem 8(b).

(c). Solve the autocatalator IVP (17) for the four values of $\omega(0)$: 50, 250, 500, 800. Since $\omega(t) = \omega_0 e^{0.002t}$, the first term on the right of the x' ODE is $0.002\omega_0 e^{0.002t} = 0.1 e^{-0.002t}$, $0.5 e^{-0.002t}$, $e^{-0.002t}$, $1.60 e^{-0.002t}$, respectively. Figures 8(c) show the time-state curves produced by our solver (your solver may generate somewhat different graphs—depending on the axis scales and viewpoint). The graphs suggest that some minimal value of $\omega(0)$ is needed to trigger the oscillations and this value is larger than 50 (Graph 1) but less than 250 (Graph 2). If $\omega(0)$ is too large (Graph 4) the time of the onset of the oscillations may be considerably delayed (note the new time scale in Graph 4) and their duration and amplitude diminished. You might want to plot component curves and see what information they give you about these questions.

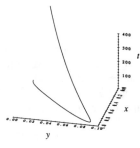

Problem 8(c), Graph 1.

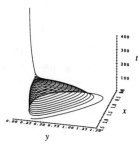

Problem 8(c), Graph 2.

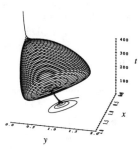

Problem 8(c), Graph 3.

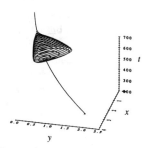

Problem 8(c), Graph 4.

9. Group project.

10. Group project.

5.2 Properties of Systems

Suggestions for preparing a lecture

Topics: We have enough material in this section that you will want to pick and choose topics and examples to support what you want to emphasize. You might give a brief mention of the Fundamental Theorem (Theorem 5.2.1), a discussion (again) of autonomous systems with a graphical example based on one of the autonomous damped spring models of Sections 3.1 and 3.2, or on the simple pendulum of Section 4.1, and a brief comment on sensitivity of solutions to a parameter change. Introduce direction fields for planar autonomous systems, but be prepared to answer questions about the difference between the direction fields here and the direction fields of Section 1.2. The latter may be interpreted as direction fields for the system $y' = f(t, y)$, $t' = 1$.

Remarks: Either spend two class sessions on the material, or else select some of the material for one lecture and bring in the remainder in later lectures or other sections, which is what we do. When we talk about systems in our courses, we often treat this material rather hastily, but then keep referring back to it. Eventually, students seem to get it.

Making up a problem set

The problems for this section are something of a grab bag. None is especially deep or abstract (except for Problem 5, which is both). One aim is to get students working with systems and seeing how the solutions change when parameters change. Two or three parts of 1 and 2, 4, 6, and possibly 7. If you haven't given a group problem for a while, assign Problem 8—scientists and engineers like this problem, maybe because pendulum motion is easy to picture. Problem 5 might be given for the proof-oriented students.

Comments

Here we are with another Fundamental Theorem (Theorem 5.2.1). This time the theorem is so inclusive it contains the Fundamental Theorem of Section 2.3. Since we work mostly with systems of first-order ODEs in Chapters 5, 7, 8, and 9, the theorem is formulated for these systems. There is quite a bit of basic material for the study of systems of ODEs in this section besides the Fundamental Theorem: autonomous systems, equilibrium points and cycles, maximal extension of solutions of autonomous systems, planar autonomous systems and their direction fields and nullclines, sensitivity to parameter changes.

One very important point may get buried in the mass of detail: parameters in the rate functions can be treated as additional state variables whose rates of change are all zero. The text above Example 5.2.7 shows how this is done. Applying the Fundamental Theorem to the extended system, we conclude that solutions are continuous in all the parameters as long as the hypotheses of the theorem are satisfied by the rate functions.

---

1. The hypotheses of the Fundamental Theorem are satisfied in each part for all points in state space and for all t since the rate functions in each case are polynomials in the state variables, and all polynomials are continuous and continuously differentiable. So, each of the initial value problems has a unique maximally extended solution defined on some time interval containing $t_0 = 0$. The solutions are defined for all real t unless specifically noted otherwise. The origin in state space is the only equilibrium point since the right-hand sides vanish only at that point. The x- and y-nullclines are the dashed curves. They are not always visible because sometimes orbits lie on top of them. Some of the nine orbits coincide with one another, and so fewer than nine orbits may be visible in the pictures. The 9 initial points to use are $(-1, -1)$, $(-1, 0)$, $(-1, 2)$, $(0, -1)$, $(0, 0)$, $(0, 2)$, $(2, -1)$, $(2, 0)$, and $(2, 2)$.

   (a). The system is $x' = y$, $y' = -x - 2y$. The corresponding scalar IVP, $x'' + 2x' + x = 0$, $x(0) = a$, $x'(0) = y(0) = b$, has characteristic polynomial $r^2 + 2r + 1$ with -1 as a double root. The general solution has the form $x = C_1 e^{-t} + C_2 t e^{-t}$. Applying the initial conditions, we obtain $x = e^{-t}(a + (a + b)t)$. Then $y = x' = e^{-t}(b - (a + b)t)$. See Fig. 1(a), where the x-nullcline is $y = 0$, and the y-nullcline is $y = -x/2$; $(0, 0)$ is the equilibrium point.

(b). The ODEs, $x' = 2x$, $y' = -4y$, are already uncoupled; the solutions are $x = ae^{2t}$, $y = be^{-4t}$. See Fig. 1(b), where the x-nullcline is $x = 0$ and the y-nullcline is $y = 0$; $(0, 0)$ is an equilibrium point.

(c). The system is $x' = y$, $y' = -9x$. The equivalent scalar IVP $x'' + 9x = 0$, $x(0) = a$, $x'(0) = y(0) = b$ has characteristic polynomial $r^2 + 9$ with roots $\pm 3i$. The general solution has the form $x = C_1 \cos 3t + C_2 \sin 3t$. Applying the initial conditions, we obtain $x = a \cos 3t + (b \sin 3t)/3$, and so $y = x' = -3a \sin 3t + b \cos 3t$. All orbits aside from the equilibrium point $(0, 0)$ are cycles. See Fig. 1(c), where the x- and y-nullclines are, respectively, $y = 0$ and $x = 0$.

(d). From the system $x' = y^3$, $y' = -x^3$ we have that $dy/dx = -x^3/y^3$, $x^3 dx + y^3 dy = 0$, or upon solving, $x^4 + y^4 = C$. Since $x(0) = a$ and $y(0) = b$, the equation of the orbit is $x^4 + y^4 = a^4 + b^4$. Note that we have not found $x(t)$ and $y(t)$, but solutions are defined for all t since the orbits are cycles. See Fig. 1(d), where the x- and y-nullclines are, respectively, $y = 0$ and $x = 0$.

(e). The ODEs $x' = -x^3$, $y' = -y$ are uncoupled and may be solved directly. Separating variables, we have $x^{-3} dx = -dt$, $x(0) = a$, which leads to $a^{-2} - x^{-2} = -2t$, or $x = a(1 + 2a^2 t)^{-1/2}$, where $t > -1/2a^2$ (or if $a = 0$, $|t| < \infty$). In addition, the solution of the IVP, $y' = -y$, $y(0) = b$, is $y = be^{-t}$ for the same t-interval. See Fig. 1(e), where the x- and y-nullclines are, respectively, $y = 0$ and $x = 0$; $(0, 0)$ is an equilibrium point.

(f). The third ODE, $z' = -z/2$, is uncoupled from the others and has solution $z = e^{-t/2}$. The scalar ODE equivalent to the system $x' = y$, $y' = -26x - 2y$ is $x'' + 2x' + 26x = 0$, where $x(0) = x'(0) = y(0) = 1$. The characteristic polynomial $r^2 + 2r + 26$ has roots $-1 \pm \sqrt{5}$. The general solution has the form $x = C_1 e^{-t} \cos 5t + C_2 e^{-t} \sin 5t$. Applying the initial conditions, we obtain $x = e^{-t}[\cos 5t + (2/5) \sin 5t]$; then $y = e^{-t}[\cos 5t - (27/5) \sin 5t]$. So, all state variables tend to 0 as $t \to \infty$, x and y in an oscillatory manner, z monotonically decreasing. See Figs. 1(f) for the respective xy, xz, yz, and xyz graphs; $(0, 0, 0)$ is an equilibrium point.

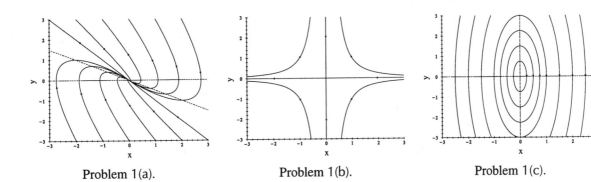

Problem 1(a).   Problem 1(b).   Problem 1(c).

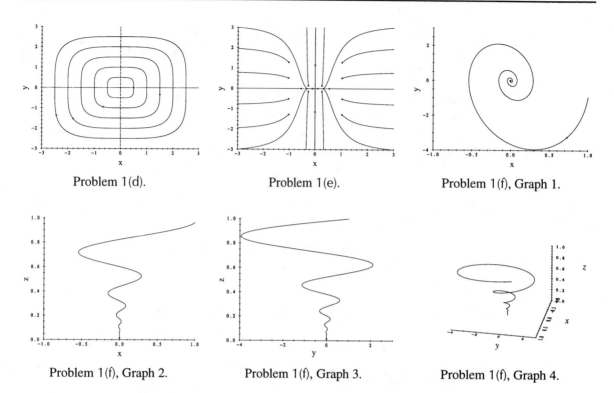

Problem 1(d).   Problem 1(e).   Problem 1(f), Graph 1.

Problem 1(f), Graph 2.   Problem 1(f), Graph 3.   Problem 1(f), Graph 4.

2.   The equilibrium points are found by setting the right-hand sides of each system equal to 0 and solving simultaneously for x and y. The x- and y-nullclines (dashed, unless along an axis) are determined by setting each rate function equal to zero.

(a). The system is $x' = x - y^2$, $y' = x - y$. Since $x = y$ and $x = y^2$ at an equilibrium point, we have $y = y^2$ or $y = 0, 1$. The equilibrium points are $(0,0)$ and $(1,1)$. See Fig. 2(a), where the x- and y-nullclines (dashed) are, respectively, $x = y^2$ and $x = y$. There seems to be a family of cycles enclosing $(1, 1)$, while other orbits head toward $(0, 0)$, but then veer away on one side or the other.

(b). The system is $x' = y \sin x$, $y' = xy$. Since either $x = 0$ or $y = 0$ makes $x' = 0$ and $y' = 0$, the two coordinate axes are lines of equilibrium points: $(a, 0)$, $(0, b)$, where a and b are arbitrary constants. See Fig. 2(b), where the x-nullclines are the lines given by $y = 0$ and by $x = n\pi$ and the y-nullclines are the x- and y- axes. Orbits seem to emerge from the positive x- and y-axes as t increases from $-\infty$. They seem to approach the negative x- or y-axes as $t \to +\infty$. The axes are lines of equilibrium points.

(c). The system is $x' = 2 + \sin(x + y)$, $y' = x - y^3 + 27$. There are no equilibrium points since $2 + \sin(x + y)$ is never smaller than $+1$; x' is always positive, so no cycles. See Fig. 2(c). The y-nullcline is given by $x = y^3 - 27$. Note that segments of orbits approach this curve asymptotically, hiding it from view in Fig. 2(c).

(d). The system is $x' = y + 1$, $y' = \sin^2 3x$. The equilibrium points are $(n\pi/3, -1)$, $n = 0, \pm1, \pm2, \ldots$, since $\sin^2 3x$ is 0 if $x = n\pi/3$. See Fig. 2(d), where the x- and y-

nullclines (dashed) are, respectively, $y = -1$ and $x = n\pi/3$. Orbits seem to rise upward since y' is positive (except at $x = n\pi/3$) toward the equilibrium points and then veer around them on one side or the other.

(e). The system is $x' = 3(x - y)$, $y' = y - x$. The x- and y-nullclines are the common line $y = x$, which is a line of equilibrium points. See Fig. 2(e). The orbits are rays which emerge from the equilibrium points on $y = x$ as t increases from $-\infty$.

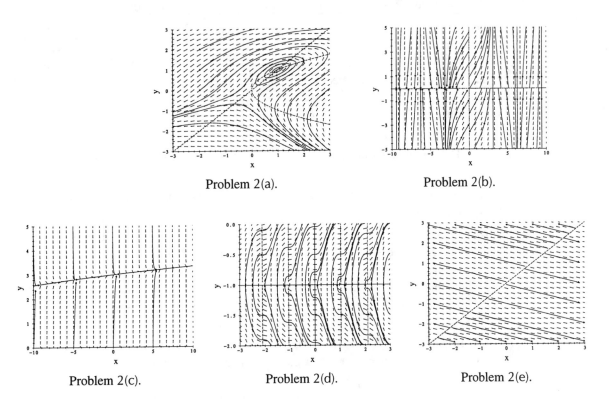

Problem 2(a).   Problem 2(b).

Problem 2(c).   Problem 2(d).   Problem 2(e).

3. The ODEs in parts **(a)**–**(c)** are uncoupled and may be solved separately. Notice that translating solutions by T just amounts to changing the values of the integration constants.

(a). Since the ODEs, $x' = 3x$, $y' = -y$, are uncoupled, the equations are easily solved, $x(t) = C_1 e^{3t}$, $y(t) = C_2 e^{-t}$, $-\infty < t < \infty$. Since $x(t - T) = C_1 e^{3(t-T)}$, we have that $dx(t - T)/dt = 3C_1 e^{3(t-T)} = 3x(t - T)$, and so $x(t - T)$ is a solution of $x' = 3x$ if $x(t)$ is a solution. Similarly, $dy(t - T)/dt = -C_2 e^{-(t-T)} = -y(t - T)$, so $y(t - T)$ is a solution. The solution $x = x(t - T)$, $y = y(t - T)$ is defined for all t.

(b). Separating variables and solving the ODE $x' = 1/x$ for x we have that $x = \pm(2t - 2C_1)^{1/2}$, $t > C_1$. Solving the second ODE, $y' = -y$, we have that $y = C_2 e^{-t}$, $t > C_1$. [Note that all the components of a solution must be defined on the same t-interval.] To show that $x(t - T) = (2(t - T) - 2C_1)^{1/2}$, $t > C_1 + T$, is a solution, calculate $dx(t - T)/dt = (2(t - T) - 2C_1)^{-1/2} = 1/x(t - T)$. Note that $x(t - T)$ is defined on a different t-interval,

$t > C_1 + T$, from $x(t)$. Similarly, $dy(t - T)/dt = -C_2 e^{-(t-T)} = -C_2 y(t - T)$, so $x(t - T)$, $y(t - T)$, $t > C_1 + T$ is a solution of the system.

(c). Separating variables and solving the first ODE, $x' = -x^3$, for x, we have that $x = C_1(1 + 2C_1^2 t)^{-1/2}$, $t > -1/2C_1^2$ while the second ODE, $y' = 1$, has the solutions $y(t) = t + C_2$, $t > -1/2C_1^2$. Since $dx(t - T)/dt = -C_1^3(1 + 2C_1^2(t - T))^{-3/2} = -x^3(t - T)$, $x(t - T) = C_1(1 + 2C_1^2(t - T))^{-1/2}$, $t > -1/2C_1^2 + T$ is a solution. Similarly, $dy(t - T)/dt = 1$, so $x(t - T)$, $y(t - T)$, $t > -1/2C_1^2 + T$ is a solution of the system.

(d). Substitute the solution $y = C_2 e^{-t}$ of the second ODE, $y' = -y$, into $x' = x^2(1 + y)$, and then solve $x^{-2}dx = (1 + C_2 e^{-t})dt$ to obtain $x = -(t + C_1 - C_2 e^{-t})^{-1}$, where t lies in an interval I for which $t + C_1 - C_2 e^{-t} \neq 0$. I depends on the values of C_1 and C_2 and cannot be explicitly determined. Suppose that $t - T + C_1 - C_2 e^{-(t-T)} \neq 0$. Since $dx(t - T)/dt = (1 + C_2 e^{-(t-T)})(t + T + C_1 - C_2 e^{-(t-T)})^{-2} = x^2(1 + y)$, $x(t - T)$ is a solution. Similarly, $dy(t - T)/dt = -C_2 e^{-(t-T)} = -y(t - T)$, so $x(t - T)$, $y(t - T)$, $t - T + C_1 - C_2 e^{-(t-T)} \neq 0$ is a solution of the system.

4.   **(a).** The four equilibrium points of $x' = 1 - y^2$, $y' = 1 - x^2$ have coordinates $x = \pm 1$, $y = \pm 1$.

     (b). See Fig. 4(b). Note the cycles going clockwise around the equilibrium point $(1, -1)$ and counterclockwise around $(-1, 1)$.

     (c). See Fig. 4(c) for the direction field. It certainly appears that the line $y = x$ consists of orbits. Since the two equilibrium points $(1, 1)$ and $(-1, -1)$ lie on the line, and orbits cannot touch, there must be at least three other orbits on the line—one between the equilibrium points, one at the lower left, and another at the upper right. Since there are no other equilibrium points on the line $y = x$, it appears that these five orbits are the only ones on the line.

     (d). Substitute $x = f(t)$ and $y = f(t)$ into the system ODEs to obtain a single first-order ODE $f' = 1 - f^2$. Separating the variables and integrating, we obtain

$$\ln \left| \frac{1 + f}{1 - f} \right|^{1/2} = t + C, \quad \text{where } C \text{ is an arbitrary constant}$$

Exponentiating, squaring, and dropping the absolute value signs we have

$$\frac{1 + f}{1 - f} = ce^{2t}, \quad \text{where } c \text{ is an arbitrary constant.}$$

Imposing the initial conditions $x(0) = y(0) = a$, we find that $c = (1 + a)/(1 - a)$, and solving for $f(t)$ we have

$$f(t) = \frac{c - e^{-2t}}{e^{-2t} + c}$$

Now, if $|a| < 1$, then $c > 0$ and in that case $f(t) \to 1^-$, as $t \to +\infty$. If $a > 1$, then $c < -1$, and so $f(t) \to 1^+$, as $t \to +\infty$. Finally, take the case where $a < -1$. In that case $-1 < c < 0$, and we see that $f(t)$ reaches $-\infty$ as t increases to the finite value t_0 where $e^{-2t_0} + c = 0$, or $t_0 = (-1/2) \ln(-c)$.

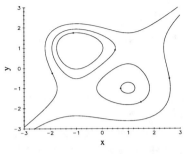

Problem 4(b).

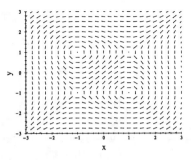

Problem 4(c).

5. (a). Suppose that P is an equilibrium point of $x' = f(x)$, and that P is inside a region S in which the conditions of the Fundamental Theorem hold. Then $x = P$ is a solution. Suppose $x = x^1(t)$, $a < t < b$, is another maximally extended solution of the system and that $x^1(t) \to P$ as $t \to T$. By the maximal extension property for the orbits of an autonomous system, and the fact that $x^1(t)$ is a continuous vector function of t, we have that either $a < T < b$ and $x^1(T) = P$, or else $T = a = -\infty$, or else $T = b = +\infty$, or else $T = a$ (or b) and $x^1(t)$ tends to the boundary of S as $t \to a^+$ (or b^-). The first alternative is forbidden by the uniqueness property since otherwise there would be two different maximally extended orbits through P, $x = P$ for all t and $x = x^1(t)$. The last alternative would imply that P lies on the edge of S, but P is assumed to lie inside S. So one of the two middle alternatives must hold, $T = \infty$ or $T = -\infty$, and a solution reaches a constant equilibrium state only as $t \to \infty$, or $t \to -\infty$.

(b). Suppose that $x = x(t)$ is a nonconstant solution for which $x(t + T) = x(t)$ for all t, and suppose that T is the smallest positive number with this property. We must show that the orbit of $x = x(t)$ does not cross or touch itself (except after each period). It is enough to show that if $0 < t_1 \le T$ and $x(t_1) = x(0)$, then $t_1 = T$. Suppose that $y(t) = x(t + t_1)$. Since the system is autonomous, $y(t)$ is a solution of the system with the same orbit as $x = x(t)$. Since $x(0) = x(t_1) = y(0)$, we have that $x(t) = x(t + t_1)$ by uniqueness. So $x(t)$ has period t_1. Since T was the smallest positive period, $t_1 = T$. So the orbit must be a simple closed curve, closed since $x(T) = x(0)$ and $x(t + T) = x(t)$ for all t, simple since for no t_1, $0 < t_1 < T$, does $x(t_1) = x(T)$.

6. (a). First we write the Cartesian rate functions in terms of r and θ using (13)

$$f = x - y - x(x^2 + y^2) = r\cos\theta - r\sin\theta - (r\cos\theta)r^2$$
$$g = x + y - y(x^2 + y^2) = r\cos\theta + r\sin\theta - (r\sin\theta)r^2$$

Using formula (15), we have

$$r' = (\cos\theta)f + (\sin\theta)g$$
$$= r\cos^2\theta - r\cos\theta\sin\theta - r^3\cos^2\theta + r\cos\theta\sin\theta + r\sin^2\theta - r^3\sin^2\theta$$
$$= r - r^3 = r(1 - r^2)$$

$$\theta' = (r^{-1}\cos\theta)g - (r^{-1}\sin\theta)f$$
$$= \cos^2\theta + \cos\theta\sin\theta - r^2\cos\theta\sin\theta - \cos\theta\sin\theta + \sin^2\theta + r^2\cos\theta\sin\theta$$
$$= 1$$

Because $r = 1$, all t, solves the rate equation for r, the unit circle is an orbit (or union of orbits) for the original system. Since $\theta' = 1$, we see that $\theta = t + C_1$ and so the unit circle is a cycle of period 2π and is traced out counterclockwise in time. We see from the original Cartesian ODEs that the origin $x = 0$, $y = 0$ is an equilibrium point since $f(0, 0) = g(0, 0) = 0$. There are no other equilibrium points because we see from the rate equations for r and θ that for all other points either r' or θ' is non-zero.

(b). We have seen in part **(a)** that $\theta = t + C_1$ for any constant C_1. The ODE, $r' = r(1 - r^2)$, is separable. Separating the variables, using partial fractions, and integrating we have that

$$\frac{1}{r} \cdot \frac{1}{1 - r^2}\, dr = \left[\frac{1}{r} - \frac{1/2}{1 + r} + \frac{1/2}{1 - r}\right] dr = dt$$

$(0 < r < 1)$
$$\ln\frac{r}{(1 - r^2)^{1/2}} = t + C$$

$$\frac{r}{(1 - r^2)^{1/2}} = Ke^t$$

$$r^2 = K^2 e^{2t}(1 - r^2)$$

$$r^2(1 + K^2 e^{2t}) = K^2 e^{2t}$$

$$r = \frac{Ke^t}{(1 + K^2 e^{2t})^{1/2}}, \qquad \text{where } K \text{ is any positive constant}$$

$(r > 1)$
$$\ln\frac{r}{(r^2 - 1)^{1/2}} = t + C$$

$$r^2(1 - K^2 e^{2t}) = -K^2 e^{2t}$$

$$r = \frac{Ke^t}{(K^2 e^{2t} - 1)^{1/2}}, \qquad \text{where } K \text{ is any constant}$$

In either case as $t \to +\infty$, $r \to +1$, and all nonconstant orbits approach the cycle $r = 1$ as $t \to +\infty$.

(c). See Fig. 6(c) for orbits starting at $(0, 0)$, $(-0.001, 0)$, $(-3, -1)$, and $(3, 2)$, where $0 \le t \le 20$. From the formula for $r(t)$, $0 < r < 1$, in part **(b)** we see that as $t \to -\infty$, $r(t) \to 0$, and so nonconstant orbits inside the unit circle "emerge from the origin" as t increases from $-\infty$. So the origin is a repeller. As shown in part **(b)** all nonconstant orbits tend to the unit circle as $t \to \infty$, and this circle is an attractor. The x- and y-nullclines are the graphs (dashed) of $x - y - x(x^2 + y^2) = 0$ and $x + y - y(x^2 + y^2) = 0$, respectively.

(d). We see from the component graphs in Fig. 6(d) that the cycle on the unit circle has period 2π.

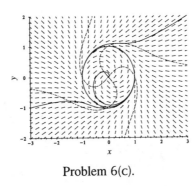

Problem 6(c).

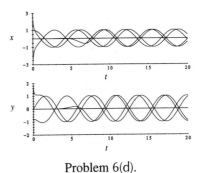

Problem 6(d).

7. The nonconstant vector functions $\begin{bmatrix} \sin t \\ t \end{bmatrix}$ and $\begin{bmatrix} te^t \\ t^3 \end{bmatrix}$ are both the zero vector at $t = 0$. So there are *no* continuously differentiable functions $f(x, y, t)$, $g(x, y, t)$ such that both vector functions are solutions of $x' = f$, $y' = g$, because this would violate uniqueness if there were.

8. Group project.

5.3 Models of Interacting Species

Suggestions for preparing a lecture

Topics: Discussion of 2-species population models in general, and any of the specific models that catches your fancy. Since the Volterra predator-prey model is done in Section 5.4, we barely mention it here.

Remarks: If you only have time for one of these two sections, you might want to consider doing 5.4 rather than 5.3.

Making up a problem set

Problems 1**(c)**, 2**(a)**, 3**(a)**, 4, 5 would be good choices.

Comments

Models of the population of interacting species have been popular for some time. These models (particularly the oversimplified models introduced here) don't have much predictive value, but they are guides to thinking about how species interact. Some would call these models caricatures, rather than models. A caricature is a distortion of reality, but the original is still recognizable—and that is just what these models are like in comparison to the real thing. We restrict attention to two interacting species whose model rate equations are autonomous. That means that time variations in the rate functions have been averaged out—a dubious proposition at best, but the modeling process usually begins with oversimplifications. We give a catalogue of possible interactions between two populations, ranging from the now classical Volterra model for predator-prey interactions to assorted models of cooperation, competition, harvesting, and satiation. No deep results here—just some computer simulations of orbits in the population quadrant $x \geq 0$, $y \geq 0$. We have not explicitly scaled the population levels or time; in fact, we don't even mention units for measuring time or populations. But the variables are all scaled down to convenient levels for computation and calculation.

1. **(a).** Each species modeled by the system $x' = (\alpha - by)x$, $y' = (\beta - cx)y + H$ would grow exponentially in isolation (the rate terms αx, βy), but when occupying the same ecological region (or *niche*, to use the technical term) they compete for common resources (the terms $-bxy$ and $-cxy$). The y-species is restocked at rate H.

    (b). Each species modeled by the system $x' = (\alpha - ax - by)x$, $y' = (\beta - cx - dy)y$ would grow exponentially in isolation (rate terms αx and βy) except that overcrowding implies logistic growth (rate terms $-ax^2$ and $-dy^2$). When occupying the same ecological niche, the two species compete for common resources (the terms $-bxy$ and $-cxy$).

    (c). The system is $x' = (\alpha + by)x$, $y' = (-\beta + cx - dy)y$. The x-species would grow exponentially in isolation (the rate term αx), while the y-species would decline exponentially (the rate term $-\beta y$) with the decline accelerated by overcrowding (the term $-dy^2$). The coupling terms bxy and cxy indicate that the two species somehow cooperate, the presence of each being favorable for the growth of the other.

    (d). The system is $x' = (\alpha - ax - by)x$, $y' = (-\beta + cx - dy)y + 2 + \cos t$. The x-species would grow exponentially in isolation (the rate term αx) except that overcrowding implies logistic growth (rate term $-ax^2$), while the y-species would decline exponentially (the rate term $-\beta y$), with the decline accelerated by overcrowding (the term $-dy^2$). The y-species is a predator of the prey x-species (the terms $-bxy$ and cxy) and is restocked at the periodic rate $2 + \cos t$.

    (e). The system is $x' = (\alpha + by)x - H\,\mathrm{sqw}(t, 25, 1)$, $y' = (\beta + cx - dy)y$. Each species would grow exponentially in isolation (rate terms αx and βy), except that overcrowding implies logistic change for the y-species (the rate term $-dy^2$). When occupying the same ecological niche the two species cooperate, the presence of each being favorable for the other (the terms bxy and cxy). The x-species is seasonally harvested at the rate H, but only for the first 25% of the period 1. This is a model of cooperation (the terms bxy and cxy).

    (f). The system is $x' = (\alpha - ax)x$, $y' = (-\beta + cx/(m + kx))y + H$. The x-species follows a model of logistic change and is unaffected by the y-species. The y-species would decline exponentially in the absence of x (the term $-\beta y$), but the presence of x contributes to y's growth. So x is the "prey" and y is the predator. The fact that the x-species is both the prey of, and apparently unaffected by, the y-species suggests that the y-species lives off a product of the x-species. The prey's contribution to the growth of the predator's population is modeled by the satiation term $cxy(m + kx)^{-1}$, which is positive inside the population quadrant. Note that $cx/(m + kx) \to c/k$ as $x \to \infty$, which means that the positive effect of lots of prey on the predator's growth rate levels off as the number of prey gets large. The y-species is restocked at constant rate H.

    (g). The system is $x' = (\alpha - bz)x$, $y' = (\beta - my - kz)y$, $z' = (-\gamma + az + cy)z - Hz$.

The y-species would change logistically if isolated (the term $\beta y - dy^2$), but the x-species would increase exponentially if isolated (the term dx), and the z-species would decline exponentially if isolated (the term $-\gamma z$). The z-species is a predator on the prey species x and y (the terms $-bzx$ and axz, $-kzy$ and cyz). The z-species is harvested with constant effort harvesting (the term $-Hz$).

2. Direction fields, nullclines (dashed), and orbits are shown in Figs. 3(a), 3(b), 3(c), 3(d).Although the graphs shown here were produced by a numerical solver, the questions can also be answered using only a straightedge and some sketching by hand. In each case the x-nullclines are given by setting $x' = 0$, and the y-nullclines are found by setting $y' = 0$. In every case the x- and y-axes are among the nullclines; the axes are also unions of orbits. The equilibrium points are where the x-nullclines meet the y-nullclines.

(a). The system is $x' = (5 - x + y)x$, $y' = (10 + x - 5y)y$. The x-nullclines are defined by $(5 - x + y)x = 0$. That is, the x-nullclines are the lines $x = 0$ and $y = x - 5$. The y-nullclines are defined by $(10 + x - 5y)y = 0$. That is, the y-nullclines are the lines $y = 0$ and $y = x/5 + 2$. The equilibrium points are the intersection points of the x-nullclines with the y-nullclines. For example, if $x = 0$, then we require also that $y = 0$ or $y = x/5 + 2$. This gives us the two equilibrium points $(0, 0)$ and $(0, 2)$. If $y = x - 5$, then we require (as before) that $y = 0$ (in which case $x = 5$) or $y = x/5 + 2$ (in which case $x/5 + 2 = y = x - 5$, so $x = 35/4$, and $y = 15/4$). Now we get two new equilibrium points $(5, 0)$ and $(35/4, 15/4)$. The terms $+xy$ in each rate equation show that the two species cooperate. From the behavior of the orbits (see Fig. 3(a)) the populations seem to approach the equilibrium point $(35/4, 15/4)$ as $t \to \infty$.

(b). The system is $x' = (10 - x + 5y)x$, $y' = (5 + x - y)y$. The x-nullclines are the lines $x = 0$ and $y = x/5 - 2$. The y-nullclines are the lines $y = 0$ and $y = x + 5$. The intersection points of the nullclines are found as follows: If $x = 0$, then $y = 0$ or $y = x + 5 = 5$. If $y = x/5 - 2$, then $0 = y = x/5 - 2$ (so $x = 10$) or $x + 5 = y = x/5 - 2$ (so $x < 0$, which is outside the population quadrant). The three equilibrium points in the quadrant are $(0, 0)$, $(0, 5)$, and $(10, 0)$. The terms $+5xy$ and $+xy$ in the rate equations show that the species cooperate. From the behavior of the orbits [see Fig. 3(b)] the populations of both species seem to explode as time goes on.

(c). The system is $x' = (5 - x - y)x$, $y' = (10 - x - 2y)y$. The x-nullclines are the lines $x = 0$ and $y = -x + 5$. The y-nullclines are the lines $y = 0$ and $y = -x/2 + 5$. The intersection points of the nullclines are found as follows: If $x = 0$, then $y = 0$ or $y = -x/2 + 5 = 5$. If $y = -x + 5$, then $-x + 5 = y = 0$ (so $x = 5$) or $-x + 5 = y = -x/2 + 5$ (so $x = 0$, $y = 5$). The equilibrium points are $(0, 0)$, $(0, 5)$, and $(5, 0)$. The terms $-xy$ in the rate equations indicate that the species are competitors. From the behavior of the orbits [see Fig. 3(c)], it seems that $x \to 0$ and $y \to 5$ as time increases. So the y-species wins the competition.

(d). The system is $x' = (10 - x - 5y)x$, $y' = (5 - x - y)y$. The x-nullclines are the lines $x = 0$ and $y = -x/5 + 2$. The y-nullclines are the lines $y = 0$ and $y = -x + 5$. The intersection points of the nullclines are found as follows: If $x = 0$, then $y = 0$ or $y =$

$-x+5=5$. If $y=-x/5+2$, then $-x/5+2=y=0$ (so x=-10) or $-x/5+2=y=-x+5$ (so x=15/4). The equilibrium points are $(0,0)$, $(0,5)$, $(10,0)$, and $(15/4, 5/4)$. The terms $-5xy$ and $-xy$ indicate that the species are competitors. From the behavior of the orbits [see Fig. 3(d)], it appears that this is a model of competitive exclusion. The population quadrant is divided into two subregions, one in the lower right where orbits approach the equilibrium point $(10, 0)$ with x winning as $t \to \infty$, and the other region in the upper left where orbits tend to $(0, 5)$ with y winning as $t \to \infty$.

3.   The dashed lined are the nullclines.

(a). See Fig 3(a) for orbits of $x' = (5 - x + y)x$, $y' = (10 + x - 5y)y$. Each population tends to a positive equilibrium value as $t \to +\infty$.

(b). See Fig 3(b) for orbits of $x' = (5 - x - y)x$, $y' = (10 - x - 2y)y$. Each population seems to explode and to tend to infinity as $t \to +\infty$.

(c). See Fig 3(c). The y-species wins the competition as $x \to 0$ and $y \to 5$ as $t \to \infty$.

(d). See Fig 3(d). The interior of the population quadrant is divided into two regions. In one region y wins as $y \to 5$ and $x \to 0$ as $t \to \infty$. In the other x wins as $x \to 10$ and $y \to 0$ as $t \to \infty$.

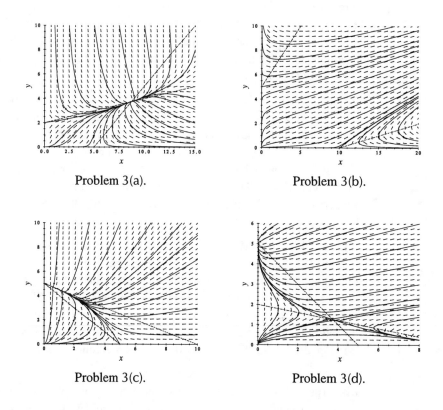

Problem 3(a).   Problem 3(b).

Problem 3(c).   Problem 3(d).

4.   The system is $x' = (2 - x + 2y)x$, $y' = (2 + 2ax - y)y$.

(a). The coefficient $2a$ measures the effectiveness of the y-species in using the x-species to promote the growth of y.

(b). We find the value of a where the x-nullcline $y = x/2 - 1$ and the y-nullcline $y = 2ax + 2$ are parallel; the lines are parallel if $2a = 1/2$, so $a = 1/4$. For $0 < a < 1/4 = a_0$ the graphs seem to show a stable equilibrium where the x- and y-nullclines cross at the point $(6/(1 - 4a), (2 + 4a)/(1 - 4a))$, where $x/2 - 1 = 2ax + 2 = y$. See Fig. 4(b), Graph 1, where a is 0.05. In Fig. 4(b), Graph 2 at the critical value $a_0 = 1/4$, there is no internal equilibrium point, and (apparently) both species grow without bound. In Fig. 4(b), Graph 3 a is 0.5, above the critical value, and both species grow without bound.

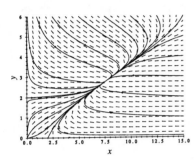
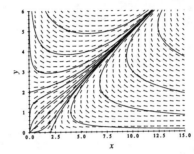
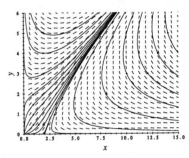

Problem 4(b), Graph 1.   Problem 4(b), Graph 2.   Problem 4(b), Graph 3.

5. The system is $x' = (2 + H - x - 2y)x$, $y' = (2 - 2x - y)y$. The equilibrium point determined by the intersection of the x-nullcline $2 + H - x - 2y = 0$ and the y-nullcline $2 - 2x - y = 0$ has coordinates $x = (2 - H)/3$, $y = 2(1 + H)/3$. This point exits the first quadrant as H increases through the value 2. The other equilibrium points are $(0, 0)$, $(0, 2)$, and $(2 + H, 0)$. For $0 < H < 2$, it seems that we have a model of competitive exclusion (see Fig. 5, Graph 1 where $H = 1$). At $H = 2$, the x-species seems to win the competition (see Fig. 5, Graph 2 where $H = 2$). For $H > 2$, the x-species wins (see Fig. 5, Graph 3 where $H = 3$).

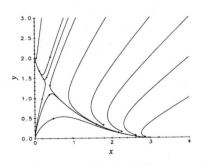
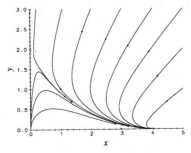
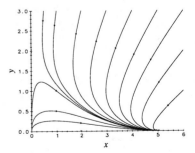

Problem 5, Graph 1.   Problem 5, Graph 2.   Problem 5, Graph 3.

6. Group project.

7. Group project.

5.4 Predator-Prey Models

Suggestions for preparing a lecture

Topics: Tell the story of D'Ancona and Volterra, and give Volterra's three laws with as many examples and graphs as possible. If you want to spend a second lecture on all this, you might discuss the proof that the predator-prey orbits are simple closed curves (cycles); see Problem 6. On the other hand the graphical evidence that the orbits are all cycles is so strong that most students won't appreciate the rather lengthy mathematical proof.

Remarks: Students who saw Green's Theorem in a multivariable calculus course like the proof of Volterra's Law of Averages, and so do we!

Making up a problem set

One part of 1 and the corresponding part of 2, possibly part of 3, 5; 7 makes an excellent project for a group.

Comments

This section is devoted to one topic: the Lotka-Volterra models of predator-prey dynamics, and the story of how Volterra came to consider the question. The model isn't very realistic, although (as we point out) it is based on data from real fish populations. Anyway, Volterra derived three laws from the model equations, and all three are presented and discussed. The implicit formula for the population cycles is derived, but the proof that the formula defines a simple closed curve is not at all easy (see Problem 6). As in Section 5.3, we assume that time and populations have been scaled to appropriate levels for ease in calculation and computation.

---

1. **(a).** The system is $x' = -x + xy$, $y' = y - xy$. We see that $x(t)$ is the predator population and $y(t)$ is the prey population, as we can tell from the signs of the interaction terms xy in the rate equations. The average predator and prey populations are 1 since the coordinates of the internal equilibrium point are $x = 1$, $y = 1$ (see Theorem 5.4.2). Because $x' < 0$ if $x > 0$, $y = 0$, and $y' > 0$ if $y > 0$, $x = 0$, the orbits turn clockwise about $(1, 1)$ as time increases.

(b). The system is $x' = 0.2x - 0.02xy$, $y' = -0.01y + 0.001xy$. We see that $y(t)$ is the predator population and $x(t)$ is the prey population at time t (look at the signs of the xy terms). The value of the average predator and prey populations is 10 because the internal equilibrium point is $x = 10$, $y = 10$. Because $x' > 0$ if $x > 0$, $y = 0$, and $y' < 0$ if $y > 0$, $x = 0$, the orbits turn counterclockwise about $(10, 10)$ as time increases.

(c). The system is $x' = (-1 + 0.09y)x$, $y' = (5 - x)y$. We see that $x(t)$ is the predator population and $y(t)$ is the prey population at time t (look at the signs of the xy terms). The average predator and prey populations are 5 and 100/9, respectively, because $(5, 100/9)$ is the internal equilibrium point (see Theorem 5.4.2). Because $x' < 0$ if $x > 0$, $y = 0$, and $y' > 0$ if $y > 0$, $x = 0$, the orbits turn clockwise about $(5, 100/9)$ as time increases.

2. In each part, the orbits are on the left and the x-component and y-component graphs are, respectively, at the upper and lower right. We can estimate the period of an orbit by looking at the times between the peaks of a corresponding component curve. In every case the low-amplitude cycles have the shorter periods.

    (a). See Fig. 2(a) for the graph of a cycle and the internal equilibrium point. The estimated period of the nonconstant orbit is about 17. Note that in this case $(5, 10)$ and $(10, 5)$ are points on the same orbit.

    (b). See Fig. 2(b) for the graphs of the cycles through $(5, 10)$ and $(10, 5)$, and the equilibrium point $(10, 10)$. The estimated periods of the cycles are about 145 and 190.

    (c). See Fig. 2(c), Graph 1 for the orbit with initial point $(5, 10)$ and Fig. 2(c), Graph 2 for the orbit with initial point $(10, 5)$. The equilibrium point $(5, 100/9)$ and its component graphs are also plotted. The estimated periods of the cycles are about 3 and 4.

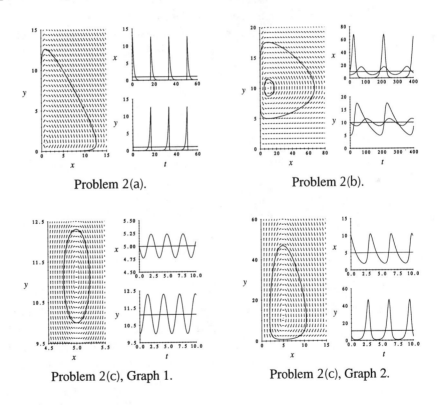

Problem 2(a).   Problem 2(b).

Problem 2(c), Graph 1.   Problem 2(c), Graph 2.

3. The xy-system is $x' = f(x, y) = -ax + bxy$, $y' = g(x, y) = cy - dxy$. Let $x_1 = c/d$ and $y_1 = a/b$ so (x_1, y_1) is the equilibrium point.

    (a). To linearize we first find the partial derivatives of $f(x, y)$ and $g(x, y)$ with respect to both x and y. These are $\partial f/\partial x = -a + by$, $\partial f/\partial y = bx$, $\partial g/\partial x = -dy$, and $\partial g/\partial y = c - dx$.

So with $x_1 = c/d$ and $y_1 = a/b$, we have $x' = \partial f/\partial x(x_1, y_1)(x - x_1) + \partial f/\partial y(x_1, y_1)(y - y_1) = (-a + b(a/b))(x - c/d) + b(c/d)(y - a/b) = bcy/d - ac/d$ and $y' = \partial g/\partial x(x_1, y_1)(x - x_1) + \partial g/\partial y(x_1, y_1)(y - y_1) = -d(a/b)(x - c/d) + (c - d(c/d))(y - a/b) = -adx/b + ac/b$. The linearized system is $x' = bcy/d - ac/d$, $y' = -adx/b + ac/b$.

(b). To show $x = c/d + A\cos\omega t$, $y = a/b + B\sin\omega t$ solves the linearized system, first solve for x'. We see that $x' = -A\omega\sin\omega t$. Substituting the value for y into the linearized system gives $x' = bc/d(a/b + B\sin\omega t) - ac/d = bcB/d\sin\omega t$. We get $-A\omega\sin\omega t = x' = bcB/d\sin\omega t$ so, $-A\omega = bcB/d$. Solving for ω gives $\omega = -Bbc/Ad$. Now $y' = B\omega\cos\omega t$. Substituting the value for x into the linearized system gives $y' = -ad(c/d + A\cos\omega t)/b + ac/b = -ad(A\cos\omega t)/b$. We get $B\omega\cos\omega t = -ad(A\cos\omega t)/b$, so $B\omega = adA/b$. Solving for ω gives $\omega = -Aad/Bb$. By setting the two expressions for ω equal to each other we get $-Bbc/Ad = -Aad/Bb$, so $A^2/B^2 = b^2c/ad^2$. So $A/B = \pm b/d\sqrt{c/a}$. Substituting this back in to either of the expressions for ω gives $\omega = \pm\sqrt{ac}$.

(c). See Fig. 3(c), Graph 1 for the graphs of the predator-prey orbits and component graphs of the nonlinear Lotka-Volterra system, $x' = (-1 + y)x$, $y' = (1 - x)y$. Graph 2 shows the orbits and component graphs for the corresponding linearized system. Note from the x-component graphs how the period of the predator-prey cycles of the nonlinear system increases with amplitude, but the period of the linearized cycles is fixed at $2\pi/\sqrt{ac}$ $(= 2\pi$ if $a = c = 1)$. The initial points used for the graphs in both figures are $x_0 = 1$, $y_0 = 1, 1.1, 1.3, 1.5,$ and 1.9.

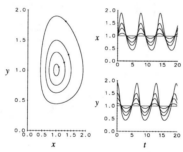
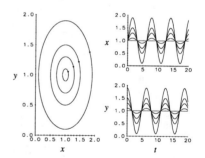

Problem 3(c), Graph 1.   Problem 3(c), Graph 2.

4.   Rewriting the harvested system $x' = -ax + bxy - H_1x$, $y' = cy - dxy - H_2y$, we have $x' = -(a + H_1)x + bxy$, $y' = (c - H_2)y - dxy$. By the Law of the Averages, we know the average predator population is $\bar{x} = (c - H_2)/d$ and the average prey population is $\bar{y} = (a + H_1)/b$. The predator fraction F of the total average catch is

$$F = \frac{\bar{x}}{\bar{x} + \bar{y}} = \frac{(c - H_2)/d}{(c - H_2)/d + (a + H_1)/b} = \frac{1}{1 + d(a + H_1)/b(c - H_2)}$$

The quantity $d(a + H_1)/b(c - H_2)$ increases as H_1 increases (because the numerator increases), and also as H_2 increases but stays smaller than c (because the denominator decreases). Since this ratio is in the denominator of F, F itself must decrease as H_1 and H_2 increase.

5. **(a).** Figure 5.4.3 graphs the system $x' = -x + xy/10 - Hx$, $y' = y - xy/5 - Hy$ with initial conditions $x(0) = 8$, $y(0) = 15$ for $H = 0, 2/5, 1, 5$. We see two extinction orbits in the graph. When $H_1 = H_2 = 1.0$, the predator species vanishes, but not the prey. When $H_1 = H_2 = 5.0$, the heavy harvesting kills off both species. The other orbits correspond to lower values of $H_1 = H_2$, where neither population becomes extinct.

(b). The equilibrium points are the origin and the point $x = (c - H_2)/d$, $y = (a + H_1)/b$ inside the population quadrant. The internal equilibrium point approaches the point $(0, (a + H_1)/b)$ on the y-axis as $H_2 \to c^-$. If $H_2 = c$, system (9) becomes $x' = (-a - H_1 + by)x$, $y' = -dxy$. So, all points on the y-axis are equilibrium points of system (9) since $x' = y' = 0$ if $x = 0$. For more information, refer to the answer to part **(d)**.

(c). If $H_2 = c$, then replacing the coefficient d by d^* to avoid confusing it with the "d" of the differential,

$$\frac{dy}{dx} = \frac{d^* xy}{(a + H_1)x - bxy} = \frac{d^* y}{a + H_1 - by}$$

After separating variables, we have

$$\left(\frac{a + H_1 - by}{d^* y} \right) dy = dx$$

So, $(a + H_1)/d^* \cdot (1/y)dy - (b/d^*)dy = dx$. Integrating, $((a + H_1)/d^*) \ln(y/y_0) - (b/d^*)(y - y_0) = x - x_0$; that is,

$$x = x_0 + \frac{a + H_1}{d} \ln \frac{y}{y_0} - \frac{b}{d}(y - y_0)$$

(d). We see from the rate equations, $x' = (-a - H_1 + by)x$, $y' = -dxy$, that $y(t)$ decreases along every orbit inside the population quadrant. Moreover, $x(t)$ increases along an orbit if $by(t) > a + H_1$, but as time goes on eventually $by(t) < a + H_1$ and $x(t)$ decreases. In fact, each orbit inside the population quadrant moves away from an equilibrium point $(0, y_1)$, $y_1 > (a + H_1)/b$, as t increases from $-\infty$, and moves toward an equilibrium point $(0, y_2)$, $0 < y_2 < (a + H_1)/b$, as $t \to \infty$. See Fig. 5(d) for several of these orbits in the case $a = b = c = d = 1$, $H_1 = 0.1$. The dashed line is the line of equilibrium points on the y-axis.

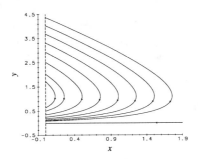

Problem 5(d).

6. **(a).** Let $f(x) = x^c e^{-dx}$, $g(y) = y^a e^{-by}$; according to text formula (6), each orbit is defined by $f(x)g(y) = K$, where $K_0 = (x_0^c e^{-dx_0})(y_0^a e^{-by_0})$ and (x_0, y_0) is a point on the orbit. We have that $f(0) = g(0) = 0$. Moreover, $f(x)$ and $g(y)$ are positive for $x > 0$ and $y > 0$, and tend to 0 as $x \to \infty$, $y \to \infty$, respectively. To show, for example, that $f(x) = x^c/e^{dx} \to \infty$ as $x \to \infty$, apply l'Hôpital's Rule, taking just enough derivatives of numerator and of denominator (say n) so that the power $c - n$ of x in the numerator is zero or negative. Then $\lim_{x \to \infty} f(x) = \lim_{x \to \infty} c(c-1)\cdots(c-n+1)x^{c-n}/(d^n e^{dx}) = 0$. Since $f'(x) = cx^{c-1}e^{-dx} - dx^c e^{-dx} = x^{c-1}e^{-dx}(c - dx)$, we see that $f(x)$ increases with x to the maximum value of $M_1 = f(c/d) = c^c(de)^{-c}$ at $x = c/d$, and then decreases as x increases beyond c/d. Similar results hold for $g(y)$, whose maximum value is $g(a/b) = a^a(be)^{-a} = M_2$ at $y = a/b$.

(b). If $K > M_1 M_2 = [\max f(x)] \cdot [\max g(y)]$, then the equation $K = f(x)g(y)$ holds for no positive values of x and y. If $K = K^* = M_1 M_2$ then the equation has the unique solution $x = c/d$, $y = a/b$, given that $M_1 = f(c/d)$ for $x = c/d$, and $M_2 = g(a/b)$ for $y = a/b$ from part **(a)**.

(c). Since $f(x)$ is continuous, strictly increasing for $0 \le x \le c/d$, and strictly decreasing for $x \ge c/d$, then given any number Δ, $0 < \Delta < M_1$, there are precisely two values of x, x_1 and x_2, $x_1 < c/d < x_2$, such that $f(x) = \Delta$. If $0 < x < x_1$, or $x > x_2$, then $0 < f(x) < \Delta$, and there is no positive y for which $f(x)g(y) = \Delta \cdot M_2$. In other words, the graph of $f(x)g(y) = \Delta \cdot M_2$ is contained entirely in the vertical strip $x_1 \le x \le x_2$. If $x = x_1$ or x_2, then $f(x) = \Delta$ and $f(x)g(y) = \Delta \cdot M_2$ has the unique y-solution $y = a/b$, since $g(a/b) = M_2$, and $g(y) < M_2$ for all other $y > 0$. On the other hand, for each x, $x_1 < x < x_2$, we have $f(x) > \Delta$, and the equation $f(x)g(y) = \Delta \cdot M_2$ has two solutions, $y_1(x) < a/b$ and $y_2(x) > a/b$ because of the way $g(y)$ rises to its maximum value M_2 and then falls as y increases through a/b. The same analysis shows that $y_1(x)$ and $y_2(x) \to a/b$ if $x \to x_1$ or $x \to x_2$, respectively. Similar results are obtained with the roles of f and g, x and y interchanged. The analogues of x_1 and x_2 are y_1 and y_2.

(d). The orbit defined by $f(x)g(y) = K$, $0 < K < K^* = M_1 M_2$ is intersected by each horizontal line $y = y_0$, $y_1 < y_0 < y_2$, exactly twice, and by each vertical line $x = x_0$, $x_1 < x_0 < x_2$ also twice. This is because the equation $f(x) = y$ has exactly two solutions, x_1 and x_2, such that $x_1 < x < x_2$ and the equation $g(y) = yM_2/f(x)$ has exactly two solutions, y_1 and y_2, such that $y_1(x) < a/b < y_2(x)$ if $x_1 < x < x_2$. Segments of the four lines $x = x_1$, $x = x_2$, $y = y_1$, $y = y_2$ form a rectangle just touching the four extreme points of the orbit. The orbit is a simple closed curve inside the rectangle, $x_1 \le x \le x_2$, $y_1 \le y \le y_2$.

7. Group project.

5.5 The Possum Plague: A Model in the Making

Suggestions for preparing a lecture

Topics: Think of this as a model in the making, and use what is done in the text as a starting point for a general discussion and critique.

Remarks: Alternatively, do the SIR models of Problem 1 in class, and let the students read about the possum plague. Or else assign the whole section as a term project, and don't lecture on it at all.

Making up a problem set

Problems 1, 2, 3.

Comments

The models for the possum/tuberculosis problem are still in the works. So what we present in this section may very well be modified or even discarded within a few years. This material gives a glimpse at a model in the making and offers the reader a chance to play with the parameters of the model and maybe come up with some suggestions. Some may want to read the article by Wake and his colleagues cited in the footnote. We have scaled the variables and the parameters one way, but there are other ways to do the scaling, and the reader may want to try something else. We have included a traditional SIR (susceptible, infective, recovered) model in the problem set. Note that the S and I rate equations of the SIR model in Problem 1 have the same form as the rate equations for a harvested predator-prey system if the harvesting coefficient of the prey has the same magnitude as the natural growth coefficient [see Section 5.4, Problem 5(**d**)].

---

1.   **(a).**   Since the parameter b represents the reciprocal of the period of infection, the rate equation for the population of recovered is $R' = bI$. The rate equation for the population of susceptibles is $S' = -aSI$ because the spread of the disease depends upon contact between susceptibles and infectives, and so a mass action term such as $-aSI$ is appropriate, where $1/a$ measures the level of exposure of the average susceptible to the average infective. Since we assume that the total population is constant, then $S + I + R =$ constant, which implies that $I' = -(S' + R') = aSI - bI$.

(b).   Since an epidemic occurs when I' is larger than R', so we must have $aSI - bI > bI$, and so $S > 2b/a$. If more people recover each day than become infected, then $R' = bI > I' = aSI - bI$, i.e., $S < 2b/a$. So the threshold value for S is $2b/a$; below that value more people recover each day than become infected, while above the value more people are infected each day than recover.

(c).   We have $a = 0.003 \cdot (1/6) = 0.0005$. Since the duration of the illness in a person is four days, $b = 1/4 = 0.25$. It is shown in part **(b)** that the epidemic occurs if $S > 2b/a = 2 \times 0.25/0.0005 = 1000$. See Fig. 1(c) for graphs of $I(t)$, where $I(0) = 50$, $S_0 = 500, 1000, 1500$ (solid, long dashes, and short dashes, respectively). Note that an epidemic occurs only when the susceptible population is large enough (in the plotted cases, $S(0) = 1000, 1500$).

(d). See Figs. 1(d) for the respective S-, I-, and R-component graphs corresponding to $S(0) = 0$ (solid), 500 (long dashes), 1000 (short dashes), 1500 (long/short dashes), 2000 (solid); $I(0) = 100$; $R(0) = 0$. Note that the epidemic really only gets going when there is an adequate supply of susceptibles, and that the larger the number of susceptibles, the sooner the number of infectives reaches its peak.

(e). Because $dI = [b/(aS) - 1]dS$, the general solution is $I = (b/a) \ln S - S + C$. According to the initial condition, we know that $I = (b/a) \ln(S/S_0) + (S_0 - S) + I_0$. See Fig. 1(e), Graph 1 where $a = 0.001$, $b = 0.08$. The four curves (from top down) correspond to SI-initial data $(1500, 50)$, $(1250, 35)$, $(1000, 20)$, $(800, 15)$; the time span is $0 \le t \le 50$. The S-axis is a line of equilibrium points, and it seems that each SI-orbit starts (at $t = -\infty$) at a point $(S_1, 0)$ where S_1 is large, and ends (as $t \to +\infty$) at $(S_2, 0)$ where S_2 is small. This is plausible since as a disease runs its course the susceptible pool diminishes and the number of infectives first increases and then diminishes. Fig. 1(e), Graph 1 is deceptive because it suggests that as time goes on the number of susceptibles tends to 0. Fig. 1(e), Graph 2 zooms on the left-hand ends of the SI-orbits, and we see that as $t \to +\infty$ the separate orbits actually approach different points on the S-axis.

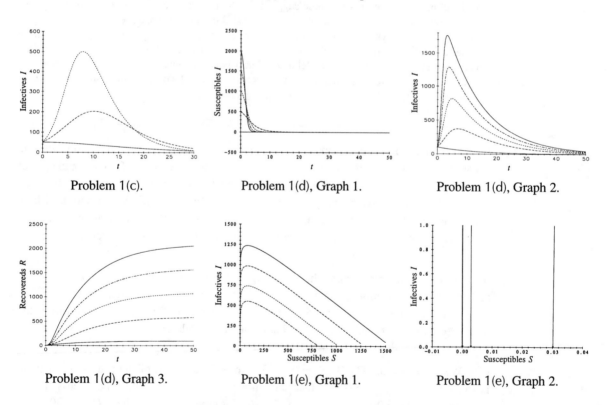

Problem 1(c).   Problem 1(d), Graph 1.   Problem 1(d), Graph 2.

Problem 1(d), Graph 3.   Problem 1(e), Graph 1.   Problem 1(e), Graph 2.

2. According to (2), we know that $x = P/h$, $y = I/j$ and $s = t/k$. If the scale constants h and j are measured in the same units of millions of animals (as P and I are) and k in years (as t is), then x, y, and s are dimensionless variables.

3. Let $S = P - I$, that is, $P = S + I$. Substituting into the possum model, we have $I' = \beta IS - (\alpha + b)I$ and $S' = (a - b)(S + I) - \alpha I - I' = (a - b)S + aI - \beta IS$. Similarly, from (5) with $z = x - y$ (so z is the scaled population of susceptible possums), we have $dy/ds = yz - ry = (z - r)y$, and $dz/ds + dy/ds = c(y + z) - y$. So the z and y rate equations are

$$dz/ds = z(c - y) - (1 - c - r)y \qquad dy/ds = (z - r)y$$

The corresponding equilibria are $(r, cr/(1 - c))$ and $(0, 0)$.

In Fig. 3, Graph 1 ($c = -1$, $r = 2$), the possums and the disease all disappear as $t \to +\infty$, basically because $c = -1$, and so the possum death rate due to natural causes (not tuberculosis) exceeds the birth rate. In Fig. 3, Graph 2 ($c = 0$, $r = 2$). The natural birth and death rates balance. In this case all points $z > 0$, $y = 0$ are equilibrium points, and every orbit tends to one of these equilibrium points as $y(t)$ diminishes and t tends to $+\infty$. Distinct orbits tend to distinct equilibrium points, although that is far from evident in Fig. 3, Graph 2. See Fig. 3, Graph 3 for a zoom of Fig. 3, Graph 2 that separates orbits near the equilibrium points. In Fig. 3, Graph 4 ($c = 1.5$, $r = 2$) the constant c is so large that the number of infectives increases without bound while the number of susceptible possums remains small. The population orbits in Fig. 3, Graph 5 ($c = 0.25$, $r = 2$) spiral toward the equilibrium point $(2, 2/3)$. The interpretation is that here the disease tends to a fairly low level endemic state. The interpretation of Fig. 3, Graph 6 ($c = 0.05$, $r = 1.1$) is similar to that of Fig. 3, Graph 5.

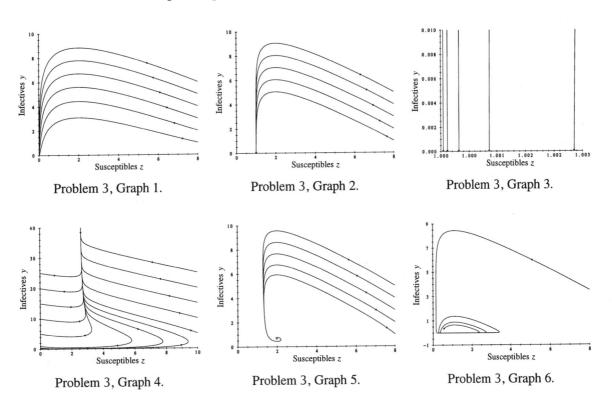

Problem 3, Graph 1.   Problem 3, Graph 2.   Problem 3, Graph 3.

Problem 3, Graph 4.   Problem 3, Graph 5.   Problem 3, Graph 6.

4. The vector $\mathbf{i} - \mathbf{j}$ is perpendicular to the line $y = x$ and is directed from the line into the region $y < x$. So orbits of $dx/ds = cx - y$, $dy/ds = y(x - y) - ry$ that touch the line $y = x$ stay on the line or move across it into the region $y < x$ if the dot product of $\mathbf{i} - \mathbf{j}$ and the field vector (evaluated for $y = x$) is non-negative. Note that

$$(\mathbf{i} - \mathbf{j}) \cdot ((cx - x)\mathbf{i} - rx\mathbf{j}) = cx - x + rx = x(c + r - 1) \geq 0$$

since $x > 0$ inside the first quadrant and by assumption, $c + r \geq 1$. Note that if $c + r = 1$, then on $y = x$, we have $dx/ds = (c - 1)x$ and $dy/ds = -rx = (c - 1)x$. So $dy/dx = 1$ and it follows that $y = x$ is an orbit.

5. Group project.

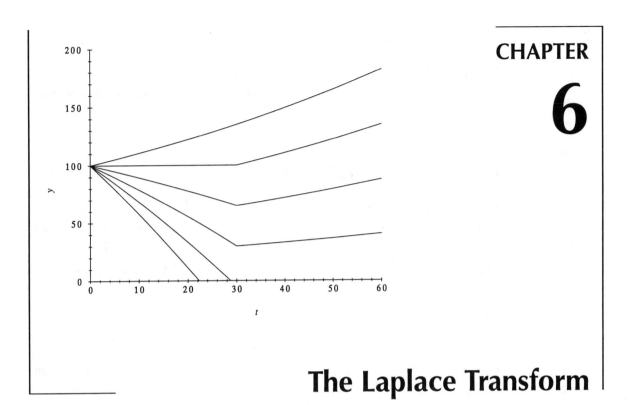

The Laplace Transform

There are differences of opinion about the place of Laplace transforms in an ODE course. Institutions with an engineering program often offer a "baby systems" course which contains a healthy dose of Laplace transforms. This reflects the central role played by the Laplace (and Fourier) transform in the work of practicing engineers. Because of this, some feel that the time in an ODE course can be better spent in doing some state plane analysis, stability, and bifurcations, subjects that engineers may not otherwise see in the classroom. On the other hand, institutions without an engineering program of their own may very well want to put Laplace transforms in their ODE course. This chapter has been included just in case it is needed, but no harm is done in skipping the chapter altogether.

With the above remarks out of the way, we make a few general observations about the contents of Chapter 6. The chapter is quite demanding technically, requiring, as it does, a great deal of algebraic pyrotechnics and much skill in using tables and the calculus of transforms. For convenience in calculating Laplace transforms, we have put a table giving the transforms of commonly used functions at the end of the chapter. Partial fractions play a big role in using Laplace transforms to solve IVPs, and the basic expansions are given in Table 6.2.1.

In using Laplace transforms to solve IVPs, the attempt will fail unless the ODEs are linear, and will be very difficult even if linear if the coefficients are not constant. The big advantage of transforms lies in their ability to handle discontinuous driving terms such as square waves or sawtooth waves and in the simple way the transform involves initial data right from the start of the solution process. The disadvantage (other than the restriction to linear ODEs with constant coefficients) is that the formula for the inverse transform is beyond the scope of this book. And so we must be satisfied with inversion by pattern matching from tables and by the calculus of transforms. It is a remarkable fact, however, that Laplace transforms can be used to solve constant-coefficient linear operator equations other than linear ODEs. An example of the transform technique to difference equations is given in Problem 10, Section 6.2, and to an unusual car-following model that involves differential-delay equations in Sections 6.1 and 6.3.

6.1 Introduction to the Laplace Transform

Suggestions for preparing a lecture

Topics: The idea of the Laplace transform, the transform of some specific functions (including the square pulse of Example 6.1.6), and how to use the transform to solve an IVP. We like to outline the car-following model.

Remarks: When we do teach this subject, we like to emphasize the linear operator properties on the transform.

Making up a problem set

The problems for this section are all straightforward—none are particularly exciting, but who is looking for transform excitement? Problems 1(**d**) and 1(**f**) are somewhat deceptive in that although the particular transforms as functions of s appear to be defined for all s, the process by which they are derived shows that, in fact, the transforms are defined only for $s > 0$. Unless the students notice this, they will inevitably say that the transform is defined for all s. At least one of the parts of Problem 2 should be given so that students get practice in breaking integrals at points of discontinuity of the integrand. Since IVPs for linear ODEs are the reason we are doing the transforms at all, at least one of the parts of Problem 3 should be assigned. The instructor might mention that one big advantage of using the transform to solve an IVP is that you don't have to find the general solution first and then impose the initial conditions. The transform does the whole thing in one fell swoop. We like Problem 3(**c**) because students must use the tables in the endpapers of the chapter to invert $1/(s+2)(s+3)$, unless they use partial fractions.

Comments

The purpose of this short section is to give a quick introduction to the transform, the ability to calculate the transform of some elementary functions, and a glimpse at applications of Laplace transforms to solving IVPs. The curious car-following process is modeled here, but the model equations are only solved in Section 6.3. The inverse transform is introduced (but not by formula). We note that although distinct continuous functions have distinct transforms, two different discontinuous functions may have the same transform. Logically, we should introduce equivalence classes of piecewise-continuous functions at this point so that the inverse transform $\mathcal{L}^{-1}$ is well-defined, but that level of abstraction seems out of place at the introductory level. How do you set about finding transforms—brain only, calculus book, tables of integrals, or tables of Laplace transforms? Our preference is to work out a few transforms from scratch, but after that to use tables, which is why we have included a set of transform tables at the end of the chapter. It is not a good idea to spend so much time working out integrals that you lose sight of what the transform is all about.

---

1.   The transform $\mathcal{L}[f](s) = \int_0^\infty e^{-st} f(t)\, dt$ may be computed by consulting tables of integrals or directly through integration by parts. The details of (**a**) are given, but answers only are listed for the other parts.

(a).

$$L[3t - 5] = \int_0^\infty e^{-st}(3t - 5)\,dt = \left[3\frac{e^{-st}}{s^2}(-st - 1) + \frac{5}{s}e^{-st}\right]\Big|_{t=0}^{t=\infty} = \frac{3}{s^2} - \frac{5}{s}$$

since $\lim_{t\to+\infty} e^{-st} = 0$ and $\lim_{t\to+\infty} te^{-st} = 0$ if (and only if) $s > 0$. And so, $0 < s < \infty$ is the largest interval on which the transform is defined.

(b). $L[t^2] = 2/s^3$, $s > 0$ (using tables or two integrations by parts).

(c). $L[t^n] = n!/s^{n+1}$, $s > 0$ (using tables or repeated integration by parts).

(d). $L[\cos at] = s/(s^2 + a^2)$, $s > 0$ (using tables or two integrations by parts). As a function of s, $s/(s^2 + a^2)$ is defined for all s. However, the derivation of the transform is invalid if $s \le 0$ because at some point the expression $\lim_{t\to+\infty} e^{-st}\cos at$ must be evaluated. This limit exists if and only if $s > 0$.

(e). $L[te^{at}] = 1/(s - a)^2$, $s > a$ (using tables or integration by parts).

(f). $L[t\sin at] = 2as/(s^2 + a^2)^2$, $s > 0$ (using tables or integration by parts). The transform variable s must be positive for much the same reason as in part **(d)**.

2. **(a).** If $f(t) = \sin t$, $0 \le t \le \pi$, but $f(t) = 0$ for $t > \pi$, then (using integral tables)

$$L[f](s) = \int_0^\pi e^{-st}\sin t\,dt = \frac{e^{-st}}{1 + s^2}(-s\sin t - \cos t)\Big|_{t=0}^{t=\pi} = \frac{1 + e^{-s\pi}}{1 + s^2}, \qquad -\infty < s < \infty$$

(b). If $f(t) = 0$, $0 \le t \le 1$, $f(t) = t$ for $t > 1$, then (using integral tables and L'Hôpital's Rule for $\lim_{t\to+\infty} st/e^{st}$)

$$L[f](s) = \int_1^\infty te^{-st}\,dt = \frac{e^{-st}}{s^2}(-st - 1)\Big|_{t=1}^{t=\infty} = \frac{(1 + s)e^{-s}}{s^2}, \qquad s > 0$$

(c). If $f(t) = t$, $0 \le t \le 1$, $f(t) = 0$ for $t > 1$, then (using integral tables)

$$L[f](s) = \int_0^1 te^{-st}\,dt = \frac{e^{-st}}{s^2}(-st - 1)\Big|_{t=0}^{t=1} = \frac{1}{s^2}[1 - (1 + s)e^{-s}], \qquad s > 0$$

(d). If $f(t) = t$, $0 \le t \le 1$, $f(t) = 2 - t$ for $t \ge 1$, then (splitting the interval of integration into $[0, 1]$ and $[1, \infty)$)

$$L[f](s) = \int_0^1 te^{-st}\,dt + 2\int_1^\infty e^{-st}\,dt - \int_1^\infty te^{-st}\,dt$$

$$= \frac{e^{-st}}{s^2}(-st - 1)\Big|_{t=0}^{t=1} - \frac{2}{s}e^{-st}\Big|_{t=1}^{t=\infty} + \frac{e^{-st}}{s^2}(st + 1)\Big|_{t=1}^{t=\infty}$$

$$= \frac{1}{s^2}[1 - (1 + s)e^{-s}] + \frac{2}{s}e^{-s} - \frac{(1 + s)e^{-s}}{s^2}$$

$$= \frac{1}{s^2}(1 - 2e^{-s}), \qquad s > 0$$

3. Apply the operator L to each side of the ODE and use the formula $L[y'](s) = sL[y](s) - y(0)$.

(a). The IVP is $y' + 2y = 0$, $y(0) = 1$. Applying $\mathcal{L}$, we have that $s\mathcal{L}[y] - 1 + 2\mathcal{L}[y] = 0$. We have $\mathcal{L}[y] = 1/(s+2)$. Since $\mathcal{L}[e^{-2t}] = 1/(s+2)$ (see Example 6.1.3 or II.3 in the tables), we have that $y = e^{-2t}$.

(b). The IVP is $y' + 2y = 0$, $y(0) = 1$. Applying $\mathcal{L}$, we have $s\mathcal{L}[y] - 5 + 2\mathcal{L}[y] = 1/(s+3)$ (see Example 6.1.3 or II.3 in the tables). We have that $\mathcal{L}[y] = 1/[(s+2)(s+3)] + 5/(s+2)$. By II.9 and II.3 in the tables, $y = e^{-2t} - e^{-3t} + 5e^{-2t} = 6e^{-2t} - e^{-3t}$.

(c). The IVP is $y' + 2y = e^{it}$, $y(0) = 0$. Applying $\mathcal{L}$, we have $s\mathcal{L}[y] + 2\mathcal{L}[y] = 1/(s-i)$, where II.3 in the tables has been used with $a = i$. So, $\mathcal{L}[y] = 1/[(s+2)(s-i)]$. By II.9 with $a = -2$, and $b = i$, we have $y = (e^{it} - e^{-2t})/(2+i)$.

4. Group project

6.2 Calculus of the Transform

Suggestions for preparing a lecture

Topics: Work out several IVPs (including at least one that requires a partial fraction expansion, and another that uses a shifting theorem), and show students how to use the tables in the entries in Table 6.2.1.

Remarks: There is a lot of material in this section, and you probably won't cover it all in the lecture hall, even if you have a lecture and a half or two lectures available.

Making up a problem set

Here is a collection of problems that pretty much covers the most important techniques: 1**(d)**, **(g)**, **(i)**, and 3**(e)**. Problems 6–8 show how to use transforms in solving IVPs that model physical phenomena. It is a good idea to assign at least one of these Problems—maybe 6 and 7**(c)**. You might want to assign Problem 10 (for a different kind of equation, that is *not* an ODE).

Comments

Laplace transforms are a great tool for solving IVPs, but, as with any tool, you have to learn the operational rules first—and there are a lot of them. We have put the most important of the rules in this section—how do you transform a derivative, an integral, or an on-off function? Going backward, how do you inverse transform a quotient of polynomials? An important part of the answer to the last question is the algebraic technique of partial fractions (see Table 6.2.1). The Shifting Theorem (the three formulas in Theorem 6.2.4) are essential when dealing with functions where the time scale (or the s-variable scale) has been shifted. You may want to use a CAS to avoid what are often lengthy and exacting algebraic manipulations.

1. The transforms can be obtained by straightforward integration, integration by parts, tables of integrals, or the tables at the chapter's end. For brevity, we refer to the tables or to formulas derived in this section. The basic partial fraction expansions are given in Table 6.2.1; every expansion needed here can be reduced to one in the table.

(a). Since $\sinh at = (e^{at} - e^{-at})/2$, $L[\sinh at] = a/(s^2 - a^2)$ by II.7 (or II.3).

(b). Since $\cosh at = (e^{at} + e^{-at})/2$, $L[\cosh at] = s/(s^2 - a^2)$ by II.7 (or II.3).

(c). By Eq. (11), $L[t^2 e^{at}] = L[t^2](s - a) = 2/(s - a)^3$, where we have used II.1.

(d). By Eq. (11), $L[(1 + 6t)e^{at}] = L[1 + 6t](s - a) = 1/(s - a) + 6/(s - a)^2$, where we have used II.1.

(e). Using Eq. (11) and Eq. (4), $L[te^{2t} f(t)] = L[tf(t)e^{2t}] = L[tf(t)](s - 2) = -\phi'(s - 2)$, where $L[f](s)$ is denoted by $\phi(s)$.

(f). $L[(D^2 + 1)f] = L[f'' + f] = s^2 L[f] - sf(0) - f'(0) + L[f]$ by Eq. (2).

(g). $L[(t + 1)\,\text{step}(t - 1)] = L[(t + 2 - 1)\,\text{step}(t - 1)] = e^{-s} L[t + 2] = e^{-s}(1/s^2 + 2/s)$ by Eq. (12) and II.3.

(h). $L[te^{at}\,\text{step}(t - 1)] = L[t\,\text{step}(t - 1)](s - a) = e^{-(s-a)}L[t + 1] = e^{-(s-a)}[1/(s - a)^2 + 1/(s - a)]$, where we have used Eq. (11) and Eq. (13) and II.3.

(i). $L\big[e^{at}[\text{step}(t - 1) - \text{step}(t - 2)]\big] = e^{-s}L[e^{a(t+1)}] - e^{-2s}L[e^{a(t+2)}] = e^{-(s-a)}/(s - a) - e^{-2(s-a)}/(s - a)$, where we have used Eq. (13).

(j). $L[(t - 2)\big[\text{step}(t - 1) - \text{step}(t - 3)\big]] = L[((t - 1) - 1)\,\text{step}(t - 1)] - L[((t + 1) - 3)\,\text{step}(t - 3)] = e^{-s}L[t - 1] - e^{-3s}L[t + 1] = e^{-s}(1/s^2 - 1/s) - e^{-3s}(1/s^2 + 1/s)$ by Eq. (12) and II.3.

2. In each case the appropriate formula number in Section 6.2 or in the Transform Tables is given.

(a). $f(t) = 3\,\text{step}(t - 2) - 3\,\text{step}(t - 5)$, and so $L[f] = 3(e^{-2s} - e^{-5s})/s$, where (14) and (15) have been used.

(b). $f(t) = 2\sin t\,\text{step}(\pi - t)$, and so $L[f] = 2(1 + e^{-\pi s})/(1 + s^2)$, where integral tables (or integration by parts) is used to evaluate $2\int_0^{\pi} e^{-st}\sin t\,dt$.

(c). II.3: $f(t) = t^{12}e^{5t}$, and so $L[f] = 12!/(s - 5)^{13}$.

(d). II.12: $f(t) = 6t\sin 3t$, and so $L[f] = 36s/(s^2 + 9)^2$.

(e). II.13: $L[f] = \sqrt{s + 2} - \sqrt{s}$, and so $f = (1 - e^{-2t})/(2\sqrt{\pi t^3})$.

(f). II.17: $L[f] = \ln(s + 5) - \ln(s - 2)$, and so $f = (e^{2t} - e^{-5t})/t$.

3. Apply the operator L to each side of the ODE in the given IVP, and use formula (2) with $n = 2$. See Table 6.2.1 for partial fraction expansions.

(a). The IVP is $y'' - y' - 6y = 0$, $y(0) = 1$, $y'(0) = -1$, so $s^2 L[y] - s + 1 - sL[y] + 1 - 6L[y] = 0$. Using partial fractions

$$L[y] = \frac{s - 2}{s^2 - s - 6} = \frac{s - 2}{(s - 3)(s + 2)} = \frac{1}{5(s - 3)} + \frac{4}{5(s + 2)}$$

So, $y(t) = (e^{3t} + 4e^{-2t})/5$ by II.3.

(b). The IVP is $y'' + y = \sin t$, $y(0) = 0$, $y'(0) = 1$, so $s^2 L[y] - 1 + L[y] = 1/(s^2 + 1)$

by II.5:

$$L[y] = \frac{1}{(s^2+1)^2} + \frac{1}{s^2+1}$$

Using II.5 and II.11, we have that $y = (3\sin t - t\cos t)/2$.

(c). The IVP is $y'' - 2y' + 2y = 0$, $y(0) = 0$, $y'(0) = 1$, so $s^2 L[y] - 1 - 2sL[y] + 2L[y] = 0$:

$$L[y] = \frac{1}{s^2-2s+2} = \frac{1}{(s-1)^2+1}$$

By Eq. (11) and II.5, $y(t) = e^t \sin t$.

(d). The IVP is $y'' + 4y' + 4y = e^t$, $y(0) = 1$, $y'(0) = 1$, so $s^2 L[y] - s - 1 + 4sL[y] - 4 + 4L[y] = 1/(s-1)$ by II.3. Then

$$L[y] = \frac{s+5}{s^2+4s+4} + \frac{1}{(s-1)(s^2+4s+4)} = \frac{s^2+4s-4}{(s-1)(s+2)^2}$$

$$= \frac{1/9}{s-1} + \frac{8/9}{s+2} + \frac{24/9}{(s+2)^2}$$

where we have used partial fractions. So $y = (e^t + 8e^{-2t} + 24te^{-2t})/9$ by II.3.

(e). The IVP is $y'' - 2y' + y = \text{step}(t-1)$, $y(0) = 1$, $y'(0) = 0$, so $s^2 L[y] - s - 2sL[y] + 2 + L[y] = L[\text{step}(t-1)] = e^{-s}/s$ by (14). Using partial fractions, we have

$$L[y] = \frac{s-2}{s^2-2s+1} + \frac{e^{-s}}{s(s^2-2s+1)} = \frac{1}{s-1} - \frac{1}{(s-1)^2} + e^{-s}\left(\frac{1}{s} - \frac{1}{s-1} + \frac{1}{(s-1)^2}\right)$$

Then using II.3 and (14), we have

$$y = e^t(1-t) + \text{step}(t-1)[1 - e^{t-1} + (t-1)e^{t-1}] = e^t(1-t) + \text{step}(t-1)[1 + (t-2)e^{t-1}]$$

(f). The IVP is $y'' + 2y' - 3y = \text{step}(1-t)$, $y(0) = 1$, $y'(0) = 0$, so $s^2 L[y] - s + 2sL[y] - 2 - 3L[y] = (1-e^{-s})/s$ by Eq. (15). Using partial fractions, we have

$$L[y] = \frac{s+2}{s^2+2s-3} + \frac{1-e^{-s}}{s(s^2+2s-3)} = \frac{s^2+2s+1}{s(s+3)(s-1)} - \frac{e^{-s}}{s(s+3)(s-1)}$$

$$= \frac{-1/3}{s} + \frac{1/3}{s+3} + \frac{1}{s-1} - e^{-s}\left(\frac{-1/3}{s} + \frac{1/3}{s+3} + \frac{1}{s-1}\right)$$

By II.3 and (12), we have

$$y(t) = -\frac{1}{3} + \frac{1}{3}e^{-3t} + e^t - \text{step}(t-1)\left[-\frac{1}{3} + \frac{1}{3}e^{-3(t-1)} + e^{(t-1)}\right]$$

4. Use formula (8) in every case. Since $a = 0$, the final integral in (8) is 0.

(a). The transform is (by II.3)

$$L\left[\int_0^t (x-1)e^x \, dx\right] = \frac{1}{s}L[(t-1)e^t] = \frac{1}{s}\left[\frac{1}{(s-1)^2} - \frac{1}{(s-1)}\right]$$

(b). The transform is (by II.1)

$$L\left[\int_0^t (x^2 - 2x)\,dx\right] = \frac{1}{s}L[t^2 - 2t] = \frac{1}{s}\left[\frac{2}{s^3} - \frac{2}{s^2}\right]$$

(c). Since $\sin(t - \pi/4) = (\sin t - \cos t)/\sqrt{2}$ by a trigonometric identity, the transform of $\int_0^t \sin(x - \pi/4)e^x\,dx$ is [using II.5, II.6, and (8)]

$$\frac{1}{\sqrt{2}s}L[e^t \sin t - e^t \cos t] = \frac{1}{\sqrt{2}s}\left[\frac{1}{(s-1)^2+1} - \frac{s-1}{(s-1)^2+1}\right]$$

(d). From (8) and II.6 the transform is

$$L\left[\int_0^t e^{(x-a)}\cos x\,dx\right] = \frac{1}{s}L[e^{-a}e^t \cos t] = \frac{1}{s}e^{-a}\frac{s-1}{(s-1)^2+1}$$

5. Since

$$\text{step}(t-a) = \begin{cases} 1, & t \geq a \\ 0, & t < a \end{cases} \quad \text{and} \quad \text{step}(t-b) = \begin{cases} 1, & t \geq b \\ 0, & t < b \end{cases}$$

and $b > a$, we have $\text{step}(t-a) - \text{step}(t-b) = 1$ for $a \leq t < b$, and $= 0$ for all $t < a$ or $t \geq b$. This is a square pulse over the interval (a, b), and is denoted by the engineering function $\text{sqp}(t-a, b-a)$. Using the above, $L[\text{sqp}(t-1, 3)] = L[\text{step}(t-1) - \text{step}(t-4)]$, which, using II.19, equals $(e^{-s} - e^{-4s})/s$.

6. **(a).** The constant input rate over the time span $a \leq t \leq b$ is k_2, while k_1 is the basic radioactive decay coefficient. The input to the sample may be described as a square pulse composed of two step functions as shown in Problem 5. The Balance Law [Net Rate = Rate in − Rate out] provides: $y'(t) = -k_1 y(t) + k_2[\text{step}(t-a) - \text{step}(t-b)]$, where $y(t)$ is the amount present at time t.

(b). Let $y(0) = y_0$ and apply L to each side of the ODE in (a): $sL[y] - y_0 = -k_1 L[y] + k_2(e^{-as} - e^{-bs})/s$, where we have used (1), (14):

$$L[y] = \frac{y_0}{s+k_1} + \frac{k_2}{s(s+k_1)}[e^{-as} - e^{-bs}]$$

Since using partial fractions

$$\frac{1}{s(s+k_1)} = \frac{1}{k_1}\left[\frac{1}{s} - \frac{1}{s+k_1}\right]$$

we have that

$$L[y] = \frac{y_0}{s+k_1} + \frac{k_2}{k_1}\left[\frac{1}{s} - \frac{1}{s+k_1}\right][e^{-as} - e^{-bs}]$$

By II.3 and formula (12), we have

$$y(t) = y_0 e^{-k_1 t} + \frac{k_2}{k_1}\text{step}(t-a)(1 - e^{-k_1(t-a)}) - \frac{k_2}{k_1}\text{step}(t-b)(1 - e^{-k_1(t-b)})$$

See Fig. 6(b) for $y(t)$ with $k_1 = 1$, $k_2 = 0.5$, $a = 1$, $b = 10$, $y_0 = 1$, and $0 \leq t \leq 20$.

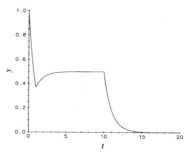

Problem 6(b).

7. The answers to parts **(a)** and **(b)** are obtained from **(c)** and **(d)** by setting the resistance $R = 0$. The answers have the same form as in **(c)** and **(d)** because the roots r_1 and r_2 of the characteristic polynomial $s^2 + Rs/L + 1/LC$ are given symbolically.

(c). Applying the transform to each side of the ODE, $Lq'' + Rq' + \frac{1}{C}q = E_0 \operatorname{step}(t - a)$, rearranging terms, and using (14), we obtain

$$\mathcal{L}[q] = \frac{E_0}{Ls^2 + Rs + 1/C} \frac{e^{-as}}{s} = \frac{E_0/L}{(s - r_1)(s - r_2)s} e^{-as}$$

where r_1 and r_2 are assumed to be distinct roots of $s^2 + Rs/L + 1/(LC)$, but they need not be real. Since we have the partial fraction decomposition

$$\frac{1}{(s - r_1)(s - r_2)s} = \frac{1}{r_1 r_2 (r_1 - r_2)} \left[\frac{r_2}{s - r_1} - \frac{r_1}{s - r_2} + \frac{r_1 - r_2}{s} \right]$$

we can use II.3 and formula (12) and the fact that $r_1 r_2 = 1/LC$ to obtain

$$q_a(t) = \operatorname{step}(t - a) C E_0 \left[1 + \frac{r_2}{r_1 - r_2} e^{r_1(t-a)} + \frac{r_1}{r_2 - r_1} e^{r_2(t-a)} \right]$$

where $q_a(t)$ is the charge at time $t \geq 0$ if the switch is turned on at time $t = a$.

(d). Referring to the solution to part **(c)**, we see that $q(t) = q_a(t) - q_b(t)$, where $q_a(t)$ is given in part **(c)**'s solution and $q_b(t)$ is as well (replacing a by b).

8. From Eq. (18), we see that we must have $A = y(360) = (A - H/k)e^{360k} + He^{330k}/k$, assuming, of course, that the population does not become extinct during the harvest season, i.e., assuming that $H < kA(1 - e^{-30k})^{-1}$.

9. **(a).** The IVP is $x_1' = 3x_1 - 2x_2 + t$, $x_2' = 5x_1 - 3x_2 + 5$, $x_1(0) = x_2(0) = 0$. Using the initial data and applying the Laplace transform to each term in each equation, we have that $s\mathcal{L}[x_1] = 3\mathcal{L}[x_1] - 2\mathcal{L}[x_2] + s^{-2}$, $s\mathcal{L}[x_2] = 5\mathcal{L}[x_1] - 3\mathcal{L}[x_2] + 5s^{-1}$. Solving these equations for the transforms $\mathcal{L}[x_1]$ and $\mathcal{L}[x_2]$, we have (using a partial fractions decomposition)

$$\mathcal{L}[x_1] = \frac{3 - 9s}{s^2(s^2 + 1)} = \frac{-9}{s} + \frac{3}{s^2} + \frac{9s - 3}{s^2 + 1}$$

$$\mathcal{L}[x_2] = \frac{5s^2 - 15s + 5}{s^2(s^2 + 1)} = \frac{-15}{s} + \frac{5}{s^2} + \frac{15s}{s^2 + 1}$$

Using the Laplace transform tables, we can invert the transforms to obtain $x_1(t) = -9 + 3t + 9\cos t - 3\sin t$, $x_2(t) = -15 + 5t + 15\cos t$.

(b). The IVP is $x_1' = x_1 + 3x_2 + \sin t$, $x_2' = x_1 - x_2$, $x_1(0) = 0$, $x_2(0) = 1$. Proceeding as in **(a)**, we have that $sL[x_1] = L[x_1] + 3L[x_2] + (s^2 + 1)^{-1}$, $sL[x_2] - 1 = L[x_1] - L[x_2]$. Solving these equations for the transforms $L[x_1]$ and $L[x_2]$ and using partial fractions, we have

$$L[x_1] = \frac{3s^2 + s + 4}{(s^2 + 1)(s^2 - 4)} = \frac{-1/5}{s^2 + 1} + \frac{-s/5}{s^2 + 1} + \frac{9/10}{s - 2} - \frac{7/10}{s + 2}$$

$$L[x_2] = \frac{s^3 - s^2 + s}{(s^2 + 1)(s^2 - 4)} = \frac{-1/5}{s^2 + 1} + \frac{3/10}{s - 2} + \frac{7/10}{s + 2}$$

Using the Laplace transform tables, we have

$$x_1(t) = -\frac{1}{5}\sin t - \frac{1}{5}\cos t + \frac{9}{10}e^{2t} - \frac{7}{10}e^{-2t}$$

$$x_2(t) = -\frac{1}{5}\sin t + \frac{3}{10}e^{2t} + \frac{7}{10}e^{-2t}$$

10. The difference equation is $3x(t) - 4x(t - 1) = 1$, where $x(t) = 0$ for $t \le 0$. First, we find $L[x(t - 1)]$. We have, by setting $u = t - 1$, that

$$L[x(t - 1)] = \int_0^\infty e^{-st} x(t - 1)\, dt = \int_{-1}^\infty e^{-s(u+1)} x(u)\, du = \int_0^\infty e^{-s(u+1)} x(u)\, du$$

$$= e^{-s} \int_0^\infty e^{-su} x(u)\, du = e^{-s} L[x]$$

where we have used the fact that $x(t) = 0$, $t \le 0$. Applying L to each side of the difference equation, we then have that $3L[x] - 4e^{-s} L[x] = 1/s$. So,

$$L[x] = \frac{1}{s(3 - 4e^{-s})} = \left[3s\left(1 - \frac{4}{3}e^{-s}\right) \right]^{-1}$$

For s large enough, we have that $4e^{-s}/3 < 1$ and a geometric series expansion of $(1 - 4e^{-s}/3)^{-1}$ can be used, which gives

$$L[x] = \frac{1}{3s}\left[1 + \frac{4}{3}e^{-s} + \frac{16}{9}e^{-2s} + \cdots + \left(\frac{4}{3}\right)^n e^{-ns} + \cdots \right]$$

(Geometric series are found in Appendix B.2, item 9.)

By (14) the solution of the difference equation is

$$x(t) = \frac{1}{3}\text{step}(t) + \frac{4}{9}\text{step}(t - 1) + \frac{16}{27}\text{step}(t - 2) + \cdots + \frac{1}{3}\left(\frac{4}{3}\right)^n \text{step}(t - n) + \cdots$$

6.3 Applications of the Transform: Car Following

Suggestions for preparing a lecture

Topics: The transform of a periodic function and some IVP with an on-off driving function.

Remarks: We have done a driven LC circuit in the text. You may want to redo the RC-communication circuit of Example 2.2.7, or the model of antihistamine in the GI-tract (the first ODE in Example 1.8.5), or (our favorite) the car-following model and its solution.

Making up a problem set

A selection of problems from 1, 2, 3, and 4, should be assigned so that students have experience with a variety of transforms and can use them in solving IVPs. Our favorite problems, though, are not these, but the much harder (and more intriguing) Problems 5, 7, 9, 10, 12, 13 because they have engineering functions as inputs or else involve time delays. Particularly interesting are 5(**b**), 5(**c**), 7(**b**) (with period $T = 3$), and 12. In 5(**b**) and 7(**b**) the Fourier series expansion of the engineering function input has a sinusoidal term whose frequency exactly matches or is close to the actual frequency of the ODE—and this means resonance or beats. The instructor may want to work up something along these lines for the lecture. Problem 6 or 8 should be assigned so that students see how to use geometric series expansions, which are of central importance in working with transforms of the periodic engineering functions.

Comments

The main topic of this section is the Laplace transform of a periodic function. Of course, we have already shown how to transform sinusoids, but now we derive a formula for the transform of any periodic function. That formula is then applied to an LC circuit modeled by an ODE in the current $I(t)$ with a square wave driving function and to the car-following model of Section 6.1. The details are all worked out, and they are far from trivial since geometric expansions and partial fractions are required. Partial fractions formulas were introduced in Table 6.2.1. As is usually the case when working with transforms, there is a lot of algebra. The geometric series expansion $(1-x)^{-1} = 1+x+x^2+\cdots+x^n+\cdots$ for $|x| < 1$ is used throughout the section and its problems. It is valid for any x whose magnitude is less than 1; for example, it is valid for $x = e^{-s}$, $s > 0$.

Traffic models are widely used these days to predict what will happen in congested areas when certain conditions occur. As always when human factors are involved, the models are more suggestive than exact. Nevertheless, they do give a way of thinking about traffic flow, and they do include certain kinds of human behavior that most of us can immediately recognize. The modeling outlined in Section 6.1 and in this section is straightforward. However, solving the model equations by using transform techniques is anything but straightforward. On the other hand, formulas (22)–(24) show that the solution is not that hard to understand and use.

---

1. Since a periodic function appears as a factor in each case, formula (2) may be used at some point in the calculation. On the other hand, it is often simpler to use integration tables (or integration by parts) or transform formulas from Sections 6.1 or 6.2 if the periodic functions involve sines and cosines. We use both approaches for (**a**), but give answers only for the rest.

(a). $\mathcal{L}[e^t \sin t](s) = \mathcal{L}[\sin t](s - 1)$ by (11) of the Shifting Theorem (Theorem 6.2.4). By formula (2) and integral tables (or a double integration by parts)

$$L[\sin t](s) = \frac{1}{1 - e^{-2\pi s}} \int_0^{2\pi} e^{-st} \sin t \, dt = \frac{1}{1 - e^{-2\pi s}} \left[\frac{1 - e^{-2\pi s}}{1 + s^2} \right] = \frac{1}{1 + s^2}$$

So, $L[e^t \sin t] = 1/(1 + (s - 1)^2)$. Alternatively, by integral tables (or a double integration by parts) we have that

$$L[e^t \sin t](s) = \int_0^\infty e^{-st} e^t \sin t \, dt = \int_0^\infty e^{-(s-1)t} \sin t \, dt$$

$$= \frac{1}{1 + (s - 1)^2} \left[e^{-(s-1)t} (-(s-1) \sin t - \cos t) \right]_{t=0}^{t=\infty} = \frac{1}{1 + (s - 1)^2}$$

(b). $L[\sin^2 t](s) = 2/[s(s^2 + 4)]$ by integration tables.

(c). Using the result [from part **(b)**] that $L[\sin^2 t](s) = 2/[s(s^2 + 4)]$, we have

$$L[\cos^2 t](s) = L[1 - \sin^2 t](s) = \frac{1}{s} - \frac{2}{s(s^2 + 4)} = \frac{s^2 + 2}{s(s^2 + 4)}$$

(d). $L[t \sin at](s) = -dL[\sin at](s)/ds$ by formula (4) in Section 6.2. Since $L[\sin at](s) = a/(a^2 + s^2)$, we have that $L[t \sin at](s) = 2as/(a^2 + s^2)^2$.

(e). By formula (4) in Section 6.2 [and part **(d)**] $L[t \sin at](s) = 2as/(a^2 + s^2)^2$,

$$L[t^2 \sin at] = -\frac{d}{ds} L[t \sin at] = -\frac{d}{ds} \left[\frac{2as}{(a^2 + s^2)^2} \right] = -\frac{2a(3s^2 - a^2)}{(a^2 + s^2)^3}$$

(f). $L[\cos^3 t] = (s^3 + 7s)/[(s^2 + 9)(s^2 + 1)]$ by integration tables.

(g). Using a trigonometric identity, we have

$$L[e^{-3t} \cos(2t + \pi/4)](s) = \frac{1}{\sqrt{2}} L[e^{-3t} \cos(2t)](s) - \frac{1}{\sqrt{2}} L[e^{-3t} \sin(2t)](s)$$

$$= \frac{1}{\sqrt{2}} L[\cos(2t)](s + 3) - \frac{1}{\sqrt{2}} L[\sin(2t)](s + 3)$$

by (11) of the Shifting Theorem (Theorem 6.2.4). And so by Example 6.2.1,

$$L[e^{-3t} \cos(2t + \pi/4)] = \frac{1}{\sqrt{2}} \left[\frac{s + 3}{(s + 3)^2 + 4} - \frac{2}{(s + 3)^2 + 4} \right] = \frac{1}{\sqrt{2}} \frac{s + 1}{(s + 3)^2 + 4}$$

(h). Using formulas (4) and (11) in Section 6.2, we have

$$L[t^2 e^t \cos t](s) = \frac{d^2}{ds^2} \{ L[e^t \cos t](s) \} = \frac{d^2}{ds^2} \{ L[\cos t](s - 1) \}$$

We have by Example 6.2.1 that

$$L[t^2 e^t \cos t](s) = \frac{d^2}{ds^2} \left[\frac{s - 1}{(s - 1)^2 + 1} \right] = \frac{2(s - 1)(s^2 - 2s - 2)}{((s - 1)^2 + 1)^3}$$

2.   **(a).** $L^{-1}[1/s + e^{-s}/s] = 1 + \text{step}(t - 1)$ by (14) in Section 6.2 and Example 6.2.5.

(b). Using formula (12) of Section 6.2, we have that

$$\mathcal{L}^{-1}\left[\frac{3e^{-2s}}{3s^2+1}\right] = \mathcal{L}^{-1}\left[e^{-2s}\frac{1}{s^2+1/3}\right] = \text{step}(t-2)f(t-2)$$

where $\mathcal{L}[f] = 1/(s^2+1/3)$. By Example 6.2.1, $f(t-2) = \sqrt{3}\sin((t-2)/\sqrt{3})$ and so

$$\mathcal{L}^{-1}\left[\frac{3e^{-2s}}{3s^2+1}\right] = \sqrt{3}\,\text{step}(t-2)\sin((t-2)/\sqrt{3})$$

(c). If $\mathcal{L}[f](s) = 1/s^n$, then by (11) of the Shifting Theorem (Theorem 6.2.4) $\mathcal{L}[e^{at}f(t)] = 1/(s-a)^n$. On the other hand, by formula (4) of Section 6.2,

$$s^{-n} = \frac{(-1)^{n-1}}{(n-1)!}\frac{d^{n-1}}{ds^{n-1}}(s^{-1}) = \frac{(-1)^{n-1}}{(n-1)!}\mathcal{L}[(-1)^{n-1}t^{n-1}]$$

since $\mathcal{L}[1] = s^{-1}$. So,

$$\mathcal{L}^{-1}\left[\frac{1}{(s-a)^n}\right] = \frac{t^{n-1}e^{at}}{(n-1)!}$$

(d). $\mathcal{L}^{-1}[1/s^2 - e^{-s}/s^2] = t - (t-1)\,\text{step}(t-1)$ by Example 6.1.6 and (12) of the Shifting Theorem (Theorem 6.2.4).

(e). As in **(a)** and **(b)** we have that

$$\mathcal{L}^{-1}\left[\frac{(s-1)e^{-s}+1}{s^2}\right] = \mathcal{L}^{-1}\left[\frac{1}{s}e^{-s} - \frac{1}{s^2}e^{-s} + \frac{1}{s^2}\right]$$

$$= \text{step}(t-1) - (t-1)\,\text{step}(t-1) + t = t + (2-t)\,\text{step}(t-1)$$

(f). By formula (4) in Section 6.2, and Example 6.1.6, we have that

$$\mathcal{L}^{-1}\left[\frac{d}{ds}\ln\frac{s+3}{s+2}\right] = \mathcal{L}^{-1}\left[\frac{1}{s+3} - \frac{1}{s+2}\right] = e^{-3t} - e^{-2t} = -t\mathcal{L}^{-1}\left[\ln\frac{s+3}{s+2}\right]$$

So,

$$\mathcal{L}^{-1}\left[\ln\frac{s+3}{s+2}\right] = \frac{1}{t}(e^{-2t} - e^{-3t})$$

3.   The partial fraction decompositions and the answers $f(t)$ are listed, but the detailed calculations are omitted. The partial fraction decompositions follow from Table 6.2.1.

(a).

$$\frac{1}{s(s+1)} = \frac{1}{s} - \frac{1}{s+1}, \qquad f(t) = 1 - e^{-t}$$

(b).

$$\frac{1}{s(s+2)^2} = \frac{1/4}{s} - \frac{1/2}{(s+2)^2} - \frac{1/4}{s+2}, \qquad f(t) = \frac{1}{4} - \frac{1}{2}te^{-2t} - \frac{1}{4}e^{-2t}$$

(c).

$$\frac{1}{(s-a)(s-b)} = \frac{1}{a-b}\left[\frac{1}{s-a} - \frac{1}{s-b}\right], \qquad f(t) = \frac{1}{a-b}[e^{at} - e^{bt}]$$

(d).

$$\frac{s^2 + 3}{(s-1)^2(s+1)} = \frac{1}{s+1} + \frac{2}{(s-1)^2}, \qquad f(t) = e^{-t} + 2te^t$$

(e).

$$\frac{3s+1}{(s^2+2s+2)(s-1)} = -\frac{1}{5} \cdot \frac{4s-3}{(s+1)^2+1} + \frac{4}{5} \cdot \frac{1}{s-1}$$

$$f(t) = -\frac{4}{5}e^{-t}\cos t + \frac{3}{5}e^{-t}\sin t + \frac{4}{5}e^t$$

(f).

$$\frac{s+1}{(s-2)(s^2+9)} = \frac{1}{13}\left(\frac{3}{s-2} - \frac{3s}{s^2+9} + \frac{7}{s^2+9}\right)$$

$$f(t) = \frac{1}{13}(3e^{2t} - 3\cos 3t + \frac{7}{3}\sin 3t)$$

4. Apply $\mathcal{L}$ to each side of the ODE and use partial fractions (see Table 6.2.1) and the appropriate transform formulas from Sections 6.1-6.3 and in the tables. Use the formula $\mathcal{L}[y'' + ay' + by] = (s^2 + as + b)\mathcal{L}[y] - (s+a)y(0) - y'(0)$. The detailed calculations are omitted. See Figs. 4(a)–4(g) for plots of the solution curves.

(a). The IVP is $y'' + 6y' + 5y = t$, $y(0) = y'(0) = 0$.

$$\mathcal{L}[y] = \frac{1}{s^2(s^2+6s+5)} = \frac{1}{s^2(s+1)(s+5)} = \frac{-6/25}{s} + \frac{1/5}{s^2} + \frac{1/4}{s+1} - \frac{1/100}{s+5}$$

$$y = -\frac{6}{25} + \frac{1}{5}t + \frac{1}{4}e^{-t} - \frac{1}{100}e^{-5t}$$

(b). The IVP is $y'' + 2y' + y = e^t$, $y(0) = y'(0) = 0$.

$$\mathcal{L}[y] = \frac{1}{(s+1)^2(s-1)} = \frac{-1/4}{s+1} - \frac{1/2}{(s+1)^2} + \frac{1/4}{s-1}$$

$$y = \frac{1}{4}[-e^{-t} - 2te^{-t} + e^t]$$

(c). The IVP is $y'' - 2y' + 2y = \sin t$, $y(0) = y'(0) = 0$.

$$\mathcal{L}[y] = \frac{1}{(s^2+1)((s-1)^2+1)}$$

$$= \frac{2}{5} \cdot \frac{s}{s^2+1} + \frac{1}{5} \cdot \frac{1}{s^2+1} - \frac{2}{5} \cdot \frac{s-1}{(s-1)^2+1} + \frac{1}{5} \cdot \frac{1}{(s-1)^2+1}$$

$$y = (2\cos t + \sin t)/5 + e^t(-2\cos t + \sin t)$$

(d). The IVP is $y'' - 4y' + 4y = 2e^t + \cos t$, $y(0) = 3/25$, $y'(0) = -4/25$.

$$\mathcal{L}[y] = \frac{2}{(s-1)(s-2)^2} + \frac{s}{(s^2+1)(s-2)^2} + \frac{3s}{25(s-2)^2} - \frac{4}{25(s-2)^2}$$

$$= \frac{2}{s-1} - \frac{2}{s-2} + \frac{2}{(s-2)^2} + \frac{3s}{25(s^2+1)} - \frac{4}{25(s^2+1)}$$

$$y = 2e^t + 2e^{2t}(t-1) + (3\cos t - 4\sin t)/25$$

(e). The IVP is $y'' - 2y' + y = e^t \sin t$, $y(0) = y'(0) = 0$.

$$\mathcal{L}[y] = \frac{1}{(s-1)^2} \cdot \frac{1}{(s-1)^2 + 1} = \frac{1}{(s-1)^2} - \frac{1}{(s-1)^2 + 1}$$

$$y = e^t(t - \sin t)$$

(f). The IVP is $y'' + 2y' + y = te^{-t}$, $y(0) = 1$, $y'(0) = -2$.

$$\mathcal{L}[y] = \frac{1}{(s+1)^4} + \frac{s}{(s+1)^2} = \frac{1}{(s+1)^4} + \frac{1}{s+1} - \frac{1}{(s+1)^2}$$

$$y = t^3 e^{-t}/6 + e^{-t}(1 - t)$$

(g). The IVP is $y''' + y'' + 4y' + 4y = -2$, $y(0) = 0$, $y'(0) = 1$, $y''(0) = -1$. Applying the Laplace transform operator $\mathcal{L}$ to each side of the ODE and using the initial data, we have that $(s^3 + s^2 + 4s + 4)\mathcal{L}[y] - s = -2/s$, and so

$$\mathcal{L}[y] = \frac{s^2 - 2}{s(s+1)(s^2+4)} = \frac{1}{5(s+1)} + \frac{3s}{10(s^2+4)} + \frac{6}{5(s^2+4)} - \frac{1}{2s}$$

$$y = e^{-t}/5 + (3/10)\cos 2t + (3/5)\sin 2t - 1/2$$

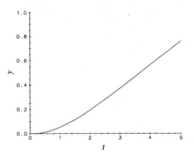

Problem 4(a).

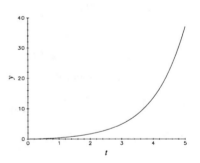

Problem 4(b).

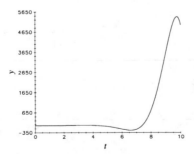

Problem 4(c).

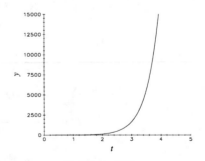

Problem 4(d).

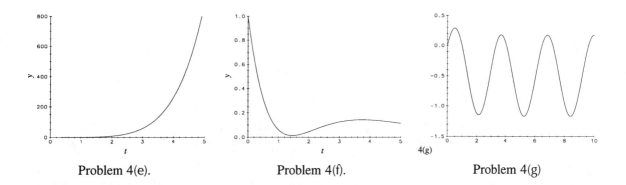

| Problem 4(e). | Problem 4(f). | Problem 4(g) |

5. The three graphs in each case show (from top down) the square wave input and the component graphs of I and I'. The IVP is $I'' + \omega^2 I = 4\,\text{sqw}(t, 50, T) - 2$, $I(0) = 0$, $I'(0) = 0$.

(a). ($\omega^2 = 1$, $T = 5\pi$) The natural period is 2π since $I'' + I = 0$ has solutions $\sin t$ and $\cos t$ of period 2π. The period of the square wave input is 5π, so the common period is 10π. One expects to see a periodic response of period $10\pi \approx 31.4$, and it is visible in the graphs of I and I' vs. t in Fig. 5(a).

(b). ($\omega^2 = 20$, $T = 5\pi$) The natural period is $2\pi/\sqrt{20} = \pi/\sqrt{5}$, and the period of the square wave input is 5π. Since the ratio of the two periods is the irrational number $\sqrt{5}$, the response is not expected to be periodic, and indeed it is not. See Fig. 5(b).

(c). ($\omega^2 = 1$, $T = 2\pi$) The natural period and the period of the square wave input are both 2π. One might expect to see resonance, i.e., a response of growing amplitude (see Section 4.2). The graphs of Fig. 5(c) show that this is the case.

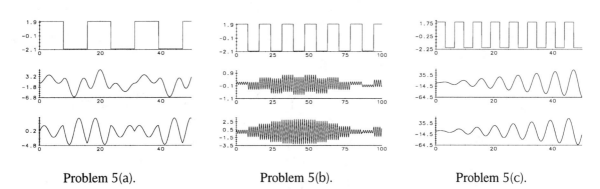

| Problem 5(a). | Problem 5(b). | Problem 5(c). |

6. Following the technique of Example 6.3.2, we see that if $g = 2\,\text{sqw}(t, 50, 2a) - 1$, then $\mathcal{L}[g] = s^{-1}\tanh(as/2)$. If $f = A\,\text{sqw}(t, 50, 2a) - B$, then $f(t) = A(g(t) + 1)/2 - B = Ag(t)/2 + A/2 - B$. Since $\mathcal{L}$ is a linear operator, $\mathcal{L}[f] = A\mathcal{L}[g]/2 + \mathcal{L}[A/2 - B] = A\tanh(as/2)/2s + (A/2 - B)/s$. Since $\tanh(as/2) = (e^{as/2} - e^{-as/2})/(e^{as/2} + e^{-as/2}) = (1 - e^{-as})/(1 + e^{-as})$, we have [using a geometric series expansion for $(1 + e^{-as})^{-1}$]:

$$\tanh(as/2) = \left(1 - e^{-as}\right)\left(1 - e^{-as} + e^{-2as} - e^{-3as} + \cdots + (-1)^n e^{-nas} + \cdots\right)$$

$$= 1 + 2 \sum_{n=1}^{\infty} (-1)^n e^{-nas}$$

So,

$$\mathcal{L}[f] = \frac{A}{2s} \left[1 + 2 \sum_{n=1}^{\infty} (-1)^n e^{-nas} \right] + \frac{A}{2s} - \frac{B}{s} = \frac{A}{s} \left[1 + \sum_{n=1}^{\infty} (-1)^n e^{-nas} \right] - \frac{B}{s}$$

7.   **(a).** The IVP is $y' + y = \text{sqw}(t, 50, T)$, $y(0) = 0$. $\mathcal{L}[y' + y] = s\mathcal{L}[y] + \mathcal{L}[y] = \mathcal{L}[\text{sqw}(t, 50, T)]$. So,

$$\mathcal{L}[y] = \frac{1}{s+1} \cdot \frac{1}{s} \left[1 + \sum_{n=1}^{\infty} (-1)^n e^{-nTs/2} \right]$$

where we have used the expansion in Problem 6 of the square wave with $T = 2a$. Since

$$\frac{1}{(s+1)s} = \frac{1}{s} - \frac{1}{s+1}$$

we have

$$\mathcal{L}[y] = \frac{1}{s} - \frac{1}{s+1} + \sum_{n=1}^{\infty} (-1)^n e^{-nTs/2} \left[\frac{1}{s} - \frac{1}{s+1} \right]$$

Inverting, we have

$$y = \mathcal{L}^{-1} \left[\frac{1}{s} \right] - \mathcal{L}^{-1} \left[\frac{1}{s+1} \right] + \sum_{n=1}^{\infty} (-1)^n \mathcal{L}^{-1} \left\{ e^{-nTs/2} \left[\frac{1}{s} - \frac{1}{s+1} \right] \right\}$$

$$= 1 - e^{-t} + \sum_{n=1}^{\infty} (-1)^n \text{step}(t - nT/2)[1 - e^{-(t - nT/2)}]$$

where we have used formula (12) in Theorem 6.2.4. See Graph 1 where the period of the driving term is $T = 30$, and Graph 2 for $T = 1$. In each case, the response tends to a forced oscillation of period T. Graph 1 and Graph 2 show (from top down) the square wave input and the output $y(t)$.

If the ODE models a communication channel that receives square waves of various periods as inputs, then the long period (i.e., low frequency) input comes out of the channel [as $y(t)$] with much less distortion and amplitude loss (Graph 1 where $T = 30$) than the short period and high-frequency input (Graph 2 where $T = 1$).

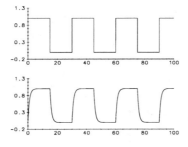

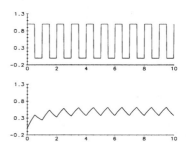

Problem 7(a), Graph 1.   Problem 7(a), Graph 2.

(b). The IVP is $y'' + 4y = \text{sqw}(t, 50, T)$, $y(0) = 0$, $y'(0) = 0$. $L[y'' + 4y] = s^2 L[y] + 4L[y] = L[\text{sqw}(t, 50, T)]$. As in part **(a)**,

$$L[y] = \frac{1}{s^2 + 4} \cdot \frac{1}{s} \left[1 + \sum_{n=1}^{\infty} (-1)^n e^{-nTs/2} \right]$$

Since

$$\frac{1}{s(s^2 + 4)} = \frac{1}{4} \left(\frac{1}{s} - \frac{s}{s^2 + 4} \right)$$

we have

$$L[y] = \frac{1}{4} \left(\frac{1}{s} - \frac{s}{s^2 + 1} \right) + \frac{1}{4} \sum_{n=1}^{\infty} (-1)^n e^{-nTs/2} \left[\frac{1}{s} - \frac{s}{s^2 + 4} \right]$$

So,

$$y = \frac{1}{4} L^{-1} \left[\frac{1}{s} \right] - \frac{1}{4} L^{-1} \left[\frac{s}{s^2 + 4} \right] + \frac{1}{4} \sum_{n=1}^{\infty} (-1)^n L^{-1} \left\{ e^{-nTs/2} \left[\frac{1}{s} - \frac{s}{s^2 + 4} \right] \right\}$$

$$= \frac{1}{4} - \frac{1}{4} \cos 2t + \frac{1}{4} \sum_{n=1}^{\infty} (-1)^n \text{step}(t - nT/2)[1 - \cos(t - nT/2)]$$

where we have used formula (12) in Theorem 6.2.4. See Graph 1 (for $T = 7$) and Graph 2 (for $T = 3$) where the graphs of the input and $y(t)$ and $y'(t)$ are shown. The natural period is π, and the driving period is 7 in the first case. Since $\pi/7$ is not a rational number, one would not expect to see a periodic response, and indeed Graph 1 does not show one. However, if $T = 3$, that period is close to the natural period of π, and one might expect to see beats. Graph 2 shows one "beat." The three graphs in each case show (from top down) the square wave input and the component graphs of y and y'.

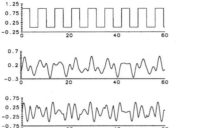
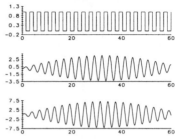

Problem 7(b), Graph 1.   Problem 7(b), Graph 2.

8. Let $g = \text{trw}(t, 100, a)$, and so $f = Ag - B$. Then by the linearity of L, we have $L[f] = AL(g) - BL[1] = (2A/as^2)\tanh(as/4) - B/s$. Since

$$\tanh(as/4) = (e^{as/4} - e^{-as/4})/(e^{as/4} + e^{-as/4})$$

$$= (1 - e^{-as/2})/(1 + e^{-as/2})$$

we use a geometric series expansion for $(1 + e^{-as/2})^{-1}$ to obtain

$$\tanh(as/4) = (1 - e^{-as/2})\left[1 - e^{-as/2} + e^{-as} - e^{-3as/2} + \ldots (-1)^n e^{-nas/2} + \ldots\right]$$

$$= 1 + 2\sum_{n=1}^{\infty}(-1)^n e^{-nas/2}$$

So, we have that

$$L[A\,\text{trw}(t, 100, a) - B] = \frac{2A}{as^2}\left[1 + 2\sum_{n=1}^{\infty}(-1)^n e^{-nas/2}\right] - B/s$$

9. (a). Using the expansion of a triangular wave given in Problem 8, we see that

$$L[y' + y] = sL[y] + L[y] = L[\text{trw}(t, 100, T)]$$

$$= \frac{2}{Ts^2}\left[1 + 2\sum_{n=1}^{\infty}(-1)^n e^{-nTs/2}\right]$$

and so

$$L[y] = \frac{2}{T(s+1)s^2}\left[1 + 2\sum_{n=1}^{\infty}(-1)^n e^{-nTs/2}\right]$$

Since

$$\frac{1}{(s+1)s^2} = \frac{1}{s+1} - \frac{1}{s} + \frac{1}{s^2}$$

we have

$$L[y] = \frac{2}{T}\left(\frac{1}{s+1} - \frac{1}{s} + \frac{1}{s^2}\right) + \frac{4}{T}\sum_{n=1}^{\infty}(-1)^n e^{-nTs/2}\left(\frac{1}{s+1} - \frac{1}{s} + \frac{1}{s^2}\right)$$

Inverting, we have

$$y = \frac{2}{T}\left(e^{-t} - 1 + t\right) + \frac{4}{T}\sum_{n=1}^{\infty}(-1)^n \operatorname{step}(t - \frac{nT}{2})\left[e^{-(t-nT/2)} - 1 + (t - \frac{nT}{2})^2\right]$$

See Graph 1 for the input and response when $T = 30$ and Graph 2 for the case $T = 1$. This is a model for a low-pass communication channel [see also Problem 7(a)]. The channel transmits the long-period (low-frequency) triangular wave input without much distortion, but badly distorts the short-period (high-frequency) triangular wave input. The two graphs in each case show (from top down) the triangular wave input and the output $y(t)$.

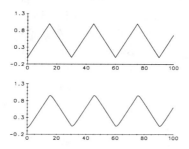

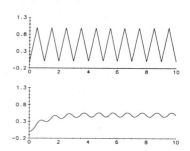

Problem 9(a), Graph 1.   Problem 9(a), Graph 2.

10.   Group project.

11.   (a), (b),(c). Using the last inequality in (25), we see that a collision occurs in every case, because $\lambda T > 12 + 5/T^2$ for the values $T = 1$, $\lambda T = 20$[part (a)] and $T = 2$, $\lambda T = 14$[part (c)].

12.   (a). The given equation for $y_3(t)$, $0 \leq t \leq 3T$, follows directly from formula (21) by setting $j = 3$. Only the first term of the series in (21) needs to be used.

(b). The given equation for $y_2(t) - y_3(t)$, $0 \leq t \leq 3T$, is immediate from formula (23) and the formula in part (a).

(c). The condition that the second and third cars collide at some time t, $2T \leq t \leq 3T$ is that $y_2(3T) - y_3(3T) = 20 + \alpha\lambda(2T)^3/6 - \alpha\lambda^2T^4/12 \leq 15$. In terms of T and λT with $\alpha = 6$ ft/sec^2, this can be written as $10 \leq T^2(\lambda T - 16)\lambda T$. At the equality, we can solve for λT in terms of T: $\lambda T = 8 \pm (64 + 10/T^2)^{1/2}$.

(d). Following the derivation of condition (25) from formula (24), we see that the first two cars do *not* crash during the time span $0 \leq t \leq 2T$ if

$$\lambda T < 12 + 5/T^2 \tag{i}$$

If the two cars don't crash for $2T \leq t \leq 3T$, then at $t = 3T$ [using the last equation of formula (24)] we must have

$$3(3T)^2 + 20 - \lambda(2T)^3 + \lambda^2T^4/4 > 15$$

which can be written in terms of a quadratic in λT

$$(\lambda T)^2 - 32(\lambda T) + 108 + 20/T^2 > 0$$

Equating the left side to 0, solving the quadratic in λT as a function of T, and then requiring that λT exceed the larger of the two roots, we obtain

$$\lambda T > 16 - 16(1 + (20/T^2 + 108)/256)^{1/2} \qquad \text{(ii)}$$

The inequalities (i), (ii) and that derived in part (c) bound a region of points $(T, \lambda T)$ in the $T, \lambda T$-plane corresponding to a crash between the second and third cars during the time span $0 \le t \le 3T$, but no crash between the first two cars during the same time interval. The region can be most easily seen by replacing the three inequalities by equalities and plotting λT as a function of T. The "acceptable" region R is below the graph of $\lambda T = 12 + 5/T^2$ [the bold curve in Fig. 12(d)], but above the other two graphs. In Fig. 12(d) it is the region between the bold curve and the long-dashed curve [the graph of $\lambda T = 16 - 16(1 - (108 + 20/T^2)/256)^{1/2}$]. Note that the response time T is very short in this region, while the value of λT is quite large. This means that the values of λ, the sensitivity factor, are comparatively large, maybe too large to be realistic.

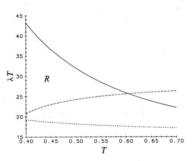

Problem 12(d).

13. (a). The model is much like the one in system (13) in the text, but separation of cars, rather than the difference in velocities, is the stimulus. If the acceleration of the driver $j + 1$ is α_{j+1} and the response time and sensitivity are T seconds and λ (seconds)$^{-1}$, respectively, then $\alpha_{j+1}(t + T) = \lambda[y_j(t) - y_{j+1}(t)]$, $j \ge 1$. As before, $v_1(t) = f(t)$ for the lead car, so $y_1(t) = \int_0^t f(s)\, ds$.

(b). Combining the model of (a) with that of system (13) in the text, we have that $\alpha_{j+1}(t + T) = C_1\lambda_1[v_j(t) - v_{j+1}(t)] + C_2\lambda_2[y_j(t) - y_{j+1}(t)]$, where C_1 and C_2 are constants.

(c). We have that $y_1(t) = \alpha_0 t^2/2$, $y_2''(t + T) = \lambda[\alpha_0 t^2/2 - y_2(t)]$. Since $\mathcal{L}[y'(t + T)] = e^{Ts}\mathcal{L}[y(t)]$ [see formula (17) in the text], we have that $\mathcal{L}[\alpha_2(t + T)] = \mathcal{L}[y_2''(t + T)] = e^{Ts}\mathcal{L}[y_2'(t)] = e^{Ts}\{s\mathcal{L}[y_2] - y_2(0)\} = e^{Ts}\{s\mathcal{L}[y_2] + 5\}$. On the other hand, from part (a), $\mathcal{L}[\alpha_2(t + T)] = \lambda\mathcal{L}[y_1(t) - y_2(t)] = \lambda\mathcal{L}[\alpha_0 t^2/2 - y_2] = \alpha_0\lambda/s^3 - \lambda\mathcal{L}[y_2]$. So,

$$\mathcal{L}[y_2] = \frac{\alpha_0\lambda s^{-3} - 5e^{Ts}}{se^{Ts} + \lambda} = \frac{1}{se^{Ts}}(\alpha_0\lambda s^{-3} - 5e^{Ts})\left[1 + \frac{\lambda}{s}e^{-Ts}\right]^{-1}$$

Expanding the last factor in a geometric series, we have that

$$L[y_2] = \frac{1}{se^{Ts}} \left(\alpha_0 \lambda s^{-3} - 5e^{Ts} \right) \left(1 - \frac{\lambda}{s} e^{-Ts} + \frac{\lambda^2}{s^2} e^{-2Ts} - \cdots \right)$$

$$= \left(\frac{-5}{s} + \frac{\alpha_0 \lambda e^{-Ts}}{s^4} \right) \left(1 - \frac{\lambda}{s} e^{-Ts} + \frac{\lambda^2}{s^2} e^{-2Ts} - \cdots \right)$$

$$= \frac{-5}{s} + \left(\frac{\alpha_0 \lambda}{s^4} + \frac{5\lambda}{s^2} \right) e^{-Ts} + \left(\frac{-5\lambda^2}{s^3} - \frac{\alpha_0 \lambda^2}{s^5} \right) e^{-2Ts} + \cdots$$

So

$$y_2(t) = -5 + \lambda H(t - T) [\frac{\alpha_0}{6} (t - T)^3 + 5(t - T)]$$

$$- \lambda^2 H(t - 2T) \left[\frac{\alpha_0}{24} (t - T)^4 + \frac{5}{2} (t - T)^2 \right] + \cdots$$

The first and second "point" cars collide if $y_2(t) = y_1(t)$. The values of α_0, λ, T determine whether the first two cars collide, and if they do, when the collision occurs. For example, let $\alpha_0 = 6$ ft/sec^2 and so $y_1(t) = 3t^2$. Suppose that $T = 2$ sec. Let's find a value of λ such that there is a collision at some time t, $2 < t < 4$. We have that $y_1(2) - y_2(2) = 12 + 5 = 17 > 0$, while $y_1(4) - y_2(4) = 48 - [-5 + \lambda(8 + 10)] = 53 - 18\lambda$. If the sensitivity λ exceeds $53/18 \approx 2.9$ sec^{-1}, there is a collision at some time $\bar{t}$, $2 < \bar{t} < 4$ since $y_1(t) - y_2(t)$ reverses sign on the interval $[2, 4]$.

6.4 Convolution

Suggestions for preparing a lecture

Topics: The convolution product and how it is used to solve IVPs.

Remarks: If you did the Variation of Parameters Technique (see Problem 8 in Section 3.7), you might mention that the function $g(t - u)$ in Theorem 6.4.3 is just the Green's kernel $K(t, u)$.

Making up a problem set

Three or four parts of Problems 1–4, one part of Problem 5, one of the proofs in Problem 6.

Comments

The Convolution Theorem says that the Laplace transform of a product, $L[fg]$ is equal to the convolution product of $L[f]$ adnd $L[g]$ which is denoted by $L[f] * L[g]$. So the transform of a product is *not* the product of the transforms. At the same time, the convolution product gives a practical way to construct the solution of $P(D)[y] = f$ (with zero initial data) for any constant coefficient operator $P(D)$ and integrable input f. The solution is $y = g * f$, where $g(t) = L^{-1}[1/P(s)]$. The convolution product is a neat idea and new to many readers. The more theoretically inclined like to prove its various properties, and all should appreciate its power in representing solutions of $P(D)[y] = f$ (with zero initial data). Because the approach in this section is to focus on the idea of the convolution, we have deliberately kept calculations to a minimum in the text and in the problems. This will come as a welcome relief after all of the laborious computations of the previous sections.

1. **(a).** Using the Convolution Theorem with $f(t) = t$, $g(t) = t$, we see that

$$L\left[\int_0^t (t-u)u\,du\right] = L[t]L[t] = 1/s^4$$

(b). Using the Convolution Theorem with $f(t) = 1$, $g(t) = \sin t$, we see that

$$L\left[\int_0^t \sin u\,du\right] = L[1]L[\sin t] = 1/(s^3 + s)$$

(c). Since $t^2 - 2tu + u^2 = (t-u)^2$, let $f(t) = t^2$, $g(t) = 1$. The transform of the integral is (by the Convolution Theorem)

$$L\left[\int_0^t (t^2 - 2tu + u^2)\,du\right] = L[t^2]L[1] = 2/s^4$$

(d). Here $f(t) = \sin t$ and $g(t) = \sin t$ since $\sin(t-u) = \sin t \cos u - \cos t \sin u$. Then by the Convolution Theorem

$$L\left[\int_0^t \sin(t-u)\sin u\,du\right] = L[\sin t]L[\sin t] = 1/(s^2 + 1)^2$$

2. In each case, the function of s may be written as the product of two other functions whose inverse transforms appear in Sections 6.1–6.4 or in the tables at the chapter's end. The convolution integrals are not evaluated.

(a).

$$L^{-1}\left[\frac{s}{(s^2+1)^2}\right] = L^{-1}\left[\frac{s}{s^2+1}\frac{1}{s^2+1}\right] = \cos t * \sin t = \int_0^t \cos(t-u)\sin u\,du$$

(b).

$$L^{-1}\left[\frac{s}{(s^2+10)^2}\right] = L^{-1}\left[\frac{s}{s^2+10}\frac{1}{s^2+10}\right] = \cos\sqrt{10}t * \frac{1}{\sqrt{10}}\sin\sqrt{10}t$$

$$= \frac{1}{\sqrt{10}}\int_0^t \cos(\sqrt{10}(t-u))\sin\sqrt{10}u\,du$$

(c).

$$L^{-1}\left[\frac{1}{s^2(s+1)}\right] = L^{-1}\left[\frac{1}{s^2}\frac{1}{s+1}\right] = t * e^{-t} = \int_0^t (t-u)e^{-u}\,du$$

(d).

$$L^{-1}\left[\frac{s}{(s^2+9)^3}\right] = L^{-1}\left[\frac{1}{s^2+9}\frac{s}{(s^2+9)^2}\right] = \frac{1}{3}\sin 3t * \frac{1}{6}t\sin 3t$$

$$= \frac{1}{18}\int_0^t \sin 3(t-u)(u\sin 3u)\,du$$

(e).

$$L^{-1}\left[\frac{s}{(s+1)(s+2)^3}\right] = L^{-1}\left[\frac{s}{(s+1)(s+2)}\frac{1}{(s+2)^2}\right]$$

$$= L^{-1}\left[\left(\frac{2}{s+2} - \frac{1}{s+1}\right)\cdot\frac{1}{(s+2)^2}\right]$$

$$= (2e^{-2t} - e^{-t}) * te^{-2t} = \int_0^t \left[2e^{-2(t-u)} - e^{-(t-u)}\right]ue^{-2u}\,du$$

(f).

$$L^{-1}\left[\frac{s^2+4s+4}{(s^2+4s+13)^2}\right] = L^{-1}\left[\frac{s+2}{(s+2)^2+3^2}\frac{s+2}{(s+2)^2+3^2}\right] = e^{-2t}\cos 3t * e^{-2t}\cos 3t$$

$$= \int_0^t e^{-2(t-u)}\cos 3(t-u)e^{-2u}\cos 3u\,du$$

$$= \int_0^t e^{-2t}\cos 3(t-u)\cos 3u\,du$$

(g).

$$L^{-1}\left[\frac{L[f]}{s^2+1}\right] = L^{-1}\left[L[f]\frac{1}{s^2+1}\right] = f * \sin t = \int_0^t f(t-u)\sin u\,du$$

(h). Using formula (12) of the Shifting Theorem (Theorem 6.2.4), we have

$$L^{-1}\left[e^{-3s}\frac{L[f]}{s^3}\right] = L^{-1}\left[\frac{1}{s^3}e^{-3s}L[f]\right] = \frac{1}{2}t^2 * [H(t-3)f(t-3)]$$

$$= \frac{1}{2}\int_0^t (t-u)^2 H(u-3)f(u-3)\,du$$

3. In each case the Convolution Theorem can be used to write $y(t)$ as an integral by arranging $L[y](s)$ as an appropriate product of functions of s. The transforms are given below along with the final solutions in integral form.

(a). The IVP is $y'' + y = t\,\text{step}(t-1),\ \ y(0) = y'(0) = 0$.

$$L[y] = \frac{1}{s^2+1}\cdot\frac{e^{-s}}{s^2};\quad y(t) = \int_0^t \sin(t-u)[u\,\text{step}(u-1)]\,du$$

(b). The IVP is $y'' + y = f(t),\ \ y(0) = y'(0) = 0$.

$$L[y] = \frac{1}{s^2+1}\cdot L[f];\quad y(t) = \int_0^t \sin(t-u)f(u)\,du$$

(c). The IVP is $2y'' + y' - y = f(t),\ \ y(0) = y'(0) = 0$.

$$L[y] = \frac{1}{3}\left[\frac{1}{s-1/2} - \frac{1}{s+1}\right]L[f];\quad y(t) = \frac{1}{3}\int_0^t f(t-u)\left[e^{u/2} - e^{-u}\right]du$$

(d). The IVP is $y'' + 2y' + y = f(t)$, $y(0) = y'(0) = 0$.

$$\mathcal{L}[y] = \frac{1}{(s+1)^2}\mathcal{L}[f]; \quad y(t) = \int_0^t f(t-u)ue^{-u}\,du$$

(e). The IVP is $y'' + 4y = \begin{cases} 0, & 0 \le t \le 1 \\ t-1, & 1 \le t \le 2 \\ 1, & t > 2 \end{cases}$, $\quad y(0) = y'(0) = 0$.

$$\mathcal{L}[y] = \frac{1}{s^2+4}\mathcal{L}\left[(t-1)\,\text{step}(t-1) - (t-2)\,\text{step}(t-2)\right] = \frac{1}{s^2+4}\cdot\frac{1}{s^2}\left(e^{-s} - e^{-2s}\right)$$

$$y(t) = \frac{1}{2}\int_0^t \sin 2(t-u)\left[(u-1)\,\text{step}(u-1) - (u-2)\,\text{step}(u-2)\right]\,du$$

4. In each case $g(t) = \mathcal{L}^{-1}[1/P(s)]$, where $P(D)$ is the operator.

(a). $P(D) = D^2 + 6D + 13$, and so $g(t) = \mathcal{L}^{-1}[1/(s^2 + 6s + 13)] = \mathcal{L}^{-1}[1/((s+3)^2 + 2^2)] = (1/2)e^{-3t}\sin 2t$.

(b). $P(D) = D^2 + (1/3)D + 1/36$, and so

$$g(t) = \mathcal{L}^{-1}\left[\frac{1}{s^2 + s/3 + 1/36}\right] = \mathcal{L}^{-1}\left[\frac{1}{(s+1/6)^2}\right] = te^{-t/6}$$

(c). $P(D) = D^3 + 1$, and so

$$g(t) = \mathcal{L}^{-1}\left[\frac{1}{s^3+1}\right] = \mathcal{L}^{-1}\left[\frac{1}{s+1}\frac{1}{(s-1/2)^2 + 3/4}\right]$$

$$= \mathcal{L}^{-1}\left[\frac{1/3}{s+1} - \frac{1}{3}\frac{s-1/2}{(s-1/2)^2 + 3/4} + \frac{1/2}{(s-1/2)^2 + 3/4}\right]$$

$$= \frac{1}{3}e^{-t} - e^{-t/2}\left[\frac{1}{3}\cos\frac{\sqrt{3}t}{2} - \frac{1}{\sqrt{3}}\sin\frac{\sqrt{3}t}{2}\right]$$

5. In each case we write the answer as a convolution of the input function on the right side of the ODE with $\mathcal{L}^{-1}[1/P(s)]$, where $P(D)$ is the operator of the ODE. Use Theorem 6.4.3.

(a). The IVP is $y'' + 6y' + 13y = f(t)$, $y(0) = y'(0) = 0$. So, $y = [(e^{-3t}\sin 2t)/2] * f(t)$, since $\mathcal{L}^{-1}[(s^2 + 6s + 13)^{-1}] = (1/2)e^{-3t}\sin 2t$.

(b). The IVP is $y'' + y'/3 + y/36 = f(t)$, $y(0) = y'(0) = 0$. So, $y = te^{-t/6} * f(t)$.

(c). The IVP is $y''' + y = t$, $y(0) = y'(0) = y''(0) = 0$. So, the solution of this IVP is

$$y = \left[e^{-t}/3 - e^{t/2}\left(\frac{1}{3}\cos\frac{\sqrt{3}t}{2} - \frac{1}{\sqrt{3}}\sin\frac{\sqrt{3}t}{2}\right)\right] * t$$

since

$$\mathcal{L}^{-1}[1/(s^3 + 1)] = e^{-t}/3 - e^{t/2}\left(\frac{1}{3}\cos\frac{\sqrt{3}t}{2} - \frac{1}{\sqrt{3}}\sin\frac{\sqrt{3}t}{2}\right).$$

6. First let's show that $f * g$ is in **E** if f and g are (the closure property). If f and g belong to **E** then there are positive constants M_1, M_2, a_1, a_2 such that $|f(t)| \leq M_1 e^{a_1 t}$ and $|g(t)| \leq M_2 e^{a_2 t}$; in addition, f and g are assumed to be piecewise continuous on $[0, \infty)$. Then (assuming for simplicity that $a_1 \neq a_2$) we have

$$|f * g| \leq \int_0^t |f(t-u)||g(u)|\, du \leq M_1 M_2 \int_0^t e^{a_1(t-u)} e^{a_2 u}\, du$$

$$= M_1 M_2 e^{a_1 t} \int_0^t e^{(a_2-a_1)u}\, du \leq \frac{M_1 M_2}{|a_2 - a_1|} e^{a_2 t}$$

which therefore belongs to **E**. If $a_1 = a_2$, then $|f * g| \leq M_1 M_2 t e^{a_1 t} \leq M_1 M_2 e^{(a_1+1)t}$, which belongs to **E**.

To show that $(f * g) * h = f * (g * h)$ (associativity), we proceed as follows:

$$f * (g * h) = \int_0^t f(t-u)(g * h)(u)\, du = \int_0^t f(t-u) \left\{ \int_0^u g(u-v)h(v)\, dv \right\} du$$

$$= \int_0^t \left\{ \int_v^t f(t-u)g(u-v)\, du \right\} h(v)\, dv$$

$$= \int_0^t \left\{ \int_0^{t-v} f(t-v-w)g(w)\, dw \right\} h(v)\, dv = (f * g) * h$$

where the order of integration has been interchanged and the variable $w = u - v$ has been introduced.

To show that $(f + g) * h = f * h + g * h$ (distributivity), we have

$$(f + g) * h = \int_0^t (f+g)(t-u)h(u)\, du = \int_0^t f(t-u)h(u)\, du + \int_0^t g(t-u)h(u)\, du$$

$$= f * h + g * h$$

7. We have that $e^{at}(f * g)(t) = e^{at} \int_0^t f(t-u)g(u)\, du$, while

$$e^{at} f(t) * e^{at} g(t) = \int_0^t e^{a(t-u)} f(t-u) e^{au} g(u)\, du = e^{at} \int_0^t f(t-u)g(u)\, du$$

so $e^{at}(f * g)(t) = e^{at} f(t) * e^{at} g(t)$.

6.5 Convolution and the Delta Function

Suggestions for preparing a lecture

Topics: Enough treatment of the delta function to get the idea across and then a discussion of impulsive forces (see Examples 6.5.1, 6.5.2).

Making up a problem set

Parts of Problem 1, one of 2, 3, and one of Problems 4–6.

Comments

The elusive delta function finally makes its appearance, and the we are faced with a hard choice: "Define" the delta function in a slightly shady way as we do, just "wave one's hands," or do enough with distributions to give the whole thing a mathematically respectable aura. At various times in various courses, we have used each of these approaches—and with varying success. The approach we take in this section is as good (or as bad) as any other we have tried. The essential properties of the delta function are contained in Theorem 6.5.1. The model of a vibrating spring that is subjected to an impulsive force helps to flesh out the idea of the delta function.

Experiment

If you are willing to risk the possibility of making a fool of yourself, here is a marvelous opportunity for a demonstration: ask a physicist to give you a dependable spring with a known spring constant ω^2. Then hang it up, put a weight on it, start it oscillating and then suddenly hit it and bring it to a stop. Or better yet, organize a team to do it, or do a computer simulation (see Example 6.5.2).

---

1. Apply the Laplace transform operator $\mathcal{L}$ to each side of the ODEs, using the property of the delta function that $\mathcal{L}[\delta(t-u)](s) = e^{-us}$. Also, the Shifting Theorem (Theorem 6.2.4) is sometimes used. In every case the initial values are $y(0) = y'(0) = 0$.

(a). The ODE is $y'' + 2y' + 2y = \delta(t - \pi)$. We have that

$$\mathcal{L}[y](s) = \frac{e^{-\pi s}}{s^2 + 2s + 2} = \frac{e^{-\pi s}}{(s+1)^2 + 1}$$

and so $y(t) = \text{step}(t - \pi) f(t - \pi)$, where $\mathcal{L}[f](s) = 1/[(s+1)^2 + 1]$, and $f(t) = e^{-t}\sin t$; $y(t) = \text{step}(t - \pi)e^{-(t-\pi)}\sin(t - \pi) = -\text{step}(t - \pi)e^{-(t-\pi)}\sin t$.

(b). The ODE is $y'' + 4y = \delta(t - \pi) - \delta(t - 2\pi)$. We have

$$\mathcal{L}[y](s) = \frac{e^{-\pi s}}{s^2 + 4} - \frac{e^{-2\pi s}}{s^2 + 4}$$

and so

$$y(t) = \frac{1}{2}\text{step}(t - \pi)\sin(2t - 2\pi) - \frac{1}{2}\text{step}(t - 2\pi)\sin(2t - 4\pi)$$

$$= \frac{1}{2}\sin 2t[\text{step}(t - \pi) - \text{step}(t - 2\pi)]$$

(c). The ODE is $y'' + 3y' + 2y = \sin t + \delta(t - \pi)$. Here, we have

$$\mathcal{L}[y](s) = \frac{1}{(s^2 + 1)(s + 1)(s + 2)} + \frac{e^{-\pi s}}{(s + 1)(s + 2)}$$

After using partial fractions, we have

$$y(t) = [-3\cos t + \sin t + 5e^{-t} - 2e^{-2t}]/5 + \text{step}(t - \pi)[e^{-(t-\pi)} - e^{-2(t-\pi)}]$$

(d). The ODE is $y'' + y = \delta(t - \pi)\cos t$. Use formula (10) of Section 6.4 and Theorem 6.5.2 to obtain

$$y = \int_0^t \sin(t - s)\delta(s - \pi)\cos s \, ds = \text{step}(t - \pi)\int_0^t \sin(t - s)\delta(s - \pi)\cos s \, ds$$

$$= \text{step}(t - \pi) \int_0^t \sin(s - t)\delta(\pi - s)\cos s\, ds$$

$$= \text{step}(t - \pi)\sin(\pi - t)\cos\pi = \text{step}(t - \pi)\sin t$$

since $\sin(\pi - t) = -\sin(t - \pi) = -\sin t$ and $\cos\pi = -1$.

(e). The ODE is $y'' + y = e^t + \delta(t - 1)$. In this case

$$\mathcal{L}[y] = \frac{1}{(s^2 + 1)(s - 1)} + \frac{e^{-s}}{s^2 + 1}$$

After using partial fractions, we have $y = [e^t - \cos t - \sin t]/2 + \text{step}(t - 1)\sin(t - 1)$.

2. The motion is modeled by $y'' + ky = A\sin\omega t + 2\delta(t - 2)$, $y(0) = y'(0) = 0$, $\omega^2 \neq k$. Applying $\mathcal{L}$, we have that

$$\mathcal{L}[y](s) = \frac{A\omega}{s^2 + \omega^2}\frac{1}{s^2 + k} + \frac{2e^{-2s}}{s^2 + k} = \frac{A\omega}{k - \omega^2}\left[\frac{1}{s^2 + \omega^2} - \frac{1}{s^2 + k}\right] + \frac{2e^{-2s}}{s^2 + k}$$

Inverting, we have

$$y(t) = \frac{A}{k - \omega^2}\left(\sin\omega t - \frac{\omega}{\sqrt{k}}\sin\sqrt{k}t\right) + \frac{2}{\sqrt{k}}\text{step}(t - 2)\sin\sqrt{k}(t - 2)$$

See Fig. 2 for the solution curve where $k = 1$, $A = 1$, $\omega = 2$. Note the "angle" in the curve at $t = 2$ when the blow is struck, and observe the consequences of the resultant increase in energy producing higher amplitude oscillations.

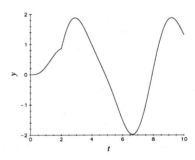

Problem 2.

3. The charge IVP is

$$Lq'' + \frac{1}{C}q = F_0 + 2F_0\delta(t - 1), \qquad q(0) = q'(0) = 0$$

Applying $\mathcal{L}$, we have that

$$\mathcal{L}[q](s) = \frac{F_0/L}{s(s^2 + 1/LC)} + \frac{2F_0 e^{-s}/L}{s^2 + 1/LC}$$

Using partial fractions and then inverting, we have

$$q(t) = CF_0\left[1 - \cos\left(\frac{t}{\sqrt{LC}}\right)\right] + 2F_0\text{step}(t - 1)\sqrt{C/L}\sin\left(\frac{t - 1}{\sqrt{LC}}\right)$$

See Fig. 3 for the solution curve where $L = 1$, $C = 1$, $F_0 = 10$. The burst of voltage at $t = 1$ introduces energy into the circuit, and the charge then oscillates with higher amplitude sinusoids.

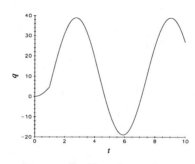

Problem 3.

4. Here's how to show that $L[\delta(t - u)](s) = e^{-us}$. $L[\delta(t - u)] = \int_0^\infty e^{-st}\delta(t - u)\,dt = \int_0^\infty e^{-st}\delta(t - u)\,\text{step}(t - u)\,dt$ [since both $\delta(t - u)$ and $\text{step}(t - u)$ have value 0 if $t < u$ and $\text{step}(t - u) = 1$ if $t \geq u$]. This quantity equals $e^{-us}L[\delta(t)](s)$ [by Theorem 6.2.4] which equals e^{-us}.

5. Let's show that $\delta(at) = (1/|a|)\delta(t)$, $a \neq 0$. Let f be any function in $\mathbf{E}_1$. We will make the substitution $v = au$ to obtain

$$\int_{-\infty}^\infty \delta(at - au)f(u)du = \frac{1}{a}\int_{-\infty}^\infty \delta(at - v)f(v/a)dv$$

$$= \frac{1}{a}f(\frac{at}{a}) = \frac{1}{a}f(t) \text{ if } a > 0$$

if $a > 0$. If $a < 0$, then we have

$$\frac{1}{a}\int_\infty^{-\infty}(\cdot) = -\frac{1}{a}\int_{-\infty}^\infty(\cdot) = \frac{1}{|a|}\int_\infty^{-\infty}(\cdot)$$

Since the above holds for all functions f, we have that $\delta(at) = |a|^{-1}\delta(t)$ if $a \neq 0$.

6. Here's why $\delta(t) * f(t) = f(t)$. $L[\delta(t) * f(t)](s) = L[\delta](s) \cdot L[f](s) = L[f](s)$ by the Convolution Theorem 6.4.2 and Theorem 6.5.1. And so, since $\delta * f$ and f have the same Laplace transform, $\delta * f$ and f are "equivalent."

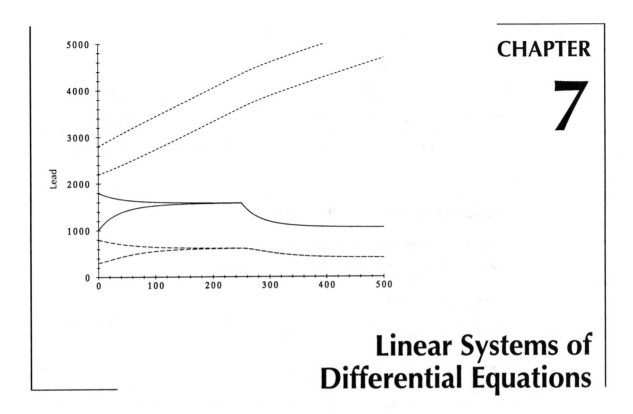

Linear Systems of
Differential Equations

It is hard to judge how much linear algebra readers of this text know before beginning a serious study of linear differential systems. There is a growing trend in many engineering curricula to drop linear algebra as a required course, expecting people to pick it up along the way in other courses. So some will have had a lot of exposure to linear algebra and a growing number may not. But our experience is that even people who have had a linear algebra course are often quite rusty when they have to use it. For these reasons, we have included several sections on matrices, systems of linear (algebraic) equations, determinants, eigenvalues, and eigenvectors. The notion of a linear space is gently described so that we can talk about linear independence. We have tried to make the matrix material as self-contained as possible, and there are examples of every kind of computation needed for the chapter. We recommend that these concepts be thoroughly reviewed by the reader.

After the sections on matrix algebra and eigenvalues, we treat undriven linear systems with constant coefficients, $x' = Ax$. The operator L with action $L[x] = x' - Ax$ is introduced to simplify the presentations of ideas and some of the computations. A gallery of orbital portraits of planar systems with constant coefficients is displayed. This is a place where people can have some fun with their solvers. Basically we just explore, via computer simulation, the behavior of the orbits of planar systems. We also introduce second-order linear systems and solve a problem of coupled springs introduced in Section 5.1. We introduce the exponential matrix as a convenient way of finding a formula for the solution of the IVP, $x' = Ax$, $x(0) = x^0$. The exponential matrix comes in handy when the method of Variation of Parameters is used to solve the IVP, $x' = Ax + F(t)$, $x(0) = 0$. An interesting feature of this chapter is our treatment of the steady-state solution where F is a constant or periodic driving term. Along with this we present some useful tests for determining if all the eigenvalues of the system matrix A have negative real parts without actually finding the eigenvalues: the Routh Test, the Gerschgorin Disks, plus other tests. The solutions of a linear system whose system matrix A has only eigenvalues with negative real parts have some very appealing features. Finally, we extend everything to the general linear system $x' = A(t)x + F(t)$.

There are three "big" models in this chapter. Lead in the human body is the subject of the entire first section and half of the next-to-the-last section. Coupled spring models are solved in Section 7.7. The low-pass filter (also used as a noise filter) is introduced and extensively studied in Section 7.10. The lead and filter models offer nice examples of the sensitivity of linear systems to changing data.

7.1 Tracking Lead Through the Body

Suggestions for preparing a lecture

Topics: The lead model, the form of a linear system and IVP.

Making up a problem set

Problem 1 or 2 or 3, 4 or 5; maybe 6 if a straightforward group problem that mostly involves computer simulations is appropriate at this point.

Comments

This section is devoted to a single model as the title indicates. It provides a nice example of applying the Balance Law to a three-compartment model to obtain a linear differential system. People seem to enjoy this model, perhaps because it shows the vulnerability of living in Southern California (at least in the bad old days of leaded gasoline and leaded paint). The model provides an excellent backdrop for illustrating the asymptotic approach to equilibrium, the effect of a discontinuous driving term, and the sensitivity of the system as parameters in the model change. Turn ahead to Section 7.10 and the corresponding part of the Resource Manual for more on the lead model and a discussion of general compartment models. Here are two more references on the lead model: *Getting the Lead Out*, Irene Kessel and John T. O'Connor [Plenum, New York, 1997]; "Biological Modeling for Predictive Purposes," Naomi H. Harley, pp. 27–36 in *Methods for Biological Monitoring*, T.J. Kneip and J.V. Crable (editors) [American Public Health Association, Washington, 1988].

---

1.   We want to find the equilibrium points of IVP (4):

$$x_1' = -0.0361x_1 + 0.0124x_2 + 0.000035x_3 + 49.3, \qquad x_1(0) = 0$$
$$x_2' = 0.0111x_1 - 0.0286x_2, \qquad x_2(0) = 0$$
$$x_3' = 0.0039x_1 - 0.000035x_3, \qquad x_3(0) = 0$$

(a). Set the rate equations of IVP (4) equal to 0 and solve simultaneously for x_1, x_2, x_3, e.g., by using elementary row operations (see Section 7.2). The equilibrium levels are found to be $x_1 = 1800$, $x_2 = 699$, and $x_3 = 200583$.

(b). Using the answers to part **(a)**, we see that 85% of the respective equilibrium levels of lead in the blood, tissues, and bones is 1530, 594, and 170496. If we plot each of the x_1, x_2, and x_3 component curves over an appropriate time span, and then zoom in on each curve over a time interval enclosing the time when the 85% level is attained, we get the graphs shown in Fig. 1(**b**), Graphs 1, 2, and 3. Reading from those graphs, we see that the 85% level is reached in the blood, tissues, and bones after (approximately) 147 days, 206 days, and 61500 days respectively. **Warning!** These are only approximations because the time interval is so long. More accurate answers would be obtained if the problem were first rescaled to dimensionless form. **Second warning!** Don't be fooled by the apparent leveling off of your component curves long before the equilibrium levels are reached. In exponential decay phenomena, the approach to equilibrium can be extremely slow near the equilibrium. We have chosen to use zoom graphs in order to show that the curves are actually still rising and to make it easier to find the desired times. **Third warning!** Your solver may crash trying to solve over the long time span of 75000 days (205 years) of lead accumulation. In any case, the subject of this case study would hardly be alive by then.

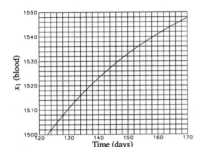

Problem 1(b), Graph 1.

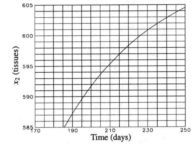

Problem 1(b), Graph 2.

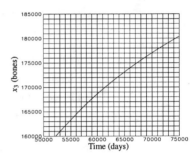

Problem 1(b), Graph 3.

(**c**). Replace 49.3 by I_1 in the above rate equations. The equilibrium levels, $x_1 = 36.5I_1$, $x_2 = 14.16I_1$, and $x_3 = 4068.6I_1$, are obtained by setting the right-hand sides of the ODEs in (4) (with 49.3 replaced by I_1) equal to 0, and solving for x_1, x_2, and x_3. Since we are solving a linear algebraic system of the form $Ax = F$ with solution $x = \hat{x}$, say, then $Ax = F/2$ has solution $x = \hat{x}/2$ by linearity.

(**d**). See Fig. 1(**d**), Graphs 1–5. From these graphs, we see that it takes approximately 850, 440, 305, 237, and 197 days, respectively, for the lead level in the bones to reach 1000 micrograms if $I_1 = 10, 20, 30, 40, 50$ micrograms/day.

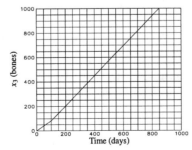

Problem 1(d), Graph 1.

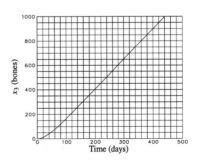

Problem 1(d), Graph 2.

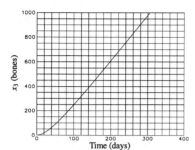

Problem 1(d), Graph 3.

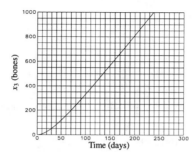

Problem 1(d), Graph 4.

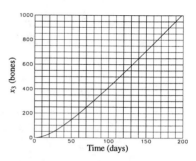

Problem 1(d), Graph 5.

2.   Here are the rate equations if all lead intake stops for $t \geq 400$ days:

$$x_1' = -0.0361x_1 + 0.0124x_2 + 0.000035x_3 + 49.3\,\text{step}(400 - t), \qquad x_1(0) = 0$$
$$x_2' = 0.0111x_1 - 0.0286x_2, \qquad\qquad\qquad\qquad\qquad\qquad x_2(0) = 0$$
$$x_3' = 0.0039x_1 - 0.000035x_3, \qquad\qquad\qquad\qquad\qquad\quad x_3(0) = 0$$

Notice that inserting $\text{step}(400 - t)$ models the situation where the subject is removed to a lead-free environment after 400 days. See Graph 1 for the graph of the bone lead levels over $0 \leq t \leq 400$. From this figure we estimate that $x_3(400) = 2215$. From Graph 2 we estimate that it takes 22500 (62 years), 45000 (123 years), and 75000 days(205 years), respectively, for the lead levels in the bones to decline to 50%, 25%, and 10% of $x_3(400)$. These numbers show that we have extended the lead model so far into the future that it no longer has much use.

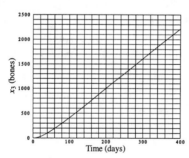

Problem 2, Graph 1.

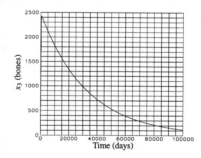

Problem 2, Graph 2.

3.   Begin with IVP (4):

$$x_1' = -0.0361x_1 + 0.0124x_2 + 0.000035x_3 + 49.3, \qquad x_1(0) = 0$$
$$x_2' = 0.0111x_1 - 0.0286x_2, \qquad\qquad\qquad\qquad\qquad x_2(0) = 0$$
$$x_3' = 0.0039x_1 - 0.000035x_3, \qquad\qquad\qquad\qquad\quad x_3(0) = 0$$

(a). Replacing 49.3 by $49.3\,\text{step}(400 - t) + 24.65\,\text{step}(t - 400)$ models cutting the intake rate in half because at $t = 400$ days the second step function turns on, and the first one turns off; for $t \geq 400$ the intake rate is 24.65 micrograms/day. Fig. 3(a) shows that by $t = 800$ days, the lead levels in the blood and the tissues have been cut in half.

(b). Replacing 0.0361 by $0.0361\,\text{step}(400 - t) + 0.0572\,\text{step}(t - 400)$ corresponds to doubling k_{01} from 0.0211 [see the data in (3)] to 0.0422 since $0.0361 = k_{01} + k_{21} + k_{31}$. The increase occurs at $t = 400$ and continues thereafter. Fig. 3(b) shows that the lead levels in the blood and the tissues drop significantly, but not quite as much as in **(a)**.

(c). Fig. 3(c) shows the significant changes in blood and tissues lead levels if the intake rate is cut in half after 400 days [part **(a)**] and medication doubles the clearance coefficient k_{01} [part **(b)**]. The new rate equation for x_1 is $x_1' = -[0.0361\,\text{step}(400 - t) + 0.0572\,\text{step}(t - 400)]x_1 + 0.0124x_2 + 0.000035x_3 + 49.3\,\text{step}(400 - t) + 24.65\,\text{step}(t - 400)$.

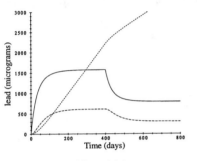
Problem 3(a).

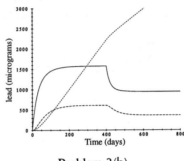
Problem 3(b).

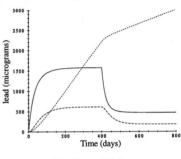
Problem 3(c).

4. Here is the data for the problem:

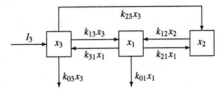

$$I_3 = 1, \quad k_{31} = 3, \quad k_{21} = 1, \quad k_{01} = 16, \quad k_{13} = 1, \quad k_{03} = 5, \quad k_{23} = 2, \quad k_{12} = 4$$

(a). The rate equations for $x_1(t)$, $x_2(t)$, $x_3(t)$ are obtained from the Balance Law applied to each compartment: $x_1'(t) = -20x_1 + 4x_2 + x_3$, $x_2'(t) = x_1 - 4x_2 + 2x_3$, and $x_3'(t) = 3x_1 - 8x_3 + 1$. The coefficient -20 is given by $-(k_{01} + k_{21} + k_{31})$, the coefficient $-4 = -k_{12}$, and the coefficient $-8 = -(k_{03} + k_{13} + k_{23})$.

(b). The equilibrium levels of Q are $x_1 = 3/143 \approx 0.021$, $x_2 = 41/572 \approx 0.072$, and $x_3 = 19/143 \approx 0.133$. They are obtained by setting the three rate functions equal to zero and solving for x_1, x_2, and x_3.

(c). 90% of the equilibrium values of x_1, x_2, and x_3 are about 0.019, 0.065, and 0.120 respectively. See Fig. 4(c) for the tx_1 (solid), tx_2 (long dashes), and tx_3 (short dashes) component curves. The levels all reach 90% of equilibrium within two time units.

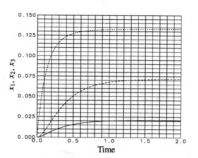

Problem 4(c).

5.   **(a).** Here is the compartment diagram for the flow of inulin through the bloodstream (compartment 1) and the intercellular areas (compartment 2):

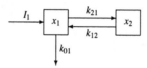

Suppose that $x_1(t)$, $x_2(t)$ are the respective amounts of inulin at time t in the blood and in the intercellular area. The Balance Law (1) implies that $x_1' = I_1 - (k_{21} + k_{01})x_1 + k_{12}x_2$, $x_2' = k_{21}x_1 - k_{12}x_2$.

(b). The equilibrium levels of inulin in the blood and intercellular areas are found by setting $x_1' = 0$ and $x_2' = 0$ and solving for x_1 and x_2. This gives $x_1 = I_1/k_{01}$ and $x_2 = (k_{21}/k_{12})(I_1/k_{01})$.

(c). See Fig. 5(c), where the respective equilibrium levels of inulin from the formulas of part **(b)** are $x_1 = 200$, $x_2 = 100$. The x_1 curve (solid) and x_2 curve (dashed) are graphed over the time span $0 \le t \le 1500$, but each reaches the 90% equilibrium level long before $t = 1500$ (around $t = 750$).

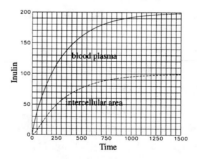

Problem 5(c).

6. Group project.

7.2 Overview of Vectors and Matrices

Suggestions for preparing a lecture

Topics: After discussing the algebra of matrices and vectors, you might want to jump immediately to the next section and concentrate on how to solve linear systems of equations via examples (for instance Example 7.3.1). Then, if you have time, go into invertibility and its connection with determinants, again via examples. Say something about the vector spaces $\mathbb{R}^n$ and $\mathbb{C}^n$, and matrices of functions, especially the function vector space $\mathbf{C}^n(I)$. This is a lot of material for one lecture so you'll have to ask your students to fill in the details by reading the relevant examples in the text.

Remarks: You will have to judge how much linear algebra your students have picked up one way or another before enrolling in your ODE class. In any event, students need to pick up everything necessary for finding eigenvalues and eigenvectors of $n \times n$ matrices because the rest of the chapter depends on having these skills in hand. Sections 7.2–7.4 have all the linear algebra needed for the rest of the text (that's why there is no appendix on linear algebra). Leave the notion of linear independence and dimension to the next section. We recommend leaving the Vector/Matrix Inequalities until you actually need them.

Making up a problem set

Problems 1, 2, 3, 5 **(b)**, 6, 8. (This covers most of the concepts in this section.)

Comments

We avoid the axiomatic definition of a vector space. Instead, we define a vector space as the collection of all column vectors with n entries along with matrix addition and multiplication by a scalar. It is important to distinguish the case where all entries in the vectors (and the scalars) are reals (i.e., the case $\mathbb{R}^n$) from the case where the entries (and the scalars) are complex numbers (i.e., the case $\mathbb{C}^n$). We use the term function vector space to mean the collection of all n-entry column vectors with function entries (the notion $\mathbf{C}^n(I)$ is used again, leaving the number of entries to the context).

It is important to take a look at matrices with complex entries and the corresponding linear systems with complex coefficients. Ditto for function vector spaces with entries which are complex-valued functions of a real variable.

1. Here's the matrix equation we want to solve for A:

$$2A + \begin{bmatrix} 0 & 1 \\ 1 & -1 \end{bmatrix} A = \begin{bmatrix} 1 & -2 \\ 0 & 1 \end{bmatrix}$$

Put $A = \begin{bmatrix} a & b \\ c & d \end{bmatrix}$ into the given equation to obtain

$$\begin{bmatrix} 2a & 2b \\ 2c & 2d \end{bmatrix} + \begin{bmatrix} c & d \\ a-c & b-d \end{bmatrix} = \begin{bmatrix} 1 & -2 \\ 0 & 1 \end{bmatrix}$$

Comparing entries, we conclude that

$$2a + c = 1, \quad a + c = 0, \quad 2b + d = -2, \quad b + d = 1$$

Solving this system, we obtain the unique solution $a = 1$, $b = -3$, $c = -1$, $d = 4$.

2. The proof is by direct computation from the definition of matrix product. From the definition of matrix products note that the j-th column of AB is exactly A applied to the j-th column of B. We leave the examples to the reader.

3. The proof is by direct computation. The i-th entry in the column vector $x_1 a^1 + \cdots + x_n a^n$ is $\sum_{j=1}^{n} a_{ij} x_j$, but this is also the i-th entry of the matrix product Ax. We leave the examples to the reader.

4. Verify by direct computation. In each case calculate $A + B$, $B + A$, $A + (B + C)$, $(A + B) + C$, $A(BC)$, $(AB)C$, $A(B + C)$, $AB + AC$, $(A^T)^T$, $(A + B)^T$, $(AB)^T$, $B^T A^T$.

5. **(a).** Let v^1 and v^2 be any two vectors common to the subspaces U and W of the space V, and let α, β be any two scalars. Then $\alpha v^1 + \beta v^2$ must belong to U and also to W because they are subspaces so each contains all finite linear combinations of its own elements. Therefore, $\alpha v^1 + \beta v^2$ is common to U and W, and the vectors common to U and W form a subspace of V (by the definition of subspace).

    (b). A vector $x = [x_1 \quad x_2 \quad x_3]^T$ in $\mathbb{R}^3$ lies in $U = \text{Span}\{[2 \quad 1 \quad 0]^T, [1 \quad 0 \quad -1]^T\}$ if and only if there are scalars α, β such that $x_1 = 2\alpha + \beta$, $x_2 = \alpha$, $x_3 = -\beta$, so if and only if $x_1 = 2x_2 - x_3$. Similarly, $[x_1 \quad x_2 \quad x_3]^T$ lies in $W = \text{Span}\{[-1 \quad 1 \quad 1]^T, [0 \quad -1 \quad 1]^T\}$ if and only if there exist scalars γ, δ such that $x_1 = -\gamma$, $x_2 = \gamma - \delta$, $x_3 = \gamma + \delta$, so if and only if $2x_1 + x_2 + x_3 = 0$. The subspace of all vectors in $\mathbb{R}^3$ common to U and W must consist of all vectors $x = [x_1 \quad x_2 \quad x_3]^T$ which satisfy the two conditions: $x_1 - 2x_2 + x_3 = 0$, $2x_1 + x_2 + x_3 = 0$. Using elementary row operations we find that the solution set of this pair of linear equations is given by $x = \alpha[-3 \quad 1 \quad 5]^T$, where α is an arbitrary constant. So the subspace of vectors common to U and W has dimension $= 1$.

6. **(a).** From the definition of matrix products and addition and the product of a matrix with a scalar, we have $A(\alpha x + \beta y) = A(\alpha x) + A(\beta y) = \alpha Ax + \beta Ay$.

    (b). If $Ax^d = y$, then to solve $Ax = y$ let $v = x - x^d$, so $x = x^d + v$ and $Ax = A(x^d + v)$. Part **(a)** shows that $A(x^d + v) = Ax^d + Av$ which equals $y + Av$. But we want $Ax = y$. This is possible if and only if $Av = 0$.

7. **(a).** The given system is linear and has the standard form

$$\begin{bmatrix} x_1 \\ x_2 \\ x_3 \end{bmatrix}' = \begin{bmatrix} 1 & 1 & -2 \\ -5 & -1 & 2 \\ -1 & -2 & 1 \end{bmatrix} \begin{bmatrix} x_1 \\ x_2 \\ x_3 \end{bmatrix} + \begin{bmatrix} 0 \\ e^{-t} \\ \sin t \end{bmatrix}$$

    (b). This system is not linear because of the $x_1 x_3$ term in the second equation.

8. Group problem.

7.3 Systems of Linear Equations

Suggestions for preparing a lecture

Topics: Using elementary row operations on the augmented matrix $[A|b]$ to solve the matrix equation $Ax = b$, an example where some entries of A or b are complex numbers, invertibility of a matrix, determinant of a matrix, linear independence and basis, Theorem 7.3.1 (Solvability of Linear Equations).

Making up a problem set

Problems 1(b), 2(a) (b) (c), 3, 4, 8(a). These problems cover most of the concepts in this section. A favorite of ours is Problem 13 (the Fredholm Alternative) because it puts such a nice finish on the solvability of linear systems. You might assign it if your class has already had a linear algebra course.

Comments

There are some subtle ideas here, and it may take some time for everything to sink in. Many readers have already seen determinants and linear systems before in some math course, but they might be a bit rusty on these topics. The main point is how to find *all* solutions of a linear system, even when complex numbers are involved (we will need this later). The other main point is linear independence and how to find a basis for a subspace. Work as many examples as you can and read over the many examples in this section.

---

1. **(a).** Write the system in standard matrix form $\begin{bmatrix} 1 & -2 \\ -2 & 4 \end{bmatrix} \begin{bmatrix} x_1 \\ x_2 \end{bmatrix} = \begin{bmatrix} 0 \\ 0 \end{bmatrix}$ and use elementary row operations on the augmented matrix $\begin{bmatrix} 1 & -2 & 0 \\ -2 & 4 & 0 \end{bmatrix}$ to obtain $\begin{bmatrix} 1 & -2 & 0 \\ 0 & 0 & 0 \end{bmatrix}$.

So the solution set of our system is given by the single equation $x_1 - 2x_2 = 0$. This equation determines a line through the origin in the $x_1 x_2$-plane. To describe this line assign the arbitrary value s to x_2. Then $x_1 = 2s$, and the solution set is given by

$$x = \begin{bmatrix} x_1 \\ x_2 \end{bmatrix} = \begin{bmatrix} 2s \\ s \end{bmatrix} = s \begin{bmatrix} 2 \\ 1 \end{bmatrix}$$

(b). Write the system in standard augmented matrix form

$$\begin{bmatrix} 1 & -2 & -1 & 0 \\ 0 & 1 & 1 & 0 \\ 1 & 1 & 2 & 0 \end{bmatrix}$$

and use elementary row operations to reduce the matrix to the form

$$\begin{bmatrix} 1 & -2 & -1 & 0 \\ 0 & 1 & 1 & 0 \\ 0 & 0 & 0 & 0 \end{bmatrix}$$

So the solution set of our system is given by the two equations $x_1 - 2x_2 - x_3 = 0$, $x_2 + x_3 = 0$. Assigning the arbitrary value s to x_3 we see from the second equation that $x_2 = -s$, and from the first equation that $x_1 = -s$. So the solution set is

$$x = \begin{bmatrix} x_1 \\ x_2 \\ x_3 \end{bmatrix} = \begin{bmatrix} -s \\ -s \\ s \end{bmatrix} = s \begin{bmatrix} -1 \\ -1 \\ 1 \end{bmatrix}$$

which is a line through the origin in $x_1x_2x_3$-space parallel to the vector $-\mathbf{i} - \mathbf{j} + \mathbf{k}$.

(c). Writing the system in standard augmented matrix form we have

$$\left[\begin{array}{ccc|c} 1 & 1 & 2 & 1 \\ 3 & 4 & -1 & -1 \end{array} \right]$$

and using elementary row operations we have

$$\left[\begin{array}{ccc|c} 1 & 1 & 2 & 1 \\ 0 & 1 & -7 & -4 \end{array} \right]$$

which yields the two equations $x_1 + x_2 + 2x_3 = 1$, and $x_2 - 7x_3 = -4$. Assigning the arbitrary value s to x_3, the second equation reduces to $x_2 = -4 + 7s$, and the first equation becomes $x_1 = 5 - 9s$. So the solution set of our system is

$$x = \begin{bmatrix} x_1 \\ x_2 \\ x_3 \end{bmatrix} = \begin{bmatrix} 5 - 9s \\ -4 + 7s \\ s \end{bmatrix} = \begin{bmatrix} 5 \\ -4 \\ 0 \end{bmatrix} + s \begin{bmatrix} -9 \\ 7 \\ 1 \end{bmatrix}$$

which is a line in $x_1x_2x_3$-space passing through the point $(5, -4, 0)$ and parallel to the vector $-9\mathbf{i} + 7\mathbf{j} + \mathbf{k}$.

2. (a). Forming the augmented matrix

$$[A|b] = \left[\begin{array}{ccc|c} 1 & -1 & 0 & 1 \\ -1 & 3 & -2 & -1 \\ 1 & 3 & -4 & 1 \end{array} \right]$$

and applying elementary row operations, we obtain

$$\left[\begin{array}{ccc|c} 1 & -1 & 0 & 1 \\ 0 & 2 & -2 & 0 \\ 0 & 4 & -4 & 0 \end{array} \right] \quad \begin{array}{l} \text{(Row 1)} \\ \text{(Row 2 + Row 1)} \\ \text{(Row 3 - Row 1)} \end{array}$$

$$\left[\begin{array}{ccc|c} 1 & -1 & 0 & 1 \\ 0 & 2 & -2 & 0 \\ 0 & 0 & 0 & 0 \end{array} \right] \quad \begin{array}{l} \text{(Row 1)} \\ \text{(Row 2)} \\ \text{(Row 3 - 2 × Row 2)} \end{array}$$

which, when converted back into equation form, becomes

$$x_1 - x_2 = 1$$

$$2x_2 - 2x_3 = 0$$

$$0 = 0$$

This system is consistent. Assigning x_3 to have the arbitrary value s, the second equation above yields that $x_2 = s$, and the first equation above yields that $x_1 = 1 + s$. All solutions of the original system are given by

$$x = \begin{bmatrix} x_1 \\ x_2 \\ x_3 \end{bmatrix} = \begin{bmatrix} 1+s \\ s \\ s \end{bmatrix} = \begin{bmatrix} 1 \\ 0 \\ 0 \end{bmatrix} + s \begin{bmatrix} 1 \\ 1 \\ 1 \end{bmatrix}$$

where s is an arbitrary real. Taking $s = 0$ shows that $x^d = [1 \quad 0 \quad 0]^T$ is a particular solution of $Ax = [1 \quad -1 \quad 1]^T$.

(b). Repeating the above procedure for the homogeneous system $Ax = 0$, we find that all solutions are given by $x^h = s[1 \quad 1 \quad 1]^T$, where s is an arbitrary real number.

(c). Set up the augmented matrix

$$[A|y] = \begin{bmatrix} 1 & -1 & 0 & a \\ -1 & 3 & -2 & b \\ 1 & 3 & -4 & c \end{bmatrix}$$

and perform the same elementary row operations as in part **(c)** above to obtain

$$\begin{bmatrix} 1 & -1 & 0 & a \\ 0 & 2 & -2 & a+b \\ 0 & 0 & 0 & c-3a-2b \end{bmatrix}$$

So for the system to be consistent we must have $c - 3a - 2b = 0$. So the range of the operator $L[x] = Ax$ is exactly the collection of all points $(a, b, c,)$ on the plane through the origin $c - 3a - 2b = 0$. That is, $Ax = y$ is solvable if and only if y has the form $[a \ b \ 3a + 2b]^T$, where a and b are any real numbers.

3. **(a).** We omit the calculations that carry us to the reduced system from which the solution can be found. Applying elementary row operations to the augmented matrix of the system, we obtain

$$\begin{bmatrix} 3 & -1 & 1 & 2 & -2 & 1 \\ 0 & -7 & 1 & 5 & -5 & 3 \\ 0 & 0 & 15 & -30 & 9 & 17 \end{bmatrix}$$

which translates to the system

$$3x_1 - x_2 + x_3 + 2x_4 - 2x_5 = 1$$

$$-7x_2 + x_3 + 5x_4 - 5x_5 = 3$$

$$15x_3 - 30x_4 + 9x_5 = 17$$

Putting $x_4 = a$, $x_5 = b$ for arbitrary reals a and b, we see that $x_3 = 17/15 + 2a - (3/5)b$, $x_2 = 2/35 + a - (4/5)b$, and $x_1 = -8/315 - a + (3/5)b$. So all solutions of the system are given by

$$x = \begin{bmatrix} x_1 \\ x_2 \\ x_3 \\ x_4 \\ x_5 \end{bmatrix} = \begin{bmatrix} -8/315 - a + (3/5)b \\ 2/35 + a - (4/5)b \\ 17/15 + 2a - (3/5)b \\ a \\ b \end{bmatrix} = \begin{bmatrix} -8/315 \\ 2/35 \\ 17/15 \\ 0 \\ 0 \end{bmatrix} + a \begin{bmatrix} -1 \\ 1 \\ 2 \\ 1 \\ 0 \end{bmatrix} + b \begin{bmatrix} 3/5 \\ -4/5 \\ -3/5 \\ 0 \\ 1 \end{bmatrix}$$

for arbitrary reals a and b.

(b). Applying elementary row operations (which we omit) to the augmented matrix of the system, we obtain

$$\begin{bmatrix} 1 & -2 & 4 & -1 & -1 \\ 0 & 5 & -9 & 3 & 3 \\ 0 & 0 & 0 & 0 & -2 \end{bmatrix}$$

whose last equation, $0 = -2$, can never be satisfied for any choice x_1, x_2, x_3, x_4. So the given system has no solutions.

4. The augmented 2×3 matrix for the system is

$$[A|y] = \begin{bmatrix} 1 & 2 & -i \\ -i & i & 1 \end{bmatrix}$$

Adding i times the first row to the second row, we obtain

$$\begin{bmatrix} 1 & 2 & -i \\ 0 & 3i & 2 \end{bmatrix}$$

which translates to the system

$$x_1 + 2x_2 = -i$$

$$3ix_2 = 2$$

Solving this reduced system, we have the unique solution $x_2 = 2/(3i) = -(2/3)i$, and $x_1 = -i - 2x_2 = -i + (4/3)i = (1/3)i$.

5. The upper triangular matrix $\begin{bmatrix} a & b \\ 0 & c \end{bmatrix}$ is invertible if and only if there exists a matrix $\begin{bmatrix} A & B \\ C & D \end{bmatrix}$ such that

$$\begin{bmatrix} A & B \\ C & D \end{bmatrix} \begin{bmatrix} a & b \\ 0 & c \end{bmatrix} = \begin{bmatrix} a & b \\ 0 & c \end{bmatrix} \begin{bmatrix} A & B \\ C & D \end{bmatrix} = \begin{bmatrix} 1 & 0 \\ 0 & 1 \end{bmatrix}$$

So, in particular, $Ca = 0$, $Aa = 1$, and $cC = 0$, $cD = 1$. So $a \neq 0$, and $c \neq 0$, which implies that $C = 0$, and we are done.

6. The matrix A is invertible if and only if there is a matrix M such that $MA = AM = I$. Taking the transpose of these relations, we obtain that $A^T M^T = M^T A^T = I$. So, A^T is also invertible and $(A^T)^{-1} = M^T = (A^{-1})^T$.

7.   **(a).** If a matrix A is invertible, then there is a matrix M such that $MA = AM = I$. Say that there is another matrix N such that $NA = AN = I$. Then $M = MI = MAN = IN = N$. So inverses are unique.

(b). Let A^{-1} be the inverse of A, and B^{-1} the inverse of B. Then since (using the associativity of the matrix product)

$$AB(B^{-1}A^{-1}) = A(BB^{-1})A^{-1} = AIA^{-1} = AA^{-1} = I$$

it follows that $(AB)^{-1} = B^{-1}A^{-1}$.

8.   **(a).** Expanding the given 4×4 determinant along the third row via cofactors (we chose the third row because of all the zero entries), the determinant becomes

$$-2\det\begin{bmatrix} 5 & 7 & 2 \\ 4 & 1 & 1 \\ 1 & 3 & 4 \end{bmatrix} = -2\left(5\det\begin{bmatrix} 1 & 1 \\ 3 & 4 \end{bmatrix} - 7\det\begin{bmatrix} 4 & 1 \\ 1 & 4 \end{bmatrix} + 2\det\begin{bmatrix} 4 & 1 \\ 1 & 3 \end{bmatrix} \right)$$

$$= -2(5 \cdot 1 - 7 \cdot 15 + 2 \cdot 11) = 156$$

where we have expanded the 3×3 determinant via cofactors along the top row.

(b). Before using cofactors, first introduce as many zeros as possible on some row or column by using some of the techniques listed in Properties of $\det A$. We just give the answer here. The final value of this determinant turns out to be 118.

9.   **(a).** The set $\{e^{-t}, 3e^t, \cosh t\}$ is dependent, since $\cosh t = \frac{1}{2}(e^{-t}) - \frac{1}{6}(-3e^t) = \frac{1}{2}(e^t + e^{-t})$.

(b). The set $\{e^t, te^t, -t^2e^t\}$ is independent, for assume that for some scalars a, b, c we have that $ae^t + bte^t - ct^2e^t = 0$ for all t. Then, dividing out e^t, it would follow that $a + bt - ct^2 = 0$ for all t, but this cannot happen unless $a = b = c = 0$.

(c). The set $\{e^{-t}\cos t, e^{-t}\sin t\}$ is independent, for let scalars a, b be such that $ae^{-t}\cos t + be^{-t}\sin t = 0$ for all t. Then it would follow (after dividing by e^{-t}) that $a\cos t + b\sin t = 0$ for all t. Substituting $t = 0$ and $t = \pi/2$ into this statement we obtain that $a = b = 0$, so the set is independent.

(d). Put $p_0 = 1$, $p_1 = t - 1$, $p_2 = 1 - t^2$, $p_3 = 3t^2 + t - 1$, and note that $p_1 = t - p_0$, $p_2 = p_0 - t^2$. So $t = p_1 + p_0$ and $t^2 = p_0 - p_2$, and substituting we see that $p_3 = 3(p_0 - p_2) + (p_1 + p_0) - p_0 = -3p_2 + p_1 + 3p_0$. So the set $\{p_0, p_1, p_2, p_3\}$ is dependent.

10.   **(a).** The general solution of $y'' = 0$ is $y = At + B$, for arbitrary constants A, B. Imposing the boundary conditions $2y(0) + y'(0) = 0$, $2y(1) - y'(1) = 0$, we see that the constants must satisfy the equation $2B + A = 0$. So the given set is $\{y = B(-2t + 1):$ for all B in $\mathbb{R}\}$ which is a subspace of $\mathbf{C}^2(\mathbb{R})$ of dimension 1.

(b). The general solution of $y'' = 2$ is $y = t^2 + At + B$ for all constants A, B. The boundary conditions $y(0) = 0$, $y(1) = 0$ imply that $B = 0$ and $1 + A + B = 0$. So the given subset of $\mathbf{C}^2(\mathbb{R})$ consists of the single element $y = t^2 - t$ which is *not* a subspace of $\mathbf{C}^2(\mathbb{R})$ since, for example, $2y(t)$ is not in the subset.

(c). Again, $y'' = 2$ has the general solution $y = t^2 + At + B$. The boundary conditions $2y(0) + y'(0) = 0$, $2y(1) - y'(1) = 0$ imply that $2B + A = 0$, whose solutions are $A =$

$-2B$, for all real B. The given set has the form $\{y = t^2 - 2Bt + B: B \text{ an arbitrary constant}\}$ which is not a subspace of $\mathbf{C}^2(\mathbb{R})$, since $2y(t)$ is *not* in the subset if $y(t)$ *is* in the subset.

11. **(a).** The set of polynomials of degree ≥ 2 is *not* a subspace since it doesn't contain the zero function.

(b). The set of polynomials of degree ≥ 3 is *not* a subspace since the zero function is not included.

(c). Since any linear combination of odd functions is odd, it follows that the odd functions in $\mathbf{C}^0[-1, 1]$ form a subspace.

(d). The set of nonnegative functions is *not* a subspace since the product of the scalar -1 with a nonnegative function is not in general a nonnegative function.

12. Group project.

13. With the matrix A in Problem 2 we see from Problem 2 **(c)** that $Ax = y$ is solvable if and only if $c - 3a - 2b = 0$. Now let's find $N(A^T)$, that is, the set of all vectors $v = [v_1 \ v_2 \ v_3]^T$ such that $A^T v = 0$. Setting up the augmented matrix

$$[A^T|0] = \begin{bmatrix} 1 & -1 & 1 & | & 0 \\ -1 & 3 & 3 & | & 0 \\ 0 & -2 & -4 & | & 0 \end{bmatrix}$$

and applying elementary row operations, we obtain

$$\begin{bmatrix} 1 & -1 & 1 & | & 0 \\ 0 & 2 & 4 & | & 0 \\ 0 & 0 & 0 & | & 0 \end{bmatrix}$$

So the system $A^T v = 0$ translates into the system

$$v_1 - v_2 + v_3 = 0$$

$$2v_2 + 4v_3 = 0$$

Assigning the arbitrary real value of s to v_3, we see that $v_2 = -2s$, and $v_1 = -3s$. So $A^T v = 0$ if and only if $v = s[-3 \ \ -2 \ \ 1]^T$. Setting $y = [a \ \ b \ \ c]^T$, we see that "$y^T v = 0$ for all v in $N(A^T)$" just means that $-3a - 2b + c = 0$ because $v = [-3 \ \ -2 \ \ 1]^T$ is a basis for $N(A^T)$.

7.4 Eigenvalues and Eigenvectors of Matrices

Suggestions for preparing a lecture

Topics: Eigenvalues and eigenvectors, the characteristic polynomial of a matrix, a little bit on generalized eigenvectors.

Remarks: The discussion just before Example 7.5.1 motivates why in an ODE course we are interested in eigenvalues and eigenvectors of a matrix.

Making up a problem set

Problems 1(a), one of 1(b)–(d), 1(e) or 1(j), 1(g) [to cover all of the cases]; 2, 5, 6, 7.

Comments

Look ahead to Section 7.5 and read the discussion just before Example 7.5.1 about the need to do eigen-analysis in order to generate the solution set of $x' = Ax$, where A is a constant matrix. However, most of the section is devoted to the process of finding eigenvalues and eigenvectors, maybe even using a CAS for larger matrices. Everyone needs to understand about eigenspaces and multiple eigenvalues, and at least have a basic idea about generalized eigenvectors. The results of Problems 5–7 are important in later sections. People sometimes have trouble spanning an eigenspace, probably because of having to deal with $(A - \lambda I)v = 0$, where $A - \lambda I$ is singular. Several examples will usually clear this up. A worked-out example also helps with complex eigenvalues or eigenvectors.

1. The eigenvalues of a matrix A are the roots of the characteristic polynomial $p(\lambda) = \det[A - \lambda I]$. In each solution below we begin with the calculation of $p(\lambda)$, followed by the listing of the eigenvalues (i.e., the roots of $p(\lambda)$ along with their multiplicities). Any nonzero constant multiple of an eigenvector is also an eigenvector, so the eigenvectors given below are not unique. If λ_j is a root of multiplicity m_j, then the corresponding eigenspace has dimension d_j, $1 \le d_j \le m_j$. A basis for the eigenspace consists of d_j independent vectors v such that $(A - \lambda_j I)v = 0$. In each case the spanning vectors are not unique. Details are given in only a few cases.

(a). $p(\lambda) = \det \begin{bmatrix} 1 - \lambda & 0 \\ 2 & 1 - \lambda \end{bmatrix} = (1 - \lambda)^2$; $\lambda = 1$ is a double eigenvalue. The corresponding eigenspace V_1 may have dimension 1 or 2. In this case the eigenvector equation $(A - I)v = 0$ reduces to the single equation $2v_1 = 0$. V_1 is spanned by the eigenvector $[0\ 1]^T$ and is one-dimensional.

(b). $p(\lambda) = \det \begin{bmatrix} 1 - \lambda & 1 \\ 1 & 1 - \lambda \end{bmatrix} = (1 - \lambda)^2 - 1 = \lambda^2 - 2\lambda$, so the eigenvalues 0 and 2 are simple. V_0 is spanned by $[1\ -1]^T$, V_2 by $[1\ 1]^T$.

(c). $p(\lambda) = \det \begin{bmatrix} -\lambda & 3 \\ 3 & -\lambda \end{bmatrix} = \lambda^2 - 9 = (\lambda - 3)(\lambda + 3)$. The eigenvalues 3 and -3 are simple. V_3 is spanned by $[1\ 1]^T$, V_{-3} by $[1\ -1]^T$.

(d). $p(\lambda) = \det \begin{bmatrix} 6 - \lambda & -7 \\ 1 & -2 - \lambda \end{bmatrix} = (6 - \lambda)(-2 - \lambda) + 7 = \lambda^2 - 4\lambda - 5 = (\lambda + 1)(\lambda - 5)$. The eigenvalues -1 and 5 are simple. V_{-1} is spanned by $[1\ 1]^T$ because the eigenvector equation $(A + I)v$ reduces to $7v_1 - 7v_2 = 0$. V_5 is spanned by $[7\ 1]^T$.

(e). $p(\lambda) = \det \begin{bmatrix} 3 - \lambda & -5 \\ 5 & 3 - \lambda \end{bmatrix} = (3 - \lambda)^2 + 25 = \lambda^2 - 6\lambda + 34$. The simple eigenvalues are the complex conjugates $3 + 5i$ and $3 - 5i$. V_{3+5i} is spanned by $[1\ -i]^T$ and V_{3-5i} by the conjugate $[1\ i]^T$.

(f). $p(\lambda) = \det \begin{bmatrix} 1-\lambda & 0 & 1 \\ 0 & 1-\lambda & 0 \\ 1 & 0 & 1-\lambda \end{bmatrix} = (1-\lambda)^3 - (1-\lambda) = (1-\lambda)(\lambda^2 - 2\lambda).$ The eigenvalues 0, 1, 2 are simple. V_0 has basis vector $[1\ 0\ -1]^T$, V_1 has basis vector $[0\ 1\ 0]^T$, and V_2 has the basis vector $[1\ 0\ 1]^T$.

(g). Expanding $\det(A - \lambda I) = \det \begin{bmatrix} 5-\lambda & -6 & -6 \\ -1 & 4-\lambda & 2 \\ 3 & -6 & -4-\lambda \end{bmatrix}$ by cofactors (see Section 7.2), we obtain $p(\lambda) = -\lambda^3 + 5\lambda^2 - 8\lambda + 4 = -(\lambda - 1)(\lambda - 2)^2$, so 1 is a simple eigenvalue and 2 is a double eigenvalue. V_1 is spanned by $[3\ -1\ 3]^T$ and V_2 has basis $\{[2\ 1\ 0]^T,\ [2\ 0\ 1]^T\}$. Any two independent vectors in V_2 form a basis, so there is nothing unique about the given basis.

(h). $p(\lambda) = \det \begin{bmatrix} \cos\theta - \lambda & -\sin\theta \\ \sin\theta & \cos\theta - \lambda \end{bmatrix} = (\cos\theta - \lambda)^2 + \sin^2\theta = \lambda^2 - (2\cos\theta)\lambda + 1;$ $\cos\theta \pm i\sin\theta$ are the simple eigenvalues. The corresponding eigenspaces are spanned by $[i\ 1]^T$ and $[-i\ 1]^T$, respectively.

(i). $p(\lambda) = \det \begin{bmatrix} -1-\lambda & 36 & 100 \\ 0 & -1-\lambda & 24 \\ 0 & 0 & 5-\lambda \end{bmatrix} = (-1-\lambda)^2(5-\lambda);\ -1$ is a double eigenvalue and 5 is a simple eigenvalue. V_{-1} is spanned by $[1\ 0\ 0]^T$ and V_5 is spanned by $[262\ 27\ 6]^T$ since the eigenvector equation $(A - 5I)v = 0$ reduces to $-6v_1 + 36v_2 + 100v_3 = 0,\ -6v_2 + 27v_3 = 0.$

(j). $p(\lambda) = \det \begin{bmatrix} a-\lambda & b \\ -b & a-\lambda \end{bmatrix} = (a-\lambda)^2 + b^2 = \lambda^2 - 2a\lambda + a^2 + b^2;$ the simple complex conjugate eigenvalues are $a \pm ib$. V_{a+ib} is spanned by $[1\ i]^T$, and V_{a-bi} by $[1\ -i]^T$.

2. If A is upper triangular, then

$$A - \lambda I = \begin{bmatrix} a_{11}-\lambda & a_{12} & \cdots & a_{1n} \\ 0 & a_{22}-\lambda & \cdots & a_{2n} \\ \vdots & & \ddots & \vdots \\ 0 & \cdots & 0 & a_{nn}-\lambda \end{bmatrix}$$

Since the determinant of a triangular matrix is the product of the diagonal elements (see Properties of det A on page 357), we see that $\det(A - \lambda I) = p(\lambda) = (a_{11} - \lambda) \cdots (a_{nn} - \lambda)$, and the eigenvalues are the diagonal entries of A, each eigenvalue having multiplicity equal to the number of times the value appears on the diagonal.

3. If $Av = \lambda v$ for some vector $v \neq 0$, then $A^k v = A^{k-1}(Av) = A^{k-1}(\lambda v) = \lambda A^{k-1}(v) = \cdots = \lambda^k v.$ So λ^k is an eigenvalue of A^k if λ is an eigenvalue of A. The other way around, however, is not always true. For example, $A = \begin{bmatrix} 2 & 0 \\ 0 & -3 \end{bmatrix}$ has eigenvalues 2 and -3, while $A^2 = \begin{bmatrix} 4 & 0 \\ 0 & 9 \end{bmatrix}$ has eigenvalues 4 and 9, but $3 = 9^{1/2}$ is *not* an eigenvalue of A.

4. First recall from Section 7.2 that a matrix and its transpose have the same determinant. Since $\det(A^T - \lambda I) = \det(A^T - (\lambda I)^T) = \det(A - \lambda I)^T = \det(A - \lambda I)$, A and A^T have a common characteristic polynomial.

5. We have $\det[A - \lambda I] = p(\lambda) = (-1)^n(\lambda - \lambda_1) \cdots (\lambda - \lambda_n)$ if $\lambda_1, \ldots, \lambda_n$ are the eigenvalues of A, each repeated according to its multiplicity (see "Polynomials" in Appendix B.4). So $p(0) = (-1)^n(-\lambda_1) \cdots (-\lambda_n) = \lambda_1 \cdots \lambda_n = \det(A - 0I) = \det A$. On the other hand, the constant term of $p(\lambda) = \det(A - \lambda I)$ is $p(0) = \det A$, so $\det A$ is the product of its eigenvalues.

6. **(a).** We know that $p(\lambda) = (-1)^n(\lambda - \lambda_1) \cdots (\lambda - \lambda_n) = \det(A - \lambda I)$. Equate the coefficients of λ^{n-1} on both sides. First note that $p(\lambda) = (-1)^n \lambda^n + (-1)^{n-1}(\lambda_1 + \lambda_2 + \cdots + \lambda_n)\lambda^{n-1} + \cdots$. Next, note that the coefficient of λ^{n-1} in $\det(A - \lambda I)$ is $(-1)^{n-1}(a_{11} + a_{22} + \cdots + a_{nn})$. So $\operatorname{tr} A = \lambda_1 + \lambda_2 + \cdots + \lambda_n = a_{11} + a_{22} + \cdots + a_{nn}$.

(b). Since $\operatorname{tr} A = 1 - 7 + 2 = -4 = \lambda_1 + \lambda_2 + \lambda_3$, where λ_1, λ_2 and λ_3 are the three eigenvalues of A, at least one of the three must have a negative real part.

7. By Problem 5, $\det A = \lambda_1 \cdots \lambda_n$, where $\lambda_1, \ldots, \lambda_n$ are the eigenvalues of A. Since A is nonsingular if and only if its determinant is nonzero (see Theorem 7.3.1), we see that A is nonsingular if and only if none of its eigenvalues are zero.

8. **(a).** First read the theorem to see just what it is that has to be shown. Then suppose that each B_i contains the single element v^i. Then $Av^i = \lambda_i v^i$, for each i, and the eigenvalues λ_i are all distinct. The anchor step (for $p = 1$) follows immediately from the definitions. To show the induction step, assume that the Eigenspace Property holds for $B = \{v^1, v^2, \cdots, v^p\}$ for $p \leq k - 1$; we shall show that it also holds when $p = k$, i.e., for $B = \{v^1, v^2, \cdots, v^k\}$. Now assume that there exist scalars $a_1, a_2, \ldots, a_k$, such that

$$a_1 v^1 + a_2 v^2 + \cdots + a_k v^k = 0 \tag{i}$$

Multiplying both sides of this equation with the matrix $A - \lambda_k I$, we see that

$$a_1(\lambda_1 - \lambda_k)v^1 + a_2(\lambda_2 - \lambda_k)v^2 + \cdots + a_k(\lambda_{k-1} - \lambda_k)v^{k-1} = 0$$

But from the induction hypothesis the set $\{v^1, v^2, \cdots, v^{k-1}\}$ is linearly independent, so the coefficients $a_i(\lambda_i - \lambda_k) = 0$, for all $i = 1, 2, \ldots, k - 1$, or, since $\lambda_i \neq \lambda_k$, $a_i = 0$, for all $i = 1, 2, \ldots, k - 1$. Inserting this fact into (i) we see that $a_k = 0$ as well, ending the proof.

(b). We now lift the restriction that the eigenspaces V_{λ_i} have to be one-dimensional [as was done in part **(a)**] in order for the Eigenspace Property to hold. So the B_i may contain more than just one vector. Now say that the collection B of all vectors in $B_1, B_2, \ldots, B_p$, is dependent so there exists a linear combination over B which adds up to the zero vector. But that would mean that there are nonzero vectors w^i in B_i, $i = 1, 2, \ldots, p$ such that

$$w^1 + w^2 + \cdots + w^p = 0$$

From part **(a)** we know that this is impossible since the set $\{w^1, w^2, \ldots, w^p\}$ is linearly independent. This contradiction shows that our assumption that B was dependent was false, ending our proof.

9. First we find the eigenvalues of the given matrix:

$$p(\lambda) = \det \begin{bmatrix} -2-\lambda & -2 & 0 \\ 2 & 3-\lambda & 0 \\ 0 & 0 & -1-\lambda \end{bmatrix} = (\lambda^2 - \lambda - 2)(-1 - \lambda) = -(\lambda - 2)(\lambda + 1)^2;$$

$\lambda = 2$ is a simple eigenvalue and $[1 \quad -2 \quad 0]^T$ is a corresponding eigenvector; $\lambda = -1$ is a double eigenvalue, and V_{-1} is 2-dimensional and is spanned by $[2 \quad -1 \quad 0]^T$ and $[2 \quad -1 \quad 1]^T$. So the basis of $\mathbb{R}^3$ consists of three eigenvectors, and we don't need any generalized eigenvectors.

10. Group project. The verification is by induction on k, but we do not carry out all the details:

I. (Anchor Step) Assertion holds when $k = 1$.

(1.) Assume first that B_1 is a simple chain of generalized eigenvectors $u^1, u^2, \ldots, u^r$, corresponding to the eigenvalue λ. We claim that $\{u^1, u^2, \cdots, u^r\}$ is independent. Say the constants $a_1, \ldots, a_r$ are such that $a_1 u^1 + a_2 u^2 + \cdots + a_r u^r = 0$. Apply $(A - \lambda_1 I)^{r-1}$ to this identity to obtain $a_r u^1 = 0$. Since u^1 is an eigenvector, $u^1 \neq 0$, and $a_r = 0$. Repeating this argument with $(A - \lambda_1 I)^\alpha$, for $1 \leq \alpha \leq r - 2$, yields that $a_1 = a_2 = \cdots = a_r = 0$. So the chain is linearly independent.

(2.) Say, now, that B_1 consists of two chains: $u^1, u^2, \ldots, u^r$, and $v^1, v^2, \ldots, v^s$, where the base $\{u^1, v^1\}$ of the chains is linearly independent. Say there are constants $a_1, \ldots, a_r$ and $b_1, \ldots, b_s$ such that

$$a_1 u^1 + \cdots + a_r u^r + b_1 v^1 + \cdots b_s v^s = 0$$

Then using step (1.) above, applying $A - \lambda_1 I$ the appropriate number of times (and using the fact that $\{u^1, v^1\}$ is independent), we can show that $a_1 = a_2 = \cdots = a_r = b_1 = b_2 = \cdots = b_s = 0$. B_1 is linearly independent in this case as well.

(3.) Finally, induction can be used to show that when B_1 consists of finitely many chains (whose base elements form an independent set of eigenvectors), then B_1 is linearly independent.

II. (Induction Step) Assume that the assertion holds for all $k \leq K - 1$, and then show that it holds also for $k = K$. We will deal with the case $k = 1$; the general case follows in a similar manner. Let B_1 and B_2 correspond to λ_1 and λ_2, respectively. We see from part I. above that B_1 and B_2 are each linearly independent. Now consider a linear combination over the union of B_1 and B_2 which adds up to the zero vector. Say that the length of the largest chain in B_2 is m, then applying $(A - \lambda_2 I)^{m+1}$ to the linear combination will eliminate all the vectors in B_2. But because B_1 is linearly independent it follows that the constants multiplying vectors in B_1 are zero. The constants multiplying vectors in B_2 are also zero because B_2 is an independent set.

7.5 Undriven Linear Systems with Constant Coefficients

Suggestions for preparing a lecture

Topics: Linear systems of ODEs, solutions, Linearity and Closure Properties, general solution of $x' = Ax$ when all eigenvalues are real and eigenspaces nondeficient, what to do when eigenvalues are deficient (example only).

Remarks: Complex eigenvalues are taken up in Section 7.6.

Making up a problem set

You might choose a small number of crank-type problems from the various parts of Problems 1–4 so that students get practice finding solution formulas. These could be done by a CAS, but we prefer students to do the problems by hand (that's why we usually pick problems where the algebra isn't hard). We also like Problem 7, so that students have to think hard about how to start a chain of generalized eigenvectors. If you assign Problem 7, you may want to give some suggestions.

Comments

With the matrix and eigen machinery of Sections 7.2–7.4 in hand, we can solve undriven, constant coefficient linear systems, and that is what we do in this section. Operator notation is used first to simplify matters, and to help explain why the process of finding all solutions of $x' = A(t)x + F(t)$ splits two subproblems: find all solutions of $L[x] = 0$, where L is the linear operator $L = d/dt - A$ [i.e., first find the general solution of $x' = A(t)x$], and then find a particular solution of $Lx = f$. In this section we show how the first problem can be solved if A is a matrix of constants which has all real eigenvalues and the eigenvalues of A (and their multiplicities) are known, as well as the corresponding eigenspaces. Wronskians and fundamental sets are introduced. The only real algebraic problem in all this occurs when the dimension of the eigenspace corresponding to a multiple eigenvalue is less than the multiplicity of the eigenvalue. The generalized eigenvectors introduced in Section 7.4 come to the rescue in this case. We work out a few examples using a chain of generalized eigenvectors. We don't dwell on this awkward case, but focus on the more common situation of full eigenspaces (we call them nondeficient in the text).

1. In each case we write out the system matrix rather than the full system of ODEs. That's all that is needed to construct the solutions (using Theorem 7.5.3).

(a). The system matrix is $A = \begin{bmatrix} 5 & 3 \\ -1 & 1 \end{bmatrix}$. Since $\det(A - \lambda I) = p(\lambda) = \lambda^2 - 6\lambda + 8 = (\lambda - 2)(\lambda - 4)$, the eigenvalues are $\lambda_1 = 2$ and $\lambda_2 = 4$. To find the eigenspace for λ_1, note that $(A - 2I)v = 0$ reduces to $3v_1 + 3v_2 = 0$, so V_2 is spanned by $[1 \ \ -1]^T$. For λ_2, note that $(A - 4I)v = 0$ reduces to $v_1 + 3v_2 = 0$ so V_4 is spanned by $[3 \ \ -1]^T$. The general real solution of the given system is

$$x = c_1 \begin{bmatrix} 1 \\ -1 \end{bmatrix} e^{2t} + c_2 \begin{bmatrix} 3 \\ -1 \end{bmatrix} e^{4t} \qquad c_1, c_2 \text{ arbitrary reals}$$

(b). The characteristic polynomial of the system matrix $\begin{bmatrix} 5 & -2 \\ -2 & 8 \end{bmatrix}$ is $p(\lambda) = \lambda^2 -$

$13\lambda + 36 = (\lambda - 4)(\lambda - 9)$. The eigenvalues are $\lambda_1 = 4$, $\lambda_2 = 9$; the corresponding eigenspaces are respectively spanned by $[2 \ 1]^T$ and $[1 \ -2]^T$. The general real-valued solution is $x = c_1 \begin{bmatrix} 2 \\ 1 \end{bmatrix} e^{4t} + c_2 \begin{bmatrix} 1 \\ -2 \end{bmatrix} e^{9t}$, c_1, c_2 arbitrary reals.

(c). The system matrix is $A = \begin{bmatrix} 1 & -1 \\ 1 & 3 \end{bmatrix}$ and since $\det(A - \lambda I) = p(\lambda) = (\lambda - 2)^2$, $\lambda_1 = 2$ is the only eigenvalue and it has multiplicity $m_1 = 2$. To find the eigenspace for λ_1, note that $(A - 2I)v = 0$ becomes $v_1 + v_2 = 0$. So the eigenspace for λ_1 is spanned by $[1 \ -1]^T$. So $x^1 = e^{2t}[1 \ -1]^T$ is one solution of the given system. Since the eigenspace is deficient, we need to find a generalized eigenvector in order to construct a fundamental set. Any vector w which solves $(A - 2I)w = [1 \ -1]^T$ is a generalized eigenvector. Solving, we have $w = [0 \ -1]^T$ as one such solution. Now using the notation of (16) for a chain of eigenvectors, we put $u^1 = [1 \ -1]^T$ and $u^2 = [0 \ -1]^T$, so from (18), $x^2 = ([0 \ -1]^T + t[1 \ -1]^T)e^{2t}$ is another solution of the given system. Also, $\{x^1, x^2\}$ is a fundamental set, so $x = c_1 x^1 + c_2 x^2$, c_1 and c_2 arbitrary reals, is the general solution of the given system.

2. **(a).** Imposing the initial conditions $x_1(0) = 1$, $x_2(0) = 1$ on the general solution given in Problem 1(a), we have that c_1 and c_2 must satisfy the condition

$$\begin{bmatrix} 1 \\ 1 \end{bmatrix} = c_1 \begin{bmatrix} 1 \\ -1 \end{bmatrix} + c_2 \begin{bmatrix} 3 \\ -1 \end{bmatrix}$$

or in component form, $1 = c_1 + 3c_2$, $1 = -c_1 - c_2$. Solving this system, we obtain $c_1 = -2$ and $c_2 = 1$, and so the unique solution of the given IVP is given by $x = -2[1 \ -1]^T e^{2t} + [3 \ -1]^T e^{4t}$.

(b). Imposing the initial conditions in the general solution given in Problem 1(b), we obtain

$$\begin{bmatrix} 1 \\ 1 \end{bmatrix} = c_1 \begin{bmatrix} 2 \\ 1 \end{bmatrix} + c_2 \begin{bmatrix} 1 \\ -2 \end{bmatrix}$$

or in component form, $1 = 2c_1 + c_2$, $1 = c_1 - 2c_2$. Solving we obtain $c_1 = 3/5$, $c_2 = -1/5$, and so the unique solution of the given IVP is

$$x = (3/5)e^{4t}[2 \ 1]^T + (-1/5)e^{9t}[1 \ -2]^T$$

(c). Imposing the initial conditions on the general solution derived in Problem 1(c), we obtain

$$\begin{bmatrix} 1 \\ 1 \end{bmatrix} = c_1 \begin{bmatrix} 1 \\ -1 \end{bmatrix} + c_2 \begin{bmatrix} 0 \\ -1 \end{bmatrix}$$

or in component form $1 = c_1$, $1 = -c_1 - c_2$. Solving, we have $c_1 = 1$ and $c_2 = -2$, and so the unique solution of the given IVP is

$$x = e^{2t}[1 \ -1]^T - 2e^{2t}([0 \ -1]^T + t[1 \ -1]^T)$$

3. **(a).** The eigenvalues of the system matrix $\begin{bmatrix} 0 & 1 \\ 8 & -2 \end{bmatrix}$ are $\lambda_1 = -4$, $\lambda_2 = 2$. The eigenspace for λ_1 is spanned by $[1 \quad -4]^T$, and the eigenspace for λ_2 is spanned by $[1 \quad 2]^T$. The general real-valued solution of the system is $x = c_1 e^{-4t}[1 \quad -4]^T + c_2 e^{2t}[1 \quad 2]^T$, where c_1 and c_2 are arbitrary reals.

(b). The eigenvalues of the system matrix $\begin{bmatrix} 3 & -2 \\ 2 & -2 \end{bmatrix}$ are $\lambda_1 = 2$, $\lambda_2 = -1$ and the corresponding eigenspaces are spanned, respectively, by $[2 \quad 1]^T$ and $[1 \quad 2]^T$. So the general real-valued solution is $x = c_1 e^{2t}[2 \quad 1]^T + c_2 e^{-t}[1 \quad 2]^T$, where c_1 and c_2 are arbitrary reals.

(c). The eigenvalues of the system matrix $\begin{bmatrix} 2 & 1 \\ -3 & 6 \end{bmatrix}$ are $\lambda_1 = 5$ and $\lambda_2 = 3$, and the corresponding eigenspaces are spanned, respectively, by $[1 \quad 3]^T$ and $[1 \quad 1]^T$. So the general real-valued solution is $x = c_1 e^{5t}[1 \quad 3]^T + c_2 e^{3t}[1 \quad 1]^T$, where c_1 and c_2 are arbitrary reals.

(d). The eigenvalues of the system matrix $\begin{bmatrix} -1 & 4 \\ 2 & 1 \end{bmatrix}$ are $\lambda_1 = 3$ and $\lambda_2 = -3$, and the corresponding eigenspaces are spanned, respectively, by $[1 \quad 1]^T$ and $[2 \quad -1]^T$. The general real-valued solution is $x = c_1 e^{3t}[1 \quad 1]^T + c_2 e^{-3t}[2 \quad -1]^T$, where c_1 and c_2 are arbitrary reals.

(e). The eigenvalues of the system matrix $\begin{bmatrix} 3 & 2 & 2 \\ 1 & 4 & 1 \\ -2 & -4 & -1 \end{bmatrix}$ are $\lambda_1 = 1$, $\lambda_2 = 2$, $\lambda_3 = 3$, and the corresponding eigenspaces are spanned by $[1 \quad 0 \quad -1]^T$, $[2 \quad -1 \quad 0]^T$ and $[0 \quad 1 \quad -1]^T$, respectively. So the general real-valued solution is

$$x = c_1 e^t[1 \quad 0 \quad -1]^T + c_2 e^{2t}[2 \quad -1 \quad 0]^T + c_3 e^{3t}[0 \quad 1 \quad -1]^T$$

where c_1, c_2, and c_3 are arbitrary reals.

(f). The characteristic polynomial of the system matrix $\begin{bmatrix} 2 & 2 & 1 \\ 1 & 3 & 1 \\ 1 & 2 & 2 \end{bmatrix}$ is $p(\lambda) = -(\lambda - 1)^2(\lambda - 5)$, and so the eigenvalue $\lambda_1 = 1$ has multiplicity 2, and $\lambda_2 = 5$ is a simple eigenvalue. The eigenspace corresponding to $\lambda_1 = 1$ has dimension 2 (and is nondeficient) and has the basis $\{[1 \quad 0 \quad -1]^T, \ [2 \quad -1 \quad 0]^T\}$. The eigenspace corresponding to $\lambda_2 = 5$ is spanned by $[1 \quad 1 \quad 1]^T$. The general real-valued solution is

$$x = c_1[1 \quad 0 \quad -1]^T e^t + c_2[2 \quad -1 \quad 0]^T e^t + c_3[1 \quad 1 \quad 1]^T e^{5t}$$

where c_1, c_2, c_3 are arbitrary reals.

4. **(a).** Imposing the initial conditions on the general solution in Problem 3(a), we have $[-1 \quad 1]^T = c_1[1 \quad -4]^T + c_2[1 \quad 2]^T$, or in component form $-1 = c_1 + c_2$, $1 = -4c_1 + 2c_2$. Solving, we have $c_1 = c_2 = -1/2$, and so the unique solution of the IVP is $x = (-1/2)(e^{-4t}[1 \quad -4]^T + e^{2t}[1 \quad 2]^T)$.

(b). Imposing the initial conditions on the general solution given in Problem 3(b), we obtain $-1 = 2c_1 + c_2$, $1 = c_1 + 2c_2$. Solving, we have $c_1 = -1$, $c_2 = 1$. The unique solution of the given IVP is $x = -e^{2t}[2 \quad 1]^T + e^{-t}[1 \quad 2]^T$.

(c). Imposing the initial conditions on the general solution given in Problem 3(c), we obtain $-1 = c_1 + c_2$, $1 = 3c_1 + c_2$. Solving, we have $c_1 = 1$, $c_2 = -2$. The unique solution of the given IVP is $x = e^{5t}[1 \ \ 3]^T - 2e^{3t}[1 \ \ 1]^T$.

(d). Imposing the initial conditions on the general solution given in Problem 3(d), we obtain $-1 = c_1 + 2c_2$, $1 = c_1 - c_2$. Solving, we have $c_1 = 1/3$, $c_2 = -2/3$. The unique solution of the given IVP is $x = (1/3)e^{3t}[1 \ \ 1]^T + (-2/3)e^{-3t}[2 \ \ -1]^T$.

(e). Imposing the initial conditions on the general solution given in Problem 3(e), we obtain $1 = c_1 + 2c_2$, $1 = -c_2 + c_3$, and $1 = -c_1 - c_3$. Solving, we have $c_1 = -5$, $c_2 = 3$, and $c_3 = 4$. The unique solution of the given IVP is

$$x = -5e^t[1 \ \ 0 \ \ -1]^T + 3e^{2t}[2 \ \ -1 \ \ 0]^T + 4e^{3t}[0 \ \ 1 \ \ -1]^T$$

(f). Imposing the initial conditions on the general solution given in Problem 3(f), we obtain $1 = c_1 + 2c_2 + c_3$, $1 = -c_2 + c_3$, and $1 = -c_1 + c_3$. Solving, we have $c_1 = 0$, $c_2 = 0$, and $c_3 = 1$. The unique solution of the given IVP is $x = e^{5t}[1 \ \ 1 \ \ 1]^T$.

5. **(a).** Note that $x_i = y^{(i-1)}$, and so $x_i' = y^{(i)} = x_{i+1}$, for $i = 1, 2, \ldots, n-1$. Finally, using the ODE,

$$x_n' = y^{(n)} = -a_0 y - a_1 y' - \cdots - a_{n-1} y^{(n-1)}$$

$$= -a_0 x_1 - a_1 x_2 - \cdots - a_{n-1} x_n$$

The vector $[x_1' \ \ x_2' \cdots x_n']^T$ takes the form $A[x_1 \ \ x_2 \cdots x_n]^T$, where the matrix has the asserted form.

(b). We use column operations on the matrix $A - \lambda I$ before calculating its determinant. Add λ^{j-1} times the jth column to the first column, for each $j = 2, 3, \ldots, n$. The new matrix looks like this: the first column has all zeros, except for the bottom entry which is $-p(\lambda)$, where $p(\lambda) = \lambda^n + a_{n-1}\lambda^{n-1} + a_{n-2}\lambda^{n-2} + \cdots + a_1\lambda + a_0$. The $(n-1) \times (n-1)$ matrix formed by deleting the first column and last row of the full matrix is a triangular matrix with 1's on the diagonal. Expanding the determinant by cofactors on the first column, we obtain $\det(A - \lambda I) = (-1)^{n+1}(-p(\lambda)) \cdot 1 = (-1)^n p(\lambda)$, which was to be shown.

6. **(a).** Referring to Examples 7.4.7 and 7.4.8, we see that the characteristic polynomial of the system matrix is $p(\lambda) = (1 - \lambda)(\lambda + 1)^2$, and so $\lambda_1 = 1$ is a simple eigenvalue and $\lambda_2 = -1$ has multiplicity 2. We also have that the eigenspace for λ_1 is spanned by $[1 \ \ -1 \ \ 1]^T$, and that the eigenspace for λ_2 is spanned by $[0 \ \ 1 \ \ -1]^T$. Thus, the eigenspace for λ_2 is deficient, and we seek a chain of eigenvectors based in that eigenspace in order to find a fundamental set of solutions: Setting $u^1 = [0 \ \ 1 \ \ -1]^T$ we see that u^1 must support a chain of generalized eigenvectors of length 2. To find u^2 we must solve $(A + I)u^2 = u^1 = [0 \ \ 1 \ \ -1]^T$ (any solution u^2 will do). In Example 7.4.8 we found that $u^2 = [1 \ \ 1 \ \ 0]^T$. Now using (18) to construct solutions based on a chain of eigenvectors, we have the solutions

$$x^1 = e^t[1 \ \ -1 \ \ 1]^T$$
$$x^2 = e^{-t}[0 \ \ 1 \ \ -1]^T$$
$$x^3 = e^{-t}([1 \ \ 1 \ \ 0]^T + t[0 \ \ 1 \ \ -1]^T)$$

The Wronskian $W[x^1, x^2, x^3]$, evaluated at $t = 0$, is 1, and so $\{x^1, x^2, x^3\}$ is a basic solution set. The general real-valued solution of the system is $x = c_1 x^1 + c_2 x^2 + c_3 x^3$, where c_1, c_2, and c_3 are arbitrary reals.

(b). The characteristic polynomial of the system matrix $\begin{bmatrix} 1 & 0 & -2 \\ 0 & -1 & 1 \\ 0 & 0 & -1 \end{bmatrix}$ is $p(\lambda) =$ $-(\lambda - 1)(\lambda + 1)^2$, and so the eigenvalue $\lambda_1 = 1$ is simple and the multiplicity of the eigenvalue $\lambda_2 = -1$ is 2. The eigenspace for λ_1 is spanned by $[1 \ 0 \ 0]^T$, but the eigenspace for λ_2 is spanned by $[0 \ 1 \ 0]$, and is deficient. Put $u^1 = [0 \ 1 \ 0]^T$ and use it as a base for a chain of eigenvectors; i.e., find any solution u^2 of the equation $(A + I)u^2 = u^1 = [0 \ 1 \ 0]^T$. Solving, we find a solution $u^2 = [1 \ 0 \ 1]^T$, and the chain ends here. Using (18) to find solutions based on a chain of eigenvectors, we have the basic solution set

$$x^1 = e^t[1 \ 0 \ 0]^T$$

$$x^2 = e^{-t}[0 \ 1 \ 0]^T$$

$$x^3 = e^{-t}([1 \ 0 \ 1]^T + t[0 \ 1 \ 0]^T)$$

The general real-valued solution of the given system is $x = c_1 x^1 + c_2 x^2 + c_3 x^3$, where c_1, c_2, and c_3 are arbitrary reals.

7. Group project, but here is a sketch of how to do it.

- Because A is a triangular matrix, we immediately have

$$\det(A - \lambda I) = (-1 - \lambda)^3 = -(\lambda + 1)^3$$

and so A has only one eigenvalue $\lambda_1 = -1$; its multiplicity is $m_1 = 3$. To find the eigenspace we look for all solutions of the algebraic system $(A + I)v = 0$. All solutions of this equation have the form

$$v = \begin{bmatrix} a \\ b \\ 0 \end{bmatrix} = a \begin{bmatrix} 1 \\ 0 \\ 0 \end{bmatrix} + b \begin{bmatrix} 0 \\ 1 \\ 0 \end{bmatrix}$$

for arbitrary reals a and b. The eigenspace V_{-1} has dimension two because it is spanned by the basis $\{[1 \ 0 \ 0]^T, [0 \ 1 \ 0]^T\}$.

- Let the eigenvector u^1 have the general form $u^1 = [c_1, \ c_2, \ 0]^T$. Find constants c_1, c_2 such that the equation $(A + I)u^2 = [c_1 \ c_2 \ 0]^T$ has a solution for u^2. Writing this equation in component form (with $u^2 = [u_1^2 \ u_2^2 \ u_3^2]^T$), we have that $2u_3^2 = c_1$ and $u_3^2 = c_2$, which is solvable for any pair c_1, c_2, with $2c_2 = c_1$. So taking $c_1 = 2$, $c_2 = 1$, and solving we get $u_3^2 = 1$. There are no restrictions on u_1^2 and u_2^2, and so take them both to be zero. We have the chain

$$u^1 = [2 \ 1 \ 0]^T, \quad \text{and} \quad u^2 = [0 \ 0 \ 1]^T$$

- We have three solutions of $x' = Ax$,

$$x^1 = e^{-t}[1 \ 0 \ 0]^T$$

$$x^2 = e^{-t}[2 \ 1 \ 0]^T$$

$$x^3 = e^{-t}([0 \ 0 \ 1]^T + t[2 \ 1 \ 0]^T)$$

The Wronskian $W[x^1, x^2, x^3]$ evaluated at $t = 0$ has the value $W(0) = 1$, and so $\{x^1, x^2, x^3\}$ is a basic solution set. So the general real-valued solution of $x' = Ax$ is $x = c_1 x^1 + c_2 x^2 + c_3 x^3$, where c_1, c_2, c_3 are arbitrary reals.

8.   Group project.

7.6 Undriven Linear Systems: Complex Eigenvalues

Suggestions for preparing a lecture

Topics: Review matrices with complex entries, conjugation, real and imaginary parts of such matrices; finding a basis for an eigenspace corresponding to a complex eigenvalue, finding the general complex-valued (and real-valued) solution of $x' = Ax$ in the nondeficient case, and (example only) the deficient case; the idea of a basic solution set; what components of every real-valued solution of $x' = Ax$ look like (Theorem 7.6.4).

Making up a problem set

Problems 1, 2, 3(**a**), 4(**a**), and 5(**d**). The third matrix in Problem 5(**d**) involves a lot of work, as does Problem 6, so if you assign either of these make sure your students have enough time.

Comments

The reason why we put this material in a separate section is because people are often uncomfortable working with matrices and vectors whose entries are complex numbers or complex-valued functions. The matter-of-fact approach in Section 7.5 also works in the complex eigenvector case if the algebra of complex numbers is used. Our "proof" of solution formula (3) is by construction, once a basis of eigenvectors and generalized eigenvectors is known. We did not prove the existence of such a basis for the general case (because that would be beyond the scope of this text—or even a first course in Linear Algebra). This is not a problem because our constructive approach works just fine when $n \leq 4$, and these are the examples that occur in this text. Basic solution sets are important, so is Theorem 7.6.4 because it tells us the form of every term in every component of every solution vector of $x' = Ax$.

1.   The eigenvalues of the system matrix $A = \begin{bmatrix} 2 & -1 \\ 1 & 2 \end{bmatrix}$ are $2 \pm i$. The corresponding eigenspaces are spanned by $[1 \ \mp i]^T$. The general solution of $x' = Ax$ is

$$x = c_1[1 \ -i]^T e^{(2+i)t} + c_2[1 \ i]^T e^{(2-i)t},$$

where c_1, c_2 are arbitrary complex numbers. To find the general real-valued solutions, put $x^1 = \operatorname{Re}\left[[1 \ -i]^T e^{(2+i)t}\right] = e^{2t}[\cos t \ \sin t]^T$, and $x^2 = \operatorname{Im}\left[[1 \ -i]^T e^{(2+i)t}\right] = e^{2t}[\sin t \ -\cos t]^T$. Then $x = c_1 x^1 + c_2 x^2$, c_1 and c_2 arbitrary reals, is the general real solution.

2. Imposing the initial conditions on the complex-valued general solution given by Problem 1, we obtain that $1 = c_1 + c_2$, $1 = i(c_2 - c_1)$. So, $c_1 = \frac{1}{2}(1 - i) = \bar{r}_2$. Substituting these values of c_1 and c_2 into the general formula, we obtain the solution $x = e^{2t}[\cos t - \sin t \quad \sin t + \cos t]^T$.

3. **(a).** The eigenvalues of the system matrix $A = \begin{bmatrix} 0 & 1 \\ -64 & 0 \end{bmatrix}$ are $\lambda_1 = 8i$, $\lambda_2 = -8i$. The eigenspaces are spanned, respectively, by $[1 \quad 8i]^T$ and $[1 \quad -8i]^T$. So the general (complex-valued) solution of $x' = Ax$ is $x = c_1 e^{8it}[1 \quad 8i]^T + c_2 e^{-8it}[1 \quad -8i]^T$. To find all real-valued solutions, put

$$x^1 = \text{Re}\left[e^{8it}[1 \quad 8i]^T\right] = [\cos 8t \quad -8\sin 8t]^T$$

$$x^2 = \text{Im}\left[e^{8it}[1 \quad 8i]^T\right] = [\sin 8t \quad 8\cos 8t]^T$$

Then the general real-valued solution is $x = c_1 x^1 + c_2 x^2$, where c_1 and c_2 are arbitrary reals.

(b). The eigenvalues of the system matrix $A = \begin{bmatrix} 2 & -1 \\ 8 & -2 \end{bmatrix}$ are $\lambda_1 = 2i$, $\lambda_2 = -2i$, and the corresponding eigenspaces are spanned, respectively, by $[1 \quad 2 - 2i]^T$ and $[1 \quad 2 + 2i]^T$. The general solution of $x' = Ax$ is $x = c_1 e^{2it}[1 \quad 2 - 2i]^T + c_2 e^{-2it}[1 \quad 2 + 2i]^T$, where c_1 and c_2 are arbitrary complex numbers. For all real-valued solutions put

$$x^1 = \text{Re}\left[e^{2it}[1 \quad 2 - 2i]^T\right] = [\cos 2t \quad 2\cos 2t + 2\sin 2t]^T$$

$$x^2 = \text{Im}\left[e^{2it}[1 \quad 2 - 2i]^T\right] = [\sin 2t \quad 2\sin 2t - 2\cos 2t]^T$$

Then $x = c_1 x^1 + c_2 x^2$, where c_1 and c_2 are arbitrary reals, defines all real-valued solutions of the given system.

4. **(a).** We have our choice of using the complex-valued general solution or the real-valued general solution. It doesn't matter which one is used. We use the complex-valued general solution. Imposing the initial condition, we have $[-1 \quad 1]^T = c_1[1 \quad 8i]^T + c_2[1 \quad -8i]^T$, or in component form, $-1 = c_1 + c_2$, $1 = 8ic_1 - 8ic_2$. Solving, we obtain $c_1 = (-8 - i)/16$, $c_2 = (-8 + i)/16$. We have the unique solution of the IVP

$$x = (-1/2 - i/16)e^{8it}[1 \quad 8i]^T + (-1/2 + i/16)e^{-8it}[1 \quad -8i]^T$$

$$= [-\cos 8t + (1/8)\sin 8t \quad \cos 8t + 8\sin 8t]^T$$

The same result can be obtained from the real-valued general solution (and more quickly!).

(b). Imposing the initial conditions on the real-valued general solution in Problem 3**(b)**, we obtain $-1 = c_1$, $1 = 2c_1 - 2c_2$. Solving, we have $c_1 = -1$, $c_2 = -3/2$. The unique solution of the given IVP is

$$x = [-\cos 2t - (3/2)\sin 2t \quad -\cos t - 5\sin 2t]^T$$

5. **(a).** Suppose that A is a real matrix with a complex eigenvalue λ. Now since $Av = \lambda v$, we see by taking conjugates that $\overline{Av} = \bar{\lambda}\bar{v}$. But since A is real we have that $\overline{Av} = A\bar{v}$, so $A\bar{v} = \bar{\lambda}\bar{v}$. So $\bar{v}$ is an eigenvector of A corresponding to the eigenvalue $\bar{\lambda}$. Now if v had only

real components then $v = \overline{v}$, so

$$\lambda v = Av = A\overline{v} = \overline{\lambda v} = \overline{\lambda} v$$

We see that $\lambda = \overline{\lambda}$, a contradiction since λ is assumed to be nonreal. So v must have at least one nonreal component.

(b). The pair of eigenvectors $\{v, \overline{v}\}$ in part **(a)** is an independent set since they arise from the distinct eigenvalues λ and $\overline{\lambda}$ (see Theorem 7.4.1).

(c). Let $z = ve^{\lambda t}$, where v and λ are as in part **(a)**. Then z and $\overline{z} = \overline{v}e^{\overline{\lambda}t}$ are complex-valued solutions of $x' = Ax$, and from part **(b)**, $\{z, \overline{z}\}$ is an independent set. Setting $x^1 = \mathrm{Re}[z]$ and $x^2 = \mathrm{Im}[z]$, observe that

$$x^1 = \frac{1}{2}(z + \overline{z}) \text{ and } x^2 = \frac{-i}{2}(z - \overline{z})$$

so $\{x^1, x^2\}$ are real-valued solutions of $x' = Ax$. Now let the constants a and b be such that $ax^1 + bx^2 = 0$. We show then that $a = b = 0$, and $\{x^1, x^2\}$ is an independent set. Indeed, replacing x^1 and x^2 by their equivalents in terms of $z, \overline{z}$, we see that $(a + b)z + (a - b)\overline{z} = 0$. But since $\{z, \overline{z}\}$ are independent we have that $a - bi = a + bi = 0$, so $a = b = 0$, finishing the proof.

(d). The eigenvalues and eigenvectors of each matrix A are calculated in order. First let $\begin{bmatrix} 1 & 1 \\ -5 & 3 \end{bmatrix}$. Then

$$p(\lambda) = \det \begin{bmatrix} 1 - \lambda & 1 \\ -5 & 3 - \lambda \end{bmatrix} = \lambda^2 - 4\lambda + 8 = (\lambda - 2)^2 + 4 = 0$$

and so $\lambda_1 = \overline{\lambda_2} = 2 + 2i$. The eigenspace for λ_1 is given by all vectors $[a \ b]$ such that $(-1 - 2i)a + b = 0$, and so

$$\begin{bmatrix} 1 & 1 \\ -5 & 3 \end{bmatrix} \begin{bmatrix} 1 \\ 1 + 2i \end{bmatrix} = (2 + 2i) \begin{bmatrix} 1 \\ 1 + 2i \end{bmatrix}$$

We have the complex-valued solution

$$\begin{bmatrix} 1 \\ 1 + 2i \end{bmatrix} e^{(2+2i)t} = \begin{bmatrix} e^{2t}(\cos 2t + i \sin 2t) \\ e^{2t}(1 + 2i)(\cos 2t + i \sin 2t) \end{bmatrix}$$

Extracting real and imaginary parts, we have that

$$x^1 = \begin{bmatrix} e^{2t} \cos 2t \\ e^{2t}(\cos 2t - 2 \sin 2t) \end{bmatrix} \text{ and } x^2 = \begin{bmatrix} e^{2t} \sin 2t \\ e^{2t}(2 \cos 2t + \sin 2t) \end{bmatrix}$$

form a fundamental set (from part **(c)**) and so $x = c_1 x^1 + c_2 x^2$, for arbitrary reals c_1, c_2, is the general real-valued solution in this case.

Now let $\begin{bmatrix} 1 & 2 \\ -2 & 1 \end{bmatrix}$. Then

$$p(\lambda) = \det \begin{bmatrix} 1 - \lambda & 2 \\ -2 & 1 - \lambda \end{bmatrix} = (1 - \lambda)^2 + 4 = 0 =$$

and so $\lambda_1 = \overline{\lambda_2} = 1 + 2i$. The eigenspace for λ_1 is given by all vectors $[a \quad b]^T$ such that $-2ia + 2b = 0$, and so

$$\begin{bmatrix} 1 & 2 \\ -2 & 1 \end{bmatrix} \begin{bmatrix} 1 \\ 2i \end{bmatrix} = (1 + 2i) \begin{bmatrix} 1 \\ 2i \end{bmatrix}$$

and a solution is

$$\begin{bmatrix} 1 \\ 2i \end{bmatrix} e^{(1+2i)t} = \begin{bmatrix} e^t(\cos 2t + i \sin 2t) \\ 2e^t i(\cos 2t + i \sin 2t) \end{bmatrix}$$

Extracting real and imaginary parts we have that

$$x^1 = \begin{bmatrix} e^t \cos 2t \\ -2e^t \sin 2t \end{bmatrix} \quad \text{and} \quad x^2 = \begin{bmatrix} e^t \sin 2t \\ 2e^t \cos 2t \end{bmatrix}$$

form a fundamental set (from part (c)) and so $x = c_1 x^1 + c_2 x^2$, for arbitrary reals c_1, c_2, is the general real-valued solution in this case.

Finally, let $A = \begin{bmatrix} -3 & 1 & -2 \\ 0 & -1 & -1 \\ 2 & 0 & 0 \end{bmatrix}$. Then

$$p(\lambda) = \det \begin{bmatrix} -3 - \lambda & 1 & -2 \\ 0 & -1 - \lambda & -1 \\ 2 & 0 & -\lambda \end{bmatrix} = -\lambda^3 - 4\lambda^2 - 7\lambda - 6 = 0$$

Use divisors of the constant term, -6, to find a root of $p(\lambda)$. Note that $\lambda_1 = -2$ is successful. To find other roots divide $p(\lambda)$ by $\lambda + 2$ to obtain the quotient $-\lambda^2 - 2\lambda - 3$ whose roots are $-1 \pm i\sqrt{2}$. So the eigenvalues of A are $\lambda_1 = -2$, $\lambda_2 = -1 + i\sqrt{2}$, and $\lambda_3 = -1 - i\sqrt{2}$. Find the eigenvector for $\lambda_1 = -2$:

$$\begin{bmatrix} -1 & 1 & -2 \\ 0 & 1 & -1 \\ 2 & 0 & 2 \end{bmatrix} \begin{bmatrix} a \\ b \\ c \end{bmatrix} = \begin{bmatrix} 0 \\ 0 \\ 0 \end{bmatrix}$$

Solving, we see that $[-1 \quad 1 \quad 1]^T$ is an eigenvector for $\lambda_1 = -2$. Find eigenvectors for $\lambda_2 = -1 + i\sqrt{2}$:

$$\begin{bmatrix} -2 - i\sqrt{2} & 1 & -2 \\ 0 & -i\sqrt{2} & -1 \\ 2 & 0 & 1 - i\sqrt{2} \end{bmatrix} \begin{bmatrix} a \\ b \\ c \end{bmatrix} = \begin{bmatrix} 0 \\ 0 \\ 0 \end{bmatrix}$$

We have $2a + (1 - i\sqrt{2})c = 0$, $i\sqrt{2}b + c = 0$, and so $[-3 \quad i\sqrt{2} - 2 \quad 2 + i2\sqrt{2}]^T$ is an eigenvector for $\lambda_2 = -1 + i\sqrt{2}$. Now we can construct a fundamental set x^1, x^2, x^3.

$$x^1 = \begin{bmatrix} -1 \\ 1 \\ 1 \end{bmatrix} e^{-2t}$$

$$x^2 = \text{Re} \left\{ \begin{bmatrix} -3 \\ i\sqrt{2}-2 \\ 2+i2\sqrt{2} \end{bmatrix} e^{(-1+i\sqrt{2})t} \right\} = e^{-t} \begin{bmatrix} -3\cos\sqrt{2}t \\ -2\cos\sqrt{2}t - \sqrt{2}\sin\sqrt{2}t \\ 2\cos\sqrt{2}t - 2\sqrt{2}\sin\sqrt{2}t \end{bmatrix}$$

$$x^3 = \text{Im} \left\{ \begin{bmatrix} -3 \\ i\sqrt{2}-2 \\ 2+i2\sqrt{2} \end{bmatrix} e^{(-1+i\sqrt{2})t} \right\} = e^{-t} \begin{bmatrix} -3\sin\sqrt{2}t \\ \sqrt{2}\cos\sqrt{2}t - 2\sin\sqrt{2}t \\ 2\sqrt{2}\cos\sqrt{2}t + 2\sin\sqrt{2}t \end{bmatrix}$$

The general real-valued solution in this case is $x = c_1 x^1 + c_2 x^2 + c_3 x^3$, where c_1, c_2, c_3 are arbitrary reals.

6. (a). The system matrix A and characteristic polynomial $p(\lambda)$ for the pendulum system given in the text are

$$A = \begin{bmatrix} 0 & 1 & 0 & 0 \\ -1 & 0 & \alpha & 0 \\ 0 & 0 & 0 & 1 \\ 1 & 0 & -1 & 0 \end{bmatrix}, \quad p(\lambda) = \lambda^4 + 2\lambda^2 + 1 - \alpha$$

Solving $p(\lambda) = 0$ we have that $\lambda^2 = -1 \pm \sqrt{\alpha}$. Since $0 < \alpha < 1$, the four eigenvalues are pure imaginary:

$$\lambda_1 = i\sqrt{1+\sqrt{\alpha}} = \bar{\lambda}_2 = \beta i, \quad \text{where } \beta = \sqrt{1+\sqrt{\alpha}}$$

$$\lambda_3 = i\sqrt{1-\sqrt{\alpha}} = \bar{\lambda}_4 = \gamma i, \quad \text{where } \gamma = \sqrt{1-\sqrt{\alpha}}$$

Corresponding eigenvectors, v^1, $v^2 = \bar{v}^1$, v^3, and $v^4 = \bar{v}^3$, have complex components, but we can follow the technique of this section to generate the real solution space by taking the real and the imaginary parts of $e^{\lambda_1 t} v^1$ and $e^{\lambda_3 t} v^3$. Note that the double pendulum has the two *natural frequencies*, β and γ.

After carrying out the lengthy calculations, we find that the general real solution of the system is given by

$$\begin{bmatrix} x_1 \\ x_2 \\ x_3 \\ x_4 \end{bmatrix} = c_1 \begin{bmatrix} (1+\lambda_1^2)\cos\beta t \\ -\beta(1+\lambda_1^2)\sin\beta t \\ \cos\beta t \\ -\beta\sin\beta t \end{bmatrix} + c_2 \begin{bmatrix} (1+\lambda_1^2)\sin\beta t \\ \beta(1+\lambda_1^2)\cos\beta t \\ \sin\beta t \\ \beta\cos\beta t \end{bmatrix}$$

$$+ c_3 \begin{bmatrix} (1+\lambda_3^2)\cos\gamma t \\ -\gamma(1+\lambda_3^2)\sin\gamma t \\ \cos\gamma t \\ -\gamma\sin\gamma t \end{bmatrix} + c_4 \begin{bmatrix} (1+\lambda_3^2)\sin\gamma t \\ \gamma(1+\lambda_3^2)\cos\gamma t \\ \sin\gamma t \\ \gamma\cos\gamma t \end{bmatrix}$$

$$= c_1 y^1 + c_2 y^2 + c_3 y^3 + c_4 y^4$$

where c_1, c_2, c_3, and c_4 are arbitrary real constants.

(b). The solutions y^1 and y^2 are periodic of period $2\pi/\beta$, while y^3 and y^4 have period $2\pi/\gamma$. So if $\alpha = 0.3$ and $\beta = \sqrt{1+\sqrt{\alpha}}$, we have $\beta = \sqrt{1+\sqrt{0.3}} \approx 1.24$ and $\gamma = \sqrt{1-\sqrt{0.3}} \approx$

0.67 So y^1 and y^2 have period ≈ 5.06 and y^3 and y^4 have approximate period 9.37.

(c). As in part **(b)**, if $\alpha = 0.3$, then $\beta = \sqrt{1 + \sqrt{0.3}}$, $\gamma = \sqrt{1 - \sqrt{0.3}}$, and solving the IVP (1) we find the solution $x(t) = -0.15y^1(t)$, and for IVP (2) the solution $x(t) = 0.15y^3(t)$, and finally for IVP (3) the solution $x(t) = -0.15y^1(t) + 0.15y^3(t)$.

(d). In Graph 1 we have plotted the x_1 component curves of the three solutions $-0.15y^1$ (long dashed curve), $0.15y^3$ (short dashed curve), and the sum $-0.15(y^1 - y^3)$ (solid curve). Observe that the first curve has period $2\pi/\omega_1$, the second has period $2\pi/\omega_2$, while the third is not periodic at all since it is a linear combination of solutions of period ω_1 and of period ω_2 where ω_1/ω_2 is irrational. In Graph 2 we have plotted the x_3 component curves for the same three solutions.

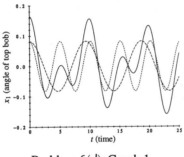 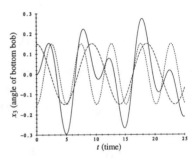

Problem 6(d), Graph 1. Problem 6(d), Graph 2.

(e). Now let's try to visualize the orbits, but since orbits live in a four-dimensional state space we will have to be content with orbital projections into lower dimensional "state spaces."

We can visualize the three solutions given in part **(d)** by projecting the three orbits from four-dimensional state space into the three-dimensional $x_1x_2x_3$-space (Graph 1) and into the two-dimensional x_1x_3-space of the two angles (Graph 2).

The periodic orbits of the solutions $-0.15y^1$ and $0.15y^3$ project into the two oval-shaped simple closed curves shown in Graph 1 and onto the crossed line segments through the origin in Graph 2. The orbit of the sum $-0.15(y^1 - y^3)$ projects onto what appears to be a curve winding around a torus (Graph 1) and onto a curve wandering through a parallelogram determined by the crossed line segments just mentioned (Graph 2). The latter curve is called a *Lissajous figure*. These projections of the third orbit look periodic, but they are not because the third solution itself is not periodic.

The orbits of the solutions $C_1y^1 + C_2y^2$ (called the ω_1-*normal modes*) project as line segments along $-0.15y^1$, but the segment length depends on C_1 and C_2. Similarly the orbits of the solutions $C_3y^3 + C_4y^4$ (the ω_2-*normal modes*) project onto the other line, a segment of which is shown in Graph 2. See the last part of Section 7.7 for another interpretation of normal modes.

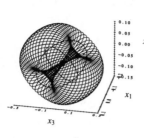   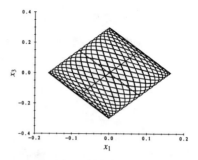

Problem 6(e), Graph 1   Problem 6(e), Graph 2

7.7 Orbital Portraits

Suggestions for preparing a lecture

Topics: The gallery of planar linear orbital portraits, sensitivity to changes in a system parameter.
Remarks: The coupled springs example is well-worth describing if there is time (maybe a second lecture?).

Making up a problem set

Two or three parts of Problem 2 (and the corresponding parts of 3), 4, 6. Maybe Problem 8. Here are our favorite problems: 4, 7 (for a change of pace), 8 (but students may find coupled spring problems hard).

Comments

The focus of this section is on the gallery of the portraits of the orbits of linear planar autonomous systems. There are six qualitatively different portraits here (more if we include system matrices with one or more zero eigenvalues). The saddle, spiral point, center, and three kinds of nodal portraits are by now visually distinct icons for the systems they portray. Then we show how a planar orbital portrait evolves as one of the coefficients in the rate function changes. Finally, we consider a very different question, the *second-order* system $z'' = Az$, where A is an $n \times n$ matrix of real constants, and what its solution set must look like. This leads us to Theorem 7.7.1, which characterizes the general solution $z(t)$ of $z'' = Az$ in terms of the eigenvectors of A and the *square roots* of the eigenvalues in the case that the eigenvalues of A are nonzero and distinct. Rather than reduce $z'' = Az$ to a system of $2n$ first order linear ODEs by introducing the n velocities $z_1', \ldots, z_n'$ as additional state variables, we treat the second-order system directly. That way we can talk about normal modes and frequencies in the way physicists, physical chemists, and mechanical engineers do. The model of the coupled springs is illustrative. It is important to realize that orbits really live in the $2n$-dimensional state space of z and z'. The normal mode oscillations and Lissajous curves that seem to cross one another in the n-dimensional z space are actually the projections of orbits in zz'-space down into z-space.

The emphasis throughout the section is on relating the orbital portraits to the nature of the eigensets of the system matrix. Incidentally, here is a good spot to focus on the art of creating good graphics—choosing the right ranges for the variables, selecting good initial points, drawing enough curves to be informative but not so many that the picture is cluttered. We have tried to follow these priniciples in our pictures. This is also a good place to look at an orbital portrait and ask yourself what the portrait implies about the eigenvalues (and eigenvectors) of the unknown system. This makes a good exam problem!

---

1. From Example 7.7.7 we know that the characteristic polynomial of the system matrix $\begin{bmatrix} \alpha & 1 \\ -1 & -1 \end{bmatrix}$ is $p(\lambda) = \lambda^2 + (1 - \alpha)\lambda + 1 - \alpha$, and the eigenvalues are

$$\lambda_{1,2} = \frac{-(1-\alpha) \pm \sqrt{(1-\alpha)^2 - 4(1-\alpha)}}{2} = \frac{-(1-\alpha) \pm \sqrt{(1-\alpha)(-\alpha-3)}}{2}$$

(a). If $\alpha > 1$, we know $\lambda_1 = [-(1-\alpha) + \sqrt{(1-\alpha)(-\alpha-3)}]/2 > 0$ and $\lambda_2 = [-(1-\alpha) - \sqrt{(1-\alpha)(-\alpha-3)}]/2 < 0$, so the equilibrium point at the origin is a saddle because the eigenvalues have opposite signs.

(b). If $-3 < \alpha < 1$, we have $(1-\alpha)(-\alpha-3) < 0$. The eigenvalues are complex conjugates with negative real parts. So the origin is a spiral point if $(1-\alpha)(-\alpha-3) < 0$.

(c). If $\alpha < -3$, $\lambda_{1,2} = -\frac{1}{2}(1-\alpha) \pm \frac{1}{2}(1-\alpha)\sqrt{1 - 4(1-\alpha)^{-1}} < 0$, and $\lambda_1 \neq \lambda_2$. So the origin is an improper node if $\alpha < -3$. Since the eigenvalues are real and distinct, a pair of eigenvectors (one for each eigenvalue) is independent.

(d). See Graphs 1, 2, and 3 for orbits in the respective cases $\alpha = 4, 0, -4$.

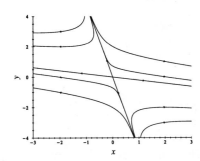
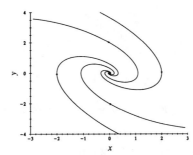
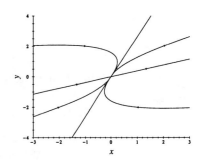

Problem 1(d), Graph 1.   Problem 1(d), Graph 2.   Problem 1(d), Graph 3.

2. Consult Table 7.7.1 for the suitable name.

(a). The characteristic polynomial of the system matrix $\begin{bmatrix} 1 & 1 \\ 4 & -2 \end{bmatrix}$ is $p(\lambda) = \lambda^2 + \lambda - 6$, and the eigenvalues are $\lambda_1 = 2$ and $\lambda_2 = -3$; corresponding eigenvectors are $v^1 = [1 \quad 1]^T$ and $v^2 = [1 \quad -4]^T$. The general real-valued solution is $[x \quad y]^T = c_1 v^1 e^{2t} + c_2 v^2 e^{-3t}$, for arbitrary reals c_1, c_2, and there is a saddle point at the origin.

(b). The characteristic polynomial of the system matrix $\begin{bmatrix} 6 & -8 \\ 2 & -2 \end{bmatrix}$ is $p(\lambda) = \lambda^2 - 4\lambda + 4$, and the double eigenvalue is $\lambda_1 = \lambda_2 = 2$. The eigenspace is one-dimensional and is spanned by $v^1 = [2 \quad 1]^T$. Solve $(A - 2I)w^2 = v^1$ for w^2 to obtain a generalized eigen-vector $w^2 = [1/2 \quad 0]^T$. The general real-valued solution is $[x \quad y]^T = c_1 v^1 e^{2t} + c_2(w^2 + tv^1)e^{2t}$ for arbitrary reals c_1, c_2: we have a deficient node.

(c). The characteristic polynomial of the system matrix $\begin{bmatrix} 7 & 6 \\ 2 & 6 \end{bmatrix}$ is $p(\lambda) = \lambda^2 - 13\lambda + 30$, and the eigenvalues are $\lambda_1 = 10$ and $\lambda_2 = 3$; corresponding eigenvectors are $v^1 = [2 \quad 1]^T$ and $v^2 = [-3 \quad 2]^T$. The general real-valued solution is $[x \quad y]^T = c_1 v^1 e^{10t} + c_2 v^2 e^{3t}$ for arbitrary reals c_1, c_2: we have an improper node.

(d). The characteristic polynomial of the system matrix $\begin{bmatrix} 0 & 4 \\ -1 & 0 \end{bmatrix}$ is $p(\lambda) = \lambda^2 + 4$, and the eigenvalues are $\lambda_1 = 2i$ and $\lambda_2 = -2i$; corresponding eigenvectors are $v^1 = [2 \quad i]^T$ and $v^2 = [2 \quad -i]^T$. The general real-valued solution is

$$\begin{bmatrix} x \\ y \end{bmatrix} = c_1 \begin{bmatrix} 2\cos 2t \\ -\sin 2t \end{bmatrix} + c_2 \begin{bmatrix} 2\sin 2t \\ \cos 2t \end{bmatrix}.$$

for arbitrary reals c_1, c_2: this is a center.

(e). The characteristic polynomial of the system matrix $\begin{bmatrix} -1 & -4 \\ 1 & -1 \end{bmatrix}$ is $p(\lambda) = \lambda^2 + 2\lambda + 5$, and the eigenvalues are $\lambda_1 = -1 + 2i$ and $\lambda_2 = -1 - 2i$; the corresponding eigenvectors are $v^1 = [2 \quad -i]^T$ and $v^2 = [2 \quad i]^T$. The general real-valued solution is

$$\begin{bmatrix} x \\ y \end{bmatrix} = c_1 e^{-t} \begin{bmatrix} 2\cos 2t \\ \sin 2t \end{bmatrix} + c_2 e^{-t} \begin{bmatrix} 2\sin 2t \\ -\cos 2t \end{bmatrix}.$$

for arbitrary reals c_1, c_2: this is a spiral point.

(f). The characteristic polynomial of the system matrix $\begin{bmatrix} -2 & 0 \\ 0 & -2 \end{bmatrix}$ is $p(\lambda) = \lambda^2 + 4\lambda + 4$, and the double eigenvalue is $\lambda_1 = \lambda_2 = -2$; V_{-2} is nondeficient, and the eigenvectors $v^1 = [1 \quad 0]^T$ and $v^2 = [0 \quad 1]^T$ constitute a basis. The general real-valued solution is $[x \quad y]^T = (c_1 v^1 + c_2 v^2)e^{-2t}$, for arbitrary reals c_1, c_2: this is a star node.

3. See the figures for orbital portraits of the systems of Problem 2.

(a). The system is $x' = x + y$, $y' = 4x - 2y$ [Fig. 3(a)].

(b). The system is $x' = 6x - 8y$, $y' = 2x - 2y$ [Fig. 3(b)].

(c). The system is $x' = 7x + 6y$, $y' = 2x + 6y$ [Fig. 3(c)].

(d). The system is $x' = 4y$, $y' = -x$ [Fig. 3(d)].

(e). The system is $x' = -x - 4y$, $y' = x - y$ [Fig. 3(e)].

(f). The system is $x' = -2x$, $y' = -2y$ [Fig. 3(f)].

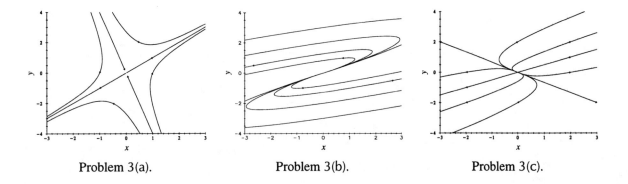

Problem 3(a).   Problem 3(b).   Problem 3(c).

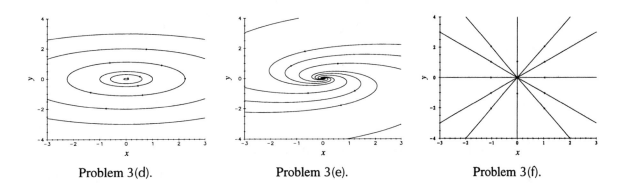

Problem 3(d).   Problem 3(e).   Problem 3(f).

4. The characteristic polynomial of the system matrix $A = \begin{bmatrix} -1 & \alpha \\ -1 & -1 \end{bmatrix}$ is $p(\lambda) = \lambda^2 + 2\lambda + 1 + \alpha$ and the eigenvalues are $\lambda_{1,2} = -1 \pm \sqrt{-\alpha}$. The orbital portrait is a saddle for $\alpha < -1$, an improper node for $-1 < \alpha < 0$, a deficient node for $\alpha = 0$, and a spiral point for $\alpha > 0$. When $\alpha = -1$, one of the eigenvalues is zero (considered a degenerate case), and the orbital portrait does not have a special name. The critical values of α, also known as bifurcation points, are -1 and 0. A *bifurcation point of the parameter* α is loosely defined as a value for which there is a qualitative change in the nature of the orbits as α passes through the value. See Sections 2.4 and 9.3 for more details on the important theory of bifurcations. See Graphs 1–4 for, respectively, a saddle ($\alpha = -2$), a line of equilibrium points and straight-line orbits approaching them ($\alpha = -1$), a deficient node ($\alpha = 0$), and a spiral point ($\alpha = 1$).

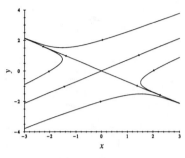

Problem 4, Graph 1.

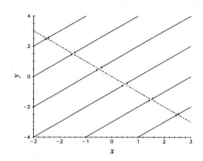

Problem 4, Graph 2.

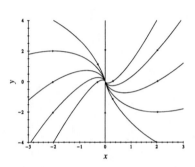

Problem 4, Graph 3.

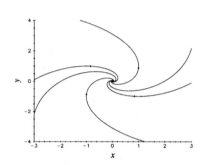

Problem 4, Graph 4.

5.   The affine system has the form $x' = Ax + [r \ s]^T$, where $x = [x \ y]^T$.

(a). We wish to find (x_0, y_0) such that $x' = 0$ and $y' = 0$. That is, we want to solve the system $Ax = -[r \ s]^T$:

$$A \begin{bmatrix} x_0 \\ y_0 \end{bmatrix} = \begin{bmatrix} a & b \\ c & d \end{bmatrix} \begin{bmatrix} x_0 \\ y_0 \end{bmatrix} = \begin{bmatrix} -r \\ -s \end{bmatrix}$$

Since $\det \begin{bmatrix} a & b \\ c & d \end{bmatrix} \neq 0$ the system has a unique solution,

$$[x_0 \ \ y_0]^T = \begin{bmatrix} a & b \\ c & d \end{bmatrix}^{-1} \begin{bmatrix} -r \\ -s \end{bmatrix}$$

(b). We have $ax_0 + by_0 + r = 0$ and $cx_0 + dy_0 + s = 0$ if (x_0, y_0) is the equilibrium point of the planar affine system. Substituting $x = u + x_0$ and $y = v + y_0$ into the affine system, we obtain the homogeneous system $u' = au + bv$, $v' = cu + dv$.

(c). This follows directly from **(b)**.

(d). Calculations show that the equilibrium point is $(-1/6, -5/6)$. The general solution of the corresponding homogeneous system is $[u \ \ v]^T = c_1 w^1 e^{2t} + c_2 w^2 e^{-3t}$, where $w^1 = [1 \ \ 1]^T$ and $w^2 = [1 \ \ -4]^T$ are eigenvectors of the homogeneous system corresponding to $\lambda_1 = 2$, $\lambda_2 = -3$, respectively. The general real-valued solution of the system $x' =$

$x + y + 1$, $y' = 4x - 2y - 1$ is $[x \quad y]^T = c_1 w^1 e^{2t} + c_2 w^2 e^{-3t} + [-1/6 \quad -5/6]^T$, for arbitrary reals c_1, c_2. The equilibrium point is a saddle. See Fig. 5(d).

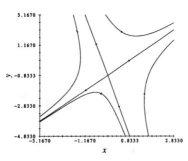

Problem 5(d).

6. The system matrix is $\begin{bmatrix} 0 & 1 \\ -b & -2a \end{bmatrix}$. The characteristic polynomial is $p(\lambda) = \lambda^2 + 2a\lambda + b$, and the eigenvalues are $\lambda_{1,2} = -a \pm \sqrt{a^2 - b}$.

    (a). If $a = 1$ and $b < 0$, then $\lambda_{1,2} = -1 \pm \sqrt{1 - b}$, and $\lambda_1 > 0$ while $\lambda_2 < 0$, and so the origin is a saddle point. If $0 < b < 1$, then $\lambda_1, \lambda_2 < 0$ and $\lambda_1 \neq \lambda_2$ so the origin is an improper node.

    (b). If $a = 1$, $\lambda_{1,2} = -1 \pm \sqrt{1 - b}$. So if $b = 1$, then $\lambda_1 = \lambda_2 = -1$ is a double eigenvalue with a one-dimensional eigenspace (spanned by $[1 \quad -1]^T$); the origin is a deficient node. If $b > 1$, then $1 - b < 0$, and the eigenvalues are complex with nonzero real part, making the origin a spiral point.

    (c). For the system to have a center, the eigenvalues must be pure imaginary. This only happens for $a = 0$, $b > 0$.

    (d). For the system to have a star node, there must be a double eigenvalue, so $a^2 = b > 0$ and $\lambda = -a$. However, the eigenspace corresponding to $\lambda = -a$ is one-dimensional (spanned by $[1 \quad -a]^T$). So the system cannot have a star node.

7. **(a).** Using the appropriate trigonometric identities from Appendix B.4, we see that $x_1 = -2\sin 2t$ and $x_2 = -\cos 2t$. Since $x_1' = 4x_2$ and $x_2' = -x_1$ for this particular solution, we already have the desired system: $x_1' = 4x_2$, $x_2' = -x_1$. Since the eigenvalues of the system matrix are $2i, -2i$, we can calculate the eigenvectors, which are, respectively, $[2 \quad i]^T$ and $[2 \quad -i]^T$. The general real-valued solution is

    $$\begin{bmatrix} x_1 \\ x_2 \end{bmatrix} = c_1 \begin{bmatrix} 2\cos 2t \\ -\sin 2t \end{bmatrix} + c_2 \begin{bmatrix} 2\sin 2t \\ \cos 2t \end{bmatrix}$$

    for arbitrary reals c_1, c_2. The solution originally given corresponds to $c_1 = 0$, $c_2 = -1$.

    (b). See Fig. 7(b) for some orbits.

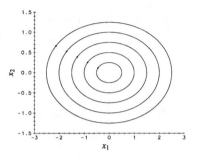

Problem 7(b).

8. From Example 7.7.8 we have that the system modeling the displacements x and y of the two weights attached to the springs is

$$x'' = -(a+b)x + by$$

$$y'' = cx - cy$$

where $a = k_1/m_1$, $b = k_2/m_1$, $c = k_2/m_2$. In this problem set $k_1 = k_2 = 1$, $m_1 = 1$. The matrix $A = \begin{bmatrix} -2 & 1 \\ c & -c \end{bmatrix}$, and its eigenvalues are $\lambda_1, \lambda_2 = -1 - c/2 \pm \sqrt{4 + c^2}/2$. An eigenvector corresponding to eigenvalue λ is $[1, \lambda + 2]^T$, or any nonzero constant multiple.

For $m_2 = 1/2, 1, 3/2$, we have $c = 2, 1, 2/3$, respectively. The eigenvalues, normal modes (eigenvectors), and normal frequencies are, respectively:

$$c = 2: \quad \lambda_1 = -2 + \sqrt{2}, \lambda_2 = -2 - \sqrt{2};$$

$$v^1 = [1, \sqrt{2}], v^2 = [1, -\sqrt{2}]; \quad \omega_1 = \sqrt{2 - \sqrt{2}}, \omega_2 = \sqrt{2 + \sqrt{2}}$$

$$c = 1: \quad \lambda_1 = -3/2 + \sqrt{5}/2, \lambda_2 = -3/2 - \sqrt{5}/2;$$

$$v^1 = [2, 1 + \sqrt{5}], v^2 = [2, 1 - \sqrt{5}]; \quad \omega_1 = \sqrt{3/2 - \sqrt{5}/2}, \omega_2 = \sqrt{3/2 + \sqrt{5}/2}$$

$$c = 2/3: \quad \lambda_1 = -4/3 + \sqrt{10}/3, \lambda_2 = -4/3 - \sqrt{10}/3;$$

$$v^1 = [3, 2 + \sqrt{10}], [3, 2 - \sqrt{10}]; \quad \omega_1 = \sqrt{4/3 - \sqrt{10}/3}, \omega_2 = \sqrt{4/3 + \sqrt{10}/3}$$

Graphs 1, 2, and 3 show normal mode oscillations and Lissajous curves corresponding to the respective initial values (x_0, x_0', y_0, y_0'):

> Graph 1 $(c = 2):$ $(1, 0, \sqrt{2}, 0), (1, 0, -\sqrt{2}, 0), (2, 0, 1, 0)$
>
> Graph 2 $(c = 1):$ $(2, 0, 1 + \sqrt{5}, 0), (1, 0, 1 - \sqrt{5}, 0), (2, 0, 1, 0)$
>
> Graph 3 $(c = 2/3):$ $(3, 0, 2 + \sqrt{10}, 0), (3, 0, 2 - \sqrt{10}, 0), (2, 0, 1, 0)$

The t-range is $0 \leq t \leq 100$ throughout.

There is nothing special about these particular initial values. The easiest way to find a normal mode oscillation is to choose $x(0)$ and $y(0)$ in the same ratio as the x-coordinate to the y-coordinate of a normal mode vector, while $x'(0)$ and $y'(0)$ must both be zero so that the orbit never gets out of the normal mode. On the other hand, a Lissajous curve is generated if the ratio of $x(0)$ to $y(0)$ does *not* meet the normal mode coordinate condition. These aren't the only ways to generate normal mode and Lissajous curves. For the given values of c, the ratio ω_1/ω_2 of the natural frequencies is irrational, and so a Lissajous curve cannot be periodic.

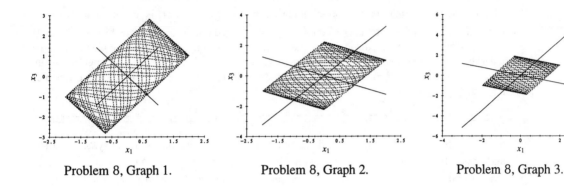

Problem 8, Graph 1.   Problem 8, Graph 2.   Problem 8, Graph 3.

9. Group project.

10. Group project.

7.8 Driven Systems and the Matrix Exponential

Suggestions for preparing a lecture

Topics: Solution matrix, transition matrix, formula (17) for the solution of the IVP $x' = Ax + F(t)$, $x(t_0) = x^0$.

Remarks: Focus on the constant system matrix case, $x' = Ax + F(t)$, $x(0) = x^0$, and the matrix exponential e^{tA}. Don't try to cover everything about the matrix exponential in one lecture!

Making up a problem set

One part of Problem 1, 2**(a)**, 2**(b)**, one part of Problem 4, 6**(a)**. Optional hard problem: 6**(b)**.

Comments

In this section we characterize the solution set of the linear system $x' = Ax + F(t)$, where A is a matrix of constants, using a variation of parameters approach. Basic matrices and transition matrices make their appearance—all quite abstract, but leading to the elegant formula (16) for the general solution and the even nicer formula (17) for the unique solution of the corresponding IVP. To find all solutions of the undriven system, $x' = Ax$, one can use any basic solution matrix $X(t)$, and write the general solution as $x(t) = X(t)c$, where c is any constant vector. On the other hand, in order to characterize the solution of the driven IVP as the sum of the responses to the initial data and the driving force as we do in formula (17), you need to use the transition matrix. Given any basic solution matrix $X(t)$, one can construct the transition matrix as $X(t)X^{-1}(t_0)$ [see formula (8)].

Finally, we get to the other topic in the title of the section, the matrix exponential. People are often mystified by the matrix exponential. Our approach is to emphasize that e^{tA} is just a name for the basic solution matrix $X(x)$ for which $X(0) = I$. Or write it as $\exp(tA)$, which seems less mysterious. Most of the properties of e^{tA} follow naturally. However, some properties depend on the series formula for e^{tA}. We don't usually prove the convergence of this series, but we do illustrate its use in calculating e^{tA} if A is diagonal or nilpotent. Incidentally, it has to be emphasized that in general the matrix exponential is *not* the matrix of the exponentials of the individual entries. In addition $\exp\left[\int_{t_0}^{t} A(s)\,ds\right]$ is *not* a solution matrix of $x' = A(t)x$ if $A(t)$ is nonconstant. We don't mention this fact in the text, but an example might be worked out. You might use the system $x' = 0$, $y' = 2tx + y$. Once you get the idea of a matrix exponential, you might try out the idea of a matrix logarithm, or a matrix sine, e.g., $\sin A = A - \frac{1}{3!}A^3 + \frac{1}{5!}A^5 - \frac{1}{7!}A^7 + \cdots$, etc., defined by means of series. Although we haven't done so, the study of matrix functions like these would make an interesting group project.

---

1. First find the eigenvalues λ_1, λ_2 and eigenvectors for the eigenspaces of the system matrix, and generalized eigenvectors if necessary. A basic solution matrix $X(t)$ can then be constructed and $e^{tA} = X(t)X^{-1}(0)$. In every case the solution of $x' = Ax$, $x(0) = [1 \quad 2]^T$ is $x = e^{tA}[1 \quad 2]^T$.

    (a). The eigenvalues of $A = \begin{bmatrix} 1 & 3 \\ 1 & -1 \end{bmatrix}$ are $\lambda_1 = -2$, $\lambda = 2$; the eigenspaces are $V_{-2} = \text{Span}[1 \quad -1]^T$, $V_2 = \text{Span}[3 \quad 1]^T$. So

    $$X(t) = \begin{bmatrix} e^{-2t} & 3e^{2t} \\ -e^{-2t} & e^{2t} \end{bmatrix}, \quad X^{-1}(0) = \frac{1}{4}\begin{bmatrix} 1 & -3 \\ 1 & 1 \end{bmatrix}$$

    $$e^{tA} = X(t)X^{-1}(0) = \frac{1}{4}\begin{bmatrix} e^{-2t} + 3e^{2t} & -3e^{-2t} + 3e^{2t} \\ -e^{-2t} + e^{2t} & 3e^{-2t} + e^{2t} \end{bmatrix}$$

    $$x(t) = \frac{1}{4}[-5e^{-2t} + 9e^{2t} \quad 5e^{-2t} + 3e^{2t}]^T$$

    (b). $\lambda_1 = 3i = \bar{\lambda}_2$, $V_{3i} = \text{Span}[1 \quad 3i]^T = \bar{V}_{-3i}$. So

    $$X(t) = \begin{bmatrix} e^{3it} & e^{-3it} \\ 3ie^{3it} & -3ie^{-3it} \end{bmatrix}, \quad X^{-1}(0) = -\frac{1}{6i}\begin{bmatrix} -3i & -1 \\ -3i & 1 \end{bmatrix}$$

    $$e^{tA} = X(t)X^{-1}(0) = \begin{bmatrix} \cos 3t & \frac{1}{3}\sin 3t \\ -3\sin 3t & \cos 3t \end{bmatrix}$$

    $$x(t) = [\cos 3t + (2/3)\sin 3t \quad 2\cos 3t - 3\sin 3t]^T$$

    (c). The eigenvalue of $A = \begin{bmatrix} -1 & 1 \\ 0 & -1 \end{bmatrix}$ is -1 and it is a double eigenvalue, but V_{-1} is only one-dimensional. Since the system is upper triangular it is simplest to solve it directly by starting with $x_2' = -x_2$, inserting its solution $c_2 e^{-t}$ into the ODE, $x_1' = -x_1 + x_2$, and solving that ODE by integrating factors. We have $x_2 = c_2 e^{-t}$, $x_1 = (c_1 + c_2 t)e^{-t}$. A fundamental matrix is

$$X(t) = \begin{bmatrix} e^{-t} & te^{-t} \\ 0 & e^{-t} \end{bmatrix}$$

obtained by first setting $c_1 = 1$, $c_2 = 0$ then $c_1 = 0$, $c_2 = 1$. Since $X(0) = I = X^{-1}(0)$, we have $X(t) = e^{tA}$, and $x(t) = e^{-t}[1 + 2t \quad 2]^T$.

(d). $\lambda_1 = 1 + i$, $\lambda_2 = 1 - i$, $V_{1+i} = \text{Span}[1 \quad i]^T$, $V_{1-i} = \text{Span}[1 \quad -i]^T$. So

$$X(t) = \begin{bmatrix} e^{(1+i)t} & e^{(1-i)t} \\ ie^{(1+i)t} & -ie^{(1-i)t} \end{bmatrix}, \quad X^{-1}(0) = \frac{1}{2}\begin{bmatrix} 1/2 & -i/2 \\ 1/2 & i/2 \end{bmatrix}$$

$$e^{tA} = X(t)X^{-1}(0) = \begin{bmatrix} e^t \cos t & e^t \sin t \\ -e^t \sin t & e^t \cos t \end{bmatrix}$$

$$x(t) = e^t[\cos t + 2\sin t \quad 2\cos t - \sin t]$$

2. Once e^{tA} has been found, the solution of the initial value problem is $x = e^{tA}[1 \quad 2 \quad 3]^T$.

(a). In this case A is nilpotent, and the series expansion for e^{tA} terminates after three terms since

$$A = \begin{bmatrix} 0 & 1 & 1 \\ 0 & 0 & 1 \\ 0 & 0 & 0 \end{bmatrix}, \quad A^2 = \begin{bmatrix} 0 & 0 & 1 \\ 0 & 0 & 0 \\ 0 & 0 & 0 \end{bmatrix}, \quad A^3 = \begin{bmatrix} 0 & 0 & 0 \\ 0 & 0 & 0 \\ 0 & 0 & 0 \end{bmatrix}$$

So

$$e^{tA} = I + tA + t^2 A^2/2 = \begin{bmatrix} 1 & t & t + t^2/2 \\ 0 & 1 & t \\ 0 & 0 & 1 \end{bmatrix}$$

$$x(t) = \begin{bmatrix} 1 + 5t + 3t^2/2 & 2 + 3t & 3 \end{bmatrix}^T$$

(b). Since A is diagonal, e^{tA} just exponentiates the diagonal entries in A:

$$e^{tA} = \begin{bmatrix} e^{2t} & 0 & 0 \\ 0 & e^{-3t} & 0 \\ 0 & 0 & e^{7t} \end{bmatrix}, \quad \text{and} \quad x(t) = \begin{bmatrix} e^{2t} & 2e^{-3t} & 3e^{7t} \end{bmatrix}^T$$

(c). The series expansion for e^{tA} terminates after three terms since

$$A = \begin{bmatrix} 0 & 0 & 0 \\ 2 & 0 & 0 \\ 3 & 4 & 0 \end{bmatrix}, \quad A^2 = \begin{bmatrix} 0 & 0 & 0 \\ 0 & 0 & 0 \\ 8 & 0 & 0 \end{bmatrix}, \quad A^k = \begin{bmatrix} 0 & 0 & 0 \\ 0 & 0 & 0 \\ 0 & 0 & 0 \end{bmatrix} \text{ for } k \geq 3$$

So

$$e^{tA} = I + tA + t^2 A^2/2 = \begin{bmatrix} 1 & 0 & 0 \\ 2t & 1 & 0 \\ 3t + 4t^2 & 4t & 1 \end{bmatrix}$$

$$x(t) = \begin{bmatrix} 1 & 2t+2 & 3+11t+4t^2 \end{bmatrix}^T$$

3. **(a).** Since A is the block matrix $\begin{bmatrix} B & 0 \\ 0 & C \end{bmatrix}$, we can write $x = [y \ z]^T$, where y and z have the appropriate number of components determined by the sizes of the blocks B and C. Then $x' = Ax$ is equivalent to the decoupled pair of systems $y' = By$ and $z' = Cz$, whose respective transition matrices are e^{tB} and e^{tC}: $\begin{bmatrix} e^{tB} & 0 \\ 0 & e^{tC} \end{bmatrix}$ is a basic solution matrix and is, in fact, e^{tA} since it reduces to the identity matrix at $t = 0$.

(b). Using part **(a)** and part of the answer to Problem 1**(a)**, we have

$$e^{tA} = \frac{1}{4} \begin{bmatrix} e^{-2t}+3e^{2t} & -3e^{-2t}+3e^{2t} & 0 \\ -e^{-2t}+e^{2t} & 3e^{-2t}+e^{2t} & 0 \\ 0 & 0 & 4e^{2t} \end{bmatrix} \quad \text{if } A = \begin{bmatrix} 1 & 3 & 0 \\ 1 & -1 & 0 \\ 0 & 0 & 2 \end{bmatrix}$$

For the second matrix observe that $C = \begin{bmatrix} 0 & 1 \\ 0 & 0 \end{bmatrix}$ is nilpotent and $e^{tC} = \begin{bmatrix} 1 & t \\ 0 & 1 \end{bmatrix}$:

$$e^{tA} = \frac{1}{4} \begin{bmatrix} e^{-2t}+3e^{2t} & -3e^{-2t}+3e^{2t} & 0 & 0 \\ -e^{-2t}+e^{2t} & 3e^{-2t}+e^{2t} & 0 & 0 \\ 0 & 0 & 4 & 4t \\ 0 & 0 & 0 & 4 \end{bmatrix} \quad \text{if } A = \begin{bmatrix} 1 & 3 & 0 & 0 \\ 1 & -1 & 0 & 0 \\ 0 & 0 & 0 & 1 \\ 0 & 0 & 0 & 0 \end{bmatrix}$$

4. In each case it is a matter of finding eigenvalues and spanning sets for eigenspaces (and possibly generalized eigenvalues), constructing from these eigensets a fundamental matrix $X(t)$, setting $e^{tA} = X(t)X^{-1}(0)$, and then solving $x(t) = Ax + F(t)$, $x(0) = x^0$ by the formula $x = e^{tA}x^0 + e^{tA}\int_0^t e^{-sA}F(s)\,ds$. Although $X(t)$ is never unique, e^{tA} is always unique (given A). One obtains e^{-sA} from e^{tA} by replacing t with $-s$. The eigenvalues, the spanning sets, a fundamental matrix $X(t)$ and e^{tA} are given below, but we haven't written out the steps of the calculations.

(a). $\lambda_1 = 2i = \bar{\lambda}_2$, $V_{2i} = \text{Span}\{[1 \ \ i]^T\} = \bar{V}_{-2i}$. So

$$X(t) = \begin{bmatrix} e^{2it} & e^{-2it} \\ ie^{2it} & -ie^{-2it} \end{bmatrix}, \qquad e^{tA} = \begin{bmatrix} \cos 2t & \sin 2t \\ -\sin 2t & \cos 2t \end{bmatrix}$$

$$e^{tA}x^0 = e^{tA}\begin{bmatrix} a \\ b \end{bmatrix} = \begin{bmatrix} a\cos 2t + b\sin 2t \\ b\cos 2t - a\sin 2t \end{bmatrix}$$

$$e^{-sA}F(s) = \begin{bmatrix} \cos 2s & -\sin 2s \\ \sin 2s & \cos 2s \end{bmatrix}\begin{bmatrix} 1 \\ 0 \end{bmatrix} = \begin{bmatrix} \cos 2s \\ \sin 2s \end{bmatrix}$$

(b). $\lambda_1 = -1$, $\lambda_2 = 1$, $V_{-1} = \text{Span}\{[1 \ \ 3]^T\}$, $V_1 = \text{Span}\{[1 \ \ 1]^T\}$. So

$$X(t) = \begin{bmatrix} e^{-t} & e^t \\ 3e^{-t} & e^t \end{bmatrix}, \qquad e^{tA} = \frac{1}{2}\begin{bmatrix} -e^{-t}+3e^t & e^{-t}-e^t \\ -3e^{-t}+3e^t & 3e^{-t}-e^t \end{bmatrix}$$

$$e^{tA}x^0 = e^{tA}\begin{bmatrix} 1 \\ 2 \end{bmatrix} = \frac{1}{2}\begin{bmatrix} e^{-t} + e^t \\ 3e^{-t} + e^t \end{bmatrix}$$

$$e^{-sA}F(s) = \frac{1}{2}\begin{bmatrix} -e^s + 3e^{-s} & e^s - e^{-s} \\ -3e^s + 3e^{-s} & 3e^s - e^{-s} \end{bmatrix}\begin{bmatrix} 3e^s \\ s \end{bmatrix} = \frac{1}{2}\begin{bmatrix} -3e^{2s} + 9 + s(e^s - e^{-s}) \\ -9e^{2s} + 9 + s(3e^s - e^{-s}) \end{bmatrix}$$

See Fig. 4(b) for the component curves.

(c). $\lambda_1 = i = \bar{\lambda}_2$, $V_i = \text{Span}\{[2+i \quad 1]^T\} = \bar{V}_{-i}$. So

$$X(t) = \begin{bmatrix} (2+i)e^{it} & (2-i)e^{-it} \\ e^{it} & e^{-it} \end{bmatrix}, \qquad e^{tA} = \begin{bmatrix} \cos t + 2\sin t & -5\sin t \\ \sin t & \cos t - 2\sin t \end{bmatrix}$$

$$e^{tA}x^0 = e^{tA}\begin{bmatrix} a \\ b \end{bmatrix} = \begin{bmatrix} a\cos t + (2a - 5b)\sin t \\ (a - 2b)\sin t + b\cos t \end{bmatrix}$$

$$e^{-sA}F(s) = e^{-sA}\begin{bmatrix} \cos s \\ 0 \end{bmatrix} = \begin{bmatrix} \cos s + 2\sin s & -5\sin 2s \\ \sin s & \cos s - 2\sin s \end{bmatrix}\begin{bmatrix} \cos s \\ 0 \end{bmatrix}$$

$$= \begin{bmatrix} \cos^2 s + 2\sin s \cos s \\ \sin s \cos s \end{bmatrix}$$

(d). $\lambda_1 = -1 + 2i = \bar{\lambda}_2$, $V_{-1+2i} = \text{Span}\{[2i \quad 1]^T\} = \bar{V}_{-1-2i}$. So

$$X(t) = \begin{bmatrix} 2ie^{(-1+2i)t} & -2ie^{(-1-2i)t} \\ e^{(-1+2i)t} & e^{(-1-2i)t} \end{bmatrix}, \qquad e^{tA} = e^{-t}\begin{bmatrix} \cos 2t & -2\sin 2t \\ (\sin 2t)/2 & \cos 2t \end{bmatrix}$$

$$e^{tA}x^0 = e^{tA}\begin{bmatrix} 0 \\ 0 \end{bmatrix} = \begin{bmatrix} 0 \\ 0 \end{bmatrix}$$

$$e^{-sA}F(s) = e^s\begin{bmatrix} \cos 2s & 2\sin 2s \\ -(\sin 2s)/2 & \cos 2s \end{bmatrix}\begin{bmatrix} e^{-3s} \\ 1 \end{bmatrix} = \begin{bmatrix} e^{-2s}\cos 2s + 2e^s\sin 2s \\ -e^{-2s}(\sin 2s)/2 + e^s\cos 2s \end{bmatrix}$$

See Fig. 4(d) for the component graphs.

(e). The eigenvalues are 1 (double) and 2 (simple). V_1 is spanned by $[1 \ 0 \ 0]^T$. We obtain a generalized eigenvector w by solving $[A - I]w = [1 \ 0 \ 0]^T$; e.g., $w = [0 \ 1 \ 0]^T$. V_2 is spanned by $[2 \ 1 \ 1]^T$. We have three independent solutions,

$$\begin{bmatrix} 1 \\ 0 \\ 0 \end{bmatrix}e^t, \qquad \begin{bmatrix} t \\ 1 \\ 0 \end{bmatrix}e^t, \qquad \text{and} \qquad \begin{bmatrix} 2 \\ 1 \\ 1 \end{bmatrix}e^{2t}$$

and

$$X(t) = \begin{bmatrix} e^t & te^t & 2e^{2t} \\ 0 & e^t & e^{2t} \\ 0 & 0 & e^{2t} \end{bmatrix}, \qquad X(0) = \begin{bmatrix} 1 & 0 & 2 \\ 0 & 1 & 1 \\ 0 & 0 & 1 \end{bmatrix}$$

$$X^{-1}(0) = \begin{bmatrix} 1 & 0 & -2 \\ 0 & 1 & -1 \\ 0 & 0 & 1 \end{bmatrix}$$

$$e^{tA} = X(t)X^{-1}(0) = \begin{bmatrix} e^t & te^t & -(2+t)e^t + 2e^{2t} \\ 0 & e^t & -e^t + e^{2t} \\ 0 & 0 & e^{2t} \end{bmatrix}$$

We omit the calculation of $e^{tA} \begin{bmatrix} a \\ b \\ c \end{bmatrix}$ and $e^{-sA} \begin{bmatrix} f_1(s) \\ f_2(s) \\ f_3(s) \end{bmatrix}$

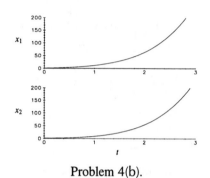

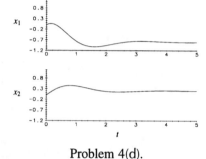

Problem 4(b).   Problem 4(d).

5.   First, find the eigenvalues and spanning sets for the eigenspaces of the system matrix; using
these, a basic solution matrix $X(t)$ can be constructed. A particular solution $x^d(t)$ is found
as explained in the hint. The general solution of $x' = Ax + F(t)$ is then $x = X(t)c + x^d(t)$,
where c is any constant column vector. Here, we calculate $X(t)$ and $x^d(t)$ only.

(a). The eigenvalues of $A = \begin{bmatrix} 2 & -1 \\ 5 & -2 \end{bmatrix}$ are $\lambda_1 = i = \bar{\lambda}_2$; the eigenspaces are $V_i =$Span$\{[1 \quad 2 -$
$i]^T\} = \bar{V}_{-i}$. So,

$$X(t) = \begin{bmatrix} e^{it} & e^{-it} \\ (2-i)e^{it} & (2+i)e^{-it} \end{bmatrix}$$

To find a particular solution $[x_1(t) \quad x_2(t)]^T$, let $x_1 = ae^t + b$, $x_2 = ce^t + d$. Then we must
have that $x_1' = ae^t = 2(ae^t + b) - (ce^t + d) + e^t$, $x_2' = ce^t = 5(ae^t + b) - 2(ce^t + d) + 1$.
Matching coefficients of corresponding terms, we have that $a = 2a - c + 1$, $c = 5a - 2c$,
and $0 = 2b - d$, $0 = 5b - 2d + 1$. Solving, we have $a = 3/2$, $c = 5/2$, $b = -1$, and
$d = -2$. So $x_1 = 3e^t/2 - 1$, $x_2 = 5e^t/2 - 2$ gives the components of a particular solution
$x^d(t)$. The general solution is $x = X(t)c + x^d(t)$, where c is any vector $[c_1 \quad c_2]^T$.

(b). $\lambda_1 = -2$, $\lambda_2 = 2$; $V_{-2} =$Span$\{[1 \quad -1]^T\}$, $V_2 = [3 \quad 1]^T$. So

$$X(t) = \begin{bmatrix} e^{-2t} & 3e^{2t} \\ -e^{-2t} & e^{2t} \end{bmatrix}$$

Assume a particular solution of the form given in the hint, insert into the ODEs and match coefficients to obtain the following two systems of equations in the unknown coefficients,

$$
\begin{aligned}
a_1 + 3a_2 + 1 &= b_1 & c_1 + 3c_2 &= d_1 \\
a_1 - a_2 &= b_2 & c_1 - c_2 &= d_2 \\
b_1 + 3b_2 &= -a_1 & d_1 + 3d_2 &= 0 \\
b_1 - b_2 &= -a_2 & d_1 - d_2 + 2 &= 0
\end{aligned}
$$

The solutions are $a_1 = a_2 = -1/5$, $b_1 = 1/5$, $b_2 = 0$, and $c_1 = 0$, $c_2 = -1/2$, $d_1 = -3/2$, $d_2 = 1/2$. So,

$$
x_1 = \frac{-\cos t + \sin t}{5} - \frac{3t}{2}, \quad x_2 = \frac{-\cos t}{5} - \frac{1}{2} + \frac{t}{2}
$$

are the components of a particular solution $x^d(t)$. The general solution is $x = X(t)c + x^d(t)$, where $c = [c_1 \; c_2]^T$ is an arbitrary vector.

6.   **(a).** With A and B as given, we find

$$
e^{tA} = \begin{bmatrix} 1 & t \\ 0 & 1 \end{bmatrix}, \quad e^{tB} = \begin{bmatrix} e^t & 0 \\ 0 & 1 \end{bmatrix}, \quad e^{tA}e^{tB} = \begin{bmatrix} e^t & t \\ 0 & 1 \end{bmatrix}, \quad e^{tB}e^{tA} = \begin{bmatrix} e^t & te^t \\ 0 & 1 \end{bmatrix}
$$

Since $A + B = \begin{bmatrix} 1 & 1 \\ 0 & 0 \end{bmatrix}$, and $\begin{bmatrix} 1 & 1 \\ 0 & 0 \end{bmatrix}^k = \begin{bmatrix} 1 & 1 \\ 0 & 0 \end{bmatrix}$ for all k, we have

$$
e^{t(A+B)} = I + \begin{bmatrix} 1 & 1 \\ 0 & 0 \end{bmatrix} t + \begin{bmatrix} 1 & 1 \\ 0 & 0 \end{bmatrix} t^2/2! + \cdots
$$

$$
= I + \begin{bmatrix} 1 & 1 \\ 0 & 0 \end{bmatrix}(e^t - 1) = \begin{bmatrix} e^t & e^t - 1 \\ 0 & 1 \end{bmatrix}
$$

Since none of the three matrices $e^{tA}e^{tB}$, $e^{tB}e^{tA}$, $e^{t(A+B)}$ match, we see that the usual multiplication rules of products of exponentials need not hold for matrix exponentials.

(b). Let's first show that if $AB = BA$, then $Ae^{tB} = e^{tB}A$. Using the series formula for e^{tB}:

$$
Ae^{tB} = A\sum t^n B^n/n! = \sum t^n AB^n/n! = \left(\sum t^n B^n/n!\right)A = e^{tB}A
$$

since $AB^n = ABB^{n-1} = BAB^{n-1} = B^2AB^{n-2} = \cdots = B^nA$.

Now, let $P(t) = e^{t(A+B)}e^{-tA}e^{-tB}$. Then since $A(A + B) = A^2 + AB = A^2 + BA = (A + B)A$ and since the derivative of a product of matrices follows the rule $(CK)' = C'K + CK'$ (but don't interchange the order of the factors because matrix multiplication need not commute), we have:

$$
P'(t) = (A + B)e^{t(A+B)}e^{-tA}e^{-tB} + e^{t(A+B)}\left(-Ae^{-tA}e^{-tB} + e^{-tA}(-B)e^{-tB}\right)
$$

$$
= (A + B)e^{t(A+B)}e^{-tA}e^{-tB} - (A + B)e^{t(A+B)}e^{-tA}e^{-tB}
$$

$$
= 0
$$

Since $P(0) = I$, $P(t) = I = e^{t(A+B)}e^{-tA}e^{-tB}$, so $e^{t(A+B)}e^{-tA}e^{-tB}e^{tB}e^{tA} = Ie^{tB}e^{tA}$, i.e.,

$e^{tB}e^{tA} = e^{t(A+B)}$. But $A + B = B + A$ and so $e^{t(A+B)} = e^{t(B+A)}$. The same kind of argument as that just given [but applied to $e^{t(B+A)}$] shows that $e^{t(B+A)} = e^{tA}e^{tB}$. So *if* A and B commute, then $e^{tB}e^{tA} = e^{tA}e^{tB}$ for all t; in particular, if A and B commute so do e^A and e^B. The dificulty in part **(a)** is that A and B do not commute.

7. **(a).** Since $L[x'] = sL[x] - x^0$, we may apply the Laplace transform operator L to every term of $x' = Ax + F$ and obtain $sL[x](s) - x^0 = AL[x](s) + L[F](s)$, where we have used the fact that $LA = AL$, so $[sI - A]L[x](s) = x^0 + L[F](s)$.

   (b). The matrix function $sI - A = -[A - sI]$ is invertible if $s > \max_{1 \le j \le n}|\lambda_j|$, where λ_j, $j = 1, 2, \ldots, n$, are the eigenvalues of A. This follows since the eigenvalues are precisely the values of s for which $A - sI$ is singular, so there is a largest $|\lambda_j|$ which is finite.

   (c). From **(a)** and **(b)**, we have for all s large enough,

   $$L[x](s) = [sI - A]^{-1}x^0 + [sI - A]^{-1}L[F](s)$$

   Applying L^{-1}, the inverse Laplace transform operator,

   $$x(t) = L^{-1}[(sI - A)^{-1}x^0] + L^{-1}[(sI - A)^{-1}L[F]](s)$$

   On the other hand, by Theorem 7.8.4, we see that $x = e^{tA}x^0 + \int_0^t e^{(t-s)A}F(s)\,ds = e^{tA}x^0 + e^{tA} * F(t)$. Upon comparison of the two formulas for $x(t)$, we see that $L^{-1}[(sI - A)^{-1}] = e^{tA}$ and $e^{tA} * F(t) = L^{-1}[(sI - A)^{-1}L[F]]$, so we have still another way to find e^{tA}.

   (d). The calculations are left to the reader, but they are straightforward. We have $L[x_1'] = sL[x_1] - 1 = L[x_2]$, while $L[x_2'] = sL[x_2] - 2 = -9L[x_1]$. We then have

   $$sL[x_1] - L[x_2] = 1$$

   $$9L[x_1] + sL[x_2] = 2$$

   a pair of linear equations that can be solved to find $L[x_1]$ and $L[x_2]$:

   $$L[x_1] = \frac{s+2}{s^2+9} = \frac{s}{s^2+9} + \frac{2}{s^2+9}, \quad L[x_2] = \frac{2s-9}{s^2+9} = 2\frac{s}{s^2+9} - \frac{9}{s^2+9}$$

   Using the transform tables at the end of Chapter 6, we see that $x_1(t) = \cos 3t + (2/3)\sin 3t$, $x_2(t) = 2\cos 3t - 3\sin 3t$.

8. Group project.

7.9 Steady States of Driven Linear Systems

Suggestions for preparing a lecture

   Topics: Constant and periodic steady states, algebraic and inequality tests that eigenvalues of system matrix have negative real parts.

Making up a problem set

Parts of 1 and 2, one of 4 and 5 and 6, parts of 7 and 8, parts of 9 and 10, 12 (if you do Gerschgorin disks).

Comments

This section has a distinctly applications-oriented slant. We focus on linear systems where all solutions tend to a constant or to a periodic steady-state and on conditions that assure us that this will happen. So we mostly require that the input vector F for the system $x' = Ax + F$ be either constant or periodic, and that all eigenvalues of the constant system matrix A have negative real parts. Under these conditions, we get a constant or a periodic steady-state that attracts all solutions as $t \to +\infty$. We go even further and state a bounded input-bounded output theorem (BIBO) that assures that if the system is driven by a bounded input $F(t)$, then the output can never be unbounded. This is a crude kind of stability result and is often termed *engineering stability*.

The second half of the section takes up the question of how we can tell whether the eigenvalues of a matrix have negative real parts without actually determining the eigenvalues. Our aim here is to find simple tests that you can carry out with pencil and paper. The first two tests are used to decide whether the roots of a monic polynomial (i.e., a polynomial with leading coefficient 1 such as $r^4 + 3r^3 + 2r^2 + r + 5$) have only negative real parts. The prime test here is the Routh Test, which is widely used by engineers. The general form of the Routh Test is given in the Background Material; the text version covers only quadratic, cubic, and quartic polynomials. For a proof, see Chapter XV, vol. 2, of *Matrix Theory*, F. R. Gantmacher, Chelsea Publ. Co., 1960. The polynomial we have in mind is the characteristic polynomial of a matrix (after the polynomial has been prepared by dividing by its leading coefficient). But who wants to find the characteristic polynomial of a big matrix? So the second set of tests works directly with the system matrix A. Aside from the trace test which is only a necessary (but not sufficient) condition that all eigenvalues have negative real parts, the Gerschgorin Disks Theorem is the main result here. This theorem is easy to use, and you can draw circles in the complex plane that tell you (more or less) just where the eigenvalues lie. We give the proof in the Background Material.

As we mention briefly in this section, there is an ambiguity in the definition of steady-state. We have taken the hard-line and defined steady-state only when the input vector is a constant or else is periodic. On the other hand, if all eigenvalues of A have negative real parts (as we require), and if $F(t)$ is any bounded column vector for $t \geq 0$, then all solutions of $x' = Ax + F(t)$ tend to one another as $t \to +\infty$. So if $x^1(t)$ and $x^2(t)$ are any two solutions, then $\|x^1(t) - x^2(t)\| \to 0$ as $t \to +\infty$, which suggests that all solutions are steady-state. This is just too inclusive for us, and that is why we have taken the hard-line. Some people define a steady-state as any bounded solution of $x' = Ax + F$, whether or not all other solutions approach it. This also is too inclusive, and certainly is contrary to common-sense applications, where steady-state means a regular and predictable state approached by all other solutions.

1. According to the rather narrow definition of steady-state that we use, and using Theorem 7.9.1, we only need to check that the eigenvalues of A have negative real parts. If that is the case, there is a steady-state; if not, then there is no steady-state. Observe that $x = 0$ is a constant solution for all systems considered.

   (a). The trace of A is $-10 + 4 + 7 = 1$, which is positive; according to Theorem 7.9.8, at least one eigenvalue has positive real part. So, $x = 0$ is a constant solution, but it is not a steady-state.

(b). The trace of A is $-10 + 4 - 40$, which is -46, but that does *not* mean that all eigenvalues have negative real parts. We shall find the characteristic polynomial $p(\lambda)$ of A and then apply the Coefficient Test or the Routh Test to $-p(\lambda)$. We have (after some calculation) that $-p(\lambda) = \lambda^3 + 46\lambda^2 + 205\lambda - 1706$. By the Coefficient Test (Theorem 7.9.6) at least one eigenvalue has nonnegative real part; $x = 0$ is not a steady-state.

(c). Since the matrix is triangular, its eigenvalues are the diagonal entries -10, -1, -1 (see Problem 2, Section 7.4). So, $x = 0$ is the unique steady-state by the Constant Steady State Theorem (Theorem 7.9.1).

2. **(a).** The characteristic polynomial of $\begin{bmatrix} \alpha & 2 \\ 3 & -4 \end{bmatrix}$ is $\lambda^2 + (4 - \alpha)\lambda - (4\alpha + 6)$. The roots have negative real parts if and only if $4 - \alpha > 0$ and $-(4\alpha + 6) > 0$ (Routh Test, Theorem 7.9.7). So, $\alpha < 4$ and $\alpha < -3/2$. We require that $\alpha < -3/2$, if we want all solutions to tend to 0 as $t \to \infty$.

(b). The system $x' = \begin{bmatrix} 0 & 1 & 0 \\ 0 & 0 & 1 \\ \alpha & -1 & -1 \end{bmatrix} x$ is equivalent to the scalar ODE, $x_1''' + x_1'' + x_1' - \alpha x_1 = 0$. The Routh Test (Theorem 7.9.7) for $r^3 + r^2 + r - \alpha$ is $-\alpha > 0$ (so $\alpha < 0$), $1 \cdot 1 > -\alpha$ (so $-1 < \alpha$). The roots have negative real parts if and only if $-1 < \alpha < 0$. If α satisfies this condition, all solutions of $x' = Ax$ decay to the zero vector as $t \to \infty$.

(c). The system $x' = \begin{bmatrix} 0 & 1 & 0 \\ 0 & 0 & 1 \\ -2 & -3 & \alpha \end{bmatrix} x$ is equivalent to the scalar ODE, $x_1''' - \alpha x_1'' + 3x_1' + 2x_1 = 0$. The Routh Test (Theorem 7.9.7) for $r^3 - \alpha r^2 + 3r + 2$ is $-\alpha > 0$, $-3\alpha > 2$, and so we require that $\alpha < 0$, $\alpha < -2/3$. All solutions of the original system decay to the zero vector as $t \to \infty$ if $\alpha < -2/3$.

3. According to the Routh Test (Theorem 7.9.7) for the characteristic polynomial $r^2 + ar + b$ of the undriven system $x' = y$, $y' = -bx - ay$, every root has negative real part if and only if $a > 0$ and $b > 0$. The eigenvalues of the system matrix for the equivalent system, $x' = y$, $y' = -bx - ay + f(t)$ have negative real parts. The input for the system is $[0 \quad f(t)]^T$, which is bounded since $\|[0 \quad f(t)]^T\| = |f(t)|$ and $f(t)$ is assumed to be bounded. The Bounded Input-Bounded Output Theorem (Theorem 7.9.5) applies, and all solutions of the system are bounded for $t \geq 0$.

4. The equivalent scalar ODE is $x'' + x = \cos t$, which has general solution $x = c_1 \cos t + c_2 \sin t + (t \sin t)/2$, and we see that $|x(t)|$ is unbounded as $t \to \infty$. The Bounded Input-Bounded Output Theorem (Theorem 7.9.5) does not apply in this case since the eigenvalues of the system matrix are $\pm i$ and so do not have negative real parts.

5. The general solution of the system $x' = -x + e^{3t}y$, $y' = -y$ is $y = c_2 e^{-t}$, $x = c_1 e^{-t} + (c_2/3)e^{2t}$, and if $c_2 \neq 0$, $|x(t)| \to \infty$ as $t \to +\infty$. The eigenvalues of the system matrix $\begin{bmatrix} -1 & e^{3t} \\ 0 & -1 \end{bmatrix}$ are -1, -1, but the Bounded Input-Bounded Output Theorem (Theorem 7.9.5) does not apply since the system matrix is *not* a matrix of constants.

6. **(a).** Suppose that A has an eigenvalue $\lambda = \alpha + i\beta$, $\alpha > 0$, and that v is a corresponding eigenvector. Then $x = e^{(\alpha+i\beta)t}v$ is a solution of $x' = Ax$, and $||x|| = e^{\alpha t}|e^{i\beta t}| \cdot ||v|| = e^{\alpha t}||v|| \to \infty$ as $t \to \infty$ since $\alpha > 0$. So $x' = Ax$ has an unbounded solution.

   (b). Let A have an eigenvalue $\lambda = i\beta$, β real, and suppose that v is a corresponding eigenvector. We want to show that it is possible to drive the system $x' = Ax$ so that the response is unbounded. Suppose that the driving force is $F(t) = e^{\lambda t}v$. Then $te^{\lambda t}v$ is an unbounded solution of $x' = Ax + e^{\lambda t}v$, for it is certainly unbounded as $t \to \infty$ since $|e^{\lambda t}| = 1$ if $\text{Re}(\lambda) = 0$. To show that $te^{\lambda t}v$ is a solution we insert it into the ODE: if $x = te^{\lambda t}v$, then $x' = e^{\lambda t}v + \lambda te^{\lambda t}Av = A(te^{\lambda t}v) + e^{\lambda t}v$, and we are done. Note that the input $e^{\lambda t}v$ is bounded since $||e^{\lambda t}v|| = |e^{\lambda t}|||v|| = |e^{i\beta t}|||v|| = ||v||$.

7. All of the polynomials [except for that in part **(c)**] have positive coefficients. So we only need to check the other Routh coefficient conditions (Theorem 7.9.7).

   (a). The Routh Test applied to $z^3 + z^2 + 2z + 1$ is that $1 \cdot 2 > 1$, which holds. So all roots have negative real parts.

   (b). The Routh Test applied to $z^4 + z^3 + 2z^2 + 2z + 3$ requires that $1 \cdot 2 \cdot 2 > 1 \cdot 3^2$ and $1 \cdot 2 > 2$. Since the tests fail, at least one root has nonnegative real part.

   (c). Since the coefficient of z^3 (and of z^2) is 0, the Routh Test implies that at least one of the roots has a nonnegative real part.

   (d). First reduce to the monic polynomial $z^3 + (11/6)z^2 + z + 1$ by dividing by 6. Then the Routh Test requires that $(11/6) \cdot 1 > 1$, and so all roots have negative real parts.

   (e). First reduce to the monic polynomial $z^3 + z^2 + (11/12)z + 1/2$ by dividing by 12. The Routh Test requires that $1 \cdot (11/12) > 1/2$, and so all roots have negative real parts.

8. Applying the Routh Test (Theorem 7.9.7) to $z^4 + 2z^3 + 3z^2 + z + a$, we see that all roots have negative real parts if and only if $a > 0$, $1 \cdot 2 \cdot 3 > 4 \cdot a + 1^2$ and $2 \cdot 3 > 1$. So, we must have a positive and also $5 > 4a$. The condition on a is that $0 < a < 5/4$.

9. In each figure the upper graphs show the x components of solutions, the lower graphs show the y components.

   (a). The system is $x' = y + 5$, $y' = -2x - 3y + 10$. The eigenvalues of the system matrix $\begin{bmatrix} 0 & 1 \\ -2 & -3 \end{bmatrix}$ are -1 and -2. By Theorem 7.9.1, the unique steady-state solution is

$$\begin{bmatrix} x \\ y \end{bmatrix} = -A^{-1}F_0 = -\begin{bmatrix} 0 & 1 \\ -2 & -3 \end{bmatrix}^{-1}\begin{bmatrix} 5 \\ 10 \end{bmatrix} = -\begin{bmatrix} -3/2 & -1/2 \\ 1 & 0 \end{bmatrix}\begin{bmatrix} 5 \\ 10 \end{bmatrix} = \begin{bmatrix} 25/2 \\ -5 \end{bmatrix}$$

   See Fig. 9(a) for graphs of the component curves of solutions as they approach the steady state.

   (b). The system is equivalent to the scalar equation $x'' + 2x' + x = \cos t$, which has the particular periodic solution $x = 0.5\sin t$. The steady-state solution of the system is the periodic forced solution, $x = 0.5\sin t$, $y = x' = 0.5\cos t$. Since the eigenvalues of the system matrix are -1 and -1, all solutions tend to the periodic forced oscillation as $t \to \infty$. See Fig. 9(b).

(c). The system is $x' = -x + y + \sin t$, $y' = x - y$. The eigenvalues of the system matrix are $-1 \pm i$. We can find the unique periodic steady-state solution (i.e., periodic forced oscillation) by Variation of Parameters or by Undetermined Coefficients. The latter process is easier here. Assume a particular solution of the form $x = A \cos t + B \sin t$, $y = C \cos t + D \sin t$. Inserting these expressions into the ODEs and matching coefficients of corresponding terms, we have that

$$-A = -B + D + 1, \quad B = -A + C, \quad -C = -B - D, \quad D = -A - C$$

Solving, we have $A = -1/5$, $B = 3/5$, $C = 2/5$, $D = -1/5$. All solutions tend to the steady-state

$$x = -(1/5) \cos t + (3/5) \sin t, \quad y = (2/5) \cos t - (1/5) \sin t$$

See Fig. 9(c).

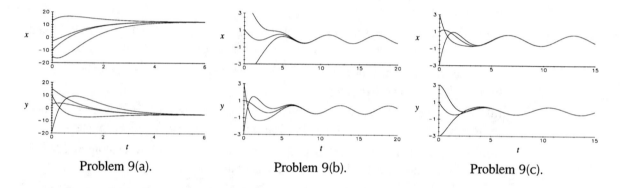

Problem 9(a).   Problem 9(b).   Problem 9(c).

10.   Each system has a periodic steady-state with the same period as the input $F(t)$ because the conditions of Theorem 7.9.3 are met. See below for graphs of the component curves of several solutions. The time interval is such that each of the components plotted essentially coincides with that component of the steady-state on the latter half of the interval. You can tell from the graphs that the steady-state has the same period as the input vector $F(t)$.

(a). The eigenvalues of the system matrix $\begin{bmatrix} -1 & 1/2 \\ -1 & -2 \end{bmatrix}$ are $-3/2 \pm i/2$, and $F(t)$ has period 2π.

(b). The eigenvalues of the system matrix $\begin{bmatrix} -2 & 5 \\ -5 & -2 \end{bmatrix}$ are $-2 \pm 5i$, and $F(t)$ has period 2.

(c). The eigenvalues The eigenvalues of the system matrix $\begin{bmatrix} 0 & 1 \\ -1 & -1 \end{bmatrix}$ are $-1/2 \pm \sqrt{3}i/2$, and $F(t)$ has period 2.

(d). The eigenvalues The eigenvalues of the system matrix $\begin{bmatrix} 0 & 1 \\ -4 & -1 \end{bmatrix}$ are $-1/2 \pm \sqrt{15}i/2$, and $F(t)$ has period 2π

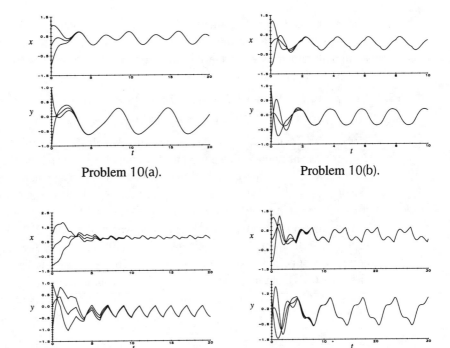

Problem 10(a).   Problem 10(b).

Problem 10(c).   Problem 10(d).

11. Suppose all eigenvalues of A have negative real parts and that $x^1(t)$ is a solution of $x' = Ax + F(t)$ which is unbounded as $t \to +\infty$. Let $x^2(t)$ be any other solution. Then $x = x^2 - x^1$ is a solution of the undriven system $x' = Ax$. Since the eigenvalues of A have negative real parts, $\|x(t)\| \to 0$ as $t \to +\infty$. So, $x^2(t) = x^1(t) + x(t)$ is the sum of a vector function which is unbounded and another which decays to 0 as $t \to +\infty$, and $x^2(t)$ is also unbounded as $t \to \infty$.

12. **(a).** The Gerschgorin row disks of $A = \begin{bmatrix} -10 & 1 & 8 \\ 1 & -3 & 2 \\ 8 & 1 & -4 \end{bmatrix}$ are given by

$$|z + 10| \le 9, \quad |z + 3| \le 3, \quad |z + 11| \le 9$$

while the column disks are defined by

$$|z + 10| \le 9, \quad |z + 3| \le 2, \quad |z + 11| \le 10$$

Since the column disks are entirely inside the left half of the complex plane, all eigenvalues have negative real parts, lying as they must inside the union of these disks.

(b). The column disks of $A = \begin{bmatrix} 10 & 1 & 10 \\ 1 & -5 & 8 \\ 1 & 1 & -20 \end{bmatrix}$ are given by

$$|z - 10| \le 2 \quad |z + 5| \le 2 \quad |z + 20| \le 18$$

Because the Gerschgorin column disk $|z - 10| \leq 2$ lies entirely to the right of the imaginary axis and does not touch the other two column disks, the Gershchorin Disks Theorem implies that there is an eigenvalue in that disk, and so there is an eigenvalue of A with positive real part.

13. **Group project.** We know $x' = Ax + F(t)$ where $F(t + T) = F(t)$ has a forced oscillation $x(t)$ of period T if $x(0) = x(T) = e^{TA} \left[x(0) + \int_0^T e^{-sA} F(s)ds \right]$, i.e., if $(I - e^{TA})x(0) = e^{TA} \int_0^T e^{-sA} F(s)ds$. Such an equation is uniquely solvable for $x(0)$ if and only if the matrix $I - e^{TA}$ is nonsingular. So we need to show that $I - e^{TA}$ is nonsingular if $2\pi i / T$ is not an eigenvalue of A. Suppose, to the contrary, that $I - e^{TA}$ is singular. Then for some nonzero vector v, $[I - e^{TA}]v = 0$, i.e., $e^{TA}v = v$. Then $y = e^{tA}v$ is a periodic solution of $y' = Ay$ of period T since the solution returns to its initial state v when $t = T$. But by Theorem 7.6.4 all components of all solutions of $y' = Ay$ are polynomial-exponentials, where the exponential factors have the form $e^{\lambda_j t}$, λ_j an eigenvalue of A. The only way any of these functions could be periodic of period T would be if some $\lambda_j = 2\pi i / T$ (leading to terms such as $\cos(2\pi t / T)$ or $\sin(2\pi t / T)$), but by assumption $2\pi i / T$ is *not* an eigenvalue of A. So the supposition that $I - e^{TA}$ is singular is wrong, and $I - e^{TA}$ is nonsingular. So there is a unique initial state $x(0) = [I - e^{TA}]^{-1} e^{TA} \int_0^T e^{-sA} F(s)ds$ which gives a periodic forced oscillation $x(t)$.

Background Material: Proofs of Root Tests

We have inserted below the proof of the Coefficient Test (Theorem 7.9.6) and the statement of the complete Routh Test (but not its proof) along with two examples. We also provide the proof of the Gerschgorin Disks Theorem (Theorem 7.9.9).

Verification of Theorem 7.9.6 (Coefficient Test).

We will prove the contrapositive, i.e., we will show that *if* all the roots of the polynomial with real coefficients $P(r) = r^n + a_{n-1}r^{n-1} + \cdots + a_0$ have negative real parts, *then* the coefficients of $P(r)$ must all be positive. Suppose the real roots (all negative) are denoted by $-s_1, \ldots, -s_j$ (so $s_1, \ldots, s_j$ are positive) with respective multiplicities $m_1, \ldots, m_j$, and the pairs of complex conjugate roots (all with negative real parts) are $-\alpha_1 \pm i\beta_1, \ldots, -\alpha_k \pm i\beta_k$, with respective multiplicities $n_1, \ldots, n_k$. Note that $(r + \alpha_p - i\beta_p)(r + \alpha_p + i\beta_p) = (r^2 + 2\alpha_p r + \alpha_p^2 + \beta_p^2)$. By the Fundamental Theorem of Algebra (Appendix B.4), we see that

$$P(r) = (r + s_1)^{m_1} \cdots (r + s_j)^{m_j} (r^2 + \alpha_1 r + \alpha_1^2 + \beta_1^2)^{n_1} \cdots (r^2 + \alpha_k r + \alpha_k^2 + \beta_k^2)^{n_k} \quad \text{(i)}$$

Since all the s_i's and α_i's are themselves positive numbers, we see that when the right side of (i) is multiplied out, the coefficients of $P(r)$ are all positive. This proves the theorem.

The Routh Test: As the example $r^3 + r^2 + r + 1 = (r + 1)(r^2 + 1) = (r + 1)(r + i)(r - i)$ shows, the positivity of all the coefficients of a polynomial does not necessarily guarantee that all the roots have negative real parts. However, the Routh Test will handle this case, and, in fact, it can be used with any polynomial whose leading coefficient is $+1$.

Routh Test

Suppose that the coefficients of $P(r) = r^n + a_{n-1}r^{n-1} + \cdots + a_0$ are real. Then all roots of $P(r)$ have negative real parts if and only if all entries in the first column of the *Routh Array* are positive:

$$\begin{bmatrix} 1 & a_{n-2} & a_{n-4} & \cdots & a_{n-2p+2} \\ a_{n-1} & a_{n-3} & a_{n-5} & \cdots & a_{n-2p+1} \\ b_{11} & b_{12} & b_{13} & \cdots & 0 \\ b_{21} & b_{22} & b_{23} & \cdots & 0 \\ \vdots & \vdots & \vdots & & \vdots \\ b_{n-1,1} & b_{n-1,2} & b_{n-1,3} & \cdots & 0 \end{bmatrix}$$

where $n = 2p - 2$ or $2p - 1$ [so, $p = (n+2)/2$, or $(n+1)/2$, whichever is an integer, a_0 is the first or second entry of the last column], $a_k = 0$ for $k < 0$, $a_n = 1$, and

$$b_{1j} = \frac{a_{n-1}a_{n-2j} - a_n a_{n-2j-1}}{a_{n-1}}, \quad b_{2j} = \frac{b_{11}a_{n-2j-1} - a_{n-1}b_{1,j+1}}{b_{11}}, \quad j < p$$

$$b_{ij} = \frac{b_{i-1,1}b_{i-2,j+1} - b_{i-2,1}b_{i-1,j+1}}{b_{i-1,1}}, \quad i > 2, \ j < p$$

For example,

$$b_{11} = \frac{a_{n-1}a_{n-2} - 1 \cdot a_{n-3}}{a_{n-1}}, \qquad b_{12} = \frac{a_{n-1}a_{n-4} - 1 \cdot a_{n-5}}{a_{n-1}}$$

$$b_{21} = \frac{b_{11}a_{n-3} - a_{n-1}b_{12}}{b_{11}}, \qquad b_{22} = \frac{b_{11}a_{n-5} - a_{n-1}b_{13}}{b_{11}}$$

$$b_{31} = \frac{b_{21}b_{12} - b_{11}b_{22}}{b_{21}}, \qquad b_{32} = \frac{b_{21}b_{13} - b_{11}b_{23}}{b_{21}}$$

The algorithm for constructing the Routh Array looks formidable, but for any particular polynomial it is easy to use. The rows are constructed from the top down and from left to right. If a zero or a negative number ever appears in the left-hand column of the Routh Array during its construction, there is no point in proceeding further since the criterion has already been violated, and the polynomial must have at least one root with nonnegative real part. The Routh Test need only be used for a polynomial with positive coefficients since the Coefficient Test handles the other cases.

The Routh Array for $r^3 + r^2 + r + 1$ is

$$\begin{bmatrix} 1 & 1 \\ 1 & 1 \\ 0 & - \\ - & - \end{bmatrix}$$

where we haven't bothered to calculate all the entries because the 0 in the first column tells us already that the polynomial has at least one root with non-negative real part. Here's a second example.

The Routh Array for the polynomial, $r^4 + 2r^3 + 3r^2 + 4r + 1/2$ is

$$\begin{bmatrix} 1 & 3 & 1/2 \\ 2 & 4 & 0 \\ 1 & 1/2 & 0 \\ 3 & 0 & 0 \\ 1/2 & 0 & 0 \end{bmatrix}$$

All roots of the given polynomial have negative real parts since all entries in the first column of the Routh Array are positive.

The Array for the polynomial $r^5 + 2r^4 + 3r^3 + 4r^2 + r/2 + 1/4$ is

$$\begin{bmatrix} 1 & 3 & 1/2 \\ 2 & 4 & 1/4 \\ 1 & 3/8 & 0 \\ 15/4 & 1/4 & 0 \\ 37/120 & 0 & 0 \\ 1/4 & 0 & 0 \end{bmatrix}$$

The roots of the polynomial all have negative real parts because the entries in the first column are positive.

Verification of Gerschgorin Theorem 7.9.9:

Suppose that λ is an eigenvalue of $A = [a_{ij}]$, $v = [v_1 \cdots v_n]^T$ a corresponding eigenvector. Since $\lambda v = Av$, we have for each i, $i = 1, \ldots, n$, that

$$\lambda v_i = \sum_{j=1}^{n} a_{ij} v_j, \quad \text{and so,} \quad (\lambda - a_{ii})v_i = \sum_{j=1, j \neq i}^{n} a_{ij} v_j \tag{ii}$$

Suppose that v_k is the component of v of maximal magnitude; i.e.,

$$|v_k| = \max_{1 \leq r \leq n} |v_r| \neq 0$$

Set $i = k$ in (ii), divide $(\lambda - a_{kk})v_k = \sum_{j=1, j \neq k}^{n} a_{kj} v_j$ by v_k, and take magnitudes:

$$|\lambda - a_{kk}| = \left| \sum_{\substack{j=1 \\ j \neq k}}^{n} a_{kj} \frac{v_j}{v_k} \right| \leq \sum_{\substack{j=1 \\ j \neq k}}^{n} |a_{kj}| \left| \frac{v_j}{v_k} \right| \leq \sum_{\substack{j=1 \\ j \neq k}}^{n} |a_{kj}| = r_k$$

where we have used the Triangle Inequality and the fact that $|v_j/v_k| \leq 1$. So, λ lies in the k-th Gerschgorin row disk. Because A and A^T have the same eigenvalues, λ also lies in at least one Gerschgorin column disk. The collection of all eigenvalues lies in the union of all the row disks and also in the union of all the column disks. Note that this does *not* imply that each disk has one or more eigenvalues in it. The proof of the final conclusion about disjoint unions of disks is omitted.

7.10 Lead Flow, Noise Filter: Steady States

Suggestions for preparing a lecture

Topics: Depending on whether Section 7.1 or Section 4.4 was covered earlier, you can cover either one of the two models (lead in the body, or low-pass filter).

Remarks: The filter model illustrates the Routh Test (Theorem 7.9.7), and the lead model uses the Gerschgorin disks (Theorem 7.9.9).

Making up a problem set

Problems 2, 4, 5 or 6.

Comments

In this section we return to the lead-level model of Section 7.1. We also introduce a two-loop electrical circuit that is an effective noise filter in a communications channel It is also a low-pass filter, because what it actually does is filter out high-freqencies, and most noise is high-frequency. Now that we have a theory of linear systems we can use it to describe theoretically (as well as by computer simulations) just how the solutions of these systems behave as parameters in the systems are changed. In particular we can use the Routh Test (Theorem 7.9.7) and the Gerschgorin Disks (Theorem 7.9.9) to conclude that the eigenvalues of the system matrix of, respectively, the circuit system and the lead-level system all have negative real parts. This puts these models squarely in the domain of systems with steady-states. And we exploit that by showing how the steady-state periodic output voltage (in the circuit system) changes in amplitude, depending on the frequency of the input voltage. We show how the lead levels in the lead system change with heavier doses of an anti-lead medication.

1. According to text equation (11), the input voltage V_1 of amplitude $|a_0|$ is changed to an output voltage V_2 of amplitude $|a_0[p(i\omega)LRC_1C_2]^{-1}|$. Since

$$-p(i\omega) = (i\omega)^3 + \frac{(i\omega)^2}{RC_1} + \frac{1}{L}\left(\frac{1}{C_1} + \frac{1}{C_2}\right)(i\omega) + \frac{1}{LRC_1C_2}$$

we see that for ω close to 0, $|p(i\omega)| \approx 1/(LRC_1C_2)$, that is, $|LRC_1C_2p(i\omega)| \approx 1$ when ω is close to zero. The low-frequency inputs pass through the circuit with little attenuation. On the other hand if $|\omega|$ is large, then the dominant term in $|p(i\omega)LRC_1C_2|^{-1}$ is $|\omega^3LRC_1C_2|^{-1}$. For ω large enough this implies that the amplitude of the response is only a tiny fraction of the amplitude of the input voltage, the circuit effectively acting to filter out high frequencies.

2. Use system (5), (6) in the text with the circuit parameters as given:

   (a). Let $L = 1$, $R = 1$, $C_1 = C_2 = 1$. In Fig. 2(a), the circuit clearly passes $\omega = 1$ (solid), acts as a partial filter for $\omega = 2$ (long dashes), and stops $\omega = 50$ (short dashes).

   (b). Let $L = 1$, $R = 0.1$, $C_1 = C_2 = 1$. In Fig. 2(b), the circuit passes (and amplifies) $\omega = 1$ (solid), passes $\omega = 1.5$ (amplitude about 0.9), and stops $\omega = 10$ (long and short dashes).

(c). Let $L = 1$, $R = 100$, $C_1 = C_2 = 0.01$. In Fig. 2(c), the circuit passes $\omega = 0.1$ (solid), acts as a partial filter for $\omega = 1.5$ (long dashes), and stops $\omega = 10$ (solid).

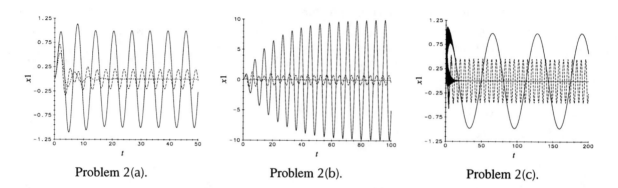

Problem 2(a).   Problem 2(b).   Problem 2(c).

3. We see from text equation (12) that the magnitude of $|V_2/V_1|$ is $= [(1 - LC_2\omega^2)^2 + R^2\omega^2(C_1 + C_2 - \omega^2 LC_1C_2)^2]^{-1/2}$, a quantity that we shall denote by $G(\omega)$.

(a). If $L = 1$, $R = 1$, $C_1 = C_2 = 1$, then

$$G(\omega) = [1 + 2\omega^2 - 3\omega^4 + \omega^6]^{-1/2}$$

So, $G(1) = 1$, $G(2) = 1/5$, $G(50) \approx 0$. The frequency 1, 2, and 50 terms are, respectively, passed with no change in amplitude, partially filtered with the amplitude reduced by 80%, and (nearly) stopped.

(b). If $L = 1$, $R = 0.1$, $C_1 = C_2 = 1$, then

$$G(\omega) = [1 - 49\omega^2/25 + 24\omega^4/25 + \omega^6/100]^{-1/2}$$

So, $G(1) = 10$, $G(2) \approx 0.33$, $G(50) \approx 0.0004$, and the frequency 1.0, 1.5, and 10 terms are, respectively, passed and amplified tenfold, passed with approximately a two-thirds drop in amplitude, and (nearly) stopped.

(c). If $L = 1$, $R = 100$, $C_1 = C_2 = 0.01$, then

$$G(\omega) = [1 + \omega^2/50 - 3\omega^4/10^4 + \omega^6/10^6]^{-1/2}$$

So, $G(1) \approx 0.99$, $G(2) \approx 0.96$, $G(50) \approx 0$, and the frequency 1, 2, and 50 terms are, respectively, passed with little change in amplitude, passed with a slight amplitude reduction, and (nearly) stopped.

4. Use system (5), (6) in the text with $L = 1$, $R = 1$, $C_1 = C_2 = 1000$ and $V_1 = 2\,\mathrm{trw}(t, 100, T) - 1$.

(a). Over the first half of a period, the triangular wave for $V_1(t)$ rises from -1 to $+1$; thus, the "rise" is 2 while the "run" is $T/2$, leading to a slope of $2/(T/2) = 4/p$. Over the second half of period, the slope is $-4/T$. The derivative of $2\mathrm{trw}(t, 100, T) - 1$ is a wave train of pulses of period T, which are at the level $4/T$ over the first half of the period, and $-4/T$ over the second half. This is just the function $(8/T)\,\mathrm{sqw}(t, 50, T) - 4/T = V_1'(t)$.

(b). From Graph 1, we see that the triangular wave $V_1(t)$ of period 2π(solid) is eventually tracked very closely by the output voltage $V_2(t)$ (dashed); the circuit passes the frequency 1 triangular wavetrain. On the other hand from Graph 2, we see that the triangular wave $V_2(t)$ of period $\pi/100$ (solid) is effectively stopped by the circuit. The output voltage $V_2(t)$ (dashed) has very low amplitude compared to the input.

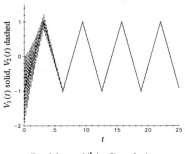
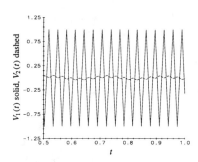

Problem 4(b), Graph 1.   Problem 4(b), Graph 2.

5. The system is text system (1).

(a). If the intake rate is lowered from I_1 to αI_1, where α is a constant between 0 and 1, then the constant steady-state (Theorem 7.9.1) is

$$\begin{bmatrix} y_1 \\ y_2 \\ y_3 \end{bmatrix} = -A^{-1}\begin{bmatrix} \alpha I_1 \\ 0 \\ 0 \end{bmatrix} = -\alpha A^{-1}\begin{bmatrix} I_1 \\ 0 \\ 0 \end{bmatrix}$$

The equilibrium levels in the compartments are lowered from the levels of $-A^{-1}[I_1 \ \ 0 \ \ 0]^T$ by the same factor α.

(b). See Graphs 1–3 for the respective graphs of the lead levels in the blood, tissues, and bones. The solid lines show the lead levels if there is no reduction in the intake, the long-dashed curves represent lead levels if the intake rate is cut in half (at day 400), and the short-dashed curves if the intake rate is cut in half again at day 800. In the first two graphs, each cut in the lead intake rate produces a corresponding cut in the lead levels in the blood and tissues as predicted in 5**(a)**. This is because the lead levels quickly reach the steady-state in the blood and the tissues. However, although the lead levels in the bones (Graph 3) drop as the intake rate is cut, we cannot tell that the new levels drop by the same ratio as the intake rate; this is because the lead levels in the bones are still far from equilibrium.

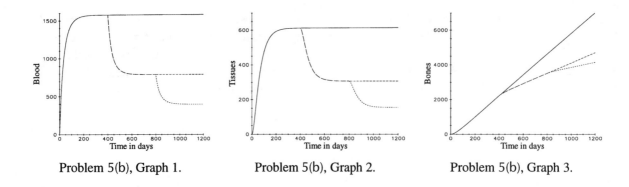

Problem 5(b), Graph 1.   Problem 5(b), Graph 2.   Problem 5(b), Graph 3.

(c). We want to find the smallest value of a, $0 < a < 0.1$ so that if the intake rate I, of lead into the blood is given by $49.3 \exp[a(400 - t)]$ for $t > 400$, then the lead level in the bones never exceeds 3000 micrograms. See Fig. 5(c). The solid line corresponds to $a = 0$, while the dashed line corresponds to $a = 0.0102$, which seems to be large enough to keep the equilibrium level less than 3000 milligrams. The value of a is determined experimentally by using a numerical solver.

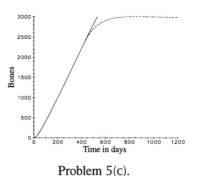

Problem 5(c).

6.   The function $[1 + 9 \sin^2(2\pi t/365)]/10$ has period 365 days, reaches its maximum value of 1 when $2\pi t/365 = \pi/2$ or $3\pi/2$, i.e., when $t = 365/4$ or $(3/4)365$; it drops to the minimum value of $1/10$ when $t = 0$ or $365/2$. This models seasonal variations in the lead intake rate, the rate ranging from 49.3 in the early spring and fall to 4.93 in the winter and summer. Graphically (see Fig. 6), the maximum level of lead in the bones appears to be between 11,000 and 12,000 micrograms, and that level is reached after about 20,000 days (or a little less than 55 years). See Fig. 6. Notice how little the seasonal variations in the lead levels in the input affect the amount of lead in the bones.

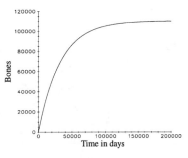

Problem 6.

Background Material: Compartment Models

The lead system is an example of a compartment model in which every compartment is "open to the environment." That is, the environment outside the three compartments of blood, tissues, and bones can be reached by a directed sequence of arrows from each compartment. Let's give a definition of a general compartmental matrix, and then specialize that to an open compartmental matrix, a particular example of which is the lead model coefficient matrix.

The square matrix $A = [a_{ij}]$ of real numbers is a *compartmental matrix* if:

- Every diagonal entry in A is nonpositive.

- Every nondiagonal entry in A is nonnegative.

- The sum of the i-th column entries, $\sum_{j=1}^{n} a_{ji}$, is nonpositive.

Here is why our experience with compartment models in Sections 1.8, 7.1, and 7.10 suggests that these three conditions are reasonable. Thinking of the matrix we would construct for an assemblage of boxes and arrows, we see that the first condition corresponds to the fact that the total rate out of each compartment and into other compartments and the environment must be nonpositive just because it is an *exit* rate. The second condition says that a_{ji}, $j \neq i$, which measures the entrance rate into compartment j from compartment i can't be negative (after all, it is an *entrance* rate). The third condition is less evident. But, recall from the lead model in Section 7.1 that $a_{ii}x_{ii}$ is the negative of the *exit* rate from compartment i, i.e., measures the rate of everything leaving compartment i. So, $a_{ii} = -\sum_{j=1, j \neq i}^{n} a_{ji} - a_{0i}$, where a_{0i} is the exit rate constant from compartment i to the environment. Since a_{0i}, a_{ji} ($j \neq i$) are all nonnegative, a_{ii} is nonpositive.

We have the following theorem about the eigenvalues of any compartmental matrix. As might be expected from the use of Gerschgorin disks to prove that all eigenvalues of the lead model system matrix (first part of Section 7.10), we validate this theorem by using Gerschgorin disks.

THEOREM

> Compartmental Matrix Theorem. All eigenvalues of a compartmental matrix have nonpositive real parts, and none is pure imaginary.

Here is how we can justify this very important result. As noted above, every diagonal entry a_{ii} is nonpositive and $a_{ii} = -\sum_{j=1,\neq i}^{n} a_{ji} - a_{0i}$. But the a_{ji} terms, $j \neq i$, are nonnegative, and so $\sum_{j=1,\neq i}^{n} a_{ji}$ is just the radius of the i-th Gerschgorin column disk whose center is at $a_{ii} \leq 0$ on the real line in the complex plane. This means that the i-th column disk lies entirely inside the left half of the complex plane, except possibly for the origin, which lies on the edge of the disk if $a_{0i} = 0$. This proves the theorem.

It is also known that if A is a compartmental matrix and $x_i(0) \geq 0$, then $x_i(t) \geq 0$ for all $t \geq 0$, $i = 1, \ldots, n$, if $x(t) = [x_1(t) \ldots x_n(t)]^T$ is a solution of $x' = Ax$.

Now let's define a compartmental system (and matrix) where every compartment is "open to the environment." If for every i there is a directed chain of arrows with positive exit rate constants which leads from compartment i to the external environment, then the system is said to be *open to the environment*. The corresponding compartmental matrix A is *open to the environment* (often called an *open compartmental matrix*) if there is a chain of compartments $j_1, \ldots, j_r$ for which $a_{j_1 i}, a_{j_2 j_1}, \ldots,$ and $a_{0 j_r}$ are positive. It isn't hard to verify that the lead matrix is open. We have the following theorem (proof omitted) for any open compartmental matrix:

THEOREM

> Washout Theorem.
>
> If A is an open compartmental matrix, then all its eigenvalues have negative real parts, so all solutions of $x' = Ax$ tend to 0 as $t \to +\infty$.

We would expect the Washout Theorem to be true because the substance being tracked through the system leads out into the environment from each compartment either directly, or indirectly through a chain of other compartments.

All of this can be extended to compartmental systems with "traps," which are compartments or groups of compartments which "absorb," but never disgorge. In a way, the outside environment is a trap, and we could add it to the set of boxes of the model as an additional compartment (the "zero-th" compartment). The enlarged model is no longer open; in fact, it is said to be a *closed compartmental model*. The corresponding matrix A is singular (i.e., 0 is an eigenvalue), and it can be shown that each solution $x(t)$ of $x' = Ax$ tends to some equilibrium point b, where $b_i \geq 0$ if $x_i(0) \geq 0$, $i = 1, \ldots, n$.

Well, that's all we have to say here about compartment models, but here are some sources for further study:

1. *Linear Models in Biology*, M.R. Cullen [Ellis Horwood Limited, Chichester, 1985].

2. *Compartmental Modeling and Tracer Kinetics*, David H. Anderson [v. 50 of *Lecture Notes in Biomathematics*, Springer-Verlag, Berlin].

3. *Compartmental Analysis in Biology and Medicine*, John A. Jacquez [Univ. of Michigan Press, Ann Arbor, 1985 (2nd ed.)].

7.11 General Linear Systems

Suggestions for preparing a lecture

Topics: The Fundamental Theorem for Linear Systems 7.11.1, Wronskians, basic solution sets, basic solution matrices, variation of parameters approach for solving driven linear systems.

Making up a problem set

Problems 2 or 3, one or more parts of 4, 5.

Comments

We avoided general linear differential systems until this final section because almost all the models we consider in this text lead to linear systems with constant coefficients. The reason for this is that there is a lot of theory for linear systems with nonconstant coefficients, but not many practical methods for finding solution formulas. Similarly, we didn't do much in Chapter 3 either with second-order linear ODEs with nonconstant coefficients (except at the end of the chapter). The one really useful solution formula technique for general second-order linear ODEs is the power series approach described in Chapter 11. If you are going to do that chapter we recommend that you first read Section 3.7.

1. The Fundamental Theorem 7.11.1 says that the IVP $x' = A(t)x$, $x(t_0) = e^j$, has a unique solution $x^j(t)$, for each $j = 1, 2, \ldots, n$. Note that the matrix $[x^1(t) \ \ldots \ x^n(t)]$ evaluated at $t = t_0$ is just the identity matrix I_n. The Wronskian

$$W(t) = \det[x^1(t) \ \ldots \ x^n(t)]$$

has the value at $t = t_0$, $W(t_0) = 1 \neq 0$ and so $\{x^1(t), \ldots, x^n(t)\}$ is a basic solution set.

2. The system is $x' = \begin{bmatrix} 0 & 1 \\ -2t^{-2} & 2t^{-1} \end{bmatrix} x$, $t > 0$.

(a). Let's denote the columns of $M(t)$ by $x^1(t) = [t^2 \ 2t]^T$, and $x^2(t) = [t \ 1]^T$ and the system matrix by A. By direct calculation, $x^1(t)$ and $x^2(t)$ are both solutions of $x' = Ax$ on the t-interval $0 < t < +\infty$. The Wronskian $W[x^1, x^2] = \det M(t)$, and since $\det M(1) = -1 \neq 0$ it follows that $\{x^1(t), x^2(t)\}$ is a basic solution set for $x' = Ax$, and so $M(t) = [x^1(t) \ x^2(t)]$ is a basic solution matrix.

(b). Using the 2×2 matrix inversion technique (7) in Section 7.3, we see that

$$M^{-1} = \frac{-1}{t^2} \begin{bmatrix} 1 & -t \\ -2t & t^2 \end{bmatrix}$$

Using (4) in this section we have that

$$\Phi(t, t_0) = M(t)M^{-1}(t_0) = \frac{-1}{t_0^2} \begin{bmatrix} t^2 - 2tt_0 & -t^2 t_0 + tt_0^2 \\ 2t - 2t_0 & -2tt_0 + t_0^2 \end{bmatrix}$$

Evidently $\Phi(t_0, t_0) = I_2$. Now using Problem 7 in Section 7.3,

$$\Phi^{-1}(t, t_0) = M(t_0)M^{-1}(t)$$

and it follows that $\Phi^{-1}(t, t_0) = \Phi(t_0, t)$, for t, $t_0 > 0$.

(c). Let $y(t)$ satisfy the Euler ODE $t^2 y'' - 2ty' + 2y = 0$ and put $x_1 = y$, $x_2 = y'$. Then the pair (x_1, x_2) solves the system

$$x_1' = x_2$$

$$x_2' = y'' = \frac{1}{t^2}(2tx_2 - 2x_1) = -(2/t^2)x_1 + (2/t)x_2$$

which is precisely the given system. On the other hand let the pair (x_1, x_2) solve the given system. Then $x_2 = x_1'$ and $x_2' = x_1'' = -(2/t^2)x_1 + (2/t)x_1'$, or

$$t^2 x_1'' - 2tx_1' + 2x_1 = 0$$

and we see that $x_1(t)$ solves the given Euler ODE.

3. The system is $x' = \begin{bmatrix} t^{-1} & 0 & -1 \\ 0 & 2t^{-1} & -t^{-1} \\ 0 & 0 & t^{-1} \end{bmatrix} x$, $t > 0$.

(a). Direct substitution shows that the columns of $M(t)$ are solutions of the given system. Since $\det M(t)$ is the Wronskian of these three solutions and since $\det M(1) = 2$, it follows that $M(t)$ is a basic solution matrix.

(b). Using equation (4) we have that $\Phi(t, t_0) = M(t)M^{-1}(t_0)$. Using the method of Example 7.3.4, let's find $M^{-1}(t)$:

$$\begin{bmatrix} t & t & -t^2 \\ 0 & t^2 & t \\ 0 & 0 & t \end{bmatrix} \begin{array}{|ccc} 1 & 0 & 0 \\ 0 & 1 & 0 \\ 0 & 0 & 1 \end{array} \rightarrow \begin{bmatrix} t & t & 0 \\ 0 & t^2 & 0 \\ 0 & 0 & t \end{bmatrix} \begin{array}{|ccc} 1 & 0 & t \\ 0 & 1 & -1 \\ 0 & 0 & 1 \end{array}$$

$$\rightarrow \begin{bmatrix} 1 & 1 & 0 \\ 0 & 1 & 1 \\ 0 & 0 & 1 \end{bmatrix} \begin{array}{|ccc} t^{-1} & 0 & 1 \\ 0 & t^{-2} & -t^{-2} \\ 0 & 0 & t^{-1} \end{array}$$

$$\rightarrow \begin{bmatrix} 1 & 0 & 0 \\ 0 & 1 & 0 \\ 0 & 0 & 1 \end{bmatrix} \begin{array}{|ccc} t^{-1} & -t^{-2} & t^{-2} + t^{-1} \\ 0 & t^{-2} & -t^{-2} - t^{-1} \\ 0 & 0 & t^{-1} \end{array}$$

So

$$M^{-1}(t) = \begin{bmatrix} t^{-1} & -t^{-2} & t^{-2} + t^{-1} \\ 0 & t^{-2} & -t^{-2} - t^{-1} \\ 0 & 0 & t^{-1} \end{bmatrix}$$

and $\Phi(t, t_0) = M(t)M^{-1}(t_0)$, t, $t_0 > 0$.

4.   **(a).** Denoting the system matrix by A, we have the IVP

$$x' = Ax + \begin{bmatrix} \cos \omega t \\ \sin \omega t \end{bmatrix}, \quad x(0) = \begin{bmatrix} a \\ b \end{bmatrix}$$

where $\omega \neq 0$, and

$$A = \begin{bmatrix} 0 & 2 \\ -2 & 0 \end{bmatrix}$$

First let's find e^{tA}. The eigenvalues of A are $\lambda = \pm 2i$, and the eigenspace corresponding to $\lambda = 2i$ has the basis $[1 \;\; i]^T$. Now put

$$x^1(t) = \mathbb{R}\left[\begin{bmatrix} 1 \\ i \end{bmatrix} e^{2it}\right] = \begin{bmatrix} 1 \\ 0 \end{bmatrix} \cos 2t - \begin{bmatrix} 0 \\ 1 \end{bmatrix} \sin 2t = \begin{bmatrix} \cos 2t \\ -\sin 2t \end{bmatrix}$$

$$x^2(t) = \text{Im}\left[\begin{bmatrix} 1 \\ i \end{bmatrix} e^{2it}\right] = \begin{bmatrix} 1 \\ 0 \end{bmatrix} \sin 2t + \begin{bmatrix} 0 \\ 1 \end{bmatrix} \cos 2t = \begin{bmatrix} \sin 2t \\ \cos 2t \end{bmatrix}$$

Notice that $M(t) = [x^1(t) \;\; x^2(t)]$ is a basic solution matrix, and since $M(0) = I_2$ we see that $M(t) = e^{tA}$. Using solution formula (10) with $\Phi(t, 0) = e^{tA}$ we have the unique solution of the IVP,

$$x(t) = e^{tA} \begin{bmatrix} a \\ b \end{bmatrix} + e^{tA} \int_0^t e^{-sA} \begin{bmatrix} \cos \omega s \\ \sin \omega s \end{bmatrix} ds$$

$$= e^{tA} \begin{bmatrix} a \\ b \end{bmatrix} + \int_0^t e^{(t-s)A} \begin{bmatrix} \cos \omega s \\ \sin \omega s \end{bmatrix} ds = e^{tA} \begin{bmatrix} a \\ b \end{bmatrix} + x^d(t)$$

where we have used the fact that $e^{tA}e^{-sA} = e^{(t-s)A}$ (see Theorem 7.8.2). Note that

$$e^{(t-s)A} \begin{bmatrix} \cos \omega s \\ \sin \omega s \end{bmatrix} = \begin{bmatrix} \cos 2(t-s) & \sin 2(t-s) \\ -\sin 2(t-s) & \cos 2(t-s) \end{bmatrix} \begin{bmatrix} \cos \omega s \\ \sin \omega s \end{bmatrix}$$

$$= \begin{bmatrix} \cos(2t - (\omega + 2)s) \\ -\sin(2t - (\omega + 2)s) \end{bmatrix}$$

where we have used some trigonometric identities in Appendix B.4. So for $\omega \neq -2$ we have that

$$x^d(t) = \int_0^t e^{(t-s)A} \begin{bmatrix} \cos \omega s \\ \sin \omega s \end{bmatrix} ds = \int_0^t \begin{bmatrix} \cos(2t - (\omega + 2)s) \\ -\sin(2t - (\omega + 2)s) \end{bmatrix} ds$$

$$= \frac{1}{\omega + 2} \begin{bmatrix} \sin \omega t + \sin 2t \\ \cos 2t - \cos \omega t \end{bmatrix}$$

Notice that if $\omega \neq -2$, then the driven solution $x^d(t)$ above is the sum

$$x^d(t) = \frac{1}{\omega + 2} \left\{ \begin{bmatrix} \sin \omega t \\ -\cos \omega t \end{bmatrix} + \begin{bmatrix} \sin 2t \\ \cos 2t \end{bmatrix} \right\}$$

Now since $[\sin 2t \quad \cos 2t]^T$ solves the undriven system $x' = Ax$, it follows that

$$\frac{1}{\omega + 2} \begin{bmatrix} \sin \omega t \\ -\cos \omega t \end{bmatrix}$$

solves the given driven system. This shows that the driven system has a unique periodic solution with the same period as the driving term, $[\cos \omega t \quad \sin \omega t]^T$, if $\omega \neq -2$ and that all solutions of the system are bounded for all t.

When $\omega = -2$ the above solution formula fails. In that case we have that

$$x^d(t) = \int_0^t e^{(t-s)A} \begin{bmatrix} \cos 2s \\ -\sin 2s \end{bmatrix} ds = \int_0^t \begin{bmatrix} \cos 2t \\ \sin 2t \end{bmatrix} ds = \begin{bmatrix} t\cos 2t \\ -t\sin 2t \end{bmatrix}$$

which is not bounded for all t. Why does this resonance phenomenon occur with $\omega = -2$ but not with $\omega = 2$? The short answer is that the driving term when $\omega = 2$ is *not* a solution of the undriven system $x' = Ax$, but when $\omega = -2$, the driving term *is* a solution of the undriven system.

(b). The IVP is $x' = \begin{bmatrix} 0 & 1 \\ -2t^{-2} & 2t^{-1} \end{bmatrix} x + \begin{bmatrix} t^3 \\ t^4 \end{bmatrix}$, $x(1) = \begin{bmatrix} 0 \\ 0 \end{bmatrix}$. Using the given solutions $[t^2 \quad 2t]^T$ and $[t \quad 1]^T$ of the undriven system, form the solution matrix

$$M(t) = \begin{bmatrix} t^2 & t \\ 2t & 1 \end{bmatrix}$$

Now $\det M(t)$ is the Wronskian of the undriven system and since $\det M(1) = -1 \neq 0$, it follows that $M(t)$ is a basic solution matrix. Using (4), we have that for t, $t_0 > 0$:

$$\Phi(t, t_0) = M(t)M^{-1}(t_0)$$

$$= \begin{bmatrix} t^2 & t \\ 2t & 1 \end{bmatrix} \begin{bmatrix} -t_0^{-2} & t_0^{-1} \\ 2t_0^{-1} & -1 \end{bmatrix}$$

$$= \begin{bmatrix} -t^2 t_0^{-2} + 2t t_0^{-1} & t^2 t_0^{-1} - t \\ -2t t_0^{-2} + 2t_0^{-1} & 2t t_0^{-1} - 1 \end{bmatrix}$$

Using solution formula (10) with $t_0 = 1$, we have the solution of the IVP:

$$x(t) = \begin{bmatrix} -t^2 + 2t & t^2 - t \\ -2t + 2 & 2t - 1 \end{bmatrix} \begin{bmatrix} 0 \\ 0 \end{bmatrix} + \int_1^t \begin{bmatrix} -t^2 s^{-2} + 2s^{-1}t & t^2 s^{-1} - t \\ -2ts^{-2} + 2s^{-1} & 2ts^{-1} - 1 \end{bmatrix} \begin{bmatrix} s^3 \\ s^4 \end{bmatrix} ds$$

$$= \int_1^t \begin{bmatrix} -t^2 s + 2s^2 t + t^2 s^3 - ts^4 \\ -2ts + 2s^2 + 2ts^3 - s^4 \end{bmatrix} ds$$

$$= \begin{bmatrix} \left(-\frac{t^2 s^2}{2} + \frac{2}{3}s^3 t + \frac{t^2 s^4}{4} - \frac{ts^5}{5} \right) \Big|_1^t \\ \left(-ts^2 + \frac{2s^3}{3} + \frac{ts^4}{2} - \frac{s^5}{5} \right) \Big|_1^t \end{bmatrix} = \begin{bmatrix} \frac{t^6}{20} + \frac{t^4}{6} + \frac{t^2}{4} - \frac{7t}{15} \\ -\frac{t^3}{3} + \frac{3t^5}{10} + \frac{t}{2} - frac715 \end{bmatrix}$$

(c). The IVP is $x' = \begin{bmatrix} t^{-1} & 0 & 0 \\ -2t^{-2} & t^{-1} & 0 \\ t^{-3} & -t^{-2} & t^{-1} \end{bmatrix} x + \begin{bmatrix} t^3 \\ t^2 \\ t^3 \end{bmatrix}$, $x(1) = \begin{bmatrix} 1 \\ 0 \\ 1 \end{bmatrix}$. Using the given solutions $[t \quad 1 \quad 0]^T$, $[0 \quad t \quad 1]^T$, $[0 \quad 0 \quad t]^T$ of the undriven system, form the solution matrix

$$M(t) = \begin{bmatrix} t & 0 & 0 \\ 1 & t & 0 \\ 0 & 1 & t \end{bmatrix}$$

Since $\det M(1) = 1 \neq 0$ it follows that $M(t)$ is a basic solution matrix. So the transition matrix is

$$\Phi(t, t_0) = M(t)M^{-1}(t_0)$$

$$= \begin{bmatrix} t & 0 & 0 \\ 1 & t & 0 \\ 0 & 1 & t \end{bmatrix} \frac{1}{t_0^3} \begin{bmatrix} t_0^2 & 0 & 0 \\ -t_0 & t_0^2 & 0 \\ 1 & -t_0 & t_0^2 \end{bmatrix}$$

$$= \frac{1}{t_0^3} \begin{bmatrix} tt_0^2 & 0 & 0 \\ t_0^2 - tt_0 & tt_0^2 & 0 \\ -t_0 + t & t_0^2 - t_0 t & tt_0^2 \end{bmatrix}$$

So using solution formula (10) with $t_0 = 1$ we have the following solution of the IVP for $t > 0$:

$$x(t) = \begin{bmatrix} t & 0 & 0 \\ 1-t & t & 0 \\ -1+t & 1-t & t \end{bmatrix} \begin{bmatrix} 1 \\ 0 \\ 1 \end{bmatrix} + \int_1^t \begin{bmatrix} ts^2 & 0 & 0 \\ s^2 - st & s^2 t & 0 \\ -s+t & s^2 - st & s^2 t \end{bmatrix} \begin{bmatrix} 1 \\ 1/s \\ 1 \end{bmatrix} ds$$

$$= \begin{bmatrix} t \\ 1-t \\ -1+2t \end{bmatrix} + \int_1^t \begin{bmatrix} ts^2 \\ s^2 \\ s^2 t \end{bmatrix} ds$$

$$= \begin{bmatrix} t \\ 1-t \\ -1+2t \end{bmatrix} + \begin{bmatrix} \frac{ts^3}{3}\Big|_1^t \\ \frac{s^3}{3}\Big|_1^t \\ \frac{s^3 t}{3}\Big|_1^t \end{bmatrix} = \begin{bmatrix} \frac{2}{3}t + \frac{1}{3}t^4 \\ \frac{2}{3} - t + \frac{t^3}{3} \\ -1 + \frac{5}{3}t + \frac{t^4}{3} \end{bmatrix}$$

5.   **(a).** Let $x(t)$ solve the system $x' = Ax$ and $z(t)$ solve the system $(z^T)' = -z^T A$. Differentiating the scalar product $z^T x$ we obtain

$$(z^T x)' = (z^T)'x + z^T x' = -z^T Ax + z^T Ax = 0$$

for all t. Therefore $z^T x = $ constant, for all t.

(b). Since $x(t)$ solves the system $x' = Ax$, we see by taking the transpose of both sides that $x^T(t)$ solves the system

$$(x^T)' = x^T A^T = -x^T A$$

Therefore by part (a), $x^T x = \|x\|^2$ is a constant for all t, but $\|x\|^2 = $ constant describes a "sphere" centered at the origin.

6.   Under the stated conditions, we know that the transition matrix $\Phi(t, t_0)$ solves the system

$x'(t) = A(t)x(t)$ for all t, t_0. So for any T we see that

$$\Phi'(t+T, t_0) = A(t+T)\Phi(t+T, t_0)$$

But we know that for some T, $A(t+T) = A(t)$ for all t. So

$$\Phi'(t+T, t_0) = A(t)\Phi(t+T, t_0)$$

for all t. So $\Phi(t+T, t_0)$ solves the system $x' = Ax$ and is also a basic solution matrix. Thus, the transition matrix $\Phi(t, t_0)$ can be "manufactured" from the basic solution matrix $\Phi(t+T, t_0)$ via (4):

$$\Phi(t, t_0) = \Phi(t+T, t_0)\Phi^{-1}(t_0+T, t_0)$$

and so the matrix C is given by

$$C = \Phi(t_0+T, t_0)$$

7. **(a).** That $X(t) = \begin{bmatrix} t^2 & t \\ 2t & 1 \end{bmatrix}$ is a solution matrix of $x' = y$, $y' = -2x/t^2 + 2y/t$, $t > 0$, follows from a straightforward calculation, i.e., show that $x = t^2$, $y = 2t$ is a solution, and then that $x = t$, $y = 1$ is also a solution.

(b). We have

$$\Phi(t, t_0) = X(t)X^{-1}(t_0) = \begin{bmatrix} t^2 & t \\ 2t & 1 \end{bmatrix} \frac{1}{t_0^2} \begin{bmatrix} -1 & t_0 \\ 2t_0 & -t_0^2 \end{bmatrix} = \frac{1}{t_0^2} \begin{bmatrix} -t^2 + 2tt_0 & (t^2 - tt_0)t_0 \\ -2t + 2t_0 & (2t - t_0)t_0 \end{bmatrix}$$

Clearly, $\Phi(t_0, t_0) = I$. A straightforward calculation (just interchange t and t_0 above) shows that

$$\Phi(t_0, t) = \frac{1}{t^2} \begin{bmatrix} -t_0^2 + 2tt_0 & (t_0^2 - tt_0)t \\ -2t_0 + 2t & (2t_0 - t)t \end{bmatrix}$$

Calculating $\Phi(t, t_0)\Phi(t_0, t)$ directly, we get I, and so $\Phi(t_0, t) = \Phi^{-1}(t, t_0)$.

(c). See Graph 1 for the orbits and Graph 2 for the component curves, given the initial data $x(1) = 1$, $y(1) = -5, -4, \ldots, 4, 5$.

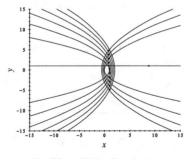

Problem 7(c), Graph 1.

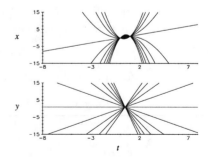

Problem 7(c), Graph 2.

8. Transform the second-order scalar IVP, $y'' + a(t)y' + b(t)y = f(t)$, $y(0) = y'(0) = 0$ to

the system IVP $u' = v$, $v' = -bu - av + f$, $u(0) = 0$, $v(0) = 0$, where $u = y$, $v = y'$. By the Variation of Parameters for Driven Systems Theorem (Theorem 7.8.3), we know the solution of the IVP is $x^p(t) = \int_{t_0}^t \Phi(t, s) F(s) \, ds$, where

$$x^p(t) = \begin{bmatrix} u(t) \\ v(t) \end{bmatrix}, \quad F(s) = \begin{bmatrix} 0 \\ f(s) \end{bmatrix}$$

So, we have

$$\Phi(t, s) F(s) = X(t) X^{-1}(s) F(s) \qquad \text{(i)}$$

$$= \frac{1}{\det X(s)} \begin{bmatrix} u_1(t) & u_2(t) \\ v_1(t) & v_2(t) \end{bmatrix} \begin{bmatrix} v_2(s) & -u_2(s) \\ -v_1(s) & u_1(s) \end{bmatrix} \begin{bmatrix} 0 \\ F(s) \end{bmatrix}$$

where $\{[u_1, v_1]^T, [u_2, v_2]^T\}$ is any fundamental set for the system. With $u_1 = y_1$ and $u_2 = y_2$, it follows that $v_1 = y_1'$ and $v_2 = y_2'$, and so

$$u(t) = y(t) = \int_{t_0}^t \frac{y_1(t) y_2(s) - y_1(s) y_2(t)}{\det X(s)} f(s) ds$$

$$= \int_{t_0}^t \frac{y_1(s) y_2(t) - y_1(t) y_2(s)}{\det \begin{bmatrix} y_1(s) & y_2(s) \\ y_1'(s) & y_2'(s) \end{bmatrix}} f(s) ds$$

where the integrands are obtained by multiplying out the matrix product in (i). So the result in Problem 7 of Section 3.7 is a special case of Theorem 7.8.4.

9. Let $x(t) = \Phi(t, t_0) c(t)$ be a solution of $x' = A(t)x + F$, where the "varied parameter" vector $c(t)$ is to be determined. We have $x' = \Phi' c + \Phi c' = A\Phi c + \Phi c' = A\Phi c + F$. So $\Phi c' = F$ and $c'(t) = \Phi^{-1}(t, t_0) F(t)$. We have $c(t) - c(t_0) = \int_{t_0}^t \Phi(t_0, s) F(s) \, ds$, where we have used the property that $\Phi^{-1}(t, t_0) = \Phi(t_0, t)$. Setting $c(t_0) = 0$ [which amounts to solving the IVP where $x(t_0) = 0$], we have that $c(t) = \int_{t_0}^t \Phi(t_0, s) F(s) \, ds$, and we have constructed a soution of the desired form $x = \Phi(t, t_0) c(t)$.

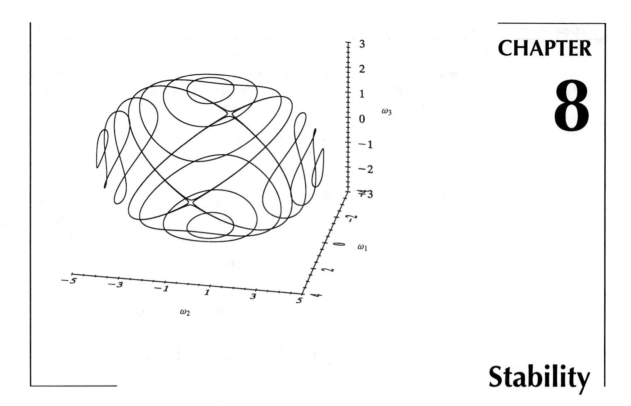

CHAPTER

8

Stability

At the end of the 19th century A. M. Lyapunov introduced a series of tests that could be used to determine the stability properties of a system of ODEs without using solution formulas. These tests involve the use of a scalar function of the state variables and its derivative following the motion. These Lyapunov functions are often based on the total energy of the system, if that can be defined, or else a quadratic form in the state variables. In Section 8.1 we define stability (both neutral and asymptotic) of a system of autonomous ODEs at an equilibrium point and illustrate the definition with many examples. We also give the criterion for stability for an autonomous linear system $x' = Ax$. In Section 8.2 we give the result that an autonomous linear system whose system matrix has only eigenvalues with negative real parts is not only asymptotically stable at the origin, but cannot be destabilized by any higher order perturbation. On the other hand, if at least one of the eigenvalues has positive real part, then the unstable linear system cannot be stabilized by any higher order perturbation. Lyapunov's result (Theorem 8.2.1) is actually phrased as follows: If the Jacobian matrix of $f(x)$ at an equilibrium point p has all eigenvalues with negative real parts, then the system $x' = f(x)$ is asymptotically stable at p. If there is at least one eigenvalue with positive real part, then $x' = f(x)$ is unstable at p. These results form the basis of the long-standing technique of linearizing a nonlinear system at an equilibrium point in order to understand the long-term behavior of the solutions of the nonlinear system near the point. In Section 8.3 we generalize the notion of exactness and the existence of an integral given in Section 1.6 and its problem set to the existence of an integral for certain systems of ODEs, which are said to be conservative. Typically, physical systems for which no energy is lost or gained have model systems of ODEs which are conservative, and this analogy is explored. The chapter cover figure displays an integral surface and several orbits of the conservative system that models the angular velocities of a whirling tennis racket. In Section 8.4 we present Lyapunov's tests and the so-called Lyapunov Method. In the process we characterize the definiteness properties of quadratic forms.

Since the material of this chapter is likely to be new and rather strange for many readers, we give lots of examples.

8.1   Stability and Linear Systems

Suggestions for preparing a lecture

Topics: The definitions of stability (asymptotic, neutral), many examples (maybe even some of the dreaded $\delta\varepsilon$-analysis!), Theorem 8.1.1 for autonomous linear systems and its implications.

Making up a problem set

Problems 1, parts of 2 and 3 and 5, 6, parts of 7.

Comments

In this section we lay out the definitions of stability and its refinements, neutral stability and asymptotic stability, for an autonomous system at an equilibrium point. Since the rest of the chapter is devoted to finding tests for stability and instability, we want to be sure the definitions are understood—and so an entire section mostly on definitions is included. If the system is linear, then Theorem 8.1.1 allows us to test the stability properties simply by inspecting the eigenvalues of the system matrix and, if an eigenvalue λ with zero real part is multiple, determining the dimension of its eigenspace. Asymptotic stability of a system at an equilibrium point requires that the system be an attractor (i.e., all orbits passing near the equilibrium point must approach it as $t \to +\infty$) *and also be stable.* At first glance, one might think that an attracting equilibrium point is automatically stable—but that need not be so (see Problem 6).

"Stability" is an everyday word that has many meanings. In science and engineering it is often used to describe something other than what we define here. For example, in engineering a system is sometimes said to be stable if it responds to bounded inputs with bounded outputs (the "BIBO" of Sections 2.3 and 7.9), but that is quite different from our definition. You might contrast stability and the continuity of a solution $x(t, x^0)$ as a function of initial position x^0. The Fundamental Theorem of Section 5.2 gives continuity in x^0, but stability requires in addition a certain "uniformity" for all $t \geq 0$ in the continuity. We give many examples here in order to help solidify understanding of what stability means.

1. Since only the orbital portraits are given (not the system itself), we use visual clues to make inferences about the unknown system.

    (a). The system is neutrally stable at the origin. Orbits starting near the origin remain near. The orbits look very much like those for the "center" of the linear system of Example 7.7.5.

    (b). The system is unstable at the origin because orbits move away as t increases and head out toward the unit circle.

    (c). The system is locally asymptotically stable at the equilibrium point $(-1, 0)$ because nearby orbits stay nearby as t increases and, in fact, are attracted to the point. For the same reason the system is also locally asymptotically stable at the equilibrium point $(1, 0)$. But the system is unstable at the equilibrium point at the origin because some nearby orbits leave the neighborhood as t increases and never come back.

2. **(a).** The system $x' = 2y$, $y' = -8x$ is linear. The system matrix has eigenvalues $\lambda_1 = 4i = \bar{\lambda}_2$. By Case 2 of Theorem 8.1.1, the system is neutrally stable at the origin. See Fig. 2(a).

    (b). Since $x' = 0$ and $y' = -y^3$, we have that $x = c_1$, and $y = c_2(1 + 2c_2^2t)^{-1/2}$, which we get by separating variables and solving. For $t \geq -1/(2c_2^2)$ ($c_2 \neq 0$) we have $\|(x(t), y(t))\| = [c_1^2 + c_2^2(1 + 2c_2^2t)^{-1}]^{1/2} \leq (c_1^2 + c_2^2)^{1/2} = [x^2(0) + y^2(0)]^{1/2}$. If $c_1 \neq 0$, then $\|(x(t), y(t))\| \to (c_1, 0) \neq (0, 0)$ and we have stability [set $\delta = \varepsilon$ in the definition of stability], but not asymptotic stability. The system is neutrally stable at the origin. See Fig. 2(b) for orbits. The dashed line, $y = 0$, is a line of equilibrium points.

    (c). The xy-subsystem $x' = y$, $y' = -x$ is linear and has the property that $x^2(t) + y^2(t) = x^2(0) + y^2(0)$ for any solution $x(t), y(t)$ because $[x^2 + y^2]' = 2xy - 2xy = 0$; the origin is *not* an attractor, and each orbit of the xy-system remains at a constant distance from the origin. On the other hand, the solutions of $z' = -z^3$ are given by $z(t) = C(1 + 2C^2t)^{-1/2}$, and $z(t) \to 0$ steadily as $t \to +\infty$. So, $x^2(t) + y^2(t) + z^2(t) \leq x^2(0) + y^2(0) + z^2(0)$ for $t \geq 0$, and the system is stable, but not asymptotically stable (so, neutrally stable) at the origin. See Fig. 2(c) for orbits; each orbit with $z_0 \neq 0$ tends to the circle in the xy-plane given by $x^2 + y^2 = x_0^2 + y_0^2$ as $t \to +\infty$.

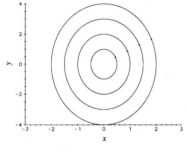

Problem 2(a).

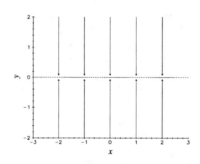

Problem 2(b).

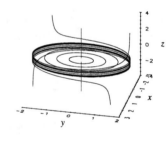

Problem 2(c).

3. If the system is linear and all the eigenvalues of the matrix of coefficients have negative real parts, then the system is asymptotically stable at the origin by Theorem 8.1.1. The systems of parts **(a)**, **(d)**,**(e)**, **(f)**, and **(h)** are linear, so we need only find the eigenvalues λ of the system matrix. The eigenvalues are calculated by the techniques of Sections 7.4 and 7.9 and are listed below (all have negative real parts). Systems **(b)**,**(c)**, **(g)**, **(i)**, and **(j)** are nonlinear and are solved explicitly in order to determine their stability properties.

    (a). The ODE $x' = -x$ is linear; the system "matrix" $[-1]$ has eigenvalue $\lambda = -1$; we have global asymptotic stability at $x = 0$. See Fig. 3(a) for some time-state curves, and the margin for the state line.

    (b). The ODE $x' = -x^3$ is nonlinear and is solved by separating variables: $x = C(1 + 2C^2t)^{-1/2}$, where C is an arbitrary constant and $t > -1/(2C^2)$ if $C \neq 0$, and $-\infty < t < \infty$

if $C = 0$. Since $|x|$ is a decreasing function of t and approaches 0 as $t \to +\infty$ (if $C \neq 0$), the ODE is asymptotically stable at the origin. In particular, we have global asymptotic stability since the origin is an attractor [$x(t) \to 0$ as $t \to +\infty$] and the system is stable [set $\delta = \varepsilon$ in the definition of stability]. See Fig. 3(b) for some time-state curves, and the margin for the state line.

(c). Separate variables in the ODE $x' = -\sin x$ and integrate to obtain $\ln|\tan(x/2)| = -t + C_1$. So, $x = 2\arctan(Ce^{-t})$, where C is an arbitrary constant. The values of the arctangent are limited to the range $(-\pi/2, \pi/2)$ since we are only interested in the ODE near $x = 0$; $|x|$ is a decreasing function of t and tends to 0 as t increases to ∞. The ODE is asymptotically stable at $x = 0$ since 0 is an attractor and we can set $\delta = \varepsilon$ in the definition of stability. We have only local asymptotic stability since, for example, equilibrium points such as $x = \pm 2\pi$ are *not* attracted to $x = 0$. See Fig. 3(c) for some time-state curves, and the margin for the state line.

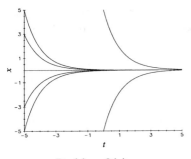

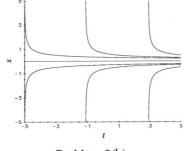

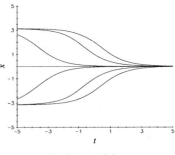

Problem 3(a).   Problem 3(b).   Problem 3(c).

(d). The system $x' = -4x$, $y' = -3y$ is linear; the system matrix has eigenvalues $\lambda_1 = -4$, $\lambda_2 = -3$. Both eigenvalues are negative, so the system is globally asymptotically stable at the origin. See Fig. 3(d) for some orbits.

(e). The system $x' = x - 3y$, $y' = 4x - 6y$ is linear; the system matrix has eigenvalues $\lambda_1 = -6$, $\lambda_2 = -1$. Both eigenvalues are negative, so the system is globally asymptotically stable at the origin. See Fig. 3(e) for some orbits.

(f). The system $x' = -x + 4y$, $y' = -3x - 2y$ is linear; the system matrix has eigenvalues $\lambda_1 = -(3 + i\sqrt{47})/2 = \bar{\lambda}_2$ with negative real parts; we have global asymptotic stability at the origin. See Fig. 3(f) for some orbits.

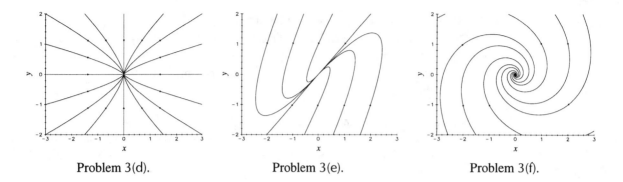

Problem 3(d).   Problem 3(e).   Problem 3(f).

(g). Solving $y' = -5y$, we have $y = c_2 e^{-5t}$, while the nonlinear ODE $x' = -2x^3$ can be solved by separating the variables to obtain $x = c_1(1 + 4c_1^2 t)^{-1/2}$, where c_1 and c_2 are arbitrary constants and $-\infty < t < \infty$ if $c_1 = 0$, and $t > -1/(4c_1^2)$ if $c_1 \neq 0$. The origin $(0,0)$ is an attractor since $(x(t), y(t)) \to (0,0)$ as $t \to +\infty$. Moreover, since $\|(x(t), y(t))\|^2 = c_2^2 e^{-10t} + c_1^2(1 + 4c_1^2 t)^{-1}$, we see that $\|(x(t), y(t))\|$ decreases steadily to 0 as t increases to $+\infty$. We may set $\delta = \varepsilon$ in the definition of stability, and the system is globally asymptotically stable at the origin. See Fig. 3(g) for some orbits.

(h). The system $x' = -x + y + z$, $y' = -2y$, $z' = -3z$ is linear; the system matrix has eigenvalues $\lambda_1 = -3$, $\lambda_2 = -2$, $\lambda_3 = -1$, and we have asymptotic stability at the origin. See Fig. 3(h) for orbits.

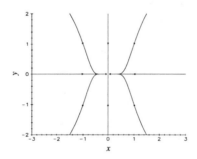

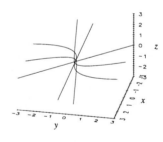

Problem 3(g).   Problem 3(h).

(i). Solving $x' = -x^3$ and $y' = -y^3$ by separating variables, we have that $x = c_1(1 + 2c_1^2 t)^{-1/2}$ and $y = c_2(1 + 2c_2^2 t)^{-1/2}$, where if $c_1, c_2 \neq 0$, then t must be greater than $-1/(2c_1^2)$ and $-1/(2c_2^2)$. Much as in the solution to Problem 1(g), we see that $(0,0)$ is an attractor since $x(t) \to 0$ and $y(t) \to 0$ as $t \to +\infty$, while $(x(t), y(t))$ decreases steadily to 0 as $t \to +\infty$ [we may set $\delta = \varepsilon$ in the definition of stability]. The system is globally asymptotically stable at the origin. See Fig. 3(i) for some orbits.

(j). The subsystem of the first two ODEs $x' = -x - y$, $y' = x - y$ is linear; the system matrix has eigenvalues $-1 \pm i$ and is globally asymptotically stable at the origin of the xy-plane. Moreover, $(x^2 + y^2)' = -2x^2 - 2y^2$ and so, except at the origin, the distance from $(0, 0)$ of any nontrivial solution of the $xy-$subsystem diminishes steadily to 0 as t increases. The solution of the ODE $z' = -x^3$ is $z = C(1 + 2C^2t)^{-1/2}$, and if $C \neq 0$, $|z| \to 0$ steadily as t increases from 0. Since $\|(x(t), y(t), z(t))\| \leq \|(x(t), y(t))\| + \|z(t)\|$ by the Triangle Inequality, $\|(x(t), y(t), z(t))\| \to 0$ steadily as $t \to +\infty$, and the system is globally asymptotically stable at the origin. See Fig. 3(j) for orbits.

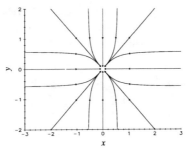

Problem 3(i).

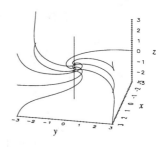

Problem 3(j).

4. The system is $x' = -x - 10y - x(x^2 + y^2)$, $y' = 10x - y - y(x^2 + y^2)$. Introduce polar coordinates, $x = r\cos\theta$ and $y = r\sin\theta$. Then by formulas (14) and (15) in Section 5.2, we have $x' = r'\cos\theta - r\theta'\sin\theta$, $y' = r'\sin\theta + r\theta'\cos\theta$. The original system becomes $r' = -r - r^3$, $\theta' = 10$. Consequently, the solution with the initial condition $r(0) = r_0$, $\theta(0) = \theta_0$ is (after separating variables in the r' equation, using integral tables or partial fractions, and solving the resulting expression for r in terms of t):

$$r = r_0 e^{-t}(1 + r_0^2 - r_0^2 e^{-2t})^{-1/2}, \qquad \theta = \theta_0 + 10t.$$

Observe that $r \to 0$ as $t \to +\infty$, regardless of the value of $r_0 > 0$. Moreover, because $r' < 0$ if $r > 0$, $r(t)$, so $\|(x(t), y(t))\|$, decreases as t increases. We may take $\delta = \varepsilon$ in the definition of asymptotic stability. So the system is asymptotically stable at the origin.

5. Systems 5(**d**)–5(**f**) are two-dimensional and linear. They are unstable because each has at least one eigenvalue with positive real part, or else a double eigenvalue with 0 real part but only a one-dimensional eigenspace [see Case 3 of Theorem 8.1.1]. Systems 5(**a**)–5(**c**) and 5(**g**) are nonlinear. They must be handled by different techniques.

(**a**). Separating variables and solving $x' = x^2$, we have $x = x_0/(1 - tx_0)$ if $x(0) = x_0 > 0$. Since $x(t) \to +\infty$ as t increases from 0 to x_0^{-1}, the ODE is unstable at 0. See Fig. 5(a) for some time-state curves, and the margin for the state line.

(**b**). Much as in Problem 3(**c**), but with a sign change, the solution of $x' = \sin x$ is found to be $x = 2\arctan(Ce^t)$, where we assume $-\pi < x < \pi$. Suppose we set $\varepsilon = 1$. Then for every $C > 0$, $x(t) \to +\pi$ as $t \to +\infty$. For no positive δ is it true that $|x(t)| < \varepsilon = 1$ for all $t \geq 0$ if $|x(0)| = |2\arctan C| < \delta$. The ODE is unstable at 0. See Fig. 5(b) for some time-state curves and the state line in the margin. Note how solution curves diverge from $x = 0$ and approach other equilibrium solutions as t increases.

x

0

(c). Suppose $x > 0$. The solutions of $x' = |x|$ are given by $x = Ce^t$, $C > 0$. Since $x(t) \to +\infty$ as $t \to +\infty$, regardless of the value of C, the ODE is unstable at 0. See Fig. 5(c) for some time-state curves, and the margin for the state line.

(d). The system $x' = 3x - 2y$, $y' = 4x - y$ is linear; the system matrix has eigenvalues λ, $\bar{\lambda}$, where $\lambda = 1 + 2i$. See Fig. 5(d) for orbits near the unstable origin [Case 3 of Theorem 8.1.1].

(e). The system $x' = 3x - 2y$, $y' = 2x - 2y$ is linear; the system matrix has eigenvalues $\lambda_1 = -1$, $\lambda_2 = 2$. One eigenvalue is positive, so the system is unstable. See Fig. 5(e) for orbits near the unstable origin [Case 3 of Theorem 8.1.1].

(f). The system $x' = x + y$, $y' = -x - y$ is linear; the system matrix has 0 as a double eigenvalue and only a one-dimensional eigenspace, which is spanned by $[1, -1]^T$. The origin is unstable by Case 2 of Theorem 8.1.1. See Fig. 5(f) for orbits; nonconstant orbits are parallel to the line $x + y = 0$ of equilibrium points. Above the line the orbits move downward and to the right since $x' > 0$ and $y' < 0$ in the region. Below the line the direction of motion is reversed.

(g). The system is $x' = x^3$, $y' = -3y$. We have $y = c_2 e^{-3t}$. Separating variables in $x' = x^3$, and solving for x in terms of t, we have that $x = c_1(1 - 2c_1^2 t)^{-1/2}$, $t < 1/(2c_1^2)$ if $c_1 \neq 0$, and $-\infty < t < \infty$ if $c_1 = 0$. Even though $y(t) \to 0$ as $t \to +\infty$, since $|x(t)| \to \infty$ as t increases from 0 to $1/(2c_1^2)$ (if $c_1 \neq 0$), the system is unstable at the origin. See Fig. 5(g) for some orbits. Note the resemblance to a saddle.

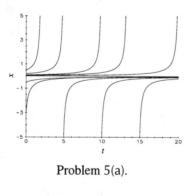

Problem 5(a).

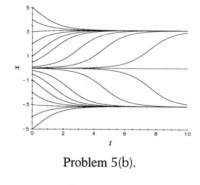

Problem 5(b).

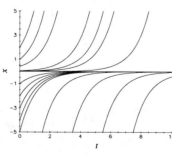

Problem 5(c).

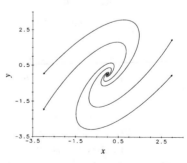

Problem 5(d).

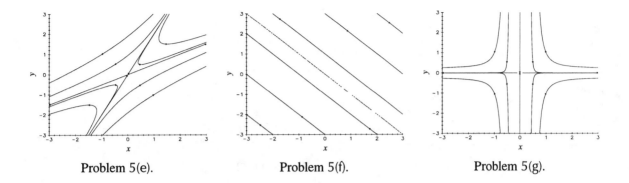

Problem 5(e).   Problem 5(f).   Problem 5(g).

6. **(a).** Using formulas (14) and (15) in Section 5.2, we see that the system $x' = x - y - x^3 - xy^2 + xy(x^2 + y^2)^{-1/2}$, $y' = x + y - x^2y - y^3 - x^2(x^2 + y^2)^{-1/2}$ can be written in polar coordinates as

$$r' = \cos\theta[r\cos\theta - r\sin\theta - r^3\cos^3\theta - r^3\cos\theta\sin^2\theta + r\cos\theta\sin\theta]$$

$$+ \sin\theta[r\cos\theta + r\sin\theta - r^3\cos^2\theta\sin\theta - r^3\sin^3\theta - r\cos^2\theta]$$

$$= r(1 - r^2)$$

where we have used $\cos^2\theta + \sin^2\theta = 1$ several times. Similarly,

$$\theta' = r^{-1}\cos\theta[r\cos\theta + r\sin\theta - r^3\cos^2\theta\sin\theta - r^3\sin^3\theta - r\cos^2\theta]$$

$$-r^{-1}\sin\theta[r\cos\theta - r\sin\theta - r^3\cos^3\theta - r^3\cos\theta\sin^2\theta + r\cos\theta\sin\theta]$$

$$= 1 - \cos\theta$$

From the given ODEs in r and θ, we see that there are two equilibrium points, $r = 0$, and $r = 1$, $\theta = 2k\pi$, [i.e., the points (0,0), (1,0) in xy-rectangular coordinates]. Since the r, θ ODEs are decoupled, they are easier to solve than the x, y ODEs. We have that $-\cot(\theta/2) = t + C$, or $\theta = 2\arctan(-t - C)^{-1} \to 2k\pi$ as $t \to \pm\infty$, and that $r(t) = r_0[r_0^2 + (1 - r_0^2)e^{-2t}]^{-1/2} \to 1$ as $t \to +\infty$ if $r_0 > 0$.

(b). We conclude that if $r_0 > 0$, then as $t \to +\infty$, we have $r \to 1$ and $\theta \to 2k\pi$; i.e., aside from the equilibrium point at the origin, all orbits tend to the point (1,0) [in rectangular coordinates], which is an attractor. This point is, however, unstable since the solution $r(t) = 1$, $\theta = -2\arctan(1/t)$ has the property that it exits the point $x = 1$, $y = 0$, (the point $r = 1$, $\theta = 0$) as t increases from $-\infty$ and turns counterclockwise around the unit circle approaching the same point (but now with $r = 1$, $\theta = 2\pi$) as $t \to +\infty$. If $\varepsilon = 1$, say, this orbit does not remain in the circular region $(x - 1)^2 + y^2 < 1$ of radius $\varepsilon = 1$ about the equilibrium point, even though it begins and ends inside that region.

(c). See Fig. 6(c) for some orbits. Note in particular the orbit on the unit circle that exits from the equilibrium point $x = 1$, $y = 0$ as t increases from $-\infty$, and then returns to the point as $t \to +\infty$. Note also how nonconstant orbits outside and inside the circle approach the point $x = 1$, $y = 0$ as $t \to +\infty$.

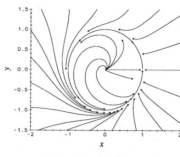

Problem 6(c).

7. In each part, use Theorem 8.1.1 to determine the stability properties of the system $x' = Ax$, where A is the given matrix.

(a). The eigenvalues of the matrix $\begin{bmatrix} -1 & 2 \\ -2 & -1 \end{bmatrix}$ satisfy $(-1 - \lambda)(-1 - \lambda) + 4 = 0$. Solving, we find $\lambda_1 = \bar{\lambda}_2 = -1 + 2i$. Since each eigenvalue has negative real part, the system is asymptotically stable.

(b). The eigenvalues of the matrix $\begin{bmatrix} -3 & 1 \\ 2 & -4 \end{bmatrix}$ satisfy $(-3 - \lambda)(-4 - \lambda) - 2 = 0$. Solving, we find $\lambda_1 = -2$ and $\lambda_2 = -5$. Since each eigenvalue is negative, the system is asymptotically stable.

(c). The eigenvalues of the matrix $\begin{bmatrix} 1 & -1 \\ 2 & -2 \end{bmatrix}$ satisfy $(1 - \lambda)(-2 - \lambda) + 2 = 0$. Solving, we find $\lambda_1 = 0$, $\lambda_2 = -1$. Since each eigenvalue is non-positive and λ_1 is zero, the system is neutrally stable.

(d). The eigenvalues of the matrix $\begin{bmatrix} 0 & 1 \\ 0 & 0 \end{bmatrix}$ satisfy $\lambda^2 = 0$. So, $\lambda = 0$ is a double eigenvalue. However, the eigenspace corresponding to $\lambda = 0$ is spanned by $[1 \ 0]^T$ and so has dimension 1. By Theorem 8.1.1, the system is unstable.

(e). The eigenvalues of the matrix $\begin{bmatrix} 5 & 1 \\ 1 & -2 \end{bmatrix}$ satisfy $(5 - \lambda)(-2 - \lambda) - 1 = 0$. Solving, we find $\lambda_1 = (3 + \sqrt{53})/2$ and $\lambda_2 = (3 - \sqrt{53})/2$. Since one of the eigenvalues is positive, the system is unstable.

(f). The eigenvalues of the matrix $\begin{bmatrix} 3 & -2 \\ 5 & -3 \end{bmatrix}$ satisfy $(3 - \lambda)(-3 - \lambda) + 10 = 0$. Solving, we have $\lambda_1 = \bar{\lambda}_2 = i$. Since both eigenvalues have zero real part, the system is neutrally stable.

(g). By Theorem 7.9.9, the eigenvalues of the matrix $\begin{bmatrix} -10 & 3 & -4 \\ 1 & -12 & 9 \\ -1 & -2 & -5 \end{bmatrix}$ all have negative real parts. To see this, note that each diagonal entry is negative and that on each row the sum of the magnitudes of the off-diagonal elements has magnitude less than the diagonal entry on that row. So using Gershgorin disks (Section 7.9), the eigenvalues lie entirely inside the left half of the complex plane, and the system is globally asymptotically stable.

(h). For the matrix $A = \begin{bmatrix} -5 & 0 & 0 \\ 0 & 0 & -3 \\ 0 & 3 & 0 \end{bmatrix}$, the first equation decouples from the last two

equations. It may be easily seen that the eigenvalues for A are $\lambda_1 = -5$, $\lambda_2 = \bar{\lambda}_3 = 3i$. Since every eigenvalue has nonpositive real part and λ_2 and λ_3 have zero real parts, the system is neutrally stable.

(i). The matrix $\begin{bmatrix} 0 & 1 & 1 \\ 0 & 0 & 1 \\ 0 & 0 & 0 \end{bmatrix}$ has $\lambda = 0$ as a triple eigenvalue. However, the eigenspace

is spanned by the vector $[1 \ 0 \ 0]^T$ and so is of dimension one. So by Theorem 8.1.1, the system is unstable since the dimension of the eigenspace is less than the multiplicity of the eigenvalue.

(j). The matrix $\begin{bmatrix} 0 & 0 & 1 & 0 \\ 0 & 0 & 0 & 0 \\ 0 & 0 & 0 & 0 \\ 0 & 0 & 0 & 0 \end{bmatrix}$ has $\lambda = 0$ as a quadruple eigenvalue. The eigenspace is

spanned by the vectors $[0 \ 1 \ 0 \ 0]^T$, $[0 \ 0 \ 0 \ 1]^T$, and $[1 \ 0 \ 0 \ 0]^T$ and so is of dimension three. By Theorem 8.1.1, the system is unstable since the dimension of the eigenspace is less than the multiplicity of the eigenvalue.

(k). For the matrix $A = \begin{bmatrix} 3 & -2 & 0 & 0 \\ 5 & -3 & 0 & 0 \\ 0 & 0 & 3 & -2 \\ 0 & 0 & 5 & -3 \end{bmatrix}$, the first two equations in the system $x' =$

Ax decouple from the last two. Looking at each pair of equations separately, we find that the eigenvalues ($\pm i$) have zero real part, so the system here is also neutrally stable.

(l). For the matrix $A = \begin{bmatrix} 0 & -3 & 1 & 0 \\ 3 & 0 & 0 & 1 \\ 0 & 0 & 0 & -3 \\ 0 & 0 & 3 & 0 \end{bmatrix}$, the eigenvalues are $\lambda_1 = \lambda_2 = 3i$ and

$\lambda_3 = \lambda_4 = -3i$. The eigenspace corresponding to $\lambda = 3i$ is spanned by the single eigenvector $[i \ 1 \ 0 \ 0]^T$, while the eigenspace corresponding to $\lambda = -3i$ is spanned by the single eigenvector $[-i \ 1 \ 0 \ 0]^T$. Since each of these eigenspaces corresponds to a double eigenvalue but the eigenspace has dimension one, the system is unstable.

8.2 Stability of a Nearly Linear System

Suggestions for preparing a lecture

Topics: Nearly linear systems and their stability properties, order of a function, Jacobian matrix, nonlinear nodes, saddles, and spiral points.

Making up a problem set

Parts of 1 and 2, the corresponding parts of 3, 4, 6, 8(**b**), group problem 12.

Comments

Theorem 8.2.1 partly justifies the often heard assertion in science and engineering that the linear terms of a system near an equilibrium point determine the stability properties without having to look at the nonlinearities. It isn't quite that simple, as Problem 4 shows, but the assertion comes close enough to the truth to be widely believed. We outline a part of the proof of Theorem 8.2.1 at the end of Section 8.4 of the Student Resource Manual. We can't do that proof here because it uses some of the ideas of Section 8.4. The Jacobian matrix is introduced here because it often is needed to determine the linearization at an equilibrium point. The table and sketches at the end of the section suggest that the terms node, saddle, and spiral point introduced in Section 7.7 for an equilibrium point of a planar autonomous linear system, still make sense for a nonlinear system. So from now on when (for example) we refer to a saddle point of a system, the system need not be linear, but the Jacobian matrix of the system at the point must have real eigenvalues of opposite sign. In order to make all of this mathematically correct, we introduce the idea of the order of a function at a point and require that the nonlinear terms of a system have order at least 2 and an equilibrium point. **Warning!** Keep your eye out for situations where the Jacobian matrix at an equilibrium point has eigenvalues that are either 0 or pure imaginary. The system may be right on the edge of a stability region and the nonlinear terms can then tip the full system either way—into stability or into instability. Problems 4 and 11 show what can happen.

---

1. Background for all of the problems here is Theorem 8.2.1.

    (a). The system $x' = -x + y^2$, $y' = -8y + x^2$ is asymptotically stable at the origin since the eigenvalues of the matrix of coefficients of the linear terms at the origin $\begin{bmatrix} -1 & 0 \\ 0 & -8 \end{bmatrix}$ are $-1, -8$ (both are negative).

    (b). The system $x' = 2x + y + x^4$, $y' = x - 2y + x^3 y$ is unstable at the origin since the eigenvalues of the Jacobian matrix $\begin{bmatrix} 2 & 1 \\ 1 & -2 \end{bmatrix}$ at the origin are $\pm\sqrt{5}$ (one is positive).

    (c). The system $x' = 2x - y + x^2 - y^2$, $y' = x - y$ is unstable at the origin since the eigenvalues of the Jacobian matrix $\begin{bmatrix} 2 & -1 \\ 1 & -1 \end{bmatrix}$ at the origin are $(1 \pm \sqrt{5})/2$ (one is positive).

    (d). The Jacobian matrix at (x, y) for the system $x' = e^{x+y} - \cos(x - y)$, $y' = -\sin x$ is
    $$\begin{bmatrix} e^{x+y} + \sin(x-y) & e^{x+y} - \sin(x-y) \\ -\cos x & 0 \end{bmatrix}$$

which evaluates at $x = 0$, $y = 0$ to $\begin{bmatrix} 1 & 1 \\ -1 & 0 \end{bmatrix}$, whose eigenvalues are $(1 \pm \sqrt{3}i)/2$ (one is negative), so the system is unstable at the origin.

2. In each case, the equilibrium points p are found by equating the right-hand sides of the ODEs to 0 and solving for x and y. Then the Jacobian matrix $A = [a_{ij}]$ is calculated, where $a_{ij} = \partial f_i / \partial x_j$, evaluated at $x = p$, and the eigenvalues of A are found. The system is asymptotically stable at p if the eigenvalues of A have negative real parts, unstable at p if an eigenvalue has a positive real part. A system may be unstable at one equilibrium point and asymptotically stable at another. See Theorem 8.2.1.

(a). The system is $x' = y$, $y' = -6x - y - 3x^2$. $P_1(0, 0)$ and $P_2(-2, 0)$ are equilibrium points. The Jacobian matrix has entries $a_{11} = 0$, $a_{12} = 1$, $a_{21} = -6 - 6x$, $a_{22} = -1$. At the point P_1, $x = 0$ and the matrix is $\begin{bmatrix} 0 & 1 \\ -6 & -1 \end{bmatrix}$, whose eigenvalues both have negative real parts since the characteristic polynomial is $\lambda^2 + \lambda + 6$, which has roots $-1/2 \pm i\sqrt{35}/2$. At P_2, $x = -2$, and the matrix is $\begin{bmatrix} 0 & 1 \\ 6 & -1 \end{bmatrix}$. The characteristic polynomial is $\lambda^2 + \lambda - 6$, whose roots are -3 and 2. The system is asymptotically stable at P_1 and unstable at P_2.

(b). The system is $x' = y^2 - x$, $y' = x^2 - y$. At an equilibrium point, $x = y^2$ and $y = x^2 = y^4$. There are two equilibrium points, $P_1(0, 0)$ and $P_2(1, 1)$. The Jacobian matrix has entries $a_{11} = -1$, $a_{12} = 2y$, $a_{21} = 2x$, $a_{22} = -1$. At P_1 we have $x = y = 0$, and the matrix is $\begin{bmatrix} -1 & 0 \\ 0 & -1 \end{bmatrix}$ with eigenvalues $-1, -1$. At P_2, we have $x = 1$, $y = 1$, and the matrix is $\begin{bmatrix} -1 & 2 \\ 2 & -1 \end{bmatrix}$ with characteristic polynomial $\lambda^2 + 2\lambda - 3$ and eigenvalues $\lambda = 1$ and -3. The system is asymptotically stable at P_1, but unstable at P_2.

(c). The system is $x' = -x - x^3$, $y' = y + x^2 + y^2$. There are equilibrium points at the origin and at $(0, -1)$. Since the Jacobian matrix is $\begin{bmatrix} -1 - 3x^2 & 0 \\ 2x & 1 + 2y \end{bmatrix}$, the matrix is $\begin{bmatrix} -1 & 0 \\ 0 & 1 \end{bmatrix}$ at the origin; there is a positive eigenvalue and the system is unstable at the origin. But the eigenvalues of the Jacobian matrix at the point $(0, -1)$ are -1 and -1; the system is asymptotically stable at the point $(0, -1)$.

(d). The system is $x' = -y - x(x^2 + y^2)$, $y' = x - y(x^2 + y^2)$. To find the equilibrium points some algebraic simplification helps: $x' = y' = 0$ implies that $xx' + yy' = 0$; but $xx' + yy' = -(x^2 + y^2)^2$, which vanishes only for $x = y = 0$. So the origin is the only equilibrium point. The Jacobian matrix at the origin is $\begin{bmatrix} 0 & -1 \\ 1 & 0 \end{bmatrix}$ with pure imaginary eigenvalues, and the approach outlined in the introduction above to the solution of this problem is inconclusive. However, note that in polar coordinates [formulas (14), (15) in Section 5.2], the system is $r' = -r^3$, $\theta' = 1$, and the origin is asymptotically stable since $r' < 0$ for all $r > 0$.

(e). The system is $x' = x + xy^2$, $y' = x$. The y-axis (i.e., $x = 0$) is a line of equilibrium points. The entries in the Jacobian matrix are $a_{11} = 1 + y^2$, $a_{12} = 2xy$, $a_{21} = 1$, $a_{22} = 0$. The characteristic polynomial when $x = 0$ is $\lambda^2 - (1 + y^2)\lambda$ with 0 and $1 + y^2$ as roots. The system is unstable at every equilibrium point $(0, y_0)$ because one of the eigenvalues is $1 + y_0^2$, which is positive.

(f). The system is $x' = -x + y^2$, $y' = x + y$. $P_1(0, 0)$ and $P_2(1, -1)$ are equilibrium points. The entries of the Jacobian matrix are $a_{11} = -1$, $a_{12} = 2y$, $a_{21} = 1$, $a_{22} = 1$. At P_1 the characteristic polynomial of this matrix is $\lambda^2 - 1$, and there is a positive eigenvalue. At P_2 the characteristic polynomial is $\lambda^2 + 1$, and $\pm i$ are the eigenvalues. P_1 is unstable, while a more detailed analysis is needed at P_2 since the eigenvalues are pure imaginary. In fact, if we set $u = x + y$, then $y' = u$ and $u' = x' + y' = y + y^2$. So $du/dy = (y + y^2)/u$. Separating variables, we have that $u^2/2 = y^2/2 + y^3/3 + C_1$, i.e., $(x + y)^2 = y^2 + 2y^3/3 + C$. A careful analysis shows that for all values of C in some interval $[-1/3, a]$, $a > -1/3$, this equation defines closed orbits surrounding $(1, -1)$, and P_2 is neutrally stable.

3. As long as no eigenvalue of the Jacobian matrix at an equilibrium point has zero real part, we can tell from the nature of the orbits near an equilibrium point whether the eigenvalues are real and negative, positive, or of opposite signs, or else are complex conjugates with negative or positive real parts.

(a). See Fig. 3(a) for some orbits of $x' = y$, $y' = -6x - y - 3x^2$. One eigenvalue of the Jacobian matrix at the apparent nonlinear saddle point at the origin must be positive, and the other must be negative. The eigenvalues of the Jacobian matrix at the apparent nonlinear spiral point must be complex conjugates with negative real parts because the orbits spiral into the point as time increases.

(b). See Fig. 3(b) for some orbits of $x' = y^2 - x$, $y' = x^2 - y$. One eigenvalue of the Jacobian matrix at the apparent nonlinear saddle point must be positive, and the other must be negative. The eigenvalues of the Jacobian matrix at the apparent nonlinear node point are negative since orbits tend to the point as t increases.

(c). See Fig. 3(c) for some orbits of $x' = -x - x^3$, $y' = y + x^2 + y^2$. One eigenvalue of the Jacobian matrix at the apparent nonlinear saddle point must be positive, and the other must be negative. The eigenvalues of the Jacobian matrix at the apparent nonlinear node point must be negative because the orbits approach the point with increasing t.

(d). See Fig. 3(d) for some orbits of $x' = -y - x(x^2 + y^2)$, $y' = x - y(x^2 + y^2)$. The eigenvalues of the Jacobian matrix at the apparent nonlinear spiral point appear to be complex conjugates with negative real parts because the direction is inward. However, the eigenvalues of the Jacobian matrix are actually $\pm i$; it is the higher order terms that pull orbits into the equilibrium point. This is one of those cases where the visual evidence is *not* sufficient to decide the nature of the eigenvalues.

(e). See Fig. 3(e) for some orbits of $x' = x + xy^2$, $y' = x$. Since the equilibrium points on the y-axis are not isolated, one of the eigenvalues at each equilibrium point must be 0. Since the nonconstant orbits move away from the equilibrium point, the other eigenvalue would seem to be positive. The y-axis through the middle of Fig. 3(e) looks like an earthquake fault line.

(f). See Fig. 3(f) for some orbits of $x' = -x + y^2$, $y' = x + y$. The eigenvalues of the Jacobian matrix at the apparent nonlinear saddle point seem to be of opposite signs. The eigenvalues of the Jacobian matrix at the apparent nonlinear center seem to be pure imaginary.

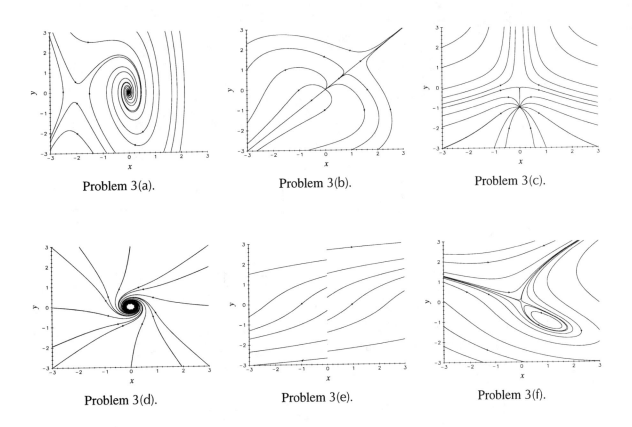

Problem 3(a).   Problem 3(b).   Problem 3(c).

Problem 3(d).   Problem 3(e).   Problem 3(f).

4. The matrix $\begin{bmatrix} -1 & 0 \\ 0 & 0 \end{bmatrix}$ of coefficients of the linear terms of $x' = -x$, $y' = ay^3$ has eigenvalues -1 and 0. Theorem 8.2.1 does not apply, and we must look at the nonlinear terms to determine the stability properties of the system. Let's find all solutions. First we see that $x = c_1 e^{-t}$, here c_1 is an arbitrary constant. Next, if $a \neq 0$, we can separate the variables in $y' = ay^3$ and solve to obtain $y = y_0(1 - 2y_0^2 at)^{-1/2}$. If $a > 0$, then $y \to \infty$ as $t \to (y_0^2 a/2)^-$, and we have instability. If $a < 0$, then $y \to 0$ as $t \to +\infty$. Since $x \to 0$ steadily as $t \to +\infty$, we have stability. If $a = 0$ then $y(t) = y_0$ for all t and we have neutral stability since $x(t) \to 0$ as $t \to +\infty$. See Figs. 4 where $a = -1, +1$, and 0. (The vertical line in Graph 3 is a line of neutrally stable equilibrium points.)

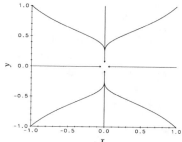

Problem 4, Graph 1.

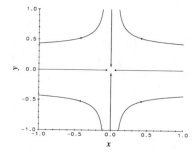

Problem 4, Graph 2.

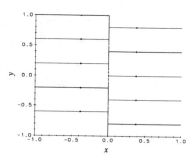

Problem 4, Graph 3.

5. The system is $x' = f(x, y) = x - y + x^2 - xy$, $y' = g(x, y) = -y + x^2$.

    (a). From $x - y + x^2 - xy = 0$ and $-y + x^2 = 0$, we have $x = 0$, $y = 0$ or $x = \pm 1$, $y = 1$. This means that the equilibrium points are $(0, 0)$, $(1, 1)$, $(-1, 1)$.

    (b). We have the Jacobian matrix

    $$J = \begin{bmatrix} \partial f/\partial x & \partial f/\partial y \\ \partial g/\partial x & \partial g/\partial y \end{bmatrix} = \begin{bmatrix} 1 + 2x - y & -1 - x \\ 2x & -1 \end{bmatrix}$$

    and so

    $$J\big|_{(0,0)} = \begin{bmatrix} 1 & -1 \\ 0 & -1 \end{bmatrix}, \quad J\big|_{(1,1)} = \begin{bmatrix} 2 & -2 \\ 2 & -1 \end{bmatrix}, \quad J\big|_{(-1,1)} = \begin{bmatrix} -2 & 0 \\ -2 & -1 \end{bmatrix}$$

    $J\big|_{(0,0)}$ has a positive and a negative eigenvalue, so the original system has an unstable nonlinear saddle at $(0, 0)$. Similarly, $J\big|_{(1,1)}$ has eigenvalues $(1 \pm \sqrt{7}i)/2$, so the original system has an unstable nonlinear spiral point at $(1, 1)$. Finally, $J\big|_{(-1,1)}$ has two negative eigenvalues, and the original system has an asymptotically stable nonlinear node at $(-1, 1)$.

    (c). See figure in the text. The linearized system at the origin is $x' = x - y$, $y' = -y$. The solid arcs are orbits of the nonlinear system, the dashed arcs are orbits of the linear system. Note that the orbits of both systems have common initial points at the upper right and lower left.

    (d). See figure in the text.

6. (a). The Jacobian matrix of $x' = y$, $y' = -2x - y - 3x^2$ is $\begin{bmatrix} 0 & 1 \\ -2 - 6x & -1 \end{bmatrix}$, and its eigenvalues at the origin are $(-1 \pm \sqrt{7}i)/2$; the system is asymptotically stable at the origin. However, the system is not globally asymptotically stable at the origin since $(-2/3, 0)$ is another equilibrium point, and so the origin is not a global attractor.

    (b). See Fig. 6(b). To find the boundaries of the region of attraction of the origin, start near the saddle point at $(-2/3, 0)$ and compute orbits backward in time.

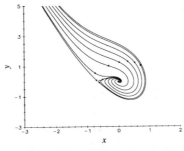

Problem 6(b).

7. The system is $x' = y$, $y' = -x/LC - ry/L - g(x, y)/L$. Since $g(x, y)$ is twice continuously differentiable in a neighborhood of $x = 0$, $y = 0$ and is also of order 2 there, $\partial g/\partial x$ and $\partial g/\partial y$ are both 0 at $x = y = 0$. So, g makes no contribution to the Jacobian matrix of the system at the origin. The eigenvalues of the Jacobian matrix at the origin are $\lambda = (-R/L \pm \sqrt{(R/L)^2 - 4/LC})/2$, so have negative real parts. The system is asymptotically stable at the origin, and the values of the nonlinear term g have no effect on the stability property, although g may severely limit the region of asymptotic stability.

8. **(a).** The system $x' = -x - x^3$, $y' = -y$ has (0,0) as its only equilibrium point. The Jacobian matrix at (0,0) is $\begin{bmatrix} -1 & 0 \\ 0 & -1 \end{bmatrix}$ with eigenvalues -1, -1. By Theorem 8.2.1 the system is asymptotically stable at the origin.

   (b). The system $x' = \alpha x - y + x^2$, $y' = x + \alpha y + x^2$, has the equilibrium points (0,0) and $(x_0, \alpha x_0 + x_0^2)$ where $x_0 = -(1 + \alpha^2)/(1 + \alpha)$, $\alpha \geq -1$. The Jacobian matrix at the origin is $\begin{bmatrix} \alpha & -1 \\ 1 & \alpha \end{bmatrix}$ with eigenvalues $\alpha \pm i$. The origin is unstable if $\alpha > 0$ and asymptotically stable if $\alpha < 0$ (Theorem 8.2.1), but Theorem 8.2.1 is *not* applicable if $\alpha = 0$. The Jacobian matrix at $(x_0, \alpha x_0 + x_0^2)$ is $\begin{bmatrix} \alpha + 2x_0 & -1 \\ 1 + 2x_0 & \alpha \end{bmatrix}$. The eigenvalues λ are the roots of quadratic $\lambda^2 - 2(\alpha + x_0)\lambda + (1 + \alpha^2) + 2x_0(1 + \alpha)$. With the value of x_0 given above and $\alpha > -1$, we see that the constant term of the quadratic is negative, and so the eigenvalues are real and of opposite signs. The system has nonlinear saddle behavior near the second equilibrium point and is unstable there.

9. The nonlinear system is $x' = z + x^2 y$, $y' = x - 4y + xz^2$, $z' = \alpha x + 2y - z + x^2$. The linearized version of the system (about the origin) is $x' = z$, $y' = x - 4y$, $z' = \alpha x + 2y - z$. The corresponding characteristic polynomial for the system matrix of this linear system is $\lambda^3 + 5\lambda^2 + (4 - \alpha)\lambda - (2 + 4\alpha) = 0$. The Routh Test (see Section 7.9) for this cubic requires that $4 - \alpha > 0$, $-(2 + 4\alpha) > 0$, $5(4 - \alpha) > -(2 + 4\alpha)$. So we must have $\alpha < 4$, $\alpha < -1/2$, and $\alpha < 22$; the requirement that α be less than $-1/2$ meets all of the conditions, so the resulting system for these values of α is (locally) asymptotically stable at the origin.

10. Group project.

11. Group project.

12. Group project.

8.3 Conservative Systems

Suggestions for preparing a lecture

Topics: Integrals, conservative systems, the "no attractors theorem," the relation between an exact system in $\mathbb{R}^2$ and a conservative system (see Problem 7) if you did exactness back in Section 1.6, Problem 10. In this regard you may also want to talk about Hamiltonian systems (Problem 6), but that would require a second lecture (or else talking fast). The tennis racket model is always fun, but does require a second lecture. This model makes a good classroom demo.

Making up a problem set

We like 1(a), (c), (d), 2, 3, 6. We also like each of the group problems, especially Problem 8 on the tennis racket model.

Comments

A system of ODEs is conservative if it has a nonconstant integral. The level sets of the integral in state space are unions of orbits, and, so, knowledge of an integral tells us a lot about orbital behavior. What kinds of ODEs have integrals? First-order exact ODEs do (see Problem 10, Section 1.6), and that is why a conservative system may be considered to be a generalization of an exact ODE. A physical system whose total energy when in state $x(t)$ is the same as when in state $x(t_0)$ is conservative; the total energy is the integral. No real physical system has this property, but for many the change in energy as time goes on is negligible and the system can be treated as if it were conservative. Undamped spring systems, undamped pendulums, resistance-free electrical circuits, the Lotka-Volterra predator-prey system, a rotating tennis racket, are all examples of (approximately) conservative systems. We note in this section that conservative systems do *not* have attracting equilibrium points. An integral of a conservative system has some of the properties of the Lyapunov functions to be introduced in Section 8.4. An integral remains constant on each orbit; any orbit touching a level set of an integral remains on the level set, and there need not be any equilibrium points at all. The reason we put this section in this chapter is that (as noted before) the construction of an integral (if there is one) gives information about orbits without having to find solution formulas.

Experiment

Problem 8. Toss a book (not this one) into the air and try to get it to rotate around a body axis. You will soon find that you can get it to rotate about the shortest axis and also the longest axis, but not about the axis of medium length. This leads to a neat analysis of the rate equations for the components of the angular velocity vector, the construction of an integral, and the detection of stable and unstable modes of rotation. The chapter cover picture is connected with this problem.

1. **(a).** Divide the second equation of the system $x' = 3x$, $y' = -y$ by the first to obtain $dy/dx = -y/3x$. Separating the variables and solving, we have $3\ln|y| = -\ln|x| + C_1$ or $y^3 x = C$, where $C = \pm e^{C_1}$. The equation $y^3 x = 0$ also defines orbits [i.e., the orbits $x = 0$

and $y = 0$]; we have the integral $K(x, y) = y^3 x$, whose level sets define the orbits. The original system is linear with a single equilibrium point at $(0, 0)$. Since the eigenvalues of the system matrix are 3 and -1, the origin is an unstable saddle point. See Fig. 1(a).

(b). The system is $x' = -y$, $y' = 25x$. Proceeding as in 1(a), we have $dy/dx = -25x/y$ or $y\,dy = -25x\,dx$; solving that equation, we obtain the level sets $K(x, y) = 25x^2 + y^2 = C$, where $K(x, y)$ is the desired integral [any nonzero constant multiple of K would do as well]. The system matrix of the original linear system has eigenvalues $\pm 5i$ and is neutrally stable at the single equilibrium point $(0,0)$. See Fig. 1(b).

(c). The system is $x' = y^3$, $y' = -x^3$. We obtain an integral by solving $dy/dx = -x^3/y^3$, or $y^3 dy + x^3 dx = 0$. $K(x, y) = \frac{1}{4}(x^4 + y^4)$ [or any nonzero constant multiple of K] is an integral, and the orbits are the "rounded-off" rectangles defined by $K(x, y) = C$ (neutral stability at the origin). See Fig. 1(c).

(d). The system is $x' = x$, $y' = x + x^2 \cos x$. There is a single equilibrium point at $(0,0)$. We have that $dy/dx = y/x + x\cos x$, which is linear in y, but may be solved most easily by noting that: $d(y/x)/dx = \cos x$ and so $y/x = C_2 + \sin x$. An integral is $K(x, y) = y/x - \sin x$, which is defined in any region not intersecting the y-axis. Since K is not defined at the origin, we must turn elsewhere to determine the stability properties of the system at the origin. Since $x = C_1 e^t$, we see that as t increases so does x; the system must be unstable at the origin. See Fig. 1(d).

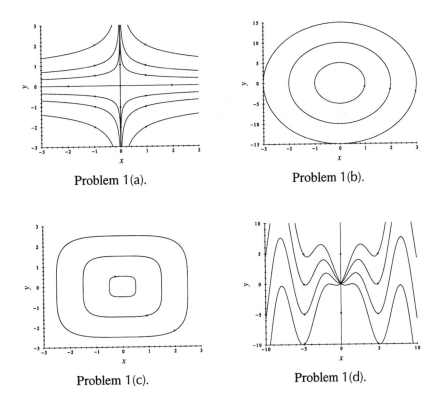

Problem 1(a).

Problem 1(b).

Problem 1(c).

Problem 1(d).

2. The system matrix of $x' = -x$, $y' = -2y$ has eigenvalues -1 and -2, and so the system is asymptotically stable at the origin. Solving $dy/dx = 2y/x$, we obtain $dy/y = 2dx/x$, $\ln|y| = 2\ln|x| + C_1$, $yx^{-2} = e^{C_1}$. $K(x, y) = yx^{-2}$ is an integral in any region not intersecting the y axis. This does not contradict the theorem that a conservative system in a region R cannot have an attractor in R since the region R in this case cannot intersect the y-axis and so does not contain the attractor at the origin.

3. The system is $x' = g(y)$, $y' = f(x)$, $z' = 0$. We have $dy/dx = f(x)/g(y)$. Separating variables and integrating, we obtain the integral $K(x, y) = F(x) - G(y)$, where F and G are any antiderivatives of $f(x)$ and $g(y)$, respectively. Note that $z(t) = C$ for all t. The system is conservative in $\mathbb{R}^3$. By Theorem 8.3.2 in this section, the system does not have an attracting equilibrium point in $\mathbb{R}^3$.

4. **(a).** Dividing the two equations of the system $x' = y$, $y' = -10x - x^3$, we have that $dy/dx = (-10x - x^3)/y$, so $ydy + (10x + x^3)dx = 0$, or $y^2/2 + 5x^2 + x^4/4 = C$. We have that the function $K = y^2/2 + 5x^2 + x^4/4$ is an integral. The level sets are oval-like curves enclosing the origin, and the system has a neutrally stable equilibrium point at the origin. This may be shown by setting $K = C$, where C is a positive constant, and solving for y: $y = \pm(2C - 10x^2 - x^4/2)^{1/2}$. The graph Γ of $K = C$ is a curve that is symmetric about both axes, cuts the y-axis at $y = \pm\sqrt{2C}$, and the x-axis at the two real roots of $2C - 10x^2 - x^4/2$, which are $\pm(-10 + (100 + 4C)^{1/2})^{1/2}$. Moreover, the upper half of the curve Γ rises steadily from 0 to $\sqrt{2C}$ as x increases from the negative root to zero and then falls back to 0 as x increases through zero to the positive root. By calculating dy/dx, we see that Γ has a vertical slope at the two extreme values of x. The origin is enclosed by these oval cycles. See Fig. 4(a).

(b). The system $x' = y$, $y' = -10x + x^3$ has the three equilibrium points $(0, 0)$, $(10^{1/2}, 0)$, $(-10^{1/2}, 0)$. Dividing the two rate equations, we have $dy/dx = (-10x + x^3)/y$, $ydy + (10x - x^3)dx = 0$, or $y^2/2 + 5x^2 - x^4/4 = C$. $K(x, y) = y^2/2 + 5x^2 - x^4/4$ is an integral for all (x, y) in R^2. The level set $K(x, y) = C$ is defined for every real constant C. For each C, the corresponding level set Γ is symmetric about each axis, but now *not* all level sets are ovals about the origin (see Fig. 4(b)). Since Γ is defined by $y = \pm(2C - 10x^2 + x^4/2)^{1/2}$, we need to analyze the behavior of $Q = 2C - 10x^2 + x^4/2$. First of all, its roots are $\pm(10 \pm (100 - 4C)^{1/2})^{1/2}$ which are not real if $C > 25$. If $C = 25$, there are just two roots, and these give the equilibrium points $(10^{1/2}, 0)$, and $(-10^{1/2}, 0)$. If $C < 25$, then the four real roots of Q may be denoted by $\pm a$, $\pm b$, where $0 < a < b$. So $y = \pm 2^{-1/2}[(x^2 - a^2)(x^2 - b^2)]^{1/2}$. Over the range $|x| \le a$, Γ resembles an oval. However, for $x \le -b$ or $x \ge b$, Γ resembles a hyperbola. Note that Γ is not defined for $a < |x| < b$ since y is imaginary for such values of x. The origin is neutrally stable, enclosed by a family of stable ovals that are components of level sets Γ defined for $C < 25$. However, the equilibrium points $(10^{1/2}, 0)$ and $(-10^{1/2}, 0)$ are unstable. This may be shown in several ways. One way is to observe that if $C = 25$, then Γ is given by $2y^2 = (10 - x^2)^2$ whose graph consists of arcs through the equilibrium points $(10^{1/2}, 0)$ and $(-10^{1/2}, 0)$. Since $y' > 0$ if $y > 0$ and $y' < 0$ if $y < 0$, we must have instability. The origin is neutrally stable, and the other two equilibrium points are unstable. See Fig. 4(b). See also Section 3.1.

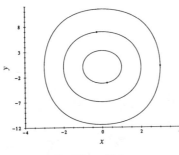

Problem 4(a).

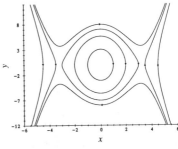

Problem 4(b).

5. The system is $x' = Ax$, where $A^T = -A$, that is, A is skew-symmetric.

(a). Let $K(x) = x^T x$. Then, $K' = (x^T)'x + x^T x' = (Ax)^T x + x^T Ax = x^T A^T x + x^T Ax = 0$ since $A^T = -A$. K remains constant along each orbit of $x' = Ax$. Since $K(x)$ is the square of the distance from the point x to the origin, K is not constant on any ball, so $K(x)$ is an integral, and the system is conservative. The level sets $K(x) = C$, C any positive constant, are the "spheres" in $\mathbb{R}^n$ of radius $\sqrt{C}$ centered at the origin: $K(x) = x^T x = x_1^2 + \cdots + x_n^2 = C$. Each orbit lies entirely on one of these spheres, and the origin is not an attractor. The system is neutrally stable at the origin with $\delta = \varepsilon$ in the definition of stability.

(b). From part (a), we know that each solution $x = x(t)$ of $x' = Ax$ lies on the surface $\|x\| = \|x(0)\|$ for all t, so $x(t)$ is bounded for all t. In particular, no component of $x(t)$ can include terms of the form $t^k \cos \omega t$ or $t^k \sin \omega t$, $k > 0$, since such terms lead to unboundedness. Since the eigenvalues have the form $\pm i\omega_1, \ldots, \pm i\omega_k$, where each ω_i is real (possibly zero), this means by Theorem 7.6.4 that each component of $x(t)$ is a linear combination of terms of the form $\cos(\omega_j t)$ or $\sin(\omega_j t)$, $j = 1, 2, \ldots, k$. Every nonconstant solution is a linear combination of constants and periodic functions, possibly of different periods.

6. The Hamiltonian system is $x_i' = \partial H / \partial y_i$, $y_i' = \partial H / \partial x_i$, $i = 1, \ldots, k$.

(a). By the Chain Rule, we have that $dH(x_1, \ldots, x_k; y_1, \ldots, y_k)/dt = \sum_{i=1}^{k}[(\partial H/\partial x_i)x_i' + (\partial H/\partial y_i)y_i'] = 0$ since $x_i' = \partial H/\partial y_i$ and $y_i' = -\partial H/\partial x_i$. Since H is assumed to be nonconstant on every region, H is an integral of the Hamiltonian system.

(b). By the No Attractors Theorem 8.3.2 in this section, a system that has an integral in a region cannot have an attracting equilibrium point in that region. By part (a), the given Hamiltonian system has an integral on R. So the system cannot be asymptotically stable at any equilibrium point.

(c). The system is $x' = y$, $y' = -f(x)$. Let $H = y^2/2 + \int_0^x f(s)ds$. Then $\partial H/\partial y = y$ and $\partial H/\partial x = f(x)$. So the system has the form $x' = y = \partial H/\partial y$, $y' = -f(x) = -\partial H/\partial x$; the system is Hamiltonian, and so cannot have an asymptotically stable equilibrium point.

(d). There is no choice of restoring force $-f(x)$ such that all solutions of $x' = y$, $y' = -f(x)$ tend to an equilibrium state, regardless of initial position and velocity. Such a restoring force would imply the existence of an attractor, but Hamiltonian systems are conservative, so cannot have any attractors.

7. Group project. See Fig. 7 for some orbits of $x' = y^2 + e^x \cos y + 2 \cos x$, $y' = -e^x \sin y + 2y \sin x$.

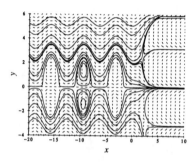

Problem 7.

8. Group project.

8.4 Lyapunov Functions

Suggestions for preparing a lecture

Topics: Quadratic forms and definiteness, the meaning of a Lyapunov function, the asymptotic stability test, examples, a final look at the spinning tennis racket.

Making up a problem set

Problems 1(**a**), (**b**), (**d**), 2(**a**), (**e**), 9, group problem 10.

Comments

Lyapunov functions provide an alternative to the ε-δ arguments of Section 8.1 to determine the stability properties of a system of ODEs at an equilibrium point. These real-valued functions of the state variables tell us a lot about what is going on in the system, but without having to find solution formulas. Given a system of ODEs with an equilibrium point, it is always possible (but *not* always easy) to find a Lyapunov function whose behavior reveals whether the system is stable, asymptotically stable, or unstable at the equilibrium point. Lyapunov functions are never unique, and this is both an advantage and a difficulty, the former because it allows a lot of flexibility, the latter because there may be no clue about how to construct a Lyapunov function. Although total energy is commonly used as a Lyapunov function for models of physical systems, quadratic forms in the state variables are usually easier to use. In this section we emphasize quadratic forms. There are many other ways to construct Lyapunov functions. Building suitable Lyapunov functions has become a cottage industry for scientists and engineers involved in designing complex systems whose stability may be a matter of life or death, systems ranging from nuclear reactors to the control systems for jets and spacecraft.

1. Since the V-functions are never unique, you may have functions that differ from those given here. However if the functions V and V' given here have certain definiteness properties, your functions must have the *same* properties. This is also the case in the other problems in this set.

   (a). The system is $x' = -4y - x^3$, $y' = 3x - y^3$. If $V = ax^2 + cy^2$, then $V' = 2ax(-4y - x^3) + 2cy(3x - y^3) = (-8a + 6c)xy - 2ax^4 - 2cy^4$. If a and c are chosen as positive constants for which $-8a + 6c = 0$ (e.g., $a = 3$, $c = 4$), then $V = 3x^2 + 4y^2$ is positive definite, and $V' = -2ax^4 - 2cy^4$ is negative definite. The system is asymptotically stable at the origin.

   (b). The system is $x' = y$, $y' = -9x$. If $V = ax^2 + cy^2$, $V' = 2axx' + 2cyy' = 2axy - 18cxy = 2(a - 9c)xy$. If we set $a = 9c > 0$ (e.g., $a = 9$, $c = 1$), then $V = 9x^2 + y^2$ is positive definite and $V' = 0$, which is (trivially) negative semidefinite. The system is stable at the origin. It is not asymptotically stable, however, since $V = 9x^2 + y^2$ is constant along each orbit (since $V' = 0$), and so $9x^2 + y^2 = C$ gives elliptical orbits, none of which tends to the origin, unless $C = 0$. Note that $9x^2 + y^2$ is an integral of the separable ODE, $9xdx + ydy = 0$, which we get from the system by dividing y' by x' and rearranging terms.

   (c). The system is $x' = -x + 3y$, $y' = -3x - y$. If $V = ax^2 + cy^2$, $V' = 2axx' + 2cyy' = -2ax^2 + 6axy - 6cxy - 2cy^2 = -2x^2 - 2y^2$ if we set $a = c = 1$. Since $V = x^2 + y^2$ is positive definite and $V' = -2x^2 - 2y^2$ is negative definite, the system is asymptotically stable at the origin.

   (d). The system is $x' = 2x + 2y$, $y' = 5x + y$. If $V = ax^2 + cy^2$, we have that $V' = 2axx' + 2cyy' = 4ax^2 + 4axy + 10cxy + 2cy^2 = 4ax^2 + (4a + 10c)xy + 2cy^2$. This V doesn't tell us much since any choice of a and c which makes V definite yields an indefinite V'. However, if we add the term $2bxy$ to V, we may be able to find a suitable test function. Since multiples of test functions are also test functions, there is no loss of freedom if we set $c = 1$. If $V = ax^2 + 2bxy + y^2$, then $V' = (2ax + 2by)x' + (2bx + 2y)y' = (4a + 10b)x^2 + (4a + 6b + 10)xy + (4b + 2)y^2$. Now if we choose $a = -2$, $b = 2$, then $V = -2x^2 + 4xy + y^2$ is indefinite since the coefficients of x^2 and y^2 have opposite signs. However, in this case, $V' = 12x^2 + 14xy + 10y^2$ is positive definite. The system is unstable at the origin (as expected, since one of the eigenvalues of the system matrix is positive).

   (e). The system is $x' = (-x + y)(x^2 + y^2)$, $y' = (x + y)(x^2 + y^2)$. With $V = ax^2 + cy^2$, we have $V' = (x^2 + y^2)(-2ax^2 + 2(a - c)xy - 2cy^2) = -2(x^2 + y^2)^2$ if we set $a = c = 1$. Since V is positive definite and V' is negative definite, the system is asymptotically stable at the origin.

   (f). The system is $x' = -2x - xe^{xy}$, $y' = -y - ye^{xy}$. If $V = ax^2 + cy^2$, $V' = 2axx' + 2cyy' = -4ax^2 - 2cy^2 - e^{xy}(2ax^2 + 2cy^2) = -4x^2 - 2y^2 - (x^2 + y^2)e^{xy}$ [if we set $a = c = 1$], which is negative definite, while V is positive definite. The system is asymptotically stable at the origin.

   (g). The system is $x' = -x^3 + x^3y - x^5$, $y' = y + y^3 + x^4$. If $V = ax^2 + cy^2$, $V' = 2axx' + 2cyy' = -2ax^4(1 + x^2) + 2cy^2(1 + y^2) + 2(c + a)x^4y$. If $a = -1$ and $c = 1$, then $V = -x^2 + y^2$ is indefinite, while $V' = 2x^4(1 + x^2) + 2y^2(1 + y^2)$ is positive definite; the system is unstable at the origin.

(h). The system is $x' = y$, $y' = -2x - 3y$. If $V = ax^2 + cy^2$, then $V' = 2(a - 2c)xy - 6cy^2$, which is indefinite (or semidefinite) for small values of $|x|$ and $|y|$. So let's try a Lyapunov function of the form $V = ax^2 + 2bxy + cy^2$ and see if we can do better. Then $V' = -4bx^2 + 2(a - 3b - 2c)xy + 2(b - 3c)y^2$. If we select $b = c = 1$, $a = 5$, then $V = 5x^2 + 2xy + y^2$ is positive definite (Theorem 8.4.2) and $V' = -4x^2 - 4y^2$ is negative definite. The system is asymptotically stable at the origin.

2. The dashed curves in the figures are the level sets of the V functions, and the solid curves are orbits of the respective systems.

(a). $x' = -4y - x^3$, $y' = 3x - y^3$; $V = 3x^2 + 4y^2 = 2, 8, 16$. See Fig. 2(a).

(b). $x' = -x + 3y$, $y' = -3x - y$; $V = x^2 + y^2 = 0.5, 2, 6$. See Fig. 2(b).

(c). $x' = (-x + y)(x^2 + y^2)$, $y' = -(x + y)(x^2 + y^2)$; $V = x^2 + y^2 = 0.5, 2, 6$. See Fig. 2(c).

(d). $x' = -2x - xe^{xy}$, $y' = -y - ye^{xy}$; $V = x^2 + y^2 = 0.5, 2, 6$. See Fig. 2(d).

(e). $x' = y$, $y' = -2x - 3y$; $V = 5x^2 + xy + y^2 = 0.25, 2, 6$. See Fig. 2(e).

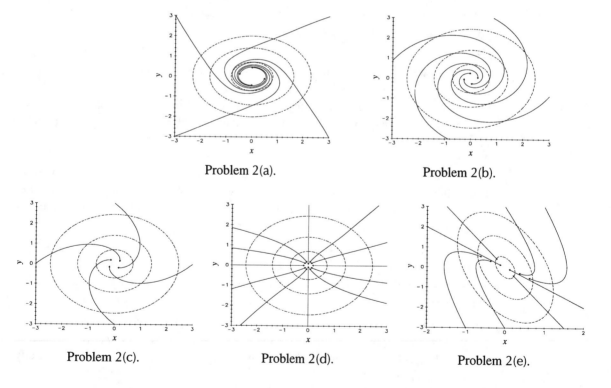

Problem 2(a).   Problem 2(b).

Problem 2(c).   Problem 2(d).   Problem 2(e).

3. **(a).** The system is $x' = -x - x^3$, $y' = -y$. If $V = x^2 + y^2$, then $V' = 2x(-x - x^3) + 2y(-y) = -2x^2 - 2y^2 - 2x^4$. Since V is positive definite and V' is negative definite, the system is asymptotically stable at the origin.

(b). The system is $x' = x + x^2$, $y' = -y$. If $V = x^2 - y^2$, then V is indefinite and $V' = 2x(x + x^2) - 2y(-y) = 2x^2 + 2y^2 + 2x^3$ is positive definite (for $|x| < 1$); the system is unstable at the origin.

(c). The system is $x' = \varepsilon x - y + x^2$, $y' = x + \varepsilon y + x^2$, $\varepsilon > 0$. If $V = x^2 + y^2$, then V is positive definite and $V' = 2x(\varepsilon x - y + x^2) + 2y(x + \varepsilon y + x^2) = 2\varepsilon(x^2 + y^2) + 2x^2(x + y)$ which is also positive definite since $\varepsilon > 0$ if $|x|$ and $|y|$ are small enough (say, $|x|$ and $|y|$ less than $\varepsilon/10$); the system is unstable at the origin.

4. **(a).** The system is $x' = -x^3 + y^3$, $y' = -x^3 - y^3$. If $V = ax^{2m} + by^{2n}$, then $V' = 2amx^{2m-1}(-x^3 + y^3) - 2bny^{2n-1}(x^3 + y^3)$. Choose $a = b = 1$ and $m = n = 2$. Then $V = x^4 + y^4$ and $V' = -4(x^6 + y^6)$. The system is asymptotically stable at the origin since V is positive definite and V' is negative definite. See Fig. 4(a) where the dashed curves are the level sets $V = 1/16, 1, 20$.

(b). The system is $x' = -2y^3$, $y' = 2x - y^3$. Let $V = ax^{2m} + by^{2n}$. Then we have that $V' = -4amx^{2m-1}y^3 + 2bny^{2n-1}(2x - y^3)$. If we choose $a = 2$, $b = 1$, $m = 1$, $n = 2$, then V is positive definite since it is the sum of positive multiples of squares, while $V' = -4y^6$ which is negative semidefinite. The system is stable, but not necessarily asymptotically stable. See Fig. 4(b) where the dashed curves are the level sets $V = 1/16, 1, 8$. Note that the orbits cross the level sets of V at points $(x, 0)$ tangentially, but still moving inward as time increases. In fact, the system is asymptotically stable at the origin, but we haven't actually shown that.

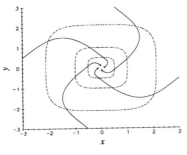

Problem 4(a).

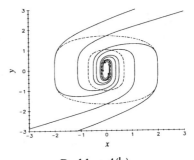

Problem 4(b).

5. The system is $x' = (x - y)^2(-x + y)$, $y' = (x - y)^2(-x - y)$. Every point on the line $y = x$ is an equilibrium point; so the origin cannot be asymptotically stable since it is not an attractor. If $V = x^2 + y^2$, then $V' = 2xx' + 2yy' = -2(x - y)^2(x^2 + y^2) \leq 0$; i.e., V' is negative semidefinite, while V is positive definite. The system is neutrally stable at the origin since it is stable, but not asymptotically stable.

6. The system is $x' = y$, $y' = -(g/L)\sin x - cy/m$, where g, L, c, and m are positive constants.

(a). Let's define V by

$$V(x, y) = \frac{4g}{L}[1 - \cos(x - 2k\pi)] + 2y^2 + \frac{2c}{m}(x - 2k\pi)y + \frac{c^2}{m^2}(x - 2k\pi)^2$$

Differentiating V following the motion of the pendulum system, and combining and canceling terms, we have that (after a lot of algebraic manipulation and replacing $\sin x$ by $\sin(x - 2k\pi)$ at one step)

$$V' = -\frac{2c}{m}y^2 - \frac{2cg}{mL}(x - 2k\pi)\sin(x - 2k\pi)$$

Now $V(x, y)$ is positive definite in the region $|x - 2k\pi| < \pi$ because it is the sum of the term $4g(1 - \cos(x - 2k\pi))/L$, which is positive except when $x = 2k\pi$ where it is zero, and a quadratic in $x - 2k\pi$ and y which is positive definite by the criterion given in Theorem 8.4.2. On the other hand, V' is negative definite if $|x - 2k\pi| < \pi$ since $-(x - 2k\pi)\sin(x - 2k\pi)$ is negative in that region except when $x = 2k\pi$, where it is zero. The Test for Asymptotic Stability (Theorem 8.4.1) shows that this system is asymptotically stable at every point $(2k\pi, 0)$.

(b). Choose $V = -y\sin(x - x_0) + c(1 - \cos(x - x_0))/m$. Then, $V' = -y'\sin(x - x_0) - yx'\cos(x - x_0) + (c/m)x'\sin(x - x_0) = -y^2\cos(x - x_0) + (g\sin^2(x - x_0))/L$, where we have replaced the term $\sin x$ in the ODE, $y' = -(g/L)\sin x - cy/m$, by $-\sin(x - x_0)$ [since $x_0 = (2k + 1)\pi$]. V' is positive definite near $((2k + 1)\pi, 0)$ since $\cos((2k + 1)\pi) = -1$. However, observe that V takes on positive values at all points $((2k + 1)\pi + \varepsilon, 0)$ close enough to $((2k + 1)\pi, 0)$. Therefore, the system is unstable at every point $((2k + 1)\pi, 0)$ by Theorem 8.4.4.

7. The system is $x' = (a - by)x$, $y' = (-c + dx)y$, where a, b, c, d are positive constants. The function Q is given by $(y^a e^{-by})(x^c e^{-dx})$. It was shown in Problem 6(a) in Section 5.4 that Q attains a strict local maximum at the equilibrium point $P(c/d, a/b)$. So $V = Q(c/d, a/b) - Q(x, y)$ is positive definite in a region containing P. Moreover, $V' = 0$ for all (x, y) in the first quadrant since $Q(x, y)$ is an integral of the equivalent first-order ODE, $(dxy - cy)\,dx + (-ax + bxy)\,dy = 0$. Therefore, the system is stable. Since the orbits are shown in Problem 6, Section 5.4 to be cycles, the system is neutrally stable, and *not* asymptotically stable.

8. Let's verify Theorem 8.4.2. Suppose $a \neq 0$. Then $V(x, y) = ax^2 + 2bxy + cy^2$ is a quadratic function in x whose roots [by the quadratic formula] are $x = a^{-1}y(-b \pm (b^2 - ac)^{1/2})$. The roots are complex if and only if $b^2 < ac$. In this case, V has a fixed sign that is positive if $a > 0$ [$V(0, 0) = 0$, however]. V is positive definite in this case. The other way around is also true since the positive definiteness of V implies that, except at the origin, $V(x, y)$ is positive, and for each $y_0 \neq 0$, $V(x, y_0) = 0$ has no real root x_0 for each value of y_0. On the other hand, if $b^2 = ac$, $a > 0$, then V vanishes along the straight line $x = -by/a$, but is otherwise positive [V is positive semidefinite]. The other way around, if V is positive semidefinite, for each $y_0 \neq 0$, the quadratic $V(x, y_0) = 0$ cannot have two distinct real roots, x_1 and x_2, because the existence of such roots implies V changes sign as x passes through each root. We must have that $b^2 \leq ac$ and $a > 0$. The criteria for negative

definiteness or semidefiniteness are shown in the same way. If none of the criteria hold, then V is indefinite and takes on both positive and negative values in every neighborhood of the origin.

9. The system is $x' = y - xf(x, y)$, $y' = -x - yf(x, y)$. Take $V(x, y) = x^2 + y^2$, which is positive definite. The derivative of V following the motion of the system is $V' = -2(x^2 + y^2)f(x, y)$. The conclusion in each part follows from the fact that $-2(x^2 + y^2)$ is negative definite, so the definiteness properties of V' are opposite to those of f.

(a). Since f is positive semidefinite, V' is negative semidefinite and the system is stable at the origin.

(b). Since f is positive definite, V' is negative definite, and the system is asymptotically stable at the origin.

(c). Since f is negative definite, V' is positive definite, and the system is unstable at the origin.

(d). See Fig. 9(d), Graph 1 where $f(x, y) = |x| + |y|$ and Fig. 9(d), Graph 2 where $f(x, y) = -\cos(x^2 + y^2)$. The level sets in Fig. 9(d), Graph 1 are $V = 0.25, 1, 4$ and in Fig. 9(d), Graph 2 are $V = 1/16, 9/16$.

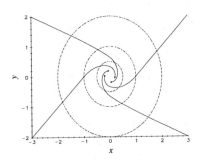

Problem 9(d), Graph 1.

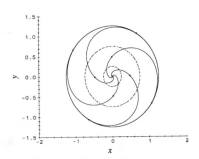

Problem 9(d), Graph 2.

10. Group project.

Background Material: Verification of Part 1 of Theorem 8.2.1

We want to show that the system $x' = Ax + P(x)$ is asymptotically stable at the equilibrium point 0 if the eigenvalues of the system matrix A have negative real parts and the "perturbation" $P(x)$ has order at least 2 at $x = 0$. We use the following fact (known as Lyapunov's Lemma) which we don't prove:

If the eigenvalues of A have negative real parts then the matrix equation

$$A^T B + BA = -I \tag{i}$$

has a matrix solution B such that the scalar function $V(x) = x^T Bx$ is positive definite.

Now we will show that $V(x)$ is a strong Lyapunov function for $x' = Ax + P(x)$ in a neighborhood of $x = 0$. By Lyapunov's Lemma we know that V is positive definite. We will show that V' is negative definite, so by Theorem 8.4.1 (Lyapunov's First Theorem) that means that the system $x' = Ax + P(x)$ is asymptotically stable at the origin. First note that because $P(x)$ has order at least 2 at the origin, then (for some positive constant c) $||P(x)|| \leq c||x||^2$ for all x in a neighborhoood of the origin.

Now let's calculate the derivative of $V(x) = x^T Bx$ following the motion of $x' = Ax + P$:

$$V' = (x')^T Bx + x' Bx' = (x^T A^T + P^T)Bx + x^T B(Ax + P)$$

$$= x^T(A^T B + BA)x + P^T Bx + x^T BP = x^T(-I)x + g(x) \qquad \text{(ii)}$$

$$= -||x||^2 + g(x)$$

where $g(x)$ is the scalar function $P^T Bx + x^T BP$ and where we have used (i). We can use the Triangle Inequality, the Cauchy-Schwarz Inequality, properties of the matrix norms, and the estimate on $||P(x)||$ to estimate $|g(x)|$:

$$|g(x)| = |P^T Bx + x^T BP| \leq |P^T Bx| + |x^T BP| = 2|P^T Bx|$$

$$\leq 2||P|| \cdot ||Bx|| \leq 2c||x||^2||Bx|| \qquad \text{(iii)}$$

$$\leq 2c||x||^2||B|| \cdot ||x|| \leq C||x||^3$$

where C is a positive constant. An upper bound for V' is found from (ii) and (iii):

$$V'(x) \leq -||x||^2[1 - C||x||] \qquad \text{(iv)}$$

If x is restricted to the open ball centered at $x = 0$ defined by $|x| < 1/C$, then (iv) implies that V' is negative definite on the open ball. $V(x)$ is a (local) strong Lyapunov function at the origin for the system $x' = Ax + P$, so that system is (locally) asymptotically stable. Note that $V(x)$ also is a (global) strong Lyapunov function for the linear system $x' = Ax$.

Set $u = x - p$ if the equilibrium point is at $x = p$ and $x' = A(x - p) + P(x)$, where $P(p) = 0$ and $P(x)$ has order at least 2 at $x = p$. Then replace the system by $u' = Au + Q(u)$ where $Q(u) = P(u + p)$, and apply the above argument.

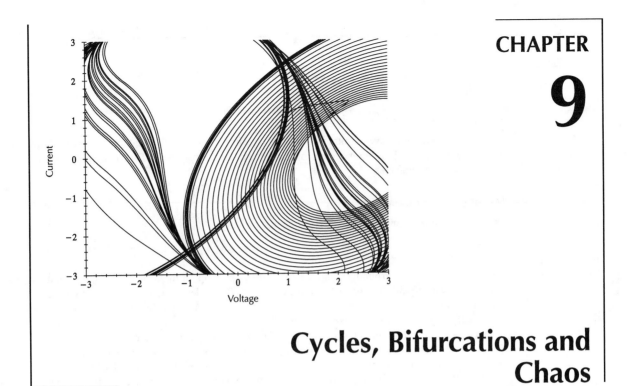

Cycles, Bifurcations and Chaos

Cycles, limit cycles, cycle-graphs, the long-term behavior of a bounded orbit of a planar autonomous system, bifurcations (including the Hopf bifurcation), and the Lorenz system and chaos are the subject of this chapter. We prove a few things, state without proof several basic results, and rely on pictures to help readers understand the ideas. Everything is carried out with just two state variables (usually) and autonomous systems (always) in the first three sections, and this is for two reasons: (1) solution and orbital behavior is easier to see, and (2) Theorem 9.2.1, which gives a precise description of the asymptotic behavior of planar orbits, fails miserably if there are more than two state variables or if the system is nonautonomous. The Lorenz system (Section 9.4) is autonomous, has three state variables, and displays apparently chaotic behavior. Some of the most interesting nonlinear systems appear in the group problems of Sections 9.2–9.4. Any one of these would make an excellent group project. The book and chapter cover figures show the apparently chaotic (but organized) wanderings of voltage and current in the nonlinear scroll circuit (see Section 9.3).

The apparently chaotic dynamics of the scroll circuit of Section 9.3 and the Lorenz system of Section 9.4 raise deep (and as yet unanswered) questions about just what chaos is. This is a wide-open area of intellectual and practical research. We expect that in a few years our approach will seem quaint and a bit old-fashioned, but that often happens when ideas are still in the development stage.

9.1   Cycles

Suggestions for preparing a lecture

Topics: Cycles, limit cycles, attracting cycles, repelling cycles, the fact that linear systems don't have attracting or repelling limit cycles, the van der Pol circuit and its mathematical model, sensitivity to a circuit parameter.

Making up a problem set

One or two parts of 1 and 2, 3, one of 4–7, one part of 8 or 9.

Comments

Cyclical behavior is all around us, and it is not hard to pick up on the idea of a periodic solution as a nonconstant equilibrium state. Limit cycles are a little harder to grasp, which is why we give several examples early on in the section. Since circular cycles are most easily described in polar coordinates, there are a lot of r's and θ's here. The van der Pol system is derived as an extension from a simple linear RLC circuit system to a circuit system with a nonlinear resistor (e.g., a diode). We introduce these circuit models as systems (rather than as second-order ODEs) because that is the most natural way to do the modeling. It is worth emphasizing that the van der Pol cycles are isolated, attracting, and with periods that change with the parameter. If you haven't done much with circuits, you can do the van der Pol system entirely graphically for various values of the parameter and various forms for the function f. Numerical orbits and component curves are pretty convincing here! The other feature is that the van der Pol cycle is a "relaxation oscillation" for large values of the parameter. We mention Hilbert's 16th problem in a footnote. You might take a look at *Mathematical Developments Arising from Hilbert Problems*, F.E. Browder (ed.), Amer. Math. Soc., 1971, for a discussion of 21 of the 23 Hilbert problems (but not the 16th!).

---

1.   The equilibrium points of these systems in r and θ are obtained by finding the points where r' and θ' are zero. If r' is zero when r is zero, then the origin in the xy-plane is an equilibrium point, regardless of the value of θ'. Elsewhere, both r' and θ' must be zero at an xy-equilibrium point. The signs of r' and θ' near a cycle or an equilibrium point will help you decide whether the cycle or point attracts or repels. The figures show arcs of orbits, with arrowheads indicating the direction of time's increase.

(a). The system is $r' = 4r(4 - r)(5 - r)$, $\theta' = 1$. The equilibrium point is $(0, 0)$, which is unstable since r' is positive for $0 < r < 1$. This system has the attracting limit cycle $r = 4$ (r' is positive for $1 < r < 4$ and negative for $4 < r < 5$) and the repelling limit cycle $r = 5$ (r' is positive for $r > 5$). See Fig. 1(a).

(b). The system is $r' = r(r - 1)(2 - r^2)(3 - r^2)$, $\theta' = -3$ The equilibrium point is $(0, 0)$, which is asymptotically stable since r' is negative for $0 < r < 1$. This system has the attracting limit cycle $r = \sqrt{2}$ ($r' > 0$ for $1 < r < \sqrt{2}$, and $r' < 0$ for $\sqrt{2} < r < \sqrt{3}$); and the repelling limit cycles $r = 1$ and $r = \sqrt{3}$ (r' is positive for $1 < r < \sqrt{2}$, negative for $0 < r < 1$ and $\sqrt{2} < r < \sqrt{3}$, and positive for $r > \sqrt{3}$). See Fig. 1(b).

(c). The system is $r' = r(1 - r^2)(4 - r^2)$, $\theta' = 1 - r^2$. The equilibrium points are the origin (unstable since r' is positive for $0 < r < 1$) and all points on the circle $r = 1$. Since $r' > 0$ for $0 < r < 1$ and $r' < 0$ for $1 < r < 2$, the circle $r = 1$ of equilibrium points attracts. In fact, each equilibrium point on the circle is stable. Here's why. Note that $dr/d\theta = r(4 - r^2)$. Separating variables and using partial fractions, we have (after a calculation which we omit) $\theta = (1/8) \ln |r^2/(4 - r^2)| + C$. So as $r \to 1$, $\theta \to (1/8) \ln(1/3) + C$; that is, by choosing C appropriately, $\theta \to$ any given number θ_0. So the circle $r = 1$ consists of equilibrium points, each of which is approached by two orbits, one with $r \to 1^+$, the other with $r \to 1^-$. So these equilibrium points are neutrally stable. This system has the repelling limit cycle $r = 2$ ($r' > 0$ for $r > 2$ and $r' < 0$ for $1 < r < 2$). See Fig. 1(c) where the circle of neutrally stable equilibrium points is dashed.

(d). The system is $r' = r(1 - r^2)(9 - r^2)$, $\theta' = 4 - r^2$. The equilibrium point is (0,0), which is unstable since r' is positive for $0 < r < 1$. This system has the attracting limit cycle $r = 1$ ($r' > 0$ for $0 < r < 1$, and $r' < 0$ for $1 < r < 3$); and the repelling limit cycle $r = 3$ ($r' < 0$ for $1 < r < 3$, and $r' > 0$ for $r > 3$). See Fig. 1(d).

(e). The system is $r' = r \cos \pi r$, $\theta' = 1$. The equilibrium point is (0,0) which is unstable since r' is positive near $r = 0$. For $n = 0, 1, 2, \ldots$, this system has the attracting limit cycles $r = 2n + 1/2$ because $r' > 0$ for r near but less than $2n + 1/2$, and $r' < 0$ for r near but larger than $2n + 1/2$; similarly, the limit cycles $r = 2n + 3/2$ are repelling. See Fig. 1(e).

(f). The system is $r' = r \sin(\pi/r)$, $\theta' = -2$. There are cycles enclosing the origin for $r = 1/n$, $n = 1, 2, \ldots$. So the equilibrium point at the origin is stable [e.g., for $\varepsilon > 0$ let $\delta = \varepsilon$ in the definition of stability], but it is not asymptotically stable. Since r' changes sign as r increases through the value $1/n$, n a positive integer, we see that the cycles defined by $r = 1/n$, $n = 1, 2, \ldots$. are alternately attracting and repelling. See Fig. 1(f). This has a center-spiral point at the origin.

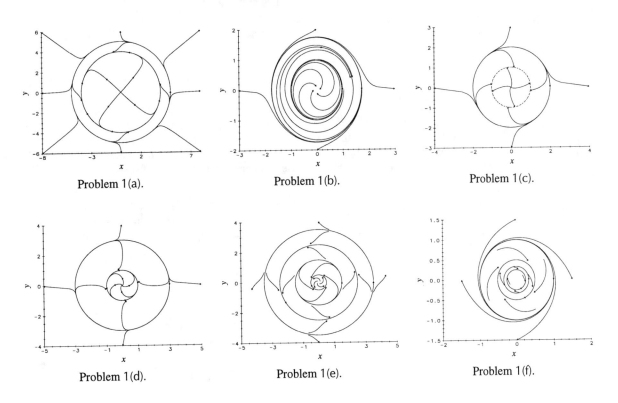

Problem 1(a). Problem 1(b). Problem 1(c).

Problem 1(d). Problem 1(e). Problem 1(f).

2. Convert the system to polar coordinates and use the sign of r' near an equilibrium point or cycle to help determine attraction or repulsion.

(a). The system is $x' = y - x(x^2 + y^2)$, $y' = -x - y(x^2 + y^2)$. Using polar coordinates, the original system becomes $r' = -r^3$, $\theta' = -1$. This system has the asymptotically stable equilibrium point $(0,0)$. There are no cycles.

(b). The system is $x' = x + y - x(x^2 + y^2)$, $y' = -x + y - y(x^2 + y^2)$. Similar to **(a)**, the original system becomes $r' = r - r^3$, $\theta' = -1$. Because $r' > 0$ if $0 < r < 1$, $r' < 0$ if $r > 1$, this system has an unstable equilibrium point at the origin, and a unique attracting limit cycle Γ of period 2π defined by $x^2 + y^2 = 1$.

(c). The system is $x' = 2x - y - x(3 - x^2 - y^2)$, $y' = x + 2y - y(3 - x^2 - y^2)$. Using polar coordinates, the original system becomes $r' = -r + r^3$, $\theta' = 1$. This system has a unique repelling limit cycle, $x^2 + y^2 = 1$, since $r' < 0$ if $0 < r < 1$, $r' > 0$ if $r > 1$.

(d). Can you see the two attracting and one repelling limit cycles in Fig. 2(d)?

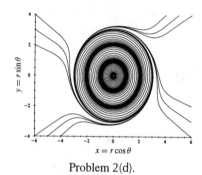

Problem 2(d).

3. The system $r' = r^3 \sin(1/r)$, $\theta' = 1$ has the repelling limit cycles $r = 1/(2n + 1)\pi$ $n = 0, 1, \dots$ and the attracting limit cycles $r = 1/(2n\pi)$, $n = 1, 2, \dots$. Because $r = 1/(2n + 1)\pi \to 0$ and $r = 1/(2n\pi) \to 0$ as $n \to \infty$, the origin is said to be a center-spiral point for this system; the origin is surrounded by a family of cycles separated from one another by annular regions of alternately inward and outward spiraling orbits, because r' changes sign across each cycle.

4. The system is $x' = y^3$, $y' = -x^3$. Dividing y' by x' and separating variables, we have that $x^3 dx + y^3 dy = 0$, which integrates to $x^4 + y^4 = C$, which defines a family of simple, closed curve orbits about the origin, each orbit corresponding to a periodic solution. See Fig. 4 for orbits and x and y component graphs; note that the periods are not constant, but seem to *decrease* as the amplitude increases.

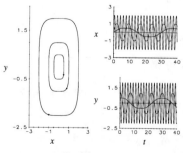

Problem 4.

5. The system $r' = r(1 - r^2)^2(4 - r^2)(9 - r^2)$, $\theta' = 1$ has a repelling limit cycle, $r = 3$, and an attracting limit cycle, $r = 2$; $r = 1$ is a semistable cycle, attracting interior orbits and repelling exterior orbits. This follows because $r' > 0$ if $0 < r < 1$, $1 < r < 2$, or $r > 3$, while $r' < 0$ if $2 < r < 3$. See Fig. 5.

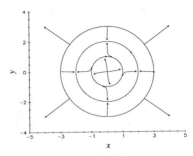

Problem 5.

6. The system is $r' = r(r - 1)^2 \sin(\pi/(r - 1))$, $\theta' = 1$. Note that as $r \to 1$, $(r - 1)^2 \sin(\pi/(r - 1)) \to 0$, since $r - 1 \to 0$ and $|\sin(\pi/(r - 1))| \le 1$ even though the sine function has no limit as $r \to 1$. So, $r = 1$ is a cycle. In addition, $r' = 0$ if $r = 1 \pm 1/n$, and so $r = 1 + 1/n$ and $r = 1 - 1/n$ are cycles for each n. But $r = 1 \pm 1/n \to 1$ as $n \to \infty$, so the cycle $r = 1$ is neither isolated from the cycles defined by $r = 1 \pm 1/n$, nor is it a limit cycle, because the cycle orbits near $r = 1$ are neither attracted to, nor repelled by $r = 1$. Note that every ring containing the circle $r = 1$ contains spiralling orbits that are *not* cycles.

7. The original system in r, θ, z coordinates is $r' = r(1 - r^2)$, $\theta' = 25$, $z' = \alpha z$. The system in xyz-coordinates is $x' = x - 25y - x(x^2 + y^2)$, $y' = 25x + y - y(x^2 + y^2)$, $z' = \alpha z$. Viewed solely from the xy-plane, the unit circle is an attracting limit cycle. It is an attracting limit cycle in $\mathbb{R}^3$ as well if $\alpha < 0$. See Fig. 7, Graph 1 (where $\alpha = -0.5$) for two nonconstant orbits, one with $x_0^2 + y_0^2 < 1$ and the other with $x_0^2 + y_0^2 > 1$. If $\alpha = 0$, then z is constant and orbits lie in planes $z = z_0$. The unit circle in each plane is a limit cycle that attracts each nonconstant orbit in the plane. However, the cycle $r = 1$, $z = 0$ is *not* an attractor because it does not attract orbits from other planes. See Fig. 7, Graph 2 for orbits in the planes $z = 10$, 40. If $\alpha > 0$ and $x = x(t)$, $y = y(t)$, $z = z(t)$ defines a nonconstant solution, then $x^2(t) + y^2(t) \to 1$ as $t \to +\infty$, while $z(t) \to +\infty$ if $z(0) > 0$. Figure 7, Graph 3 shows the orbit of $x = y = 0$, $z(t) = 0.01e^t$ and two other orbits with $z_0 > 0$, one with $x_0^2 + y_0^2 > 1$ and the other with $x_0^2 + y_0^2 < 1$.

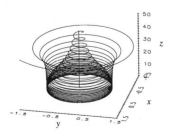

Problem 7, Graph 1.

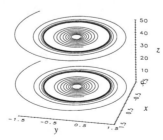

Problem 7, Graph 2.

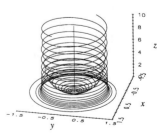

Problem 7, Graph 3.

8. After part **(a)**, the positive root a is determined for each $f(x)$ [see condition **(b)** of Theorem 9.1.1], but the other steps are omitted.

(a). The system is $x' = y - \mu(x^3 - 10x)$, $y' = -x$. Since $f(x) = x^3 - 10x$, $f(-x) = -x^3 + 10x = -f(x)$; $a = \sqrt{10}$. Then $f(x) < 0$ if $0 < x < a$ and $f(x) > 0$ if $x > a$; $f(x) \to \infty$ as $x \to \infty$. Therefore, the conditions of the van der Pol Cycle Theorem (Theorem 9.1.1) are satisfied. See Figs. 8(a), Graphs 1, 2, where $\mu = 0.1, 2$. The cycle periods are approximately 7 and 35, respectively.

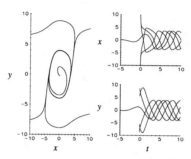

Problem 8(a), Graph 1.

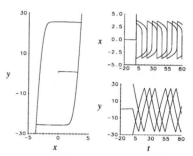

Problem 8(a), Graph 2.

(b). The system is $x' = y - \mu x(|x| - 1)$, $y' = -x$. Then $f(x) = x(|x| - 1)$ and $a = 1$. See Figs. 8(b) for $\mu = 0.5, 5, 50$. The cycle periods are approximately 7, 10, and 55, respectively.

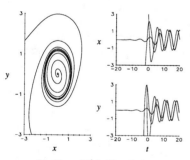

Problem 8(b), Graph 1.

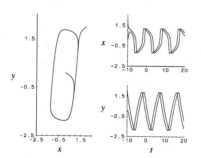

Problem 8(b), Graph 2.

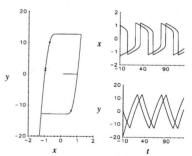

Problem 8(b), Graph 3.

(c). The system is $x' = y - \mu x(x^4 + x^2 - 1)/10$, $y' = -x$. Then $f(x) = x(x^4 + x^2 - 1)/10$ and $a = \sqrt{(-1 + \sqrt{5})/2}$. See Figs. 8(c) for $\mu = 0.1, 1$. The cycle periods are approximately 6 and 7, respectively.

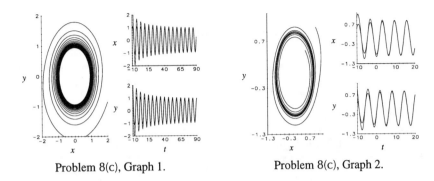

Problem 8(c), Graph 1.   Problem 8(c), Graph 2.

(d). The system is $x' = y - \mu x[2x^2 - \sin^2(\pi x) - 2]$, $y' = -x$. Then $f(x) = x[2x^2 - \sin^2(\pi x) - 2]$ and $a = 1$. See Figs. 8(d) for $\mu = 0.5, 5$. The cycle periods are approximately 8 and 30, respectively.

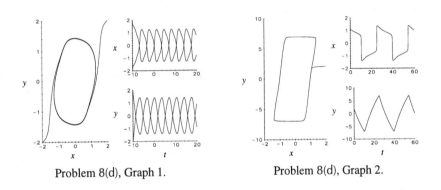

Problem 8(d), Graph 1.   Problem 8(d), Graph 2.

(e). The system is $x' = y - \mu(x - |x + 1| + |x - 1|)$, $y' = -x$. Then $f(x) = (x - |x + 1| + |x - 1|)$ and $a = 2$. See Figs. 8(e) for $\mu = 0.5, 5, 50$. The cycle periods are approximately 9, 14, and 110, respectively.

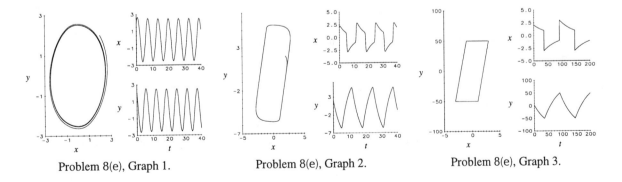

Problem 8(e), Graph 1.   Problem 8(e), Graph 2.   Problem 8(e), Graph 3.

9. In each figure, orbits in the xy-plane are at the left and x and y component graphs at upper and lower right, respectively.

(a). The system is $x' = y - \mu(x^3 - x)$, $y' = -x$. See Figs. 9(a) for $\mu = 0.1, 5, 10$. The periods appear to be about 7, 10, and 20, respectively. The period appears to increase as μ increases. The x-amplitude seems to remain constant, while the y-amplitude increases as μ increases.

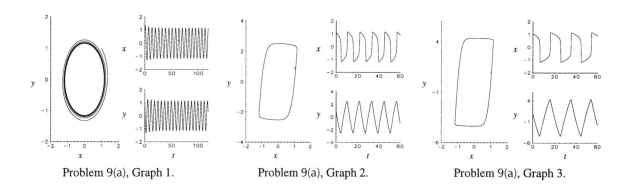

Problem 9(a), Graph 1.   Problem 9(a), Graph 2.   Problem 9(a), Graph 3.

(b). The system is $x' = y - \mu(|x|x^3 - x)$, $y' = -x$. See Figs. 9(b) for $\mu = 0.1, 5, 10$. The periods appear to be about 6, 13, and 22, respectively. The period appears to increase as μ increases. The x-amplitude seems to remain constant, while the y-amplitude increases as μ increases.

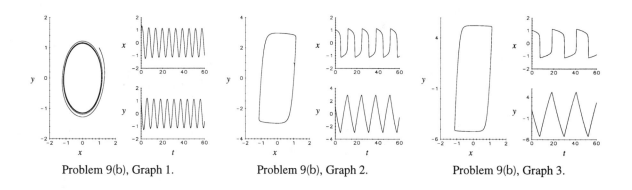

Problem 9(b), Graph 1.   Problem 9(b), Graph 2.   Problem 9(b), Graph 3.

10. The Liénard equation is $x'' + f(x)x' + g(x) = 0$.

(a). Since $y = x' + F(x)$ and $F(x) = \int_0^x f(s)\,ds$, we have $y' = x'' + f(x)x' = -g(x)$. The Liénard equation can be written in system form as $x' = y - F(x)$, $y' = -g(x)$ where $F(x) = \int_0^x f(s)\,ds$.

(b). The ODE is $x'' + (x^2 - 1)x' + x = 0$. The xx'-system is $x' = y$, $y' = -x - (x^2 - 1)y$. The Liénard system is $x' = y - (x^3/3 - x)$, $y' = -x$. See Figs. 10(b), Graphs 1, 2, for plots of the limit cycle in the xx' and Liénard planes, respectively. The graphs are obtained by selecting an initial point near the cycle, solving forward over a long enough span of time that the solution is essentially on the limit cycle, but then plotting only over a final segment of the time span.

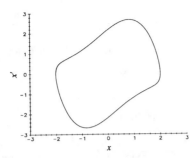

Problem 10(b), Graph 1.

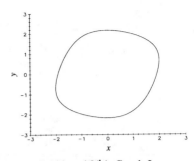

Problem 10(b), Graph 2.

(c). The Rayleigh equation is $z'' + \mu[(z')^2 - 1]z' + z = 0$. Let $x = \sqrt{3}z'$, $x' = \sqrt{3}z''$ and $x'' = \sqrt{3}z'''$. We have $x'' + \mu(x^2 - 1)x' + x = \sqrt{3}z''' + \mu(3(z')^2 - 1)\sqrt{3}z'' + \sqrt{3}z' = \sqrt{3}[z'' + \mu((z')^2 - 1)z' + z]' = 0$. Therefore, if z satisfies the Rayleigh equation, $x = \sqrt{3}\,z'$ satisfies the van der Pol equation. Let $y = x' + \mu(x^3/3 - x)$; then we have $y' = x'' + \mu(x^2 - 1)x'$. Substituting y' into the ODE, $x'' + \mu(x^2 - 1)x' + x = 0$, we see that $y' = -x$. So, we have

$$\frac{dx}{dt} = y - \mu(x^3/3 - x) \qquad \frac{dy}{dt} = -x$$

which has the same form as the van der Pol system in Liénard coordinates. So the Rayleigh equation has a unique attracting limit cycle. For orbits in the Liénard xy-plane, see Figs. 10(c), Graphs 1–4, for $\mu = 1/10$, 1, 5, and 10, respectively. Observe that the orbits in Fig. 10(c), Graph 4 would look much better if the number of points plotted per unit time had been increased.

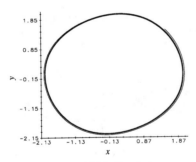

Problem 10(c), Graph 1.

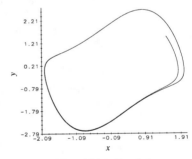

Problem 10(c), Graph 2.

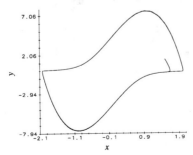

Problem 10(c), Graph 3.

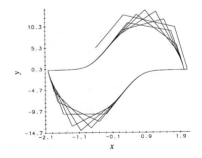

Problem 10(c), Graph 4.

Background Material: Sketch of the van der Pol Cycle Theorem (Theorem 9.1.1)

The form of the rate equation of the van der Pol system

$$x' = y - \mu f(x), \quad y' = -x \tag{i}$$

and the given conditions on μ and f imply that:

- If $x = x(t)$, $y = y(t)$ is a solution of (i), so also is $x = -x(t)$, $y = -y(t)$. Consequently, orbits of (i) in the left half-plane are simply the reflections through the origin of orbits in the right half-plane. For this reason, we only need to carry out our analysis in the right half-plane.

- If there is a cycle, it must enclose the equilibrium point at the origin.

- As noted in Section 5.2, Sign Analysis shows that each orbit Γ of (i) in the right half-plane and above the graph K of $y = \mu f(x)$ moves to the right from initial point A and falls as time increases. Such an orbit crosses K in the vertical downward direction and as τ increases moves to the left and intersects the negative y-axis at a point G. Let D be the point on K above the point $(\beta, 0)$ where $\beta > a$. Sketch a figure in the xy-plane and label the curve K and the points A, G, and D. The van der Pol graph shown uses $\mu = 1/4$, $f(x) = x^3/3 - x$. Use this graph to locate the points A, B, C, D, E, F, G as well as $x = a = \sqrt{3}$ and $x = \beta$ mentioned in the construction.

- According to the reflection property, Γ is a cycle if and only if the distance from O (the origin) to A equals the distance from O to G.

- If Γ is a cycle when $\beta = \beta_0$ then that cycle is unique and attracting if and only if for $\beta > \beta_0$ the distance $|OA|$ from O to A exceeds $|OG|$, while for $0 < \beta < a$ we have $|OA| < |OG|$. This follows (but not very easily) from the reflection properties of orbits and the fact that distinct orbits cannot intersect.

  These properties are used below.

Set $E = x^2/2 + y^2/2$; we see that $dE/dt = x\,dx/dt + y\,dy/dt = x(y - \mu f) + y(-x) = -\mu x f$, Since $dy/dt = -x$, so $-x\,dt = dy$ and we have along Γ

$$dE = \mu f(x)\,dy$$

If we set $\delta E = E_G - E_A$, the difference between the values of E at G and at A, then δE is a well-defined, continuous function of β, the x-coordinate of the point D on Γ. We have that

$$\delta E(\beta) = \frac{1}{2}|OG|^2 - \frac{1}{2}|OA|^2 = \oint_{AG} dE = \oint_{AG} \mu f(x)\,dy$$

The theorem is proven if we show that:

- $\delta E(\beta)$ is positive and bounded for $0 < \beta < a$,

- $\delta E(\beta)$ is strictly decreasing for $\beta \geq a$,

- $\delta E(\beta) \to -\infty$ as $\beta \to +\infty$.

These three properties and the continuity of $\delta E(\beta)$ imply that there is exactly one point β_0 for which $\delta E = 0$. From the formula for $\delta E(\beta)$ and the reflection property for orbits, we see that the orbit Γ corresponding to β_0 must be a cycle since $|OG| = |OA|$. Moreover, the cycle is attracting because the sign of the values of $\delta E(\beta)$ changes from $+$ to $-$ as β increases through β_0. So, we need only prove the three properties of δE listed above.

First, suppose that $0 < \beta < a$. Along the arc of Γ from A to B we have that both $f(x)$ and dy are negative because in this case the arc lies entirely in the vertical strip $0 < x < a$. Since $\mu > 0$ we have that $\delta E(\beta) > 0$. Because $\delta E(\beta)$ is continuous for all $\beta \geq 0$, the Maximum Value Theorem (Theorem B.5.1) implies that $\delta E(\beta) > 0$ has a positive upper bound E_{max} on the interval $0 \leq \beta \leq a$.

Second, suppose that $\beta > a$, and divide the line integral in the formula for $\delta E(\beta)$ into three parts:

$$\delta E(\beta) = \oint_A^G \mu f \, dy = \oint_A^B + \oint_F^G + \oint_B^F = L_1(\beta) + L_2(\beta)$$

where L_1 is the sum of the line integrals from A to B (B is the point on the orbit above $x = a$, and F is the point on the orbit below $x = a$) and from F to G, and $L_2(\beta)$ is the line integral from B to F. We shall show that $L_1(\beta)$ is a positive and decreasing function of $\beta \geq a$ and that $L_2(\beta)$ is negative and decreasing on the same interval. $L_1(\beta)$ is positive because μf and dy are negative (except possibly at A, B, F, and G) on the arcs of Γ involved in calculating the value of $L_1(\beta)$. $L_1(\beta)$ decreases as β increases because

$$\mu f \, dy = \mu f \frac{dy}{dx} dx = -\frac{x \, dx}{y - \mu x f(x)}$$

and, as β increases, the arcs of Γ from A to B and from F to G move outward. This means that as β increases the denominator $y - \mu f(x)$ increases for each fixed x, $0 < x < a$, and the quotient decreases, as must $L_1(\beta)$.

Turning to the function $L_2(\beta)$, $\beta > a$, we see that L_2 is negative because dy is negative on the arc of Γ from B to F while $\mu f(x)$ is positive (except at B and F). Finally, $L_2(\beta)$ is a decreasing function of β, $\beta \geq a$, because dy is negative (as before) and $\mu f(x)$ is positive and increasing for each x, $x \geq a$. So $\delta E(\beta)$ decreases from a positive value E_0 at $\beta = a$ as β increases, $\beta \geq a$.

$$x' = y - 0.25(x^3/3 - x), \qquad y' = -x$$

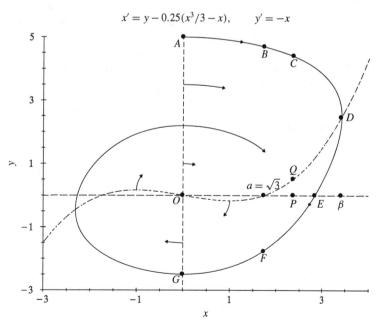

Graphs of $y = \mu f(x)$ and of van der Pol orbits.

Third, to show that $\delta E(\beta) \leq E_0$, we only need show that $L_2(\beta) \to -\infty$ as $\beta \to +\infty$. Fix the point P on the x-axis as shown in the figure. We have that

$$L_2(\beta) = \oint_B^F < \oint_C^E \mu f(x)\, dy \leq -|PQ| \cdot |PC|$$

because $\mu F\, dy$ is everywhere negative along the arc from B to F (except at B and F) and the magnitude of a line integral is at least as large as the product of the minimal magnitude of the integrand μf ($|PQ|$ in this case) along the arc from C to E and the arc length (which is at least $|PC|$ here). But $|PC| \to \infty$ as $\beta \to +\infty$. So, the above inequality for $L_2(\beta)$ implies that $L_2(\beta) \to -\infty$, and we see that the van der Pol system must have a cycle.

9.2 Long-Term Behavior

Suggestions for preparing a lecture

Topics: Cycle-graphs, positive and negative limit sets, long-term behavior in the plane, examples (such as Example 9.2.4). You might draw sketches and ask which are possible cycle-graph limit sets.

Making up a problem set

Problems 1(**d**) and (**f**), 2(**a**), 4(**c**), one or two of 7–11. Students can have fun with group problem 14.

Comments

The nature of the long-term behavior of a bounded solution of an autonomous planar system is the core of this section, but that only makes sense after the idea of a rather mysterious object called a cycle-graph has been explored. This is why we have several examples of cycle-graphs in the text and problem set. The reason for the emphasis on limit sets is that for applications in engineering and science one is often more interested in the long-term behavior of solutions than in the short-term behavior, and the limit sets show the asymptotic nature of orbits. The main result is Theorem 9.2.1 which says that a bounded orbit must tend to an equilibrium point, or to a cycle, or to a cycle-graph, and there are no other possibilities. In other words, chaotic motion *cannot* happen in this setting of a planar autonomous system. This is in marked contrast to what we will see in Section 9.4, and to some extent in Section 9.3. Bendixson's Negative Criterion (Theorem 9.2.4) is a little gem; it is useful and has an elegant proof that uses Green's theorem (see the end of this section). In doing Problem 14 people have created marvelous ODE drawings ranging from the classical kitty shown in the margin to something that might have been done by Picasso in one of his more avant-garde phases.

---

1. **(a).** The system is $x' = y$, $y' = -x$. Divide y' by x' to obtain $dy/dx = -x/y$. So, $x^2 + y^2 = C^2$ defines a circular orbit (i.e., a cycle) about the origin (which is then neutrally stable) for each value of C. Furthermore, the general solution of the original system is $x = C\cos(t + \phi)$ and $y = -C\sin(t + \phi)$, where C and ϕ are arbitrary constants. Consequently, the solutions through the points $(0,0)$ and $(1,1)$ are, respectively, $x = 0$, $y = 0$, and $x = \cos t$, $y = -\sin t$. The negative and the positive limit sets of the orbit Γ through $x = 0$, $y = 0$ are the origin, which is the only equilibrium point; so $\alpha(\Gamma) = \omega(\Gamma) = \Gamma = (0,0)$. The

negative and the positive limit sets of the second orbit are the second orbit itself, which is a cycle; so $\alpha(\Gamma) = \omega(\Gamma) = \Gamma = $ unit circle. There are no cycle-graphs.

(b). The original system $x' = y$, $y' = x$ is equivalent to $x'' - x = 0$, which has the general solution $x = c_1 e^t + c_2 e^{-t}$. Consequently, the solutions through the points (1,1) and (1, −1) are $x = e^t$, $y = e^t$ and $x = e^{-t}$, $y = -e^{-t}$, respectively. The negative and the positive limit sets of the orbit Γ through the point (1,1) are, respectively, the origin and empty (because Γ becomes unbounded as $t \to +\infty$). The negative and the positive limit sets of the orbit Γ through (1, −1) are, respectively, empty (because Γ becomes unbounded as $t \to -\infty$) and the origin. The solution through the point (1,0) is $x = (e^t + e^{-t})/2$, $y = (e^t - e^{-t})/2$, and the negative and the positive limit sets are both empty (because the orbit becomes unbounded as $t \to -\infty$ and as $t \to +\infty$). The origin is the only equilibrium point. There are no cycles or cycle-graphs.

(c). The van der Pol system is $x' = y - (x^3/3 - x)$, $y' = -x$. The negative and the positive limit sets of the orbit Γ through (0.1, 0) are, respectively, the origin and the cycle shown in Fig. 1(c); so $\alpha(\Gamma)$ is the origin and $\omega(\Gamma)$ is the cycle. For the orbit though (3, 0), the negative and positive limit sets are, respectively, the empty set (because the orbit becomes unbounded as t decreases) and the van der Pol cycle shown. The origin is the only equilibrium point, and the cycle shown in Fig. 1(c) is the only cycle. There are no cycle-graphs.

(d). The system $x' = y$, $y' = -\sin x$ has an integral $V = y^2/2 - \cos x$, whose level sets $V = C$ are unions of orbits. From Fig. 1(d), we see that the negative and positive limit sets of the cycle through (0,1) are the cycle itself, which is on the level set of V defined by $y^2/2 - \cos x = -1/2$. The orbit through $(0,\sqrt{2})$ is on the level set of V given by $y^2/2 - \cos x = 0$, and is an oval-like cycle around (0,0) similar to the first cycle; the cycle is its own negative and positive limit set. The orbit through (0,2) is on the level set of V defined by $y^2/2 - \cos x = 1$. This level set contains an oval-like curve enclosing the origin. The oval is composed of an upper arc and a lower arc bounded by the equilibrium points $(\pm\pi, 0)$ corresponding to the pendulum standing upright. The orbit through (0, 2) is the upper arc, and its negative limit set is $(-\pi, 0)$; its positive limit set is $(\pi, 0)$. The equilibrium points of the pendulum system are at $(n\pi, 0)$, $n = 0, \pm 1, \pm 2, \ldots$; each equilibrium point is surrounded by a family of cycles (which are not limit cycles). Each family of cycles is bounded by a cycle-graph (which is not a limit set).

(e). The (polar) system is $r' = r(1 - r^2)(4 - r^2)$, $\theta' = 5$. The negative and the positive limit sets of the orbit through $r = 1/2$, $\theta = 0$ are, respectively, the origin and the cycle $r = 1$, because r' is positive for $0 < r < 1$ and $r' = 0$, $\theta' \neq 0$ on $r = 1$. The negative and the positive limit sets of the orbit through $r = 3/2$, $\theta = 0$ are, respectively, the cycles $r = 2$ and $r = 1$, because $r' < 0$ for $1 < r < 2$. The negative and the positive limit sets of the orbit through $r = 5/2$, $\theta = 0$ are, respectively, the cycle $r = 2$ and the empty set (because $r(t) \to +\infty$ as t increases) since $r' > 0$ for $r > 2$. These results follow from determining the sign of r' in the regions $0 < r < 1$, $1 < r < 2$, $r > 2$. There are no cycle-graphs.

(f). The system is $x' = [x(1 - x^2 - y^2) - y][(x^2 - 1)^2 + y^2]$, $y' = [y(1 - x^2 - y^2) +$

x][$(x^2 - 1)^2 + y^2$]. The negative limit set of the orbit through (1/2,0) appears to be the equilibrium point at the origin, and the positive limit set is the unit circle, which is a cycle-graph consisting of the upper and lower semicircular arcs and the equilibrium points $(1, 0)$ and $(-1, 0)$. The semicircular orbit through (0,1) has negative and positive limit sets, respectively, the equilibrium points (1,0) and (−1,0). The negative and the positive limit sets of the orbit Γ through (3/2,0) are, respectively, empty (because Γ becomes unbounded as t increases) and the unit circle. See Fig. 1(f)

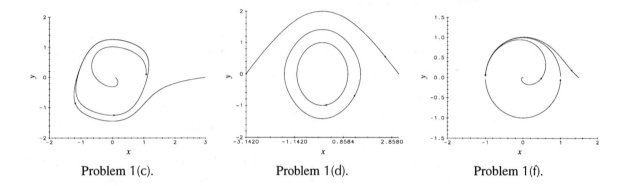

Problem 1(c).   Problem 1(d).   Problem 1(f).

2. Every system has the form $x' = f(x, y)$, $y' = g(x, y)$. Let's denote the quantity $\partial f/\partial x + \partial g/\partial y$ in Theorem 9.2.4 by the symbol Δ.

(a). The corresponding system is $x' = y$, $y' = -(g \sin x)/L - cy/m$. Then Δ is $\partial[y]/\partial x + \partial[-(g \sin x)/L - cy/m]/\partial y = -c/m < 0$. According to Bendixson's Negative Criterion, there is no cycle or cycle-graph anywhere in the xy-plane.

(b). The corresponding system is $x' = y$, $y' = -(bx + cx^3)/m - ay/m$. Then Δ is $\partial[y]/\partial x + \partial[-(bx + cx^3)/m - ay/m]/\partial y = -a/m < 0$. According to Bendixson's Negative Criterion, there is no cycle or cycle-graph anywhere in the xy-plane.

(c). The corresponding system is $x' = y$, $y' = -g(x) - (2 + \sin x)y$. Then Δ is $\partial[y]/\partial x + \partial[-g(x) - (2 + \sin x)y]/\partial y = -(2 + \sin x) < 0$. According to Bendixson's Negative Criterion, there is no cycle or cycle-graph anywhere in the xy-plane.

(d). The corresponding system is $x' = y$, $y' = -g(x) - f(x)y$. Then Δ is $\partial[y]/\partial x + \partial[-g(x) - f(x)y]/\partial y = -f(x)$, which has fixed sign (by hypothesis). According to Bendixson's Negative Criterion, there is no cycle or cycle-graph anywhere in the xy-plane.

3. Each system has the form $x' = f(x, y)$, $y' = g(x, y)$. We let the symbol Δ denote the quantity $\partial f/\partial x + \partial g/\partial y$ given in Theorem 9.2.4.

(a). The system is $x' = 2x - y + 36x^3 - 15y^2$, $y' = x + 2y + x^2y + y^5$. Here Δ is $\partial(2x - y + 36x^3 - 15y^2)/\partial x + \partial(x + 2y + x^2y + y^5)/\partial y = 4 + 109x^2 + 5y^4 > 0$. According to Bendixson's Negative Criterion, there is no cycle or cycle-graph anywhere in the xy-plane.

(b). The system is $x' = 12x + 10y + x^2y + y \sin y - x^3$, $y' = x + 14y - xy^2 - y^3$. In this case Δ is $\partial(12x + 10y + x^2y + y \sin y - x^3)/\partial x + \partial(x + 14y - xy^2 - y^3)/\partial y = 26 -$

$3x^2 - 3y^2 \geq 26 - 24 = 2 > 0$ since $3x^2 + 3y^2 \leq 24$. According to Bendixson's Negative Criterion, there is no cycle or cycle-graph anywhere in the disk $x^2 + y^2 \leq 8$.

(c). The system is $x' = x - xy^2 + y \sin y$, $y' = 3y - x^2y + e^x \sin x$, and Δ is $\partial(x - xy^2 + y \sin y)/\partial x + \partial(3y - x^2y + e^x \sin x)/\partial y = 4 - (x^2 + y^2) > 0$ if $x^2 + y^2 < 4$. According to Bendixson's Negative Criterion, there is no cycle or cycle-graph anywhere in the disk $x^2 + y^2 < 4$.

4. **(a).** The system is $x' = x - y + 10$, $y' = x^2 + y^2 - 1$. There is no real solution for the simultaneous equations, $x - y + 10 = 0$, $x^2 + y^2 - 1 = 0$, so there is no equilibrium point for this autonomous system. By the Unbounded Orbits Theorem (Theorem 9.2.2), this system has empty negative and positive limit sets because all orbits become unbounded as t increases and as t decreases. See Fig. 4(a).

(b). The system is $x' = y$, $y' = -\sin x - y + 2$. Since there is no simultaneous real solution for the equations, $y = 0$, $\sin x + y - 2 = 0$, there is no equilibrium point for this autonomous system. By the Unbounded Orbits Theorem (Theorem 9.2.2), this system has empty negative and positive limit sets because every orbit becomes unbounded as t increases and as t decreases. See Fig. 4(b).

(c). The system is $x' = x^2 + 2y^2 - 4$, $y' = 2x^2 + y^2 - 16$. Since there is no simultaneous real solution for the equations $x^2 + 2y^2 - 4 = 0$, $2x^2 + y^2 - 16 = 0$, there is no equilibrium point for the autonomous system. By the Unbounded Orbits Theorem, this system has empty negative and positive limit sets. See Fig. 4(c).

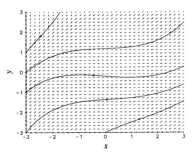
Problem 4(a).

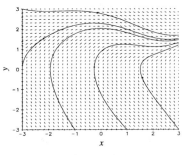
Problem 4(b).

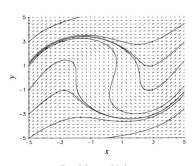
Problem 4(c).

5. **(a).** The system is $x' = e^x + y^2$, $y' = xy$. Since x' is always positive, no orbit can return to its starting point as t increases, and there are no cycles.

(b). The system is $x' = 2x^3y^4 + 5$, $y' = 2ye^x + x^3$. Since $\partial(2x^3y^4 + 5)/\partial x + \partial(2ye^x + x^3)/\partial y = 6x^2y^4 + 2e^x$ is positive for all x and y, Bendixson's Negative Criterion implies that there is no cycle in the xy-plane.

(c). The (polar) system is $r' = r \sin r^2$, $\theta' = 1$. Note that when $r = \sqrt{n\pi}$, $n = 0, 1, \ldots$, then $r' = 0$. Also, $r' > 0$ for $2k\pi < r^2 < (2k + 1)\pi$ and $r' < 0$ for $(2k + 1)\pi < r < (2k + 2)\pi$, $k = 0, 1, \ldots$. So $r^2 = 2k\pi$ defines repelling limit cycles, and $r^2 = (2k + 1)\pi$ defines attracting limit cycles, $k = 0, 1, \ldots$.

6. Suppose $x' = f$, $y' = g$ has a cycle Γ directed counterclockwise, say, with advancing time and bounding the region R whose area is the positive number A. Then, the average value of $\partial f/\partial x + \partial g/\partial y$ in R is

$$\frac{1}{A}\int\int_R (\partial f/\partial x + \partial g/\partial y)\, dx\, dy = \frac{1}{A}\oint_\Gamma (g\, dx - f\, dy) = \frac{1}{A}\int_0^T (g\frac{dx}{dt} - f\frac{dy}{dt})\, dt$$

$$= \frac{1}{A}\int_0^T (gf - fg)\, dt = 0$$

where T is the period of Γ and Green's Theorem (Theorem B.5.12) has been used to go from $\int\int_R$ to $\oint_\Gamma$..

7. The system $x' = x(1 - 0.1x - 0.1y)$, $y' = y(2 - 0.05x - 0.025y)$ is a model for competitive interaction with overcrowding. The mass-action terms $-(x^2 + xy)/10$ and $-(xy + y^2/2)/20$ model the negative effects of large populations on the species' growth rates. See Fig. 7. It appears from the figure that the y species "wins" the competition because all orbits inside the first quadrant tend to the point $x = 0$, $y = 80$, corresponding to the extinction of the x-species. So the positive limit set of every orbit inside the population quadrant appears to be the equilibrium point $(0, 80)$.

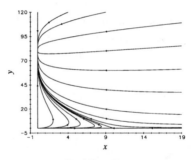

Problem 7.

8. The system is $x' = x - 10y - x(x^2 + y^2) = f(x, y)$, $y' = 10x + y - y(x^2 + y^2) = g(x, y)$, and the region R is described by $x^2 + y^2 > 0.5$, that is the region outside the circle $x^2 + y^2 = 0.5$. The region R is not simply connected, so Bendixson's Negative Criterion does not apply. In polar coordinates the system is $x' = r(1 - r^2)$, $\theta' = 10$, so $r = 1$ is a cycle that is traversed counterclockwise.

9. The system is $x' = x(1 - x - 3.75y + 2xy + y^2)$, $y' = y(-1 + y + 3.75x - 2x^2 - xy)$.

(a). The equilibrium points $(0,0)$, $(1,0)$, and $(0,1)$ appear to be saddle points, while $(1/4,1/4)$ seems to be an unstable spiral point, $(4/3,4/3)$ an unstable spiral point, and $(7/4, -3/4)$ a stable node. See Fig. 9(a).

(b). The Jacobian matrix is

$$\begin{bmatrix} \partial F_1/\partial x & \partial F_1/\partial y \\ \partial F_2/\partial x & \partial F_2/\partial y \end{bmatrix} = \begin{bmatrix} 1 - 2x - 15y/4 + 4xy + y^2 & -15x/4 + 2x^2 + 2xy \\ 15y/4 - 4xy - y^2 & -1 + 2y + 15x/4 - 2x^2 - 2xy \end{bmatrix}$$

The respective Jacobian matrices and characteristic polynomials at (0,0), (1/4,1/4), and (4/3,4/3) are

$$\begin{bmatrix} 1 & 0 \\ 0 & -1 \end{bmatrix}, \quad \begin{bmatrix} -2/16 & -11/16 \\ 10/16 & 3/16 \end{bmatrix}, \quad \begin{bmatrix} 20/9 & 19/9 \\ -35/9 & -4/9 \end{bmatrix}$$

and $\lambda^2 - 1$, $\lambda^2 - \lambda/16 + 104/256$, and $\lambda^2 - 16\lambda/9 + 585/81$. The eigenvalues of the first matrix are ± 1 [and so a nonlinear saddle at $(0,0)$]; for the second and third matrices the eigenvalues are complex conjugates with positive real parts [nonlinear unstable spiral points at $(1/4, 1/4)$, $(4/3, 4/3)$].

(c). See Figs. 9(c) where $c = -4$ (Graph 1) and $c = -3.5$ (Graph 2), respectively. A close observation of the orbits as they are being generated suggests that for $c < -3.75$ one of the spirals from the unstable spiral point at the upper right tends to the equilibrium point $(0,1)$ as $t \to \infty$. This means that the "saddle connection" on $x + y = 1$ (when $c = -3.75$) is broken. Similarly, when $c > -3.75$, an outward spiral from the unstable spiral point at the lower left approaches $(1,0)$, again breaking the original saddle connection.

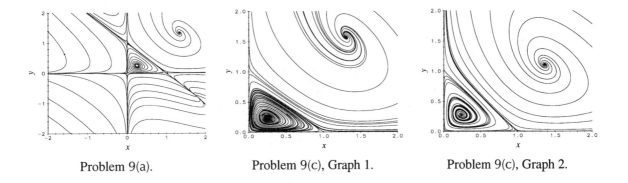

Problem 9(a).   Problem 9(c), Graph 1.   Problem 9(c), Graph 2.

10. The system is $x' = y + x(1 - x^2)(y^2 - x^2 + x^4/2)$, $y' = x - x^3 - y(y^2 - x^2 + x^4/2)$. It is not hard to show that $(0,0)$, $(1,0)$, and $(-1,0)$ are equilibrium points. See Fig. 10. The orbits outside the "lazy eight" seem to have empty negative limit sets (because the orbits become unbounded as t decreases) and the lazy eight cycle-graph as their positive limit sets. Nonconstant orbits inside the left [right] lobe have the equilibrium point $(-1, 0)$ [$(1, 0)$] as negative limit set and the left [right] lobe cycle-graph as positive limit set.

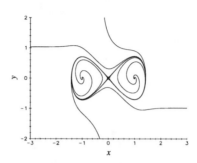

Problem 10.

11.   The system is $x' = f(x, y)$, $y' = g(x, y)$, and we will scale time by $s = s(t)$ where $dt/ds = h(x, y)$ and h is a given function with a fixed sign.

(a). Let $dt = h\,ds$. The scaled system becomes $dx/ds = fh$, $dy/ds = gh$. Since $h(x, y)$ is positive everywhere, the two systems have identical orbits and the same orientation. Since $dy/dx = g/f$ for both systems, the positive factor h affects only the rate at which orbits are generated, but not their orientation or their shape. If $h(x, y)$ is everywhere negative, the two systems have identical orbits but with opposite orientation.

(b). Suppose now that h is positive everywhere except at a point p (in the xy-plane) where the value of h is zero. Suppose p lies on an orbit Γ. The point p on Γ will be the positive limit set of one arc of Γ, the negative limit set of another adjacent arc since h is positive everywhere on Γ except at p.

(c). The system is $x' = f(x, y) = x - 10y - x(x^2 + y^2)$, $y' = 10x + y - y(x^2 + y^2)$. Let h be the function $h = 1 - \exp[-10(x - 1)^2 - 10y^2]$. Using polar coordinates, the unscaled system becomes $r' = r - r^3$, $\theta' = 10$. This system has an attracting limit cycle $r = 1$. However, the scaled system is changed to $r' = (r - r^3)\{1 - \exp[-10(r\cos\theta - 1)^2 - 10r^2\sin^2\theta]\}$, $\theta' = 10 - 10\exp[-10(r\cos\theta - 1)^2 - 10r^2\sin^2\theta]$. We see that this scaled system has an attracting cycle graph on the unit circle. See Figs. 11(c), Graph 1 and Graph 2 for an orbit and its components of the unscaled and scaled systems, respectively. The factor $h(x, y)$ in the scaled system in x, y-coordinates is positive except at the points $(\pm 1, 0)$ on the unit circle, where h is zero. The scaled system has two new equilibrium points. All other orbits are the same in the scaled and unscaled systems. However, the component graphs of solutions of the two systems are very different; the nonlinear scale factor h has dramatic effects on the "speed" at which orbits are traced out.

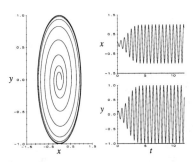
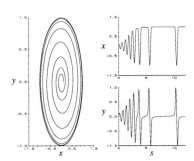

Problem 11(c), Graph 1.   Problem 11(c), Graph 2.

12. Replace f and g in the proof of Bendixson's Negative Criterion (see Background Material) by Kf and Kg and repeat the steps exactly as given.

13. The system is $x' = xF(x, y)$, $y' = yG(x, y)$, where $\partial F/\partial x$ and $\partial G/\partial y$ are negative in the population quadrant. Let $K = 1/xy$ for $x > 0$ and $y > 0$. Then $\partial(KxF)/\partial x + \partial(KyG)/\partial y = y^{-1}\partial F/\partial x + x^{-1}\partial G/\partial y < 0$ if $\partial F/\partial x < 0$ and $\partial G/\partial y < 0$, $x > 0$, $y > 0$. Now apply the Bendixson proof (see Background Material) to the scaled system $x' = xFK$, $y' = yGK$. The system $x' = xF$, $y' = yG$ has no cycles in the population quadrant.

14. Group project. In carrying out this project, you may want to search the text and the Resource Manual for pictures having bits and pieces that might help to build the face, e.g., a cycle for an eye, a line segment for a whisker, and so on. Then build a set of x and y rate functions using if... then... else statements to place these facial features where you want them.

Background Material: Proof of Bendixson's Negative Criterion (Theorem 9.2.4)

The system is $x' = f(x, y)$, $y' = g(x, y)$. Suppose that the quantity $\Delta = \partial f/\partial x + \partial g/\partial y$ is positive throughout a simply-connected region R. Suppose, contrary to the final assertion of Theorem 9.2.4, that a cycle or cycle-graph Γ lies inside R. We shall show that this leads to a contradiction. Applying Green's Theorem (Theorem B.5.12) to $\partial f/\partial x + \partial g/\partial y$ over the region S consisting of Γ and its interior, we have that

$$0 < \int_S \int (\partial f/\partial x + \partial g/\partial y)\, dx\, dy \quad \text{[since } \partial f/\partial x + \partial g/\partial y > 0 \text{ in } S\text{]}$$

$$= \oint_\Gamma (g\, dx - f\, dy) \quad \text{[Green's Theorem]}$$

But $dx = f\, dt$ and $dy = g\, dt$ along Γ because Γ is an orbit (if it is a cycle) or a union of orbits (if it is a cycle graph) of $x' = f$, $y' = g$. This implies that

$$g\, dx - f\, dy = gf\, dt - fg\, dt = 0$$

everywhere on Γ. We are left with the contradiction

$$0 < \oint_\Gamma (g\, dx - f\, dy) = 0$$

So there can be no cycle or cycle-graph with the property that it and its interior lie in the region R where the quantity Δ is positive. A similar proof applies if Δ is negative.

9.3 Bifurcations

Suggestions for preparing a lecture

Topics: Bifurcation diagrams, saddle-node bifurcation, Hopf bifurcation, a description of the scroll circuit model, examples. Alternatively, devote a separate lecture to the scroll circuit model.

Making up a problem set

Problems 1(**d**), 2(**a**), one of 3–5. One of the group problems would make a good term project.

Comments

In the last two or three decades, mathematicians, scientists, and engineers have discovered bifurcations in all sorts of dynamical systems. And it is true that bifurcation analysis often is the easiest way to handle a system where a change in a parameter leads to a sudden change in orbital or solution behavior. Nevertheless, it seems to us that it may be a case of "everything looks like a nail to a person with a hammer." With that caveat, we present in the text and in the problems some of the bifurcations most commonly observed in nonlinear systems (saddle-node, transcritical, pitchfork, Hopf). Of these, the Hopf bifurcation that spawns a limit cycle from an equilibrium point is the most interesting and useful. We present the satiable predator model, but be warned that it is not a very realistic model of population dynamics. The scroll circuit model system (see text and Problem 7) makes a nice term project for electrical engineers, who might even build and test the circuit being modeled. The scroll circuit and the model are only ten years old, and still not completely understood. Incidentally, you might also think of bifurcations in the context of elementary algebra; for example, real roots of a quadratic merging and then disappearing into the complex plane as a coefficient changes.

---

1. First find the equilibrium points, and then calculate the Jacobian matrix at each equilibrium point and its eigenvalues. Use the table after Section 8.2 to name the equilibrium points. We only describe the results and omit the details. In the bifurcation diagrams the solid curves indicate the x-coordinates of stable equilibria as functions of c, the dashed curves show the x-coordinates of unstable equilibria. The Jacobian calculations are omitted below. You might want to look back at Section 2.4.

   (**a**). Saddle-node bifurcation. The system is $x' = c + 10x^2$, $y' = x - 5y$. When c is positive, there are no equilibrium points [Fig. 1(a), Graph 1]. If $c = 0$, there is a nonelementary saddle-node equilibrium point at the origin because 0 is an eigenvalue of the Jacobian matrix of the rate function of this system at the origin [Fig. 1(a), Graph 2]. As in Example 9.3.1, the nature of each equilibrium point before and after the bifurcation is determined by the eigenvalues of the Jacobian matrix at that point. According to the table at the end of Section 8.2, the system and its linearization have the same type of equilibrium point as long as the real parts of the eigenvalues of the Jacobian matrix are nonzero. From Fig. 1(a),

Graph 3, we see that as c changes from negative values through zero and into the positive range, an asymptotically stable node at $(-\sqrt{-c/10}, -\sqrt{-c/10}/5)$ and an unstable saddle at $(\sqrt{-c/10}, \sqrt{-c/10}/5)$ move toward each other, collide (when $c = 0$), and vanish. When $c = 0$, a bifurcation occurs at the saddle-node at the origin. See Figs. 1(a), Graphs 1–3 where $c = 1, 0, -1$, respectively, and Fig. 1(a), Graph 4 for the associated bifurcation diagram.

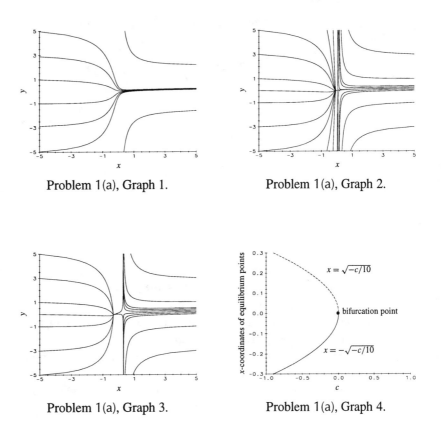

Problem 1(a), Graph 1.

Problem 1(a), Graph 2.

Problem 1(a), Graph 3.

Problem 1(a), Graph 4.

(b). Transcritical bifurcation. The system is $x' = cx - x^2$, $y' = -2y$. The system has an asymptotically stable node at $(c, 0)$ and a saddle at the origin if c is positive as the Jacobian calculations show. From Figs. 1(b), Graphs 1–3, we see that if c decreases to 0, then the two equilibrium points merge into a saddle-node at the origin. As c decreases through 0 and becomes negative, the equilibrium point $(c, 0)$ and the origin exchange orbital characters, the origin becoming the asymptotically stable node and $(c, 0)$ the saddle. A transcritical bifurcation occurs at $c = 0$. See Figs. 1(b), Graphs 1–3 where $c = -1, 0, 1$, respectively, and Fig. 1(b), Graph 4 for the associated transcritical bifurcation diagram.

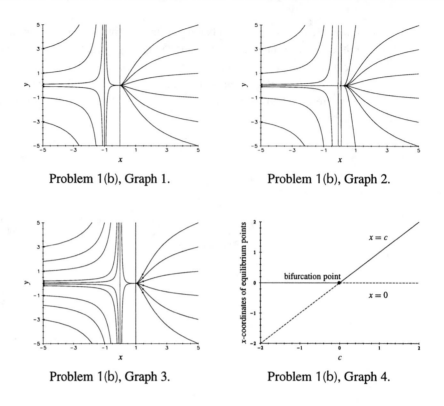

Problem 1(b), Graph 1.   Problem 1(b), Graph 2.

Problem 1(b), Graph 3.   Problem 1(b), Graph 4.

(c). Transcritical bifurcation. The system is $x' = cx + 10x^2$, $y' = x - 2y$. The system has an asymptotically stable node at the point $(-c/10, -c/20)$ and a saddle at the origin if c is positive. From Figs. 1(c), Graphs 1–3, we see that if c decreases to 0, then the two equilibrium points merge into a saddle-node at the origin. As c moves away from 0 and becomes negative, the equilibrium point $(-c/10, 0)$ and the origin exchange orbital characters, the origin becoming the asymptotically stable node and $(-c/10, -c/20)$ the saddle. A transcritical bifurcation occurs at $c = 0$. See Figs. 1(c), Graphs 1–3 where $c = -20$, 0, and 20, respectively, and Fig. 1(c), Graph 4 for the associated transcritical bifurcation diagram.

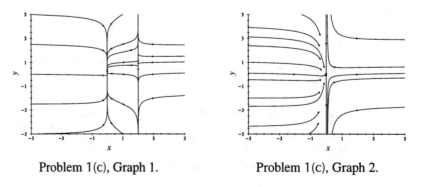

Problem 1(c), Graph 1.   Problem 1(c), Graph 2.

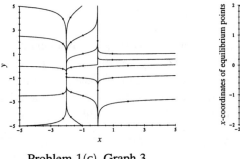

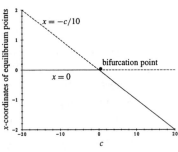

Problem 1(c), Graph 3.   Problem 1(c), Graph 4.

(d). Pitchfork bifurcation. The system is $x' = cx - 10x^3$, $y' = -5y$. The original system has asymptotically stable nodes at $(\pm\sqrt{c/10}, 0)$ and a saddle at the origin if c is positive. From Figs. 1(d), Graphs 1–3, we see that if c decreases to 0, then the three equilibrium points merge into an asymptotically stable equilibrium point at the origin. As c decreases through 0 and becomes negative, the equilibrium points $(\pm\sqrt{c/10}, 0)$ vanish. A bifurcation occurs at $c = 0$. See Figs. 1(d), Graphs 1–3 where $c = -1$, 0, and 1, respectively, and Fig. 1(d), Graph 4 for the associated pitchfork bifurcation diagram.

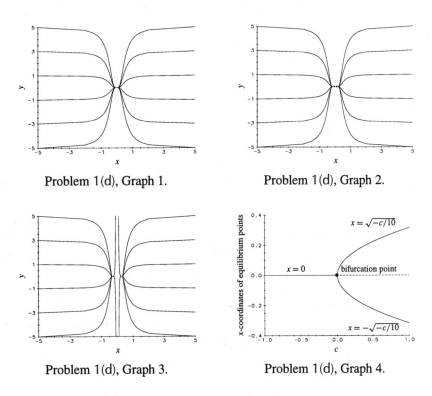

Problem 1(d), Graph 1.   Problem 1(d), Graph 2.

Problem 1(d), Graph 3.   Problem 1(d), Graph 4.

(e). Pitchfork bifurcation. The system is $x' = cx + x^5$, $y' = -y$. The system has saddles at $(\pm(-c)^{1/4}, 0)$ and an asymptotically stable node at the origin if c is negative. From

Figs. 1(e), Graphs 1–3, we see that if c increases to 0, then the three equilibrium points merge into a nonlinear saddle at the origin. Also, as c moves away from 0 and becomes positive, the equilibrium points $(\pm(-c)^{1/4}, 0)$ vanish and the origin becomes a nonlinear saddle. A bifurcation occurs if $c = 0$. See Figs. 1(e), Graphs 1–3 where $c = -1, 0$, and 1, respectively, and Fig. 1(e), Graph 4 for the associated pitchfork bifurcation diagram, where (*in this graph only*) the dashed line is the x-coordinate of a *stable* equilibrium point and the solid curves the x-coordinates of *unstable* equilibrium points.

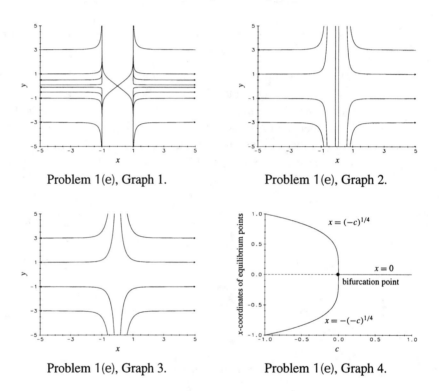

Problem 1(e), Graph 1.   Problem 1(e), Graph 2.

Problem 1(e), Graph 3.   Problem 1(e), Graph 4.

2. **(a).** The system is $x' = cx + 2y - x(x^2 + y^2)$, $y' = -2x + cy - y(x^2 + y^2)$. The rate functions are polynomials in x, y, and c. [The origin is an isolated equilibrium point for all c since in polar coordinates $r' = r(c - r^2)$, $\theta' = -2$; so the only equilibrium point is at $r = 0$.] The eigenvalues of the matrix of the linear terms of the xy-system are $c \pm 2i$. We have $\alpha(c) = c$, $\beta(c) = 2$, $\alpha(0) = 0$, $\beta(0) = 2$, $\alpha'(0) = 1$. Moreover, the system is asymptotically stable at the origin if $c = 0$, since $r' = -r^3$, and so $r = (2t + C)^{-1/2}$; $r \to 0$ as $t \to +\infty$. All the conditions of the Hopf Bifurcation Theorem are satisfied, and for $c > 0$ there is an attracting limit cycle which encloses the unstable equilibrium point at the origin and disappears into the origin as c decreases to 0. The limit cycle is given by $r = \sqrt{c}$, $\theta = -2t$, $c > 0$. See Figs. 2(a), Graphs 1–3, where $c = -1, 0$, and 1, respectively, and Fig. 2(a), Graph 4, for the associated bifurcation diagram where the solid curves indicate the r-coordinate of an attracting equilibrium point or the r-amplitude of an attracting cycle, and the dashed line is the r-coordinate of an unstable equilibrium point.

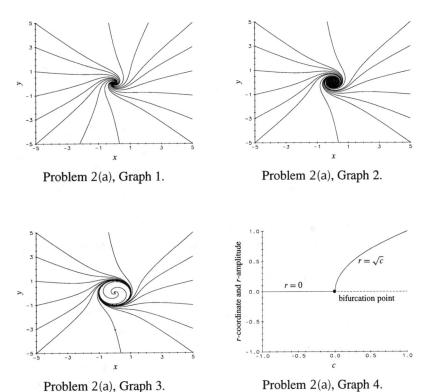

Problem 2(a), Graph 1.   Problem 2(a), Graph 2.

Problem 2(a), Graph 3.   Problem 2(a), Graph 4.

(b). The system is $x' = cx - 3y - x(x^2 + y^2)^3$, $y' = 3x + cy - y(x^2 + y^2)^3$. The argument here is exactly as in 2**(a)** except that in polar coordinates, $r' = r(c - r^6)$, $\theta' = 3$. For $c > 0$ the limit cycle is defined by $r = c^{1/6}$, $\theta = 3t$. We have $\alpha(c) = c$, $\beta(c) = 3$, $\alpha(0) = 0$, $\alpha'(0) = 1$. See Figs. 2(b), Graphs 1–3, where $c = -1$, 0, and 1, respectively, and Fig. 2(b), Graph 4, for the associated bifurcation diagram. The solid lines indicate the r-coordinate (or r-amplitude) of an attracting equilibrium point (or cycle), and the dashed line the r-coordinate of an unstable equilibrium point.

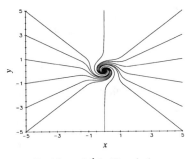

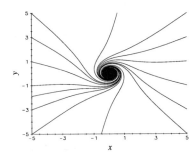

Problem 2(b), Graph 1.   Problem 2(b), Graph 2.

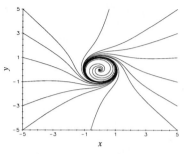

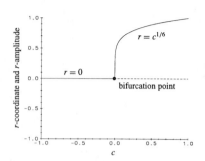

Problem 2(b), Graph 3.   Problem 2(b), Graph 4.

(c). The system is $x' = y - x^3$, $y' = -x + cy - y^3$. The linear terms of the system do not have the form given in system (5) in the text. However, the system does meet all the other conditions, as we shall show. The system has an isolated equilibrium point at the origin since in polar coordinates $r' = r[c\sin^2\theta - r^2(\cos^4\theta + \sin^4\theta)]$, $\theta' = -1 + \cos\theta\sin\theta[c + r^2(\cos^2\theta - \sin^2\theta)]$, implying that the origin is an equilibrium point for all c, but that for $|c|$ small (and $r > 0$ and small) there are no other equilibrium points. The characteristic polynomial of the coefficient matrix of the linear terms of the system [i.e., of the Jacobian matrix evaluated at the origin] is $\lambda^2 - c\lambda + 1$, whose roots are $c/2 \pm i\sqrt{1 - c^2/4}$, for all $|c| < 2$. So $\alpha(c) = c/2$ and $\beta(c) = \sqrt{1 - c^2/4}$, $\alpha(0) = 0$, $\alpha'(0) = 1/2$, $\beta(0) = 1$. The rate functions and the functions $\alpha(c)$ and $\beta(c)$ are all twice continuously differentiable in x, y, and c [$\beta(c)$ only for $|c| < 2$]. At $c = 0$, the system in polar coordinates is $r' = -r^3(\cos^4\theta + \sin^4\theta) < -r^3/2$, $\theta' = -1 + r^2\cos\theta\sin\theta(\cos^2\theta - \sin^2\theta)$, and r decreases to 0 as $t \to +\infty$, implying that the origin is asymptotically stable. The system has an attracting limit cycle enclosing an unstable equilibrium point at the origin for each $c > 0$, where c is sufficiently small. Unlike 2(a) and 2(b), there is no formula for the cycle. See Figs. 2(c), Graphs 1–3, where $c = -1$, 0, and 1.

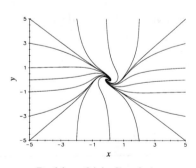
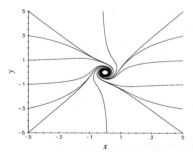
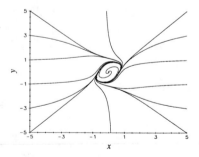

Problem 2(c), Graph 1.   Problem 2(c), Graph 2.   Problem 2(c), Graph 3.

(d). The Rayleigh system is $x' = y$, $y' = -x + cy - y^3$. The argument is the same as in part **(c)**. See Figs. 2(d), Graphs 1–3, where $c = -1$, 0, and 1.

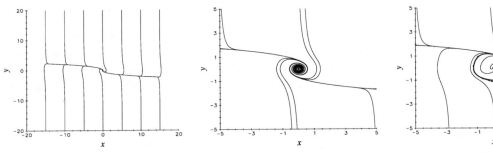

Problem 2(d), Graph 1.   Problem 2(d), Graph 2.   Problem 2(d), Graph 3.

3. The system is $x' = -cx + y + x(x^2 + y^2)$, $y' = -x - cy + y(x^2 + y^2)$. The origin is the only equilibrium point. The characteristic polynomial of the Jacobian matrix of the system at the origin is $\lambda^2 + 2c\lambda + c^2 + 1$, and the eigenvalues are $\lambda = -c \pm i$. Therefore, we see that the original system bifurcates as c increases through 0 from an unstable spiral point to a stable spiral point surrounded by a repelling limit cycle. In polar coordinates, we have $r' = r(-c + r^2)$, $\theta' = -1$. So for $c > 0$ the cycle is $r = \sqrt{c}$. See Figs. 3, Graphs 1–3 where $c = -1$, 0, and 1, respectively, and Fig. 3, Graph 4 for the associated bifurcation diagram. In this graph the solid lines indicate the r-coordinate (or r-amplitude) of an *unstable* equilibrium point (or cycle); the dashed line is the r-coordinate of a *stable* equilibrium point.

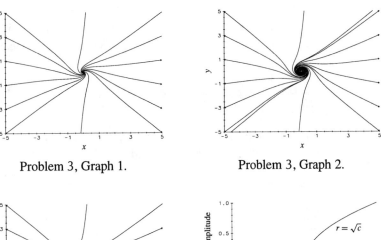

Problem 3, Graph 1.   Problem 3, Graph 2.

Problem 3, Graph 3.   Problem 3, Graph 4.

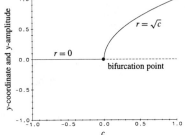

4. The system is $x' = [-0.5 + by/(0.3 + 0.9y)]x$, $y' = [1 - y - x/(0.3 + 0.9y)]y$. See Figs. 4, Graphs 1–3, where $b = 0.75$, 1, and 3, respectively. Figure 4, Graph 1 shows an asymptotically stable equilibrium point (before a Hopf bifurcation). Figure 4, Graph 2 shows the attracting limit cycle after the bifurcation. Figure 4, Graph 3 shows the expanded limit cycle which elongates in the x-direction and nearly borders the axes (one or the other species close to extinction!).

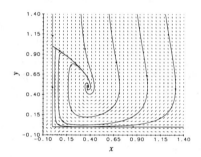

Problem 4, Graph 1.

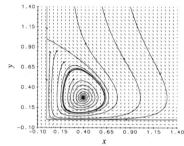

Problem 4, Graph 2.

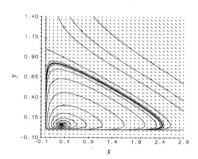

Problem 4, Graph 3.

5. The system is $x' = c(x - 5c) + (y - 5c) - (x - 5c)[(x - 5c)^2 + (y - 5c)^2]$, $y' = -(x - 5c) + c(y - 5c) - (y - 5c)[(x - 5c)^2 + (y - 5c)^2]$. The equilibrium point P has coordinates $x = y = 5c$.

(a). Let $u = x - 5c$, $v = y - 5c$. We obtain $u' = cu + v - u(u^2 + v^2)$ and $v' = -u + cv - v(u^2 + v^2)$, In polar form, $r' = cr - r^3$, $\theta' = -1$.

(b). Just as in Example 9.3.2, the original system has a supercritical Hopf bifurcation at $c = 0$. In addition, $r = \sqrt{c}$ is an orbit of the above system for $c > 0$, as we see from the rate equation for r. The original system has as an orbit the circle of radius $r = \sqrt{c}$ centered at $P(5c, 5c)$.

(c). See Figs. 5(c), Graphs 1–5, for $c = -0.5$, 0, 0.5, 1, and 1.5, respectively. See also Figures 9.3.21 and 9.3.22 in the text. The asymptotically stable equilibrium point at P [for $c < 0$] moves upward and to the right as c increases. At $c = 0$, the point is still asymptotically stable [Fig. 5(c), Graph 2, and Figure 9.3.21], but there is a Hopf bifurcation as c increases beyond 0, a bifurcation to an attracting limit cycle around the destabilized P for $c > 0$. The amplitude of the limit cycle expands as c increases, while the cycle moves upward and to the right. Figure 9.3.22 shows orbits both before and after the Hopf bifurcation. The pictures in Figure 9.3.22 resemble an expanding two-dimensional smoke ring that expands and drifts upwards and to the right as c increases.

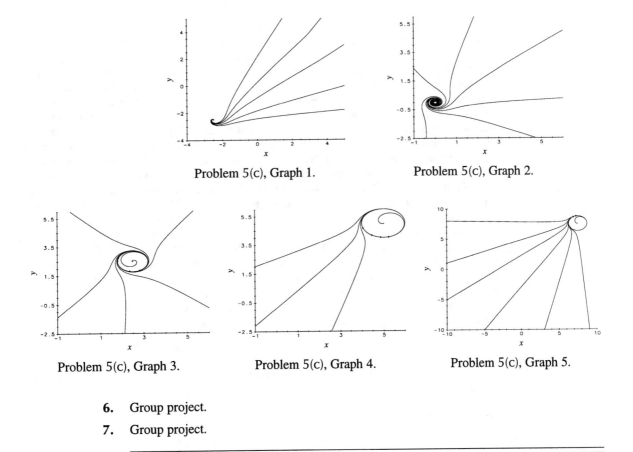

Problem 5(c), Graph 1.   Problem 5(c), Graph 2.

Problem 5(c), Graph 3.   Problem 5(c), Graph 4.   Problem 5(c), Graph 5.

6.   Group project.

7.   Group project.

9.4 Chaos

Suggestions for preparing a lecture

Topics: The Lorenz system and the numerically computed behavior of its orbits, Poincare time sections, ideas about strange attractors and chaos.

Making up a problem set

The subject is ideal for a group project (any of Problems 2–5).

Comments

Finally, the system that started it all! Lorenz's name is firmly attached to the theory of chaos, and deservedly so. It was his remarkable (but characteristically low-key) paper in 1963 that began the current multitude of studies on the implications and the very definition of chaotic dynamics. It should be noted, though, that Mary Cartwright and J. E. Littlewood did pioneering work on the subject in England in 1945. Our focus here (and in the several group projects in the problem set) is on chaotic behavior in ODEs; we have previously noted chaotic dynamics in a discrete map in

Section 2.7. We prove what we can about the behavior of the Lorenz system, but ultimately must use pictures to illustrate the chaotic motion. People often have a hard time seeing where the chaos is, because the pictures seem to show fairly regular oscillations first near one equilibrium point and then near another. However, it is nearly impossible to predict just when a given orbit will be near one equilibrium point and when it will move close to the other; moreover, the number of oscillatory whirls about an equilibrium point before moving away also seems to be unpredictable. Chaos is, as noted in the section, a feature of a set of orbits, a so-called strange attractor. These strange attractors are a generalization to higher dimensions of the attracting equilibrium point, cycle, or cycle-graph limit sets of any bounded orbit of a planar autonomous system. We only give a brief mention of what strange attractors and chaotic dynamics on a strange attractor actually mean in a mathematical (rather than an informal "hand-waving") sense. The reason is that the definitions are still in a state of flux—and the students should be so informed. **Warning**: Time-state curves and orbits of the Lorenz system, the Rössler system (Problem 3), the Duffing equation (Problem 4), and the driven and damped pendulum (Problem 5) are sensitive to changes in initial data. Since numerically computed solutions are never exact, this means that these solutions may be reasonably accurate only over a short time span. However, the orbits and solutions computed in this section are thought to be near a strange attractor that attracts all nearby solutions. So the computed solutions, orbits, and Poincaré time sections shown here are each most likely composites of several orbits or time-state curves that (at least for the end segment of the solution interval) are essentially in the strange attractor. Although we may not be looking at the actual orbit or time-state curve through the initial point, we do get a pretty good view of the strange attractor.

---

1. **(a).** The equilibrium points P_1 and P_2 are easily found by equating the right sides of the equation of the Lorenz system (1) to zero and solving for x, y, and z.

   (b). The Jacobian matrix at (x, y, z) is

   $$\begin{bmatrix} -\sigma & \sigma & 0 \\ r-z & -1 & -x \\ y & x & -b \end{bmatrix}$$

   and the eigenvalues are the roots of $\lambda^3 + (1 + \sigma + b)\lambda^2 + [b + x^2 + \sigma(1 + b - r + z)]\lambda + \sigma(-br + bz + x^2 + b + xy)$. If $x = y = z = 0$, $\lambda_3 = -b$ is easily shown to be a root; factoring the cubic, given the factor $\lambda + b$, leads to the other two roots λ_1 and λ_2 as given in formula (2) in the text. If $0 < r < 1$, λ_1 and λ_2 are both negative. If $r = 1$, λ_1 is negative but $\lambda_2 = 0$, but if $r > 1$, then λ_2 is positive.

   (c). At P_1 (and P_2), the eigenvalues of the Jacobian matrix are the roots of $\lambda^3 + (1 + b + \sigma)\lambda + (\sigma + r)b\lambda + 2\sigma b(r - 1)$. According to the Routh Test (Theorem 7.9.7) all roots of this cubic have negative real parts if and only if the coefficients are positive (which they are because b and σ are positive and $r > 1$) *and* if $(1 + b + \sigma)(\sigma + r)b > 2\sigma b(r - 1)$. This equality reduces to $\sigma(\sigma + b + 3) > r(\sigma - b - 1)$, i.e., $r < r_c = \sigma(\sigma + b + 3)/(\sigma - b - 1)$ if $\sigma > b + 1$. P_1 and P_2 are locally asymptotically stable if $1 < r < r_c$ since all the eigenvalues have negative real parts for these values of r, while if $r > r_c$ both P_1 and P_2 are unstable.

   (d). If $C = AB$, then the cubic becomes $\lambda^3 + A\lambda^2 + B\lambda + AB = (\lambda + A)(\lambda^2 + B)$. So, it has a pair of pure imaginary roots since $B > 0$ is assumed. On the other hand, if the cubic has a pair of pure imaginary roots, the cubic may be written as $(\lambda - \lambda_1)(\lambda - \bar{\lambda}_1)(\lambda - \lambda_3) =$

$\lambda^3 - \lambda_3\lambda^2 + \lambda_1\bar{\lambda}_1\lambda - \lambda_1\bar{\lambda}_1\lambda_3$. And so $A = -\lambda_3$, $B = \lambda_1\bar{\lambda}_1$, and $C = -\lambda_1\bar{\lambda}_1\lambda_3 = AB$. Applying this criterion for pure imaginary eigenvalues of the Jacobian matrix at the equilibrium point P_1 (or P_2), we must have $(1 + b + \sigma)(\sigma + r)b = 2\sigma b(r - 1)$ since according to part **(c)** the negative of the characteristic polynomial is $\lambda^3 + (1 + \sigma + b)\lambda^2 + (\sigma + r)b\lambda + 2\sigma b(r - 1)$. Solving this equation for r, we get $r = r_c$, the critical value given in formula (5).

2. Group project.

3. Group project.

4. Group project.

5. Group project.

Background Material: Verification of the Lorenz Squeeze

To show the squeeze property we argue informally along the following lines. Suppose S is a bounded region of state space with positive volume $V(S)$ and bounding surface ∂S, and suppose that ∂S is smooth, connected, and closed. For example, S could be a solid ball with spherical surface ∂S. Let ∂S_t denote the bounding surface of S_t, and let $V(S_t)$ denote the volume of the region S_t. As t increases from 0, S_t and ∂S_t will look less and less like $S = S_0$ and $\partial S = \partial S_0$. Interpreting the Lorenz equations as the equations of motion of the particles of a gas or a fluid, we have that

$$\frac{dV(S_t)}{dt} = \left\{ \begin{array}{c} \text{net volume of fluid} \\ \text{crossing } S_t \text{ in the} \\ \text{unit outward normal} \\ \text{direction per unit of} \\ \text{time} \end{array} \right\} = \iint_{\partial S_t} \mathbf{F} \cdot \mathbf{n}\, dA \qquad \text{(i)}$$

where $\mathbf{F}$ is the vector field defined by the right-hand side of the Lorenz system [text system (1)], $\mathbf{n}$ is a unit outward normal vector field to the bounding surface ∂S_t, and the integral is a surface integral. By the Divergence Theorem of vector calculus (Theorem B.5.13), the surface integral becomes the volume integral of the divergence over S_t:

$$\iint_{\partial S_t} \mathbf{F} \cdot \mathbf{n}\, dA = \iiint_{S_t} \nabla \cdot \mathbf{F}\, dV \qquad \text{(ii)}$$

The divergence of $\mathbf{F}$ is given by

$$\nabla \cdot \mathbf{F} = \frac{\partial F_1}{\partial x} + \frac{\partial F_2}{\partial y} + \frac{\partial F_3}{\partial x}$$

where F_1, F_2, F_3 are given by the right-hand sides of the equations of text system (1). We have that $\nabla \cdot \mathbf{F} = -\sigma - 1 - b$. From (i) and (ii) we have that

$$\frac{dV(S_t)}{dt} = \iiint_{S_t} -(\sigma + 1 + b)\, dV = -(\sigma + 1 + b)V(S_t) \qquad \text{(iii)}$$

Solving the scalar ODE $V' = -(\sigma + 1 + b)V$, with initial condition $V(0) = V(S)$ when $t = 0$, we have that

$$V(S_t) = V(S)e^{-(\sigma+1+b)t} \tag{iv}$$

We see that as $t \to \infty$, $V(S_t) \to 0$.

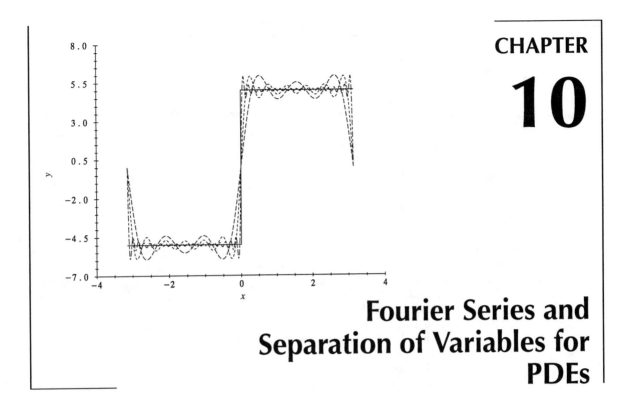

Fourier Series and Separation of Variables for PDEs

This chapter begins with the derivation of a boundary/initial value problem for the PDE which models a vibrating guitar string. Later in the chapter we develop a method for constructing a solution formula for that problem. In between we lay the groundwork for deriving solution formulas for a class of linear boundary/initial value problems. The intervening sections contain material on Fourier series and Sturm-Liouville problems that is needed to solve the boundary/initial value problems of this chapter.

Section 10.1 was written to be as self-contained as possible in order to provide a brief introduction and example of the Method of Separation of Variables. One could quit at this point and still come away with a good idea of what this method is all about. Sections 10.2 and 10.3 set the stage for Fourier Series in order to give a clean introduction to Fourier Trigonometrical Series in Sections 10.4 and 10.5. Sturm-Liouville Problems are treated in a unified way in Section 10.6. It is a short section and should be read before reading Section 10.7 on eigenfunction expansions and the Method of Separation of Variables. In Section 10.8 we discuss a model PDE for heat conduction, and apply the eigenfunction techniques for the heat equation to discuss the changing temperatures in a rod and in an underground storage cellar. In Section 10.9 we consider Laplace's PDE as a model for steady-state temperatures in a body, given the temperatures at all the boundary points. Fourier series, eigenfunction expansions, and Separation of Variables underlie all we do in this chapter.

10.1 Vibrations of a String

Suggestions for preparing a lecture

Topics: Building the model equations for the transverse vibrations u of a guitar string fastened at each end.

Remarks: You might point out the restrictions made so that all the terms $(u_x)^2$, $(u_x)^3$, . . . can be dropped. This is not easy modeling at all! Standing waves should be determined, and some time profiles of a standing wave. The various kinds of boundary conditions can be introduced and expressed in terms of linear boundary operators. Although we don't superimpose the standing waves and use Fourier methods to determine the coefficients to meet the initial conditions until later sections, it might be smart to mention the process here.

Making up a problem set

Problems 1**(a)** or **(b)**, 2.

Comments

We use the PDE of a vibrating string as the introduction to PDEs, separation of variables, and Fourier series because the twanging of a guitar string is familiar to everyone. Our approach is to start with a derivation of the wave equation for the transverse motion of the string, introduce the three different kinds of linear boundary conditions and the corresponding boundary operators, and then focus on separating the PDE to get the standing waves modulated by sinusoidal functions of time. We work out the details of a boundary problem that involves fixed endpoints (e.g., the guitar string fastened at each end). We show how the coefficients are calculated by the Fourier method, but dodge all questions of convergence.

Everything is straightforward, but the problems tend to be somewhat lengthy. If you don't have access to a CAS, use integral tables or other sources to calculate the integrals. You need to develop early on a standard format for doing a separation of variables construction. Otherwise you will lose your way in the lengthy process. Be sure to look at the steps leading up to (10) where the variables are separated, the separation constant λ is introduced, and separate ODEs are found (one with boundary conditions).

---

1. The model equations for a vibrating string with free boundary conditions are based on system (8) in the text and the condition given in (6) [with $\gamma = 0$]. We use (10) for the separated functions $X(x)$ and $T(t)$. Putting all this together, we have the model:

$$u_{tt} - c^2 u_{xx} = 0, \qquad 0 < x < L, \quad t > 0$$
$$u_x(0, t) = 0, \qquad t \geq 0$$
$$u_x(L, t) = 0, \qquad t \geq 0$$
$$u(x, 0) = f(x), \qquad 0 \leq x \leq L$$
$$u_t(x, 0) = g(x), \qquad 0 \leq x \leq L$$

(a). The standing waves $X(x)T(t)$ must satisfy the pair of equations $X'' = \lambda X$, $T'' = c^2 \lambda T$, for a constant λ. We ignore the initial conditions for $u(x, 0)$, $u_t(x, 0)$ throughout. For the standing waves to satisfy the free boundary conditions $u_x(0, t) = 0$, $u_x(L, t) = 0$, for all $t \geq 0$, we must have $X'(0) = X'(L) = 0$. Moreover, if $X(x)$ satisfies the equation $X'' = \lambda X$, for some constant λ, and the endpoint conditions $X'(0) = X'(L) = 0$, then integration by parts [see (13)] yields that $\int_0^L X(x)X''(x)dx = -\int_0^L (X'(x))^2 dx$. But since $\int_0^L X(x)X''(x)dx = \lambda \int_0^L (X(x))^2 dx$, we see that $\lambda \leq 0$. Now if $\lambda = 0$, then $X'' = 0$, so $X(x)$ has the form $Ax + B$ for some constants A, B, not both zero. To satisfy the conditions $X'(0) = X'(L) = 0$ we must take $A = 0$. So $X_0(x) = 1$, $0 \leq x \leq L$, is one of the functions that models the space dependence of a standing wave with free endpoint conditions. Similarly, if $\lambda = -k^2$, $k \neq 0$, then $X(x)$ must have the form $A \sin kx + B \cos kx$, for constants A, B, not both zero. To satisfy the free boundary conditions at $x = 0$ and $x = L$ we see that $k(A \cos kx - B \sin kx) = 0$ at $x = 0$ and $x = L$. At $x = 0$, we must have $kA = 0$, so $A = 0$, while at $x = L$, we then must have $-kB \sin kL = 0$. Since we assume that $k \neq 0$ here and $B = 0$ would give only the trivial solution, we see that $\sin kL = 0$. This implies that $kL = n\pi$, for $n = 1, 2, \ldots$, providing the further candidates for the space dependence of standing waves, $X_n = \cos n\pi x/L$, $n = 1, 2, \ldots$. So we have that the standing waves are

$$u_n(x, t) = \cos \frac{n\pi x}{L} \left(A_n \cos \frac{n\pi ct}{L} + B_n \sin \frac{n\pi ct}{L} \right), \qquad n = 0, 1, 2, \ldots$$

where A_n, B_n are arbitrary constants. Note that if $n = 0$, $u_0 = X_0(x)T_0(t) = A_0 + B_0 t$ since $X_0 = 1$ (see above) and $T''(t) = 0$. This gives all of the standing waves with free boundary conditions at the endpoints $x = 0$ and $x = L$.

(b). The model equations for this situation are

$$
\begin{aligned}
u_{tt} - c^2 u_{xx} &= 0, & 0 < x < L, \quad t > 0 \\
u_x(0, t) &= 0, & t \geq 0 \\
u_x(L, t) &= 0, & t \geq 0 \\
u(x, 0) &= f(x), & 0 \leq x \leq L \\
u_t(x, 0) &= g(x), & 0 \leq x \leq L
\end{aligned}
$$

We are looking for standing waves $u = X(x)T(t)$ that satisfy the PDE and the two boundary conditions $u(0, t) = 0$ and $u_x(L, t) = 0$; as before, we ignore the initial conditions $u(x, 0) = f$ and $u_t(x, 0) = g$. From (10) and applying the boundary conditions, we have

$$X'' = \lambda X, \qquad X(0) = 0, \quad X'(L) = 0$$

$$T'' = c^2 \lambda T$$

Let's start with the equations for $X(x)$ and find the values of λ that "work". Suppose first that $\lambda = k^2$, $k > 0$. Then the general solution of $X(x)$ is $X = c_1 e^{kx} + c_2 e^{-kx}$, where c_1 and c_2 are arbitrary constants. Applying the boundary conditions, we have

$$0 = c_1 + c_2, \qquad 0 = c_1 k - c_2 k$$

These equations imply that $c_1 = c_2 = 0$ since $k \neq 0$. Since we want nontrivial solutions, we see that $\lambda = k^2$ won't work. So could $\lambda = 0$? The general solution of $X'' = 0$ is $X = c_1 + c_2 x$. The boundary conditions imply that $0 = c_1$ and $0 = c_2$, so again we don't get anything useful.

This means that we need to take $\lambda = -k^2$, $k > 0$. Then the general solution of $X'' = -k^2 X$ is $X = c_1 \cos kx + c_2 \sin kx$, where c_1 and c_2 are any constants. Imposing the boundary conditions, we see that

$$0 = c_1, \qquad 0 = -c_1 k \sin kL + c_2 k \cos kL$$

So, since $k \neq 0$, we must have that $\cos kL = 0$ to get a nontrivial solution. Since $\cos x$ is zero only at odd multiples of $\pi/2$, we see that $kL = (2n+1)\pi/2$. So we see that

$$\lambda_n = -k^2 = -\left[\frac{(2n+1)\pi}{2L}\right]^2, \qquad n = 0, 1, 2, \ldots$$

and

$$X_n = \sin\frac{(2n+1)\pi x}{2L}$$

Turning to the equation for $T(t)$, we have

$$T'' = c^2\lambda_n T$$

whose solutions are

$$T_n = A_n\cos\frac{(2n+1)\pi ct}{L} + B_n\sin\frac{(2n+1)\pi ct}{L}$$

where A_n and B_n are arbitrary constants. The standing waves are

$$u_n = \sin\frac{(2n+1)\pi x}{2L}\left[A_n\cos\frac{(2n+1)\pi ct}{2L} + B_n\sin\frac{(2n+1)\pi ct}{2L}\right], \qquad n = 1, 2, \ldots$$

2. We will show that $X'' = \lambda X$, $X(0) = X(L) = 0$ has *no* nontrivial solutions $X(x)$ if $\lambda = 0$ or $\lambda = k^2$ ($k > 0$). We do this by supposing first that $\lambda = 0$ and then that $\lambda = k^2$, and getting contradictions. Assuming $\lambda = 0$, the equation $X'' = 0$ has the general solution $X(x) = A + Bx$ for arbitrary constants A, B. Imposing the conditions $X(0) = 0$, $X(L) = 0$, we see that the constants A, B must satisfy the conditions $A = 0$, $A + BL = 0$. This linear system has only the trivial solution $A = B = 0$. Next assume that $\lambda = k^2$, for some $k \neq 0$. The equation $X'' = \lambda X$ has the general solution $X(x) = Ae^{kx} + Be^{-kx}$ for arbitrary constants A, B. The conditions $X(0) = 0$, $X(L) = 0$ imply that $A + B = 0$, and $Ae^{kL} + Be^{-kL} = 0$. So $e^{kL} = e^{-kL}$, that is, $e^{2kL} = 1$ (if $A \neq 0$). But this implies $k = 0$ or $L = 0$, neither of which is possible. So $A = B = 0$ is the only possibility but this gives only the trivial solution, so $\lambda \geq 0$ is not possible.

3. Suppose that u and v both solve the boundary/initial problem

$$
\begin{array}{lll}
\text{(PDE)} & u_{tt} = c^2 u_{xx} + F(x,t), & 0 < x < L, \quad t > 0 \\
\text{(BC)} & u(0,t) = \alpha(t), \quad u(L,t) = \beta(t), & t \geq 0 \\
\text{(IC)} & u(x,0) = \gamma(x), \quad u_t(x,0) = \delta(x), & 0 \leq x \leq L
\end{array}
$$

We want to show $u(x,t) = v(x,t)$. Note that $U(x,t) = u(x,t) - v(x,t)$ solves the problem

$$
\begin{cases}
U_{tt} = c^2 U_{xx}, & 0 < x < L, \quad t > 0 \\
U(0,t) = U(L,t) = 0, & t \geq 0 \\
U(x,0) = U_t(x,0) = 0, & 0 \leq x \leq L
\end{cases}
$$

Defining the function

$$W(t) = \frac{1}{2}\int_0^L (U_t^2 + c^2 U_x^2)\,dx$$

we see that

$$W'(t) = \frac{1}{2} \int_0^L \frac{d}{dt} \left(U_t^2 + c^2 U_x^2 \right) dx = \int_0^L (U_t U_{tt} + c^2 U_x U_{xt}) \, dx$$

$$= \int_0^L [U_t(c^2 U_{xx}) + c^2((U_x U_t)_x - U_{xx} U_t)] \, dx$$

$$= c^2 \int_0^L (U_x U_t)_x \, dx = U_x(L, t)U_t(L, t) - U_x(0, t)U_t(0, t)$$

Now since $U(0, t) = 0$ and $U(L, t) = 0$, $t \geq 0$, we can compute $U_t(0, t)$ and $U_t(L, t)$ by just differentiating $U(0, t)$ and $U(L, t)$ with respect to t (from the definition of the partial derivative). So $U_t(0, t) = U_t(L, t) = 0$ and $W'(t) = 0$, for all $t \geq 0$; the integrand in the definition of W must vanish for all (x, t) in R (since the integrand is nonnegative), where R is the region $0 < x < L$, $t > 0$. From the Mean Value Theorem (Theorem B.5.4) we conclude that U is a constant on the region, $0 \leq x \leq L$, $t \geq 0$. Evaluating this constant on the boundary of R where $u = v$, we conclude that $U = 0$ on R, and so $u = v$ on R, as was asserted.

10.2 Orthogonal Functions

Suggestions for preparing a lecture

Topics: Orthogonal functions in a linear space, scalar products, symmetry, bilinearity, positive definiteness, distance, Euclidean space PC[a, b] (both real and complex cases); why orthogonal sets are useful.

Making up a problem set

Problems 2, one part of 3(a) (b) (c), 3(d), 5(a), 5(b), 6, 8 (hard).

Comments

A better title for this section might be "Everything You Wanted to Know about Scalar Product Spaces with Orthogonal Bases." The modern approach to Fourier Series is based on the material of this section. Without it, Fourier Series reduces to endless "turn-the-crank" computations of Fourier-Euler coefficients, and there is little chance of understanding the "how" and the "why" of it all. In the few days you have available in the course to do "all" of Fourier series, PDEs, and separation of variables you won't have the time (or the inclination) to be very mathematical. So how to solve this dilemma? Here is what you might do: focus on examples, skip theory, work problems, focus on the close relationship of all of this material to standard Euclidean concepts (e.g., the Parseval Relation (10) is just an extension of the Pythagorean theorem that the square on the hypotenuse is the sum of the squares on the sides). No harm is done by this process of "nonrigorizing" everything in your introductory studies of Fourier Analysis. In fact, as a first approach to some deep ideas, this low-level, low-key approach is the best way to do it. So don't hesitate to focus on the ideas and the examples, and not too much on the theory.

1. Using the identity $\sin \alpha \sin \beta = [\cos(\alpha - \beta) - \cos(\alpha + \beta)]/2$, we see that

    $$\int_0^L \sin \frac{n\pi x}{L} \sin \frac{m\pi x}{L} = \frac{1}{2} \int_0^L \cos \frac{(n-m)\pi x}{L} dx - \frac{1}{2} \int_0^L \cos \frac{(n+m)\pi x}{L} dx$$

    If $m = n = 0$, the integrand and the integral are zero. Now, if $n \neq m$, then the right-hand side becomes

    $$\frac{1}{2} \frac{L}{(n-m)\pi} \sin \frac{(n-m)\pi x}{L} \Big|_0^L - \frac{1}{2} \frac{L}{(n+m)\pi} \sin \frac{(n+m)\pi x}{L} \Big|_0^L = 0$$

    On the other hand, if $n = m \neq 0$, then the right-hand side of the identity becomes

    $$\frac{1}{2} \int_0^L dx - \frac{L}{4n\pi} \sin \frac{2n\pi x}{L} \Big|_0^L = \frac{L}{2}$$

    The second set of formulas in (4) is proved in the same way, using the identity $\cos \alpha \cos \beta = [\cos(\alpha - \beta) + \cos(\alpha + \beta)]/2$.

2. The solution of plucked guitar string problem

    $$\begin{aligned} u_{tt} &= c^2 u_{xx} & 0 &< x < L, \quad t > 0 \\ u(0, t) &= u(L, t) = 0, & t &\geq 0 \\ u(x, 0) &= f(x), \quad u_t(x, 0) = 0, & 0 &\leq x \leq L \end{aligned}$$

    is given by the text formula (5) where $L = \pi$, and A_n is given by text formula (6). Evaluating A_n, we have (using integral tables)

    $$A_n = \frac{2}{\pi} \int_0^\pi f(x) \sin nx \, dx = \frac{2}{\pi} \int_0^{\pi/2} x \sin nx \, dx + \frac{2}{\pi} \int_{\pi/2}^\pi (\pi - x) \sin nx \, dx$$

    $$= \frac{4}{\pi n^2} \sin \frac{n\pi}{2}$$

    Since $\sin n\pi/2 = 0$ if n is even

    $$u(x, t) = \sum_{\text{odd } n} \frac{4}{\pi n^2} \sin \frac{n\pi}{2} \sin nx \cos nt$$

    (Notice that the terms with even n drop out.) Using terms in the series up to $n = 13$ in Fig. 2, we obtain reasonably good plots for $u(x, 1)$ (the solid line), $u(x, 5)$ (the long-dashed line), and $u(x, 10)$ (the short-dashed line). In reality, these graphs are composed of straight line segments, but the "wavyness" appears because we have truncated the series.

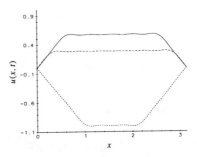

Problem 2.

3. The plucked guitar string problem in this case is

$$u_{tt} = c^2 u_{xx} \qquad\qquad 0 < x < \pi, \quad t > 0, \quad c^2 = T/\rho$$
$$u(0, t) = 0, \quad u(\pi, t) = 0, \qquad t \geq 0$$
$$u(x, 0) = f(x), \quad u_t(x, 0) = 0, \qquad 0 \leq x \leq \pi$$

According to text formulas (5) and (6), the solution is

$$u(x, t) = \sum_{n=1}^{\infty} A_n \sin nx \cos nct$$

$$A_n = \frac{2}{\pi} \int_0^\pi f(x) \sin nx\, dx, \qquad n = 1, 2, \ldots$$

So to solve parts **(a)**, **(b)**, **(c)**, all we need to do is calculate A_n, $n = 1, 2, \ldots$, for the particular $f(x)$ given.

(a). If $f(x) = \sin x$, then $A_1 = (2/\pi) \int_0^\pi \sin^2 x\, dx = 1$. $A_n = 0$ for $n > 1$ because by text formula (4) $\sin x$ and $\sin nx$, $n > 1$, are orthogonal. So $u(x, t) = \sin x \cos ct$, as one would expect.

(b). If $f(x) = 0.1 \sin 2x$, then $A_2 = (2/\pi) \int_0^\pi 0.1 \sin 2x \sin 2x\, dx = (2/\pi) \cdot 0.1 \cdot (\pi/2)$ [by text formula (4)] $= 0.1$. According to (4), $A_n = 0$ for $n = 1, 3, 4, \ldots$, so $u(x, t) = 0.1 \sin 2x \cos 2ct$.

(c). Let $f(x) = \sum_{n=1}^{10} (\sin nx)/n$. Then for $n = 1, \ldots, 10$, $A_n = (2/\pi) \int_0^\pi [\sin(nx)^2/n]\, dx$ [by (4)]. Also by (4), $A_n = 0$ for $n = 11, 12, \ldots$. So $u(x, t) = \sum_{n=1}^{10} (\sin nx \cos nct)/n$.

(d). Since c^2 is the ratio of the tension τ in the string to the density ρ of the string, c increases if τ is increased (i.e., the strip is made tauter), and c decreases if the mass (and so, presumably, the density) is increased. The frequency of a standing wave is nc, so the frequency (i.e., the pitch) is higher if the tension is increased and lower if the density is increased.

4. **(a).** We want to show that V and $\{0\}$ are subspaces of V. Let α be any scalar, and let u and v be any pair of vectors in V. Since V is a linear space, αu and $u + v$ belong to V, and so the set of all vectors in V is also a subspace of V. Clearly, $0 + 0 = 0$ and $k \cdot 0 = 0$, where k is scalar, are in V. So, the set consisting of the zero vector is a subspace of V, too.

(b). We want to show that A is a subspace of V if and only if $\text{Span}(A) = A$. Let the set A in the linear space V be a subspace. Then A is closed under the operations of addition and multiplication by a scalar, and it follows that $\text{Span}(A) = A$. Conversely, if a subset A of a linear space V is such that $\text{Span}(A) = A$, then A is a linear space since it is closed under finite linear combinations.

(c). Let $n \geq m$. We want to show that $\mathbf{C}^n(I)$ is a subspace of $\mathbf{C}^m(I)$. The set $A = \mathbf{C}^n(I)$ is contained in the linear space $V = \mathbf{C}^m(I)$, since any n-times continuously differentiable function is also m-times continuously differentiable if $m \leq n$. $\mathbf{C}^n(I)$ is also closed under finite linear combinations, and so by part **(b)**, $\mathbf{C}^n(I)$ is a subspace of $\mathbf{C}^m(I)$.

(d). Because $\mathbf{C}^\infty(I)$ is a subset of $\mathbf{C}^1(I)$ which is closed under finite linear combinations, it follows from part **(b)** that $\mathbf{C}^\infty(I)$ is a subspace of itself and hence is a linear space.

5. Define $\langle f, g \rangle = \int_0^1 f(x)g(x)e^x \, dx$.

(a). We want to show that $\langle f, g \rangle$ is a scalar product. Since $f(x)g(x) = g(x)f(x)$ for all scalar functions f and g, and since the functions in $\mathbf{C}^0[0, 1]$ are real, we have that

$$\int_0^1 f(x)g(x)e^x \, dx = \int_0^1 g(x)f(x)e^x \, dx$$

and $\langle f, g \rangle = \langle g, f \rangle$, so the first condition (8a) of a scalar product is met. The property of bilinearity [(8b)], $\langle \alpha f + \beta g, h \rangle = \alpha \langle f, h \rangle + \beta \langle g, h \rangle$ follows immediately from the corresponding property for the integral of a product. To show positive definiteness [(8c)], observe that $\langle f, f \rangle = \int_0^1 f^2(x)e^x \, dx$. If $f(x) = 0$ for all x in $[0, 1]$, then $\langle f, f \rangle = 0$. Integrals of nonnegative continuous functions satisfy the integral mean value theorem (see Theorem B.5.4), so for some x_0 in the interval $[0, 1]$, $\langle f, f \rangle = f^2(x_0)e^{x_0}$, which is certainly nonnegative. Suppose that $\langle f, f \rangle = \int_0^1 f^2(x)e^x \, dx = 0$, but that $f(x_0) \neq 0$ for some x_0, say $f(x_0) > 0$. Then there is an interval $I = [c, d]$ (maybe quite small) around x_0 and $f(x) > 0$ for all x in $[c, d]$. But then $\int_c^d f^2(x)e^x \, dx = f^2(x_1)e^{x_1}(d - c) > 0$ for some x_1 in $[c, d]$. So $\int_0^1 = \int_0^c + \int_c^d + \int_d^1 > 0$; the supposition that $f(x_0) > 0$ for some x_0 is wrong. So $\langle f, g \rangle$ meets all three conditions for an inner product.

(b). Let $f = 1 - x$, $g = e^{-x}$; we have $\langle f, g \rangle = \int_0^1 (1 - 2x)e^{-x}e^x \, dx = (x - x^2)\big|_{x=0}^{x=1} = 0$. For the second pair $f = x^2$, $g = e^x$, we have $\langle f, g \rangle = \int_0^1 x^2 e^x e^x \, dx = \int_0^1 x^2 e^{2x} \, dx = \frac{1}{4}(e^2 - 1)$, where we have used a table of integrals.

(c). Tables of integrals are used to compute the integrals, but a CAS could be used as well.

$$\langle x, 1 - x \rangle = \int_0^1 x(1 - x)e^x \, dx = 3 - e$$

$$\langle e^{-x/2}\sin(\pi x/2), e^{-x/2}\sin(3\pi x/2) \rangle = \int_0^1 \sin(\pi x/2)\sin(3\pi x/2) \, dx = 0$$

$$\langle \cos(\pi x/2), 1 \rangle = \int_0^1 \cos(\pi x/2)e^x \, dx = (2\pi e - 4)/(\pi^2 + 4)$$

6. To show that $\langle f, g \rangle = \int_a^b f(x)\bar{g}(x) \, dx$ is a scalar product for $I = [a, b]$ and all f, g in

$C^n(I)$, we need to check that (8a), (8b), and (8c) hold.

$$\langle f, g \rangle = \int_a^b f(x)\overline{g(x)}dx = \overline{\int_a^b g(x)\overline{f(x)}dx} = \overline{\langle g, f \rangle},$$

$$\langle \alpha f + \beta h, g \rangle = \int_a^b (\alpha f(x) + \beta h(x))\overline{g(x)}dx = \int_a^b \alpha f(x)\overline{g(x)}dx + \int_a^b \beta h(x)\overline{g(x)}dx$$

$$= \alpha\langle f, g \rangle + \beta\langle h, g \rangle,$$

$$\langle f, f \rangle = \int_a^b \|f\|^2 \geq 0, \quad \text{and equality holds if and only if } f = 0 \text{ since } f \text{ is continuous}$$

So $\int_a^b f\bar{g}\,dx$ is a scalar product on $C^n(I)$.

7. **(a).** Using the properties of the norm and the scalar product given in the text, we have that $\|u + v\|^2 + \|u - v\|^2 = \langle u + v, u + v \rangle + \langle u - v, u - v \rangle = \langle u, u \rangle + \langle u, v \rangle + \langle v, u \rangle + \langle v, v \rangle + \langle u, u \rangle - \langle u, v \rangle - \langle v, u \rangle + \langle v, v \rangle = 2\langle u, u \rangle + 2\langle v, v \rangle = 2\|u\|^2 + 2\|v\|^2$, which was to be shown.

    (b). We want to show that $|\langle u, v \rangle| \leq \|u\| \cdot \|v\|$. If $u = 0$ or $v = 0$, we are done. Let $u \neq 0$, $v \neq 0$. Then, for any real constants α and β, we have

    $$0 \leq \|\alpha u - \beta v\|^2 = \alpha^2\|u\|^2 + \beta^2\|v\|^2 - \alpha\beta\langle u, v \rangle - \alpha\beta\langle v, u \rangle$$

    Let $|\alpha| = 1/\|u\|$, $|\beta| = 1/\|v\|$. Then we see that $2|\alpha| \cdot |\beta| \cdot |\langle u, v \rangle| \leq |\alpha|^2\|u\|^2 + |\beta|^2\|v\|^2 = 2$ since $0 \leq \langle |\alpha|u - |\beta|v, |\alpha|u - |\beta|v \rangle = |\alpha|^2\|u\|^2 - 2|\alpha| \cdot |\beta| \cdot \langle u, v \rangle + |\beta|^2\|v\|^2$. So $|\langle u, v \rangle| \leq 1/(|\alpha| \cdot |\beta|) = \|u\| \cdot \|v\|$.

    (c). The proof that $\langle u, v \rangle = (\|u + v\|^2 - \|u - v\|^2)/4$ follows by direct computation of the right-hand side.

8. **(a).** Let $x = (x_n)$ and $y = (y_n)$ be any two elements in l^2. So $\sum |x_n|^2$ and $\sum |y_n|^2$ converge. We will use the fact that a series is convergent if it converges absolutely, that is, that $\sum a_n$ converges if $\sum |a_n|$ converges. We shall show that $\sum_{k=1}^\infty |x_k| \cdot |y_k|$ is convergent, and it will follow that $\sum_{k=1}^\infty x_k y_k$ converges, too. The partial sum $S_N = \sum_{k=1}^N |x_k| \cdot |y_k|$ can be thought of as the standard scalar product of the vectors $(|x_1|, \cdots, |x_N|)$ and $(|y_1|, \cdots, |y_N|)$ in $\mathbb{R}^N$. The Cauchy-Schwartz inequality implies that

    $$\sum_{k=1}^N |x_k| \cdot |y_k| \leq \left[\sum_{k=1}^N |x_k|^2\right]^{1/2} \left[\sum_{k=1}^N |y_k|^2\right]^{1/2} \leq \left[\sum_{k=1}^\infty |x_k|^2\right]^{1/2} \left[\sum_{k=1}^\infty |y_k|^2\right]^{1/2}$$

    Since the above inequality holds for all N, it follows that the nondecreasing sequence of partial sums $S_1, S_2, \ldots$ is bounded from above by the number $\left[\sum_{k=1}^\infty |x_k|^2\right]^{1/2} \left[\sum_{k=1}^\infty |y_k|^2\right]^{1/2}$, and the series $\sum_{k=1}^\infty |x_k| \cdot |y_k|$ is convergent.

    (b). In (a) it was shown that for any x, y in l^2 we can define the scalar $\langle x, y \rangle = \sum_{k=1}^\infty x_k y_k$. It is straightforward to verify that l^2 is a real linear space and that $\langle \cdot, \cdot \rangle$ so defined satisfies properties (8a), (8b), (8c), so defines a scalar product on l^2.

9. **(a).** For u in E, consider the vectors $v = \text{proj}_S(u)$ and $w = u - \text{proj}_S(u)$. Then $v + w = u$

and v is in S, where S is the span of the orthogonal set $\{u^1, \ldots, u^n\}$. Since

$$\langle u^k, w \rangle = \langle u^k, u - \sum_1^n [\langle u, u^i \rangle / \|u^i\|^2] u^i \rangle = \langle u^k, u \rangle - [\langle u, u^k \rangle / \|u^k\|^2] \langle u^k, u^k \rangle = 0$$

for each $k = 1, 2, \ldots, n$, it follows that w is orthogonal to S, and the assertion follows. The proof of uniqueness is omitted.

(b). We want to show that $\|u - \text{proj}_S(u)\| \le \|u - v\|$ for all v in S, equality holding if and only if $v = \text{proj}_S(u)$. First notice that for any pair of orthogonal vectors $\{x, y\}$ in E we have that $\|x + y\|^2 = \|x\|^2 + \|y\|^2$ because $\langle x, y \rangle = \langle y, x \rangle = 0$. Now let u in E and v in S be given. Then the vectors $u - \text{proj}_S(u)$, $\text{proj}_S(u) - v$ are orthogonal since $\text{proj}_S(u) - v$ is in S and $u - \text{proj}_S(u)$ is orthogonal to S by part **(a)**. We have that

$$\|u - v\|^2 = \|(u - \text{proj}_S(u)) + (\text{proj}_S(u) - v)\|^2 = \|u - \text{proj}_S(u)\|^2 + \|\text{proj}_S(u) - v\|^2$$
$$\ge \|u - \text{proj}_S(u)\|^2$$

The inequality becomes an equality if and only if $v = \text{proj}_S(u)$.

(c). We want to find the distance from $(1, 1, 1)$ to the plane $x + 2y - z = 0$. The plane $x + 2y - z = 0$ contains the two vectors $u = (1, 0, 1)$ and $v = (0, 1, 2)$. Now if v is regarded as a basis for the line S through the origin parallel to v then $w = u - \text{proj}_S(u)$ also lies in the plane and the pair $\{v, w\}$ is a basis for the plane P. We have that $w = (1, 0, 1) - 2(0, 1, 2)/5 = (1, -2/5, 1/5)$. To find the distance d between the point $q = (1, 1, 1)$ and the plane P we see from the statement of part **(b)** that this distance is given by $d = \|q - \text{proj}_S(q)\|$. Since

$$\text{proj}_S(q) = \frac{\langle q, v \rangle}{\|v\|^2} v + \frac{\langle q, w \rangle}{\|w\|^2} w = (1/3)(2, 1, 4)$$

and so $q - \text{proj}_S(q) = \frac{1}{3}(1, 2, -1)$, giving the result that $d = \|\frac{1}{3}(1, 2, -1)\| = \sqrt{6}/3$.

Background Material: Linear Spaces

$\mathbb{R}^3$ is an example of a linear space, a collection of objects, called vectors, for which vector addition and scalar multiplication are defined and satisfy the properties:

❖ **Vector Space Properties.** In the nonempty collection V of objects (called *vectors*) there are two operations defined which satisfy the properties listed below. Denote by "+" the operation of *addition* which assigns a vector $u + v$ to every pair of vectors u and v. For a given set of *scalars*, the operation of *scalar multiplication* assigns to every scalar α and every vector u the vector αu. With these operations V has the properties:

- (A1) *Addition is order independent.* A finite collection of vectors can be added together in any order without affecting the sum.

- (A2) *Existence of zero element.*[1] There is an additive *zero* element in V, denoted by 0, with the property that $u + 0 = u$ for all vectors u.

- (A3) *Existence of additive inverses.* Every vector u has an *additive inverse*, denoted by $-u$, with the property that $u + (-u) = 0$.

- (M1) *Scalar multiplication is distributive.* For any vectors u and v and any scalars α and β, we have that $\alpha(u + v) = \alpha u + \alpha v$ and $(\alpha + \beta)u = \alpha u + \beta u$.

- (M2) *Scalar multiplication is associative.* For any vectors u and any scalars α and β, we have that $\alpha(\beta u) = (\alpha\beta)u$.

- (M3) For the scalar 1 we have that $1u = u$, for all vectors u.

The Vector Space Properties define the abstract notion of a linear (or vector) space.

❖ **Linear Space.** Let V be a set of objects and **F** a set of scalars (**F** is either $\mathbb{R}$, the real numbers, or **C**, the complex numbers) with two operations, "addition" and "scalar multiplication," defined in V such that the Vector Space Properties hold. Then V is a *real linear* (or *vector*) *space* if $\mathbf{F} = \mathbf{R}$, and a *complex linear* (or *vector*) *space* if $\mathbf{F} = \mathbf{C}$. The elements of V are sometimes called *vectors*.

A feature of this abstract approach to linear spaces is that we can deduce from the defining properties (A1)–(A3) and (M1)–(M3) some additional properties which must then hold in *any* linear space. Some of them are as follows:

THEOREM

> Further Properties of Linear Spaces. Let V be any linear space.
>
> 1. There is only one zero vector in V.
>
> 2. The vector αu is zero if either the scalar $\alpha = 0$ or the vector $u = 0$. The other way around, if $\alpha u = 0$, then either $\alpha = 0$ or $u = 0$.
>
> 3. Every vector v in V has a *unique* additive inverse, $-v$; moreover $-v = (-1)v$ for any vector v.

Each assertion is verified below:

1. Suppose that 0 and $0'$ are two zero vectors for V; then by (A1) and (A2) we see that $0 = 0 + 0' = 0' + 0 = 0'$, so there is only one zero in V.

2. For any u note that $u = (1 + 0)u = u + 0u$ [using (M1) and (M3)]. Adding $-u$ to both sides of this equation, we obtain by (A3) that $0 = 0u$, as asserted. Next, we assume that $\alpha \neq 0$. Then for any vector u, $\alpha 0 + u = \alpha 0 + (\alpha \cdot \alpha^{-1})u = \alpha(0 + \alpha^{-1}u) = \alpha(\alpha^{-1}u) = 1u = u$, where we have used (M1)–(M3). So $\alpha 0$ acts

---

[1]It is traditional to use the same symbol 0 for the zero scalar and the zero vector—context distinguishes one from the other.

like a zero vector, and since there is only one such vector, we must have $\alpha 0 = 0$. Finally, suppose that $\alpha u = 0$ for some scalar α and some vector u. If $\alpha \neq 0$, then $u = \alpha^{-1}(\alpha u) = \alpha^{-1}0 = 0$.

3. Suppose that v and w are two additive inverses for the vector u. Then $v = v + 0 = v + (u + w) = (v + u) + w = 0 + w = w$ where we have used (A1) and (A2). So u has a unique inverse. Now since $0 = 0u = (1 - 1)u = 1u + (-1)u = u + (-1)u$, we see that $-u = (-1)u$, finishing the proof.

It is worth noting that there is a linear space V with only one element. In view of the result above, we see that that single element must be the zero vector; V is written as $\{0\}$ and is referred to as the *trivial* linear space. Here are two familiar examples.

EXAMPLE

Real n-tuple Space

Let $\mathbb{R}^n$ for any $n = 1, 2, \ldots$ be the collection of all ordered n-tuples of real numbers, $x = (x_1, x_2, \ldots, x_n)$. Note that x here denotes the n-tuple and the x_i are real numbers. Two elements x and y in $\mathbb{R}^n$ are identified if and only if $x_i = y_i$, $i = 1, 2, \ldots, n$. Using the real numbers $\mathbb{R}$ as scalars and defining the operations of addition and multiplication by scalars component-wise by

$$x + y = (x_1 + y_1, \ldots, x_n + y_n), \qquad \alpha x = (\alpha x_1, \ldots, \alpha x_n) \tag{i}$$

for all x, y in $\mathbb{R}^n$ and all scalars α, it is straightforward to verify that $\mathbb{R}^n$ is a real linear space. Note that $\mathbb{R}^1$ can be identified with $\mathbb{R}$, so $\mathbb{R}$ itself can be considered a real linear space.

EXAMPLE

Complex n-tuple Space

Let $\mathbf{C}^n$ for any $n = 1, 2, \ldots$ be the collection of all ordered n-tuples of complex numbers $z = (z_1, z_2, \ldots, z_n)$. Using the complex numbers $\mathbf{C}$ as scalars and defining operations of addition and multiplication by scalars as in (i), it is straightforward to show that $\mathbf{C}^n$ is a complex linear space. Note that $\mathbf{C}^1$ can be identified with $\mathbf{C}$.

10.3 Fourier Series and Mean Approximation

Suggestions for preparing a lecture

Topics: This section contains the core ideas for a Fourier expansion. If your calendar allows time for only one lecture, then it is probably best to cut right to the chase and start with the Fourier-Euler Theorem, the formula for the Fourier coefficients with respect to an orthogonal set, and the definition of a Fourier series of a function with respect to an orthogonal set (pages 535, 536). Then do an example such as Example 10.3.2, and a graph such as Figure 10.3.3. Finally mention the idea of a basis and summarize the convergence properties of a Fourier series with an emphasis on mean convergence and the Mean Convergence Theorem (Theorem 10.3.2). If there is time for a second lecture, then more examples and discussion of the various kinds of convergence are helpful.

Making up a problem set

Problems 1, one part of 2, one part of 3, one of 4–6.

Comments

This section contains the basic ideas about the Fourier expansion of a function with respect to an orthogonal set. We start with a Euclidean space and an orthogonal set in it. We show how to construct the Fourier series of a function in the Euclidean space with respect to the orthogonal set, explain mean convergence and apply that idea to the convergence of a Fourier series of a function to the function itself. The idea of a basis comes in along the way. We explain what is meant when it is said that "a partial sum of the Fourier series of a function is the best approximation in the mean to the function."

1. This problem is solved by using Theorem 10.3.1 with $\Phi_N = \{1, \cos x, \sin x, \ldots, \cos Nx, \sin Nx\}$. The element which is closest to $f(x) = x$, $-\pi < x < \pi$, in the subspace S_N of $PC[-\pi, \pi]$ is given by

$$\text{proj}_{S_N}(f) = a_0 + \sum_{k=1}^{N}(a_k \cos kx + b_k \sin kx)$$

$$a_k = \frac{\langle f, \cos kx \rangle}{\| \cos kx \|^2} = \begin{cases} (2\pi)^{-1} \int_{-\pi}^{\pi} x dx, & \text{for } k = 0 \\ (\pi)^{-1} \int_{-\pi}^{\pi} x \cos kx dx, & \text{for } k = 1, 2, \ldots, N \end{cases}$$

$$b_k = \frac{\langle f, \sin kx \rangle}{\| \sin kx \|^2} = \frac{1}{\pi} \int_{-\pi}^{\pi} x \sin kx dx, \quad \text{for } k = 1, 2, \ldots, N$$

Notice that f is odd about the origin, so $f(x) \cos kx$ is also odd about the origin for all $k = 0, 1, 2, \ldots, N$. We have $a_k = 0$ for $k = 0, 1, 2, \ldots, N$. Note that $f(x) \sin kx$ is even about the origin, so

$$b_k = \frac{2}{\pi} \int_0^{\pi} x \sin kx dx = -\frac{2}{k\pi} \int_0^{\pi} x(\cos kx)' dx$$

$$= -\frac{2}{k\pi}\left[x \cos kx \Big|_0^{\pi} - \int_0^{\pi} \cos kx dx\right] = \frac{(-1)^{k+1}2}{k}, \quad \text{for } k = 1, 2, \ldots, N,$$

So $\text{proj}_{S_n}(f) = 2\sum_{k=1}^{N}(-1)^{k+1}k^{-1}\sin kx$ is the element in S_N closest to $f(x) = x$.

2. Here the orthogonal set in PC$[0, \pi]$ is $\Phi = \{\sin x, \sin 2x, \ldots\}$.

    (a). The function $f(x)$ is x. FS$_\Phi[f]$ has the form $\sum_{k=1}^{\infty} B_k \sin kx$, where

    $$B_k = \left[\int_0^{\pi} f(x) \sin kx dx\right] \Big/ \left[\int_0^{\pi} \sin^2 kx dx\right], \quad k = 1, 2, \ldots.$$

    Note that $\int_0^{\pi} \sin^2 kx dx = \pi/2$, for all $k = 1, 2, \ldots.$ We have for any $k = 1, 2, \ldots,$

    $$B_k = \frac{2}{\pi} \int_0^{\pi} x \sin kx dx = -\frac{2}{k\pi} \int_0^{\pi} x(\cos kx)' dx$$

$$= -\frac{2}{k\pi}\left\{x\cos kx\Big|_0^\pi - \int_0^\pi \cos kx\,dx\right\}$$

$$= 2(-1)^{k+1}/k$$

So we have $FS_\Phi[f] = 2\sum_{k=1}^\infty (-1)^{k+1}(\sin kx)/k$.

(b). In this case $f(x) = 1$. So we have

$$B_k = \frac{2}{\pi}\int_0^\pi \left(1+\frac{x}{\pi}\right)\sin kx\,dx = -\frac{2}{k\pi}\int_0^\pi \left(1+\frac{x}{\pi}\right)(\cos kx)'\,dx$$

$$= -\frac{2}{k\pi}\left\{\left(1+\frac{x}{\pi}\right)\cos kx\Big|_0^\pi - \int_0^\pi \frac{1}{\pi}\cos kx\,dx\right\}$$

$$= \frac{2(\pi^{-1}+2(-1)^{k+1})}{k\pi}$$

The Fourier series of f with respect to Φ is

$$FS_\Phi[f] = \frac{2}{\pi}\sum_{k=1}^\infty \frac{\pi^{-1}+2(-1)^{k+1}}{k}\sin kx$$

(c). Here the function is $f(x) = 1 + x/\pi$, and $FS_\Phi[f] = \sum_{k=1}^\infty B_k\sin kx$, where

$$B_k = \frac{2}{\pi}\int_0^\pi \sin kx\,dx = -\frac{2}{k\pi}\cos kx\Big|_0^\pi = \frac{2(1-(-1)^k)}{k\pi} = \begin{cases} 0, & \text{even } k \\ 4/k\pi, & \text{odd } k \end{cases}$$

So

$$FS_\Phi[f] = \frac{4}{\pi}\sum_{m=0}^\infty \frac{\sin(2m+1)x}{2m+1}$$

3. Here the orthogonal set is $\Phi = \{1, \cos x, \sin x \ldots, \cos nx, \sin nx, \ldots\}$ in the Euclidean space $PC[-\pi, \pi]$, and we want to find the Fourier series of f with respect to Φ for the three engineering functions given.

(a). The sawtooth function $f(x)$ is an odd function, so $FS[f]$ contains only sine terms. Using the fact that $f(x)\sin kx$ is an even function for all k, we see that

$$B_k = \frac{2}{\pi}\int_0^\pi A(1-\frac{x}{\pi})\sin kx\,dx = \frac{-2A}{k\pi}\int_0^\pi (1-\frac{x}{\pi})(\cos kx)'\,dx$$

$$= \frac{-2A}{k\pi}\left\{\left(1-\frac{x}{\pi}\right)\cos kx\Big|_0^\pi - \int_0^\pi \left(-\frac{1}{\pi}\right)\cos kx\,dx\right\} = \frac{2A}{k\pi}$$

We have $FS_\Phi[f] = (2A/\pi)\sum_{k=1}^\infty (\sin kx)/k$.

(b). The triangular function $f(x)$ is even, so $FS[f]$ does not contain any sine terms. Also, note that $f(x)\cos kx$ is even for each $k = 0, 1, 2, \ldots$. We have

$$A_0 = \frac{2}{2\pi}\int_0^\pi A\left(1-\frac{x}{\pi}\right)dx = \frac{A}{2},$$

$$A_k = \frac{2}{\pi}\int_0^\pi A\left(1-\frac{x}{\pi}\right)\cos kx\,dx = \frac{2A}{k\pi}\int_0^\pi \left(1-\frac{x}{\pi}\right)(\sin kx)'\,dx$$

$$= \frac{2A}{k\pi} \left\{ \left(1 - \frac{x}{\pi}\right) \sin kx \Big|_0^\pi - \int_0^\pi \left(-\frac{1}{\pi}\right) \sin kx dx \right\}$$

$$= \frac{2A}{k^2 \pi^2} \left(-\cos kx \Big|_0^\pi\right) = \frac{2A}{k^2 \pi^2} (1 - (-1)^k) = \begin{cases} 0, & \text{even } k \\ \frac{4A}{k^2 \pi^2}, & \text{odd } k \end{cases}$$

We have

$$FS_\Phi[f] = \frac{A}{2} + \frac{4A}{\pi^2} \sum_{m=0}^\infty \frac{\cos(2m+1)x}{(2m+1)^2}$$

(c). The inverted triangular function $f(x)$ is even about the origin, so $FS[f]$ does not contain any sine terms.

$$A_0 = \frac{2}{2\pi} \int_0^\pi x dx = \frac{1}{2\pi} x^2 \Big|_0^\pi = \frac{\pi}{2}$$

$$A_k = \frac{2}{\pi} \int_0^\pi x \cos kx dx = \frac{2}{k\pi} \int_0^\pi x(\sin kx)' dx$$

$$= \frac{2}{k\pi} \left\{ x \sin kx \Big|_0^\pi - \int_0^\pi \sin kx dx \right\}$$

$$= -\frac{2}{k^2 \pi} (1 - (-1)^k) = \begin{cases} 0, & \text{even } k \\ -4/k^2\pi, & \text{odd } k \end{cases}$$

We have that

$$FS_\Phi[f] = \frac{\pi}{2} - \frac{4}{\pi} \sum_{m=0}^\infty \frac{\cos(2m+1)x}{(2m+1)^2}$$

4. We want to verify Theorem 10.3.3. Suppose that $f_n \to f$ and $g_m \to g$ in a Euclidean space. That means that $\|f_n - f\| \to 0$ and $\|g_m - g\| \to 0$ as $n, m \to \infty$. We must show that $\langle f_n, g_m \rangle \to \langle f, g \rangle$ as $n, m \to \infty$. Let $f_n \to f$, as $n \to \infty$, and $g_m \to g$, as $m \to \infty$ in a Euclidean space E. Put $p_n = f_n - f$ and $q_m = g_m - g$ and observe that

$$|\langle f_n, g_m \rangle - \langle f, g \rangle| = |\langle p_n + f, q_m + g \rangle - \langle f, g \rangle|$$

$$= |\langle p_n, q_m \rangle + \langle p_n, g \rangle + \langle f, q_m \rangle|$$

$$\le |\langle p_n, q_m \rangle| + |\langle p_n, g \rangle| + |\langle f, q_m \rangle|$$

$$\le \|p_n\| \|q_m\| + \|p_n\| \|g\| + \|f\| \|q_m\|$$

where we have used the Triangle inequality and the Cauchy-Schwartz inequality of Theorem 10.2.1. Now since $\|p_n\| \to 0$ as $n \to \infty$, and $\|q_m\| \to 0$ as $m \to \infty$, the desired result follows since the last item in the above inequality $\to 0$ as $n, m \to \infty$.

5. We want to show that two functions f and g with the same Fourier series with respect to an orthogonal basis Φ of a Euclidean space are "equal" in that space, where by equal we mean that $\|f - g\| = 0$. Let $w = f - g$. Since $\Phi = \{\phi_1, \phi_2, \ldots\}$ is a basis for E, the Fourier series of w over Φ must converge to w in the mean; that is,

$$w_N = \sum_{k=1}^{N} \frac{\langle w, \phi_k \rangle}{\|\phi_k\|^2} \phi_k \to w \text{ as } N \to \infty$$

By the continuity of the scalar product, $\langle w_N, w \rangle \to \|w\|^2$, as $N \to \infty$. But since $\langle w, \phi_k \rangle = 0$ for all $k = 1, 2, \ldots$, we see that $w_N = 0$ for all N, so $\langle w_N, w \rangle \to 0$ as $N \to \infty$. But $\langle w_N, w \rangle \to \langle w, w \rangle$ by Theorem 10.3.3, so $\langle w, w \rangle = \|w\|^2$, so $\|w\|^2 = 0$, which in turn implies that $f = g$.

6. E is the collection of all continuous functions $f(x)$, $|x| < \infty$, such that $\|f\|^2 = \int_{-\infty}^{\infty} |f(x)|^2 \, dx$ is finite.

(a). That $\langle f, g \rangle = \int_{\mathbb{R}} f g \, dx$, f, g in E, satisfies the defining properties of a scalar product [Section 10.2, (8a), (8b), (8c)] follows directly from properties of integrals.

(b). Notice that $0 \le f_n(x) \le 1/n^{1/2}$, for all x in $\mathbb{R}$. So $\{f_n(x)\}$ converges to the zero function uniformly on $\mathbb{R}$.

(c). Consulting a table of integrals, we see that $\langle f_n, f_n \rangle = \|f_n\|^2 = \int_{-\infty}^{\infty} (f_n)^2 dx = (\pi/2)^{1/2}$, for all $n = 1, 2, \ldots$. Now if $f_n \to g$ in E, then $\|f_n - g\|^2 = \int_{-\infty}^{\infty} (f_n - g)^2 dx \to 0$ as $n \to \infty$. This means that for any a and b, $-\infty < a < b < \infty$, we have that $\int_a^b (f_n - g)^2 dx \to 0$ as $n \to \infty$. So $g(x)$, $a < x < b$, is the limit in the mean in $PC[a, b]$ of the sequence $\{f_n(x), a < x < b\}$. Also, since f_n tends to the zero function uniformly on $\mathbb{R}$, it follows that $\{f_n(x), a < x < b\}$ converges in the mean to the zero function in $PC[a, b]$. But since limits in the mean are unique, it follows that $g(x) = 0$, $a < x < b$. Since a, b are arbitrary, it follows that $g = 0$ on $\mathbb{R}$. But this is impossible because $\|f_n - g\|^2 = \|f_n\|^2 = (\pi/2)^{1/2} \ne 0$ for all n. So $\{f_n\}$ is *not* mean convergent in E.

Background Material: Mean versus Pointwise Convergence

Mean Convergence versus Pointwise Convergence

Mean convergence of a sequence $\{f_n(x)\}$ in $PC[a, b]$ implies the existence of an element f in $PC[a, b]$ such that $\int_a^b (f_n(x) - f(x))^2 \, dx \to 0$ as $n \to \infty$. Consider the sequence g_k in $PC[0, 1]$, $k = 1, 2, \ldots$ defined in Graph 1. We claim that $\{g_k\}$ does *not* converge in the mean. Indeed, say $g_k \to g$, for some g in $PC[0,1]$. Writing $g_k = (g_k - g) + g$ and applying the Triangle Inequality, we have $\|g_k\| \le \|g_k - g\| + \|g\|$, so the sequence $\{\|g_k\|\}$ is bounded. But since $\|g_k\|^2 = \int_0^1 |g_k|^2 \, dx = 2k/3 \to \infty$, as $k \to \infty$, it follows that $\{\|g_k\|\}$ is unbounded, so $\{g_k\}$ cannot converge.

On the other hand, consider the sequence h_k in $PC[0,1]$, $k = 1, 2, \ldots$ defined in Graph 2. Note that h_k, for $k = 2^n + m$, has the value 1 inside the interval $[(m-1)/2^n, m/2^n]$ and vanishes otherwise in $[0,1]$. So $\int_0^1 |h_{2^n+m}| \, dx = 1/2^n$ for all $n = 1, 2, \ldots$, and all $m = 1, 2, \ldots, 2^n - 1$. It follows that $\|h_k\|^2 = \|h_k - 0\|^2 \to 0$ as $k \to \infty$. So $\{h_k\}$ converges in the mean to the zero function of $[0,1]$, even though for each k there is an interval on which h_k has the value one.

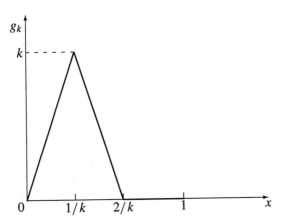

Graph 1. Sequence $\{g_k\}$ in PC[0,1].

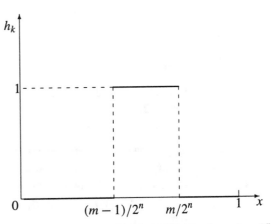

Graph 2. Sequence $\{h_k\}$ in PC[0,1], where $k = 2^n + m$.

Pointwise, Uniform, and Mean Convergence

The sequence $\{g_k\}$ above converges pointwise to the zero function on [0,1]. On the other hand, since $g_k(1/k) = k$ it is clear that the graphs of the $g_k(x)$ cannot eventually all lie in every tube about the zero function on [0,1]. So $\{g_k\}$ does *not* converge uniformly on [0,1]. It is true, however, that $\{g_k\}$ converges uniformly to the zero function on any interval $[\delta, 1]$, where $0 < \delta < 1$. The second sequence $\{h_k\}$ considered above does not converge pointwise for *any* point in [0,1]. Indeed, for any x_0 in [0,1] and any positive integer K, there are integers $r > K$, $s > K$ such that $h_r(x_0) = 0$ and $h_s(x_0) = 1$. So $\{h_k(x_0)\}$ cannot converge at all.

10.4 Fourier Trigonometric Series

Suggestions for preparing a lecture

Topics: The Fourier Trigonometric Basis Theorem and examples of Fourier trigonometric series, comments on convergence and on the decay of Fourier trigonometric coefficients. Mention the Gibbs phenomenon and the chapter cover figure.

Making up a problem set

One part of 1, one or two parts of 2, 3, 4, 5.

Comments

The basic theorems of "classical" Fourier Trigonometric Series are introduced in this section and in its problem set for the orthogonal set $\Phi = \{1, \cos \frac{\pi x}{T}, \sin \frac{\pi x}{T}, \ldots, \cos \frac{n\pi x}{T}, \sin \frac{n\pi x}{T}, \ldots\}$ in the Euclidean space PC$[-T, T]$. Pointwise convergence, uniform convergence, decay estimates—all of this is here along with the verification. This section and the previous two form a unit that can be used for self-study of the basics of the mathematical approach to Fourier Series.

1. Each of these problems can be solved without having to integrate to find Fourier coefficients.

   (a). The rearrangement $5 - \sin x - 7\sin 3x - 4\cos 6x$ already has the form of a Fourier Series (2), with $T = \pi$. So $A_0 = 5$, $B_1 = -1$, $B_3 = -7$, $A_6 = -4$, and all other coefficients are zero.

   (b). Using trigonometric identities in Appendix B.4, we see that $\cos^2 7x = (1 + \cos 14x)/2$, $2\sin x/2 \cos x/2 = \sin x$, and $2\sin x\cos 2x = \sin 3x - \sin x$. So the given function can be written as $(1 + 3\sin x - 2\sin 3x + \cos 14x)/2$ and has the form of text formula (2) with $T = \pi$. We have $A_0 = 1/2$, $B_1 = 3/2$, $B_3 = -1$, $A_{14} = 1/2$, and all other coefficients are zero.

   (c). Using trigonometric identities in Appendix B.4, we see that $\sin^3 x = \sin x(1 - \cos 2x)/2 = (\sin x - \sin x\cos 2x)/2 = (\sin x)/2 - (\sin 3x - \sin x)/4 = (3\sin x)/4 - (\sin 3x)/4$ which has the form of text formula (2) with $T = \pi$. So $B_1 = 3/4$, $B_3 = -1/4$, and all other Fourier coefficients are zero.

2. The Fourier Series for the given functions $f(x)$ have the form of text formula (2) with $T = \pi$ and coefficients given by text formulas (3). The following simple observation is often useful in calculating Fourier Series: suppose we know the coefficients in the two Fourier Series

$$FS[f] = A_0 + \sum_{k=1}^{\infty}(A_k \cos kx + B_k \sin kx)$$

and

$$FS[g] = A_0' + \sum_{k=1}^{\infty}(A_k' \cos kx + B_k' \sin kx)$$

Then for any scalars α, β it follows that

$$FS[\alpha f + \beta g] = (\alpha A_0 + \beta A_0') + \sum_{k=1}^{\infty}[(\alpha A_k + \beta A_k')\cos kx + (\alpha B_k + \beta B_k')\sin kx]$$

   (a). Here $f(x) = 2x - 2$. Notice that $FS[-2] = -2$. From this and Example 10.3.2, we have

$$FS[2x - 2] = -2 + 4\sum_{k=1}^{\infty}\frac{(-1)^{k+1}}{k}\sin kx$$

   (b). For $f(x) = x^2$, note that $A_0 = \int_{-\pi}^{\pi} x^2 dx/(2\pi) = \pi^2/3$, and for $k \geq 1$,

$$A_k = \frac{1}{\pi}\int_{-\pi}^{\pi} x^2 \cos kx dx = \frac{2}{\pi}\int_0^{\pi} x^2 \cos kx dx = \frac{4(-1)^k}{k^2}$$

   where we have used integration-by-parts twice. All the coefficients $B_k = 0$, $k \geq 1$ since

$f(x) = x^2$ is an even function on $[-\pi, \pi]$. We have

$$\text{FS}[x^2] = \frac{\pi^2}{3} + 4 \sum_{k=1}^{\infty} \frac{(-1)^k}{k^2} \cos kx$$

(c). The function $f(x)$ is $a + bx + cx^2$. Using the results of Example 10.3.2 and **(b)**, we have that

$$\text{FS}[a + bx + cx^2] = \left(a + \frac{\pi^2}{3}c\right) + \sum_{k=1}^{\infty} \left(\frac{4c(-1)^k}{k^2} \cos kx + \frac{2b(-1)^{k+1}}{k} \sin kx\right)$$

(d). Here $f(x) = \sin \pi x$. All the coefficients A_k, $k \geq 0$, vanish since $f(x) = \sin \pi x$ is an odd function on $[-\pi, \pi]$. Using an identity from Appendix B.4 for $\sin \alpha \sin \beta$, we have

$$B_k = \frac{1}{\pi} \int_{-\pi}^{\pi} \sin \pi x \sin kx \, dx = \frac{1}{\pi} \int_0^{\pi} [\cos(\pi + k)x - \cos(\pi - k)x] dx$$

$$= \frac{1}{\pi} \left[\frac{\sin(\pi + k)x}{\pi + k} - \frac{\sin(\pi - k)x}{\pi - k}\right]\bigg|_{x=0}^{x=\pi} = \frac{(-1)^{k+1} 2k \sin \pi^2}{\pi(\pi^2 - k^2)}$$

We have

$$\text{FS}[\sin \pi x] = \frac{2 \sin \pi^2}{\pi} \sum_{k=1}^{\infty} \frac{(-1)^{k+1} k}{\pi^2 - k^2} \sin kx$$

(e). Here $f(x) = |x| + e^x$. First, let $g(x) = e^x$, $-\pi \leq x \leq \pi$. To compute FS[g] note that

$$A_0 = \frac{1}{2\pi} \int_{-\pi}^{\pi} e^x dx = \frac{e^\pi - e^{-\pi}}{2\pi} = \frac{\sinh \pi}{\pi}$$

and for $k \geq 1$,

$$A_k = \frac{1}{\pi} \int_{-\pi}^{\pi} e^x \cos kx \, dx = \frac{2(-1)^k \sinh \pi}{\pi(1 + k^2)}, \qquad B_k = \frac{1}{\pi} \int_{-\pi}^{\pi} e^x \sin kx \, dx = \frac{2(-1)^{k+1} k \sinh \pi}{\pi(1 + k^2)}$$

We have that

$$\text{FS}[g] = \frac{\sinh \pi}{\pi} + \frac{2 \sinh \pi}{\pi} \sum_{k=1}^{\infty} \left[\frac{(-1)^k}{1 + k^2} \cos kx + \frac{(-1)^{k+1} k}{1 + k^2} \sin kx\right].$$

Next, from Example 10.4.1 we have that

$$\text{FS}[|x|] = \frac{\pi}{2} - \frac{4}{\pi} \sum_{m=1}^{\infty} \frac{1}{(2m+1)^2} \cos(2m+1)x$$

It follows that FS[$|x| + e^x$] has coefficients that are the sums of the corresponding coefficients in FS[$|x|$] and FS[e^x].

(f). Since $f(x) = |x^3 - x|$, $-\pi \leq x \leq \pi$, is an even function about $x = 0$, it follows that FS[f] contains only cosine terms. Note that

$$A_0 = \frac{1}{2\pi} \int_{-\pi}^{\pi} |x^3 - x| dx = \frac{1}{2\pi} \int_0^1 (x - x^3) \, dx - \frac{1}{2\pi} \int_1^\pi (x^3 - x) \, dx = \frac{2 - 2\pi^2 + \pi^4}{4\pi}$$

Similarly, splitting the integral at $x = 1$ we have that

$$A_k = \frac{2}{\pi} \int_0^\pi |x^3 - x| \cos kx \, dx$$

$$= \frac{2}{\pi k^4} \left[(12 - 4k^2) \cos k + 12k \sin k - 2k^2 - 6 + (-1)^k k^2 (3\pi^2 - 1) - 6(-1)^k \right]$$

With these values for A_0, A_k, we have

$$\text{FS}[|x^3 - x|] = A_0 + \sum_1^\infty A_k \cos kx$$

(g). Since the given rectangular pulse $f(x)$ is even, FS[f] contains only cosine terms. Observe that

$$A_0 = \frac{1}{2\pi} \int_{-B}^B A \, dx = \frac{AB}{\pi}$$

and for $k \geq 1$,

$$A_k = \frac{2}{\pi} \int_0^B A \cos kx \, dx = \frac{2A}{\pi} \frac{\sin kx}{k} \Big|_0^B = \frac{2A \sin kB}{\pi k}$$

so

$$\text{FS}[f] = \frac{AB}{\pi} + \frac{2A}{\pi} \sum_{k=1}^\infty \frac{\sin kB}{k} \cos kx$$

(h). Note that the given alternating pulse f is an odd function, so FS[f] contains only sine terms. Observe that for $k \geq 1$

$$B_k = \frac{2}{\pi} \int_0^B (-A) \sin kx \, dx = \frac{2A}{\pi} \frac{\cos kx}{k} \Big|_0^B = \frac{2A}{k\pi} (-1 + \cos kB)$$

so

$$\text{FS}[f] = \frac{2A}{\pi} \sum_{k=1}^\infty \frac{-1 + \cos kB}{k} \sin kx$$

3. We show that the function $\sin x$ has very different Fourier Series, depending on which interval $[a, b]$ we use.

(a). The function $f(x) = \sin x$, $-\pi \leq x \leq \pi$ has the Fourier Series FS[f] = $\sin x$.

(b). According to text formulas (2), (3) with $T = \pi/2$, the Fourier Series of the function $f(x) = \sin x$, $-\pi/2 \leq x \leq \pi/2$, has the form

$$\text{FS}[f] = A_0 + \sum_{k=1}^\infty [A_k \cos 2kx + B_k \sin 2kx]$$

where

$$A_0 = \frac{1}{\pi} \int_{-\pi/2}^{\pi/2} \sin x \, dx, \qquad A_k = \frac{2}{\pi} \int_{-\pi/2}^{\pi/2} \sin x \cos 2kx \, dx \quad k \geq 1,$$

$$B_k = \frac{2}{\pi} \int_{-\pi/2}^{\pi/2} \sin x \sin 2kx \, dx, \quad k \geq 1$$

Since f is odd about $x = 0$, $A_k = 0$ for all $k \geq 0$. Using an identity from Appendix B.4 for $\sin \alpha \sin \beta$, we have

$$B_k = \frac{2}{\pi} \int_0^{\pi/2} (\cos(2k-1)x - \cos(2k+1)x) \, dx$$

$$= \frac{2}{\pi} \left\{ \frac{\sin(2k-1)x}{2k-1} - \frac{\sin(2k+1)x}{2k+1} \right\} \Big|_0^{\pi/2}$$

$$= \frac{2}{\pi} \left\{ \frac{-(-1)^k}{2k-1} - \frac{(-1)^k}{2k+1} \right\} = \frac{8(-1)^k k}{\pi(1 - 4k^2)}$$

The corresponding Fourier Series is

$$FS[\sin x] = \frac{8}{\pi} \sum_{k=1}^{\infty} \frac{(-1)^k k}{1 - 4k^2} \sin 2kx$$

which is very different from the Fourier Series found in part **(a)**. The change of interval from $[-\pi, \pi]$ to $[-\pi/2, \pi/2]$ caused the orthogonal set to change [from $\sin kx$, $\cos kx$, and so on, to $\sin 2kx$, $\cos 2kx$, and so on], and this makes all the difference.

(c). The function $f(x) = \sin x$, $-3\pi/2 \leq x \leq 3\pi/2$, is odd about the origin. By (2), (3) with $T = 3\pi/2$, FS[f] has the form FS[f] $= \sum_{k=1}^{\infty} B_k \sin(2kx/3)$, where

$$B_k = \frac{2}{3\pi} \int_{-3\pi/2}^{3\pi/2} \sin x \sin \frac{2kx}{3} \, dx = \frac{2}{3\pi} \int_0^{3\pi/2} \left\{ \cos(\frac{2k}{3} - 1)x - \cos(\frac{2k}{3} + 1)x) \right\} dx$$

$$= \frac{2}{3\pi} \left\{ \frac{\sin(k\pi - 3\pi/2)}{2k/3 - 1} - \frac{\sin(k\pi + 3\pi/2)}{2k/3 + 1} \right\} = \frac{8(-1)^k k}{\pi(4k^2 - 9)}$$

So FS[$\sin x$] $= (8/\pi) \sum_{k=1}^{\infty} (-1)^k k / (4k^2 - 9) \sin(2kx/3)$, which differs from that found in part **(a)** because the intervals are different.

4. **(a).** The function is $f(x) = x$, $|x| \leq \pi$. Example 10.4.1 with $T = \pi$ gives

$$FS[|x|] = \frac{\pi}{2} - \frac{4}{\pi} \sum_{m=0}^{\infty} \frac{\cos(2m+1)x}{(2m+1)^2}$$

The Pointwise Convergence Theorem 10.4.3 implies that FS[$|x|$]$_{x=0}$ converges to $\frac{1}{2}[|0^+| + |0^-|] = 0$, and the asserted relation follows.

(b). Since the periodic extension $\tilde{f}$ of $f(x) = |x|$, $-\pi \leq x \leq \pi$, into $\mathbb{R}$ is in $\mathbf{C}^0(\mathbb{R})$ and is also piecewise smooth, the Pointwise Convergence Theorem (Theorem 10.4.3) states that FS[$|x|$] converges uniformly on $\mathbb{R}$ to $\tilde{f}$.

5. Just because a series $\sum_n A_n \cos nx + B_n \sin nx$ has the form of a Fourier Series does *not* necessarily mean that it is one. Here we look at three such series which are *not* the Fourier Series of any function in PC$[-\pi, \pi]$. By the Parseval Relation (Theorem 10.3.6), $\sum (1/(\pi n^{1/4}))^2$ must converge if $f(x) = \sum n^{-1/4} \cos nx$ is a function in PC$[-\pi, \pi]$. But

$\pi^{-2} \sum n^{-1/2}$ diverges (compare with the divergent harmonic series $\sum n^{-1}$). So, $\sum n^{-1/4} \cos nx$ is not the Fourier Series of any function in PC$[-\pi, \pi]$.

Similarly, $g = \sum (\sin n) \sin nx$ is not the Fourier Series of a function in PC$[-\pi, \pi]$ because $\sum (\sin n)^2$ doesn't converge at all [the n-th term does not $\to 0$ as $n \to \infty$].

Since $\ln(n) < n^{1/4}$ for n large enough, we have that $1/(\ln(n))^2 > 1/n^{1/2}$ for n large enough [use L'Hôpital's Theorem to show that $\lim_{n\to\infty} \ln(n)/n^{1/4} = 0$]. By the Comparison Test $\sum 1/(\ln(n))^2$ diverges since $\sum 1/n^{1/2}$ does. By Parseval's Theorem (again), $h = \sum (1/\ln(n)) \sin nx$ cannot be the Fourier Series of any function in PC$[-\pi, \pi]$.

6. The integral is the coefficient $A_n = \int_{-\pi}^{\pi} e^{\sin x}(x^5 - 7x + 1)^{52} \cos nx\, dx$ of $\cos nx$ in the Fourier Series of the function $f(x) = \pi e^{\sin x}(x^5 - 7x + 1)^{52}$, $|x| < \pi$, and so by the Decay Theorem 10.4.2 A_n must decay to zero as $n \to \infty$.

7. Let a be any given real number. We want to show that if f has period $2T$, then $\int_{-T}^{T} f(x)\, dx = \int_{a}^{a+2T} f(s)\, ds$. The interval $a \le x < a + 2T$ must contain precisely one point of the lattice $0, \pm T, \pm 3T, \dots$. Let this point be $(2k+1)T$. Then

$$\int_{a}^{a+2T} f(x)dx = \int_{a}^{(2k+1)T} f(x)dx + \int_{(2k+1)T}^{a+2T} f(x)dx$$

Making the variable substitution $x = t + 2kT$ in the first integral, and the substitution $x = t + (2k+2)T$ in the second integral, we see that

$$\int_{a}^{(2k+1)T} f(x)dx = \int_{a-2kT}^{T} f(t)dt, \qquad \int_{(2k+1)T}^{a+2T} f(x)dx = \int_{-T}^{a-2kT} f(t)dt$$

and we have that

$$\int_{a}^{a+2T} f(x)dx = \int_{-T}^{2a-2kT} f(t)dt + \int_{a-2kT}^{T} f(t)dt = \int_{-T}^{T} f(t)dt$$

8. Using a trigonometric formula from Appendix B.4, we verify that

$$2 \sin \frac{s}{2} \cos ks = \sin(k + \frac{1}{2})s - \sin(k - \frac{1}{2})s$$

for all s and all k. Sum this identity over the set $k = 1, 2, \dots, n$; observing that the right-hand side is a telescoping series, we obtain that

$$2 \sin \frac{s}{2} \sum_{k=1}^{n} \cos ks = \sin(n + \frac{1}{2})s - \sin \frac{s}{2}$$

or

$$2 \sin \frac{s}{2} \left(\frac{1}{2} + \sum_{k=1}^{n} \cos ks \right) = \sin(n + \frac{1}{2})s$$

from which Lagrange's identity

$$\frac{1}{2} + \sum_{k=1}^{n} \cos ks = \sin(n + 1/2)s / [2 \sin(s/2)]$$

follows for $s \neq 2k\pi$. It is also valid if $s = 2k\pi$, as may be verified by L'Hôpital's Rule applied to $\lim_{s \to 2k\pi} \sin((n+1/2)s)/\sin(s/2)$.

9. **(a).** By the Totality Theorem (Theorem 10.3.5) if $\{\cos nx : n = 0, 1, 2, \ldots\}$ were a basis for $PC[-\pi, \pi]$, then the *only* element in $PC[-\pi, \pi]$ orthogonal to $\cos nx$, for all $n = 0, 1, 2, \ldots$, is the zero element. But since $\langle \sin x, \cos nx \rangle = 0$, for all $n = 0, 1, 2 \ldots$, it follows that $\{\cos nx : n = 0, 1, 2, \ldots\}$ is *not* a basis for $PC[-\pi, \pi]$.

    (b). Let $\Phi_c = \{\cos nx : n = 0, 1, 2, \ldots\}$. We shall show that for every f in $PC[0, \pi]$, $FS_{\Phi_c}[f]$ converges in the mean to f, and it will follow that Φ_c is a basis for $PC[0, \pi]$. Let f be given in $PC[0, \pi]$. Extend it into $[-\pi, 0]$ as an even function f_e. So f_e is in $PC[-\pi, \pi]$ and the Fourier Trigonometrical Series of f_e, $FS[f_e]$, converges in the mean to f_e by the Fourier Basis Theorem 10.4.1. But since f_e is even, $FS[f_e]$ contains only cosine terms. So $FS[f_e] = A_0 + \sum_{k=1}^{\infty} A_k \cos kx$, where

    $$A_0 = \frac{1}{2\pi} \int_{-\pi}^{\pi} f_e(x)\,dx, \qquad A_k = \frac{1}{\pi} \int_{-\pi}^{\pi} f_e(x) \cos kx\,dx, \quad k \geq 1$$

    Note that $FS_{\Phi_c}[f] = FS[f_e]$, and that

    $$\int_{-\pi}^{\pi} \left| \left(A_0 + \sum_{k=1}^{n} A_k \cos kx \right) - f_e(x) \right|^2 dx \to 0, \text{ as } n \to \infty$$

    But since

    $$\int_{-\pi}^{\pi} \left| A_0 + \sum_{k=1}^{n} A_k \cos kx - f_e(x) \right|^2 dx \geq \int_{0}^{\pi} \left| A_0 + \sum_{k=1}^{n} A_k \cos kx - f_e(x) \right|^2 dx \geq 0$$

    it follows that $FS_{\Phi_c}[f]$ converges in the mean to f in $PC[0, \pi]$, and we are done.

    (c). The set $\{\sin x, \ldots, \sin nx\}$ is a basis for the subspace of all odd functions in $PC[-\pi, \pi]$ since $\langle f, \cos nx \rangle = 0$, for $n = 0, 1, \ldots$ if f is odd. It is *not* a basis of $PC[-\pi, \pi]$ since, for example $\langle 1, \sin nx \rangle = 0$, $n = 1, 2, \ldots$, but 1 is not the zero function.

10. We need to show that $\|f - S_n\| \leq \|f - S_m\|$ if $n \geq m$ and S_n and S_m are partial sums of $FS[f]$. For $m < n$, the Mean Approximation Theorem 10.3.1 implies that

    $$\|f - S_n\|^2 = \|f\|^2 - \frac{1}{\pi} \left[|\langle f, 1 \rangle|^2 + \sum_{k=1}^{n} \{|\langle f, \cos kt \rangle|^2 + |\langle f, \sin kt \rangle|^2\} \right]$$

    $$= \|f - S_m\|^2 - \frac{1}{\pi} \sum_{k=m+1}^{n} \{|\langle f, \cos kt \rangle|^2 + |\langle f, \sin kt \rangle|^2\} \leq \|f - S_m\|^2$$

    This completes what we were asked to do.

Background Material: Dirichlet Representation Theorem

THEOREM

> **Dirichlet Representation Theorem.** Suppose that f is in PC$[-\pi, \pi]$, and that $\tilde{f}$ is the periodic extension of f in $\mathbb{R}$ with period 2π. Suppose that $S_n(x)$ denotes the n-th partial sum of FS$[f]$: $S_n(x) = A_0 + \sum_{k=1}^{n}(A_k \cos kx + B_k \sin kx)$, where A_k, B_k are the Fourier-Euler trigonometric coefficients of f with respect to the basis $\{1, \cos x, \sin x, \ldots, \cos nx, \sin nx, \ldots\}$. Then we have the *Dirichlet Representation* for $S_n(x)$,
>
> $$S_n(x) = \int_{-\pi}^{\pi} \tilde{f}(x+s)\frac{\sin(n+\frac{1}{2})s}{2\pi \sin(s/2)}\, ds, \qquad x \text{ in } \mathbb{R}, \quad n = 0, 1, 2, \ldots \qquad \text{(i)}$$
>
> The function $D_n(s) = \sin[(n+\frac{1}{2})s]/[2\pi \sin(s/2)]$ appearing in the integrand of (i) is called the *Dirichlet kernel* and has the property that
>
> $$\int_{-\pi}^{\pi} D_n(s)\, ds = 1, \qquad \text{for all } n = 0, 1, 2, \ldots \qquad \text{(ii)}$$

To see this, use the definition of the coefficients A_k, B_k, interchanging summation and integration, and the identity $\cos(\alpha - \beta) = \cos\alpha\cos\beta + \sin\alpha\sin\beta$ to obtain

$$S_n(x) = \frac{1}{\pi}\int_{-\pi}^{\pi} f(t)\left[\frac{1}{2} + \sum_{k=1}^{n}\cos k(t-x)\right] dt, \qquad \text{for any } n = 0, 1, 2, \ldots$$

☞ Problem 8 gives a hint on how to show this identity.

Using the identity $\frac{1}{2} + \sum_{k=1}^{n}\cos kx = \sin[(n+\frac{1}{2})s]/[2\sin(s/2)]$, we see that

$$S_n(x) = \frac{1}{\pi}\int_{-\pi}^{\pi} f(t)\frac{\sin(n+\frac{1}{2})(t-x)}{2\sin\frac{1}{2}(t-x)}\, dt \qquad \text{(iii)}$$

Now after replacing f by $\tilde{f}$ in (iii), making the change of variables $t - x = s$, and applying the Integration Theorem for Periodic Functions (Theorem 10.4.2), the desired formula (i) for $S_n(x)$ is obtained. The identity (ii) results immediately from application of (i) to the function $f(x) = 1$, for all x, and we are done.

Background Material: Verification of Theorem 10.4.3

To see this, first write $S_n(x)$ in (i) above as

$$S_n(x) = \int_{-\pi}^{0} \tilde{f}(x+s)D_n(s)\, ds + \int_{0}^{\pi} \tilde{f}(x+s)D_n(s)\, ds$$

We will show that

$$\int_{0}^{\pi} \tilde{f}(x_0+s)D_n(s)\, ds \to \frac{1}{2}\tilde{f}(x_0^+), \qquad \text{as } n \to \infty$$

$$\int_{-\pi}^{0} \tilde{f}(x_0+s)D_n(s)\, ds \to \frac{1}{2}\tilde{f}(x_0^-), \qquad \text{as } n \to \infty$$

(i)

which implies the assertion. We show only the first statement in (i). Note that $D_n(s)$ is an even function, so from (ii), $\int_0^\pi D_n(s)\,ds = \int_{-\pi}^0 D_n(s)\,ds = 1/2$. Using this fact and the identity $\sin(\alpha + \beta) = \sin\alpha\cos\beta + \cos\alpha\sin\beta$, we have

$$\int_0^\pi \tilde{f}(x_0 + s)D_n(s)\,ds - \frac{1}{2}\tilde{f}(x_0^+) = \int_0^\pi [\tilde{f}(x_0 + s) - \tilde{f}(x_0^+)]D_n(s)\,ds$$

$$= \frac{1}{2\pi}\int_{-\pi}^\pi \{H(s)[\tilde{f}(x_0 + s) - \tilde{f}(x_0^+)]\}\cos ns\,ds \qquad \text{(ii)}$$

$$+\frac{1}{\pi}\int_{-\pi}^\pi \left\{ H(s)\cdot\frac{\tilde{f}(x_0 + s) - \tilde{f}(x_0^+)}{s}\frac{s/2}{\sin(s/2)}\cos(s/2)\right\}\sin ns\,ds$$

where $H(s)$ is the piecewise constant function whose value is 1, $0 < s < \pi$, and 0, $-\pi < s < 0$. Now $H(s)[\tilde{f}(x_0 + s) - \tilde{f}(x_0^+)]$ belongs to the class $PC[-\pi, \pi]$ as a function of s, so the Decay of Coefficients Theorem implies that the first integral in (ii) tends to zero as $n \to \infty$. If we can show that the function

$$H(s)\cdot\frac{\tilde{f}(x_0 + s) - \tilde{f}(x_0^+)}{s}\frac{s/2}{\sin(s/2)}\cos(s/2)$$

is also in $PC[-\pi, \pi]$ as a function of s, the Decay of Coefficients Theorem implies that this integral also tends to zero as $n \to \infty$, finishing the proof.

Observe that $H(s)$, $(s/2)\sin(s/2)$, and $\cos(s/2)$ are all piecewise continuous on $-\pi \leq s \leq \pi$. Now since $H(s) = 0$ for $-\pi \leq s < 0$, it remains only to show that $[\tilde{f}(x_0 + s) - \tilde{f}(x_0^+)]/s$ is piecewise continuous on $0 \leq s \leq \pi$. The only trouble point is $s = 0$. Since f is piecewise smooth on $[-\pi, \pi]$, it follows that $\tilde{f}(x_0 + s)$ is piecewise smooth on $-\pi \leq s \leq \pi$. So for all small enough positive s, the Mean Value Theorem implies that

☞ Theorem B.5.4 has a statement of the Mean Value Theorem.

$$\frac{\tilde{f}(x_0 + s) - \tilde{f}(x_0^+)}{s} = \tilde{f}'(x_0 + s^*) \qquad \text{for some } s^* \text{ with } 0 < s^* < s \qquad \text{(iii)}$$

Since $\tilde{f}'(x_0 + s^*) \to \tilde{f}'(x_0^+)$ as $s \to 0^+$, we see from (iii) that the function $[\tilde{f}(x_0 + s) - \tilde{f}(x_0^+)]/s$ has passed the last test to show that it is piecewise continuous on $0 \leq s \leq \pi$, and we are done.

10.5 Half-Range and Exponential Fourier Series

Suggestions for preparing a lecture

Topics: Fourier sine series, Fourier cosine series, Fourier exponential series, an example such as the circuit problem at the end of the section.

Making up a problem set

One part from each of Problems 1–4, 6.

Comments

The Fourier techniques of half-range and complex exponential expansions are "bread-and-butter" methods in doing applications. It is worthwhile pointing out the advantages of using the complex basis functions e^{ikx}: the Fourier-Euler coefficients formulas are a lot easier to remember, and one doesn't have to compute Fourier sine and cosine coefficients separately. The flip side of the coin, however, is that without the often tedious simplification at the end of the process, the Fourier Series of a real function may appear to be complex-valued.

---

1. We are looking for Fourier Sine Series here.

(a). The function $f(x) = 1$.

$$\text{FSS}[1] = \sum_{k=0}^{\infty} b \sin kx, \quad b_k = \frac{2}{\pi} \int_0^{\pi} 1 \cdot \sin kx \, dx$$

So $b_k = -2(\cos kx)/k\pi \Big|_{x=0}^{x=\pi} = 2[1 - \cos k\pi]/k\pi$, and we have $b_k = 0$ if k is even, and $b_k = 4/k\pi$ if k is odd. Then

$$\text{FSS}[1] = \frac{4}{\pi} \sum_{\text{odd } k} \frac{1}{k} \sin kx$$

We have FCS[1]=FS[1]=1 since $\Phi_c = \{1, \cos x, \ldots, \cos nx, \ldots\}$.

(b). First, we find FCS[f]. Since $f(x) = \sin x$, $0 \le x \le \pi$, we have

$$A_0 = \frac{1}{\pi} \int_0^{\pi} \sin x \, dx = -\frac{1}{\pi} \cos x \Big|_0^{\pi} = \frac{2}{\pi},$$

$$A_k = \frac{2}{\pi} \int_0^{\pi} \sin x \cos kx \, dx = \frac{1}{\pi} \int_0^{\pi} [\sin(k+1)x - \sin(k-1)x] \, dx$$

$$= \begin{cases} 0, & k = 1 \\ \dfrac{2}{\pi} \cdot \dfrac{(-1)^{k+1} - 1}{k^2 - 1}, & k \ge 2 \end{cases}$$

so

$$\text{FCS}[\sin x] = \frac{2}{\pi} - \frac{4}{\pi} \sum_{k \ge 2, \text{even}} \frac{1}{k^2 - 1} \cos kx$$

Finally, FSS[$\sin x$]=FS[$\sin x$] = $\sin x$, since $\Phi_s = \{\sin x, \ldots, \sin nx, \ldots\}$.

2. **(a).** We want to find the Fourier Exponential Series of $f(x) = \cos x$. FS[f] $= \sum_{k=-\infty}^{\infty} c_k e^{ikx}$, where

$$c_k = \frac{1}{2\pi} \int_{-\pi}^{\pi} \cos x e^{-ikx} \, dx = \frac{1}{2\pi} \int_{-\pi}^{\pi} \cos x (\cos kx - i \sin kx) \, dx$$

which is zero, unless $k = 1$ or -1. We have that $c_1 = \pi/(2\pi) = 1/2 = c_{-1}$. So FS[$f$] $= (e^{ix} + e^{-ix})/2 = \cos x$, which we should have expected in the first place.

(b). Here $f(x) = x^2$. Now, $c_0 = (2\pi)^{-1} \int_{-\pi}^{\pi} x^2 \, dx = \pi^2/3$, and using integration-by-parts twice, we have that for $k \neq 0$,

$$c_k = \frac{1}{2\pi} \int_{-\pi}^{\pi} x^2 e^{-ikx} \, dx = \frac{1}{2\pi(-ik)} \left\{ x^2 e^{-ikx} \Big|_{-\pi}^{\pi} - \int_{-\pi}^{\pi} 2x e^{-ikx} \, dx \right\}$$

$$= \frac{-2}{2\pi(-ik)^2} \left\{ x e^{-ikx} \Big|_{-\pi}^{\pi} - \int_{-\pi}^{\pi} e^{-ikx} \, dx \right\} = \frac{1}{\pi k^2} \{ 2\pi(-1)^k \} = \frac{2(-1)^k}{k^2}$$

So $FS[f] = \pi^2/3 + 2 \sum_{k=1}^{\infty} ((-1)^k/k^2)(e^{ikx} + e^{-ikx})$.

(c). Here $f(x) = |x| + ix$. We shall find FS[x] and FS[$|x|$] and combine the results to find FS[$|x| + ix$]. First, observe that $(2\pi)^{-1} \int_{-\pi}^{\pi} x \, dx = 0$, and that for $k \neq 0$,

$$\frac{1}{2\pi} \int_{-\pi}^{\pi} x e^{-ikx} \, dx = -\frac{1}{2\pi i k} \left\{ x e^{-ikx} \Big|_{-\pi}^{\pi} - \int_{-\pi}^{\pi} e^{-ikx} \, dx \right\} = \frac{i}{k}(-1)^k$$

So $FS[x] = i \sum_{k=1}^{\infty} ((-1)^k/k)(e^{ikx} - e^{-ikx})$. Next, $(2\pi)^{-1} \int_{-\pi}^{\pi} |x| \, dx = (\pi)^{-1} \int_0^{\pi} x \, dx = \pi/2$, and that for $k \neq 0$,

$$\frac{1}{2\pi} \int_{-\pi}^{\pi} |x| e^{-ikx} \, dx = -\frac{1}{2\pi} \int_{-\pi}^{\pi} \{ |x| \cos kx - i|x| \sin kx \} \, dx$$

$$= \frac{1}{\pi} \int_0^{\pi} x \cos kx \, dx = \frac{(-1)^k - 1}{\pi k^2}.$$

So, $FS[|x|] = \pi/2 + \pi^{-1} \sum_{k=1}^{\infty} [(-1)^k - 1] k^{-2} (e^{ikx} + e^{-ikx})$. Hence,

$$FS[|x| + ix] = \frac{\pi}{2} + \sum_{k=1}^{\infty} \left[\frac{(-1)^k - 1}{\pi k^2} (e^{ikx} + e^{-ikx}) - \frac{(-1)^k}{k} (e^{ikx} - e^{-ikx}) \right]$$

3. We supose that f is in PC[0, c] and that $f(x) = f(c - x)$ for $c/2 < x < c$.

(a). Let's look at FSS[f]. For $k = 2m$, $m = 1, 2, \ldots$, we see that

$$\int_0^c f(x) \sin \frac{2m\pi x}{c} \, dx = \int_0^{c/2} f(x) \sin \frac{2m\pi x}{c} \, dx + \int_{c/2}^c f(x) \sin \frac{2m\pi x}{c} \, dx \qquad \text{(i)}$$

Making the substitution $x = c - y$ in the first integral on the right side of (i) and using an identity from Appendix B.4 for $\sin(\alpha - \beta)$, we see that

$$\int_0^{c/2} f(x) \sin \frac{2m\pi x}{c} \, dx = - \int_c^{c/2} f(c - y) \sin \frac{2m\pi(c - y)}{c} \, dy$$

$$= - \int_{c/2}^c f(y) \sin \frac{2m\pi y}{c} \, dy$$

The right side of (i) adds up to zero, and we have

$$\int_0^c f(x) \sin \frac{2m\pi x}{c} \, dx = 0, \qquad m = 1, 2, \ldots$$

This leaves only the terms $\sin[(2k + 1)\pi x/c]$ in FSS[f].

(b). Reflect $\sin x$, $0 \leq x \leq \pi/4$, about $x = \pi/4$ and get a function defined on $0 \leq x \leq \pi/2$. We need to find FS[f]. Since $c = \pi/2$, FSS[f] only contains terms of the form $\sin 2kx$, where k is odd:

$$\text{FSS}[f] = \sum_{\text{odd } k} B_k \sin 2kx, \quad \text{where } B_k = \frac{4}{\pi} \int_0^{\pi/2} f(x) \sin 2kx \, dx$$

Notice that $\sin 2(2m-1)x$ is even about $x = \pi/4$ for all $m = 1, 2, \ldots$. Indeed, for any $m = 1, 2, \ldots$,

$$\sin[(4m-2)(\frac{\pi}{2} - x)] = \sin \frac{(4m-2)\pi}{2} \cos(4m-2)x - \cos \frac{(4m-2)\pi}{2} \sin(4m-2)x$$
$$= \sin(4m-2)x$$

For $m = 1, 2, \ldots$,

$$B_{2m-1} = \frac{8}{\pi} \int_0^{\pi/4} \sin x \sin(4m-2)x \, dx = \frac{4}{\pi} \int_0^{\frac{\pi}{4}} [\cos(4m-3)x - \cos(4m-1)x] \, dx$$
$$= \frac{4}{\pi} \left[\frac{\sin(4m-3)x}{4m-3} - \frac{\sin(4m-1)x}{4m-1} \right] \Big|_0^{\pi/4}$$
$$= \frac{4}{\pi} \left[\frac{\sin(m-3/4)\pi}{4m-3} - \frac{\sin(m-1/4)\pi}{4m-1} \right]$$

So $\text{FSS}[f] = \sum_{m=1}^{\infty} B_{2m-1} \sin(4m-2)x$.

4. Part **(a)** is proven in exactly the same way as in **(a)** of Problem 2. The calculation of $\text{FCS}[f]$ in **(b)** is carried out in a straightforward manner after noting the simplification resulting from **(a)**. In particular, $\text{FCS}[f] = \sum_{k=0}^{\infty} A_k \cos[(2k+1)\pi x/2]$, where $A_k = 2 \int_0^1 x^2 \cos[(2k+1)\pi x/2] \, dx = (-1)^k (2a^2 - 4)a^{-3}$, $a = (k+1/2)\pi$.

5. The circuit equation is $0.5q'' + 10q' + 10^4 q = E(t)$. From the hint we see that $\text{FS}[E] = \pi/2 + \pi^{-1} \sum_{k=1}^{\infty} (e^{ikt} + e^{-ikt})/(4k^2)$. Now if we put $\text{FS}[q] = \sum_{k=-\infty}^{\infty} c_k e^{ikt}$, then $\text{FS}[q'] = \sum_{k=-\infty}^{\infty} ik c_k e^{ikt}$, and $\text{FS}[q''] = \sum_{k=-\infty}^{\infty} (ik)^2 c_k e^{ikt}$. We can combine these results to compute $\text{FS}[0.5q'' + 10q' + 10^4 q]$ and equate it to $\text{FS}[E]$, and by comparing coefficients we obtain that

$$[0.5(ik)^2 + 10ik + 10^4]c_k = \begin{cases} \pi/2, & k = 0, \quad k \text{ odd} \\ -1/(2m^2), & k \neq 0, \quad k = 2m \end{cases}$$

Since the coefficient of c_k in the equation above is not zero for any k, we can solve the equation for c_k uniquely and if these values are inserted into $\sum_{k=-\infty}^{\infty} c_k e^{ikx}$, this series converges to the steady-state solution of the given circuit equation.

6. The spring motion is modeled by $my'' + cy' + ky = y'' + 0.2y' + 1.01y = f(t)$. The steady-state motion of the given spring-mass system can be calculated by using precisely the same method outlined in Problem 5 above. Using the fact that $f = \text{sqw}(t, 50, 2\pi)$ and

$$\text{FS}[f] = \frac{1}{2} + \frac{1}{2\pi i} \sum_{k=1}^{\infty} \frac{1 - (-1)^k}{k} (e^{ikx} - e^{-ikx})$$

we can replace $y(t)$ by $\sum_{k=-\infty}^{\infty} c_k e^{ikt}$ in the ODE to obtain

$$\sum_{k=-\infty}^{\infty} [(ik)^2 + 0.2ik + 1.01]c_k e^{ikt} = FS[f]$$

where $FS[f]$ is given above. Matching coefficients, we have

$$c_0 = \frac{1}{2.02}, \quad c_k = \begin{cases} 0, & k \neq 0, \ k \text{ even} \\ [\pi i k(-k^2 + 0.2ik + 1.01)]^{-1}, & k \text{ odd} \end{cases}$$

and the solution is $y = \sum_{-\infty}^{\infty} c_k e^{ikt}$.

10.6 Sturm-Liouville Problems

Suggestions for preparing a lecture

Topics: How to write $[P(D)]y = a_2 D^2 y + a_1 Dy + a_0 y$ in the form $M[y]$ given in (1), weighted scalar products, symmetric operators, the Symmetric Operator Theorem, and [the lecture's focus] the Sturm-Liouville Thorem. Do one or two simple examples.

Making up a problem set

Problems 1(a), 1(e).

Comments

The separation step in the technique of separation of variables for a boundary-initial value problem for a PDE always generates a Sturm-Liouville problem. This typically involves solving an ordinary differential operator equation $Ly = \lambda y$, subject to some conditions on the solution $y(x)$ at $x = a$ and $x = b$, say. More precisely, the problem is to find the values of λ so that the problem $Ly = \lambda y$ with boundary conditions has a nontrivial solution. We must find the eigenvalues of the operator L whose domain is restricted by the boundary conditions. The Sturm-Liouville Theory provides the mathematical foundations for all of this. Everything needed in Sections 10.7–10.9 and the ideas that underlie what is in Section 11.8 for the separation step in solving PDE problems can be found right here.

1.   **(a).** $L[y] = y''$, $\text{Dom}(L) = \{y \text{ in } \mathbf{C}^2[0, \pi/2]: y(0) = y(\pi/2) = 0\}$. From the Nonpositivity of Eigenvalues Theorem (Theorem 10.6.3) all the eigenvalues of L are nonpositive. First try $\lambda = 0$. The eigenfunctions (if any) must have the form $y = Ax + B$. The endpoint conditions imply that $A = B = 0$, so $\lambda = 0$ cannot be an eigenvalue. Now try $\lambda = -k^2$, $k \geq 0$. The eigenfunctions (if any) must have the form $y = A \sin kx + B \cos kx$. The condition $y(0) = 0$ implies that $B = 0$. The condition $y(\pi/2) = 0$ implies that $\sin(k\pi/2) = 0$. This can only happen if $k\pi/2 = n\pi$, $n = 1, 2, \ldots$. So $\lambda_n = -4n^2$ is an eigenvalue for each $n = 1, 2, \ldots$, and the corresponding eigenspace V_{λ_n} has $y_n = \sin 2nx$ as a basis. The Orthogonality of Eigenspaces Theorem 10.6.2 implies that the eigenspaces V_{λ_1},

$V_{\lambda_2}, \ldots$ are orthogonal with respect to the standard scalar product in $PC[0, \pi/2]$ and that $\Phi = \{\sin 2x, \cdots, \sin 2nx, \cdots\}$ is a basis for $PC[0, \pi/2]$.

(b). $L[y] = y''$, $\text{Dom}(L) = \{y \text{ in } \mathbf{C}^2[0, T]: y(0) = y'(T) = 0\}$. The eigenvalues of L are nonpositive. First, try $\lambda = 0$. The eigenfunctions (if any) must have the form $y = Ax + B$. The endpoint conditions imply that $A = B = 0$, so $\lambda = 0$ cannot be an eigenvalue. Now try $\lambda = -k^2$, $k > 0$. The eigenfunctions (if any) must have the form $y = A \cos kx + B \sin kx$. The condition $y(0) = 0$ implies that $A = 0$. The condition $y'(T) = 0$ implies that $\cos(kT) = 0$. Thus, $kT = (2n+1)\pi/2$, $n = 0, 1, 2, \ldots$, so $\lambda_n = -[(2n+1)\pi/(2T)]^2$ is an eigenvalue for L, and the eigenspace V_{λ_n} is spanned by the eigenfunction $y_n = \sin[(2n+1)\pi x/2T]$. The eigenspaces V_{λ_n} are orthogonal with the standard scalar product in $PC[0, T]$, and the orthogonal set

$$\Phi = \left\{\sin\left[(2n+1)\frac{\pi x}{2T}\right]: n = 0, 1, 2, \ldots\right\}$$

is a basis for $PC[0, T]$.

(c). $L[y] = y''$, $\text{Dom}(L) = \{y \text{ in } \mathbf{C}^2[0, T]: y'(0) = y'(T) = 0\}$. First, $\lambda = 0$ is an eigenvalue and $y = 1$ a corresponding eigenvector. As in part **(a)**, the other eigenvalues must be negative, $\lambda = -k^2$, and $y_k = A_k \cos kx + B_k \sin kx$. Imposing the boundary conditions, we see that $B_k = 0$, and the eigenvalues are $\lambda_n = -(n\pi/T)^2$, $n = 0, 1, 2, \ldots$ with each corresponding eigenspace V_{λ_n} spanned by the eigenfunction $y_n = \cos(n\pi x/T)$. The eigenspaces V_{λ_n} are orthogonal with respect to the standard scalar product in $PC[0, T]$, and the orthogonal set $\Phi = \{\cos(n\pi x/T): n = 0, 1, 2, \ldots\}$ is a basis for $PC[0, T]$.

(d). This problem is essentially worked out in Example 10.6.2. $L[y] = y''$, $\text{Dom}(L) = \{y \text{ in } \mathbf{C}^2[-T, T]: y(-T) = y(T), y'(-T) = y'(T) = 0\}$. The eigenvalues of L are $\lambda_n = -(n\pi/T)^2$, $n = 0, 1, 2, \ldots$ with each corresponding eigenspace V_{λ_n} described as follows: V_{λ_0} is one-dimensional and is spanned by the eigenfunction $y_0 = 1$; V_{λ_n}, $n > 0$, has dimension two with the eigenbasis $y_n^1 = \sin(n\pi x/T)$, $y_n^2 = \cos(n\pi x/T)$. The eigenspaces are orthogonal with respect to the standard scalar product in $PC[-T, T]$. The orthogonal set

$$\Phi = \{1, \sin(\pi x/T), \cos(\pi x/T), \cdots, \sin(n\pi x/T), \cos(n\pi x/T), \cdots\}$$

is a basis for $PC[-T, T]$.

(e). $L[y] = y''$, $\text{Dom}(L) = \{y \text{ in } \mathbf{C}^2[0, \pi]: y(0) = y(\pi) + y'(\pi) = 0\}$. First, find eigenvalues of L of the form $\lambda = k^2$, for some $k > 0$. Such eigenvalues would have eigenfunctions (if any) of the form $y = A \sinh kx + B \cosh kx$. The condition $y(0) = 0$ implies that $B = 0$. The condition $y(\pi) + y'(\pi) = 0$ implies that $\sinh k\pi + k \cosh k\pi = 0$ or $\tanh r = -\pi/r$, where $k\pi = r > 0$. The sketch in Fig. 1(e), Graph 1 indicates that there is no value $r > 0$ such that $\tanh r = -\pi/r$, so no positive eigenvalues. (Note that the Nonpositivity of Eigenvalues Theorem 10.6.3 would have given the result immediately). Next try $\lambda = 0$. The eigenfunctions (if any) must have the form $y = Ax + B$. The endpoint conditions imply that $A = B = 0$, so $\lambda = 0$ cannot be an eigenvalue. Finally, let's try $\lambda = -k^2$, $k \geq 0$. In this case any eigenfunction would have the form $y = A \cos kx + B \sin kx$. The condition $y(0) = 0$ implies that $A = 0$, and the other boundary condition implies that $\sin k\pi + k \cos k\pi = 0$ or $\tan r = -\pi/r$, for $r > 0$, where $k\pi = r$. The sketch in Fig. 1(e),

Graph 2 indicates that there is an infinite sequence $r_1, r_2, \ldots$, for which $\tan r_n = -\pi r_n$, $r_n > 0$. Note that $(n - 1/2)\pi < r_n < n\pi$, for all n, and in fact $r_n \to (n + 1/2)\pi$ as $n \to \infty$. So $\lambda_n = -k_n^2 = -(r_n/\pi)^2$ are the eigenvalues of L, and the corresponding eigenspace V_{λ_n} is spanned by $y_n = \sin k_n x = \sin(r_n x/\pi)$. Because L is a symmetric operator, the eigenspaces are orthogonal with the standard scalar product in $PC[0, \pi]$. Finally, the Sturm-Liouville Theorem 10.6.4 implies that $\Phi = \{\sin(r_n x/\pi) : n = 1, 2, \ldots\}$ is a basis for $PC[0, \pi]$.

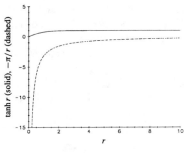

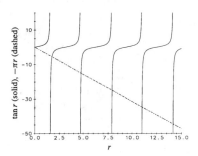

Problem 1(e), Graph 1.   Problem 1(e), Graph 2.

(f). $L = D^2 - 4D + 4$, $\text{Dom}(L) = \{y \text{ in } \mathbf{C}^2[0, \pi] : y(0) = y(\pi) = 0\}$. Note first that L has the symmetric form $L[y] = \{(e^{-4x}y')' + 4e^{-4x}y\}e^{4x}$ with $\rho = e^{-4x}$, $p = e^{-4x}$, $q = 4e^{-4x}$. If L has an eigenvalue $\lambda = k^2$, $k > 0$, then any corresponding eigenfunction must take the form $y = Ae^{(2+k)x} + Be^{(2-k)x}$. The condition $y(0) = 0$ implies that $A + B = 0$, and the condition $y(\pi) = 0$ that $Ae^{k\pi} + Be^{-k\pi} = 0$. This can only happen for $A = B = 0$ since $k > 0$; so L has no positive eigenvalues. Next try $\lambda = 0$. Then corresponding eigenfunctions must have the form $y = Ae^{2x} + Bxe^{2x}$. The endpoint conditions imply that $A = 0$, and $B\pi e^{2\pi} = 0$. So $A = B = 0$, and $\lambda = 0$ is also not an eigenvalue for L. Finally, if we try $\lambda = -k^2$, $k > 0$: $y = e^{2x}(A_k \cos kx + B_k \sin kx)$. The boundary conditions imply that $A_k = 0$ and that $\sin k\pi = 0$. Then, $\lambda_n = -n^2$, $n = 1, 2, \ldots$ are eigenvalues with corresponding eigenspaces V_{λ_n} spanned by $y_n = e^{2x} \sin nx$. The eigenspaces are orthogonal with respect to the weighted scalar product $\langle f, g \rangle = \int_0^\pi e^{-4x} f g ds$. Also, $\{e^{2x} \sin nx : n = 1, 2, \ldots\}$ is a basis for $PC[0, \pi]$ under this weighted scalar product.

10.7 Separation of Variables

Suggestions for preparing a lecture

   Topics: Go through the steps I–VII for constructing the series solution of the guitar-string problem, only mentioning step VII. Then you might talk about "shifting data" and illustrate the process by doing the vibrating string problem on page 564.

Making up a problem set

Problems 2, 4, 7 (hard).

Comments

Now we can approach a boundary/initial value problem for a PDE in a systematic way, using the techniques introduced earlier. A systematic approach is very helpful because there are so many separate steps in constructing the series solution of the problem that a "do-it-by-the-numbers" approach is preferred. This does not mean doing a problem without thought, because every new problem brings new geometry, data sets, or coefficients in the PDE. However, the template we have arranged in the form of seven steps (numbered as I–VII in the text) helps in avoiding needless errors and pointless searching for what to do next. The whole process has at its core the Method of Eigenfunction Expansions, and we make note of that important fact. Finally, we show how to solve boundary/initial value problems for PDEs when the PDE, the boundary conditions, or both, are not homogeneous. Here is where eigenfunction expansions play a central role in constructing the solution. We distinguish between a "formal solution" of a boundary/initial value problem where we find the solution as a series but don't worry about the convergence, from a "classical solution" where we add enough extra smoothness conditions on the data that all the series converge to the desired sum. On a first "read-through" you might ignore the distinction and focus on the formal solution. You might also skip the verification of the theorems.

---

1. The problems all refer to the guitar string problem (1). The initial shape $f(x)$ and velocity $g(x)$ are given.

    (a). The formal solution of this problem takes the form (8) with coefficients given by (9). Using the given data $f(x) = 0$, $g(x) = 3\sin(\pi x/L)$, we see that $A_n = 0$ for all n, and that $B_1 = (2/\pi c)\int_0^L 3(\sin(\pi x/L))^2 \, dx = 3L/\pi c$ and $B_n = 0$ for $n \geq 2$. The solution of this problem is

    $$u(x, t) = \frac{3L}{\pi c}\sin\frac{\pi ct}{L}\sin\frac{\pi x}{L}$$

    (b). Here $f(x) = g(x) = \begin{cases} x, & 0 \leq x \leq L/2 \\ L - x, & L/2 \leq x \leq L \end{cases}$. Using (8) and (9), we have that for $n \geq 1$,

    $$A_n = \frac{2}{L}\int_0^{L/2} x\sin\frac{n\pi x}{L}dx + \frac{2}{L}\int_{L/2}^{L}(L-x)\sin\frac{n\pi x}{L}dx = \frac{4L}{n^2\pi^2}\sin\frac{n\pi}{2}$$

    and

    $$B_n = \frac{2}{n\pi c}\int_0^{L/2} x\sin\frac{n\pi x}{L}dx + \frac{2}{n\pi c}\int_{L/2}^{L}(L-x)\sin\frac{n\pi x}{L}dx = \frac{4L^2}{n^3\pi^3 c}\sin\frac{n\pi}{2}$$

    The solution to the problem is

    $$u(x, t) = \sum_{n=1}^{\infty}\left(\frac{4L}{n^2\pi^2}\cos\frac{n\pi ct}{L} + \frac{4L^2}{n^3\pi^3 c}\sin\frac{n\pi ct}{L}\right)\sin\frac{n\pi}{2}\sin\frac{n\pi x}{L}$$

    (c). Here $f(x) = x(L - x) = -g(x)$. Using (8) and (9), we have that for $n \geq 1$,

    $$A_n = \frac{2}{L}\int_0^L x(L-x)\sin\frac{n\pi x}{L}dx = \frac{4L^2}{n^3\pi^3}(1 - (-1)^{n+1})$$

and

$$B_n = \frac{2}{n\pi c} \int_0^L x(x-L) \sin\frac{n\pi x}{L} dx = \frac{4L^3}{n^4\pi^4 c}((-1)^{n+1}-1) = \begin{cases} -8L^3/(n^4\pi^4 c), & \text{n even} \\ 0, & \text{n odd} \end{cases}$$

Substituting these values for A_n, B_n in the series for $u(x,t)$ in (8) yields the solution.

2. The problem is $u_{tt} - c^2 u_{xx} = 0$, $0 < x < L$, $t > 0$; $u(0,t) = 0$, $u_x(L,t) = -Lu(L,t)$, $t \geq 0$; $u(x,0) = f(x)$, $u_t(x,0) = g(x)$, $0 < x < L$.

(a). The SL problem is as follows: find all constants λ with the property that $X'' = \lambda X$ has a nontrivial solution X in $\mathbf{C}^2[0,L]$ such that $X(0) = 0$ and $X'(L) + hX(L) = 0$.

(b). First, try $\lambda = k^2$, $k > 0$. Then $X'' = k^2 X$ has the general solution $X = A\cosh kx + B\sinh kx$. The boundary conditions imply that $A = 0$ and $B(k\cosh kL + h\sinh kL) = 0$. Since B cannot vanish, it must be that $k > 0$ satisfies the equation $k\cosh kL + h\sinh kL = 0$. But this is impossible since h, $\cosh kL$, and $\sinh kL$ are all positive. There are no positive eigenvalues. Next try $\lambda = 0$. The general solution of $X'' = 0$ is $X = Ax + B$. The boundary conditions imply that $A = B = 0$, so $\lambda = 0$ is also not an eigenvalue. Now put $\lambda = -k^2$, $k > 0$. The general solution of $X'' + k^2 X = 0$ is $X = A\sin kx + B\cos kx$. The condition $X(0) = 0$ implies that $B = 0$, and the condition at $x = L$ implies that $k\cos kL + h\sin kL = 0$. Putting $s = kL$, the defining relation for k becomes the relation $s + hL\tan s = 0$ for s. Writing the relation as $\tan s = -s/(hL)$ and graphing each side against s for $s > 0$, we obtain the plot shown in Fig. 2(b) [$hL = 1$ in the plot]. From the graph we see that the relation $s + hL\tan s = 0$ has an infinite sequence of positive zeros $s_1, s_2, \ldots$. So $\lambda_n = -n^2 = -s_n^2/L^2$, $n = 1, 2, \ldots$, are the eigenvalues of our Sturm-Liouville Problem. Note that $(n-1/2)\pi < s_n < n\pi$ for all n, and $s_n - (n-1/2)\pi \to 0$ as $n \to \infty$.

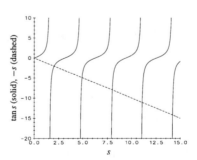

Problem 2(b).

3. A separated solution of the PDE $u_{tt} + b^2 u_t - a^2 u_{xx} = 0$ has the form $u = X(x)T(t)$ where

$$X(x)T''(t) + b^2 X(x)T'(t) - a^2 X''(x)T(t) = 0$$

Dividing by $a^2 XT$ and separating, we have that

$$\frac{X''}{X} = \frac{T'' + b^2 T'}{a^2 T} = \lambda$$

where λ is a separation constant. Imposing the boundary conditions we have the Sturm-Liouville Problem

$$X'' = \lambda X, \quad X(0) = X(L) = 0$$

and the ODE

$$T'' + b^2 T' - \lambda a^2 T = 0$$

We have that $X_n = C_n \sin(n\pi x/L)$, $\lambda_n = -n^2\pi^2/L^2$, $n = 1, 2, \ldots$, exactly as in (7). Recalling that we require that $b^2 < 2\pi a/L$, then for all integers $n = 1, 2, \ldots$, the roots of the quadratic $r^2 + b^2 r - \lambda_n a^2$ are the complex conjugates $-b^2/2 \pm i\beta_n$, where $\beta_n = \sqrt{-\lambda_n a^2 - b^4/4}$. Then we have that

$$T_n(t) = e^{-b^2 t/2} [A_n \cos \beta_n t + B_n \sin \beta_n t]$$

and the solution has the form

$$u(x, t) = e^{-b^2 t/2} \sum_1^\infty \sin \frac{n\pi}{L} x [A_n \cos \beta_n t + B_n \sin \beta_n t]$$

Now set $t = 0$ to obtain

$$u(x, 0) = f(x) = \sum_1^\infty A_n \sin \frac{n\pi}{L} x$$

and

$$u_t(x, 0) = 0 = \sum_1^\infty \sin \frac{n\pi}{L} x \left[-\frac{b^2 A_n}{2} + \beta_n B_n \right]$$

We have that

$$A_n = \frac{2}{L} \int_0^L f(x) \sin(n\pi x/L) \, dx$$

$$B_n = b^2 A_n / (2\beta_n)$$

and we are done.

4.   **(a).** Since the driving force $F(x, t) = g$, we will try to "shift" F off of the PDE onto the initial data by using a function $v(x)$ that satisfies $v_{tt} - c^2 v_{xx} = g$, that is, $-c^2 v_{xx} = g$, and the boundary conditions $v(0) = 0$, $v(a) = 0$. Solving the ODEs for $v(x)$ we have the general solution

$$v = -\frac{g}{2c^2} x^2 + Ax + B$$

If v is to satisfy the boundary conditions, then $B = 0$ and $A = gL/(2c^2)$. The function $v = gx(L - x)/(2c^2)$ satisfies all the required conditions.

(b). Let $u(x, t)$ be the solution of the given problem and put $w(x, t) = u(x, t) - v(x)$, where v is as in (a). Since $u_{tt} - c^2 u_{xx} = g$ and $v_{tt} - c^2 v_{xx} = g$, it follows that

$$w_{tt} - c^2 w_{xx} = (u_{tt} - c^2 u_{xx}) - (v_{tt} - c^2 v_{xx}) = g - g = 0$$

We have that

$$w(0, t) = u(0, t) - v(0) = 0 - 0 = 0$$

$$w(L, t) = u(L, t) - v(L) = 0 - 0 = 0$$

$$w(x, 0) = u(x, 0) - v(x) = -gx(L - x)/2c^2$$

$$w_t(x, 0) = u_t(x, 0) - 0 = 0$$

(c). Since $w = u - v$ solves the problem

$$\begin{aligned}
w_{tt} - c^2 w_{xx} &= 0, & 0 < x < L, \quad t > 0 \\
w(0, t) = w(L, t) &= 0, & t \geq 0 \\
w(x, 0) &= -gx(L - x)/2c^2, & 0 < x < L \\
w_t(x, 0) &= 0, & 0 < x < L
\end{aligned}$$

we may use (9) to find w, with $f(x) = -gx(L - x)/2c^2$ and $g(x) = 0$. We have $B_n = 0$ for all n, and

$$A_n = \frac{2}{L} \int_0^L -\frac{g}{2c^2} x(L - x) \sin\frac{n\pi x}{L} dx = \frac{2gL^2}{n^3\pi^3 c^2} ((-1)^n - 1)$$

so

$$w(x, t) = \sum_{n=1}^{\infty} A_n \cos\frac{n\pi ct}{L} \sin\frac{n\pi x}{L}$$

From the way w was constructed in parts **(a)** and **(b)**, we see that $u(x, t) = w(x, t) + v(x)$ is the solution to the original problem.

5. The problem is $u_{tt} - u_{xx} = 0$, $0 < x < 1$, $t > 0$; $u(0, t) = \sin(\pi t/2)$, $u(1, t) = 0$, $t \geq 0$; $u(x, 0) = f(x)$, $u_t(x, 0) = g(x)$, $0 \leq x \leq 1$. We need to shift the boundary data $\sin(\pi t/2)$ onto an initial condition. Recall that $X(x)T(t)$ is a solution of the wave equation $u_{tt} = u_{xx}$ if $X'' = \lambda X$ and $T'' = \lambda T$ for the same constant λ.

(a). Let $v(x, t)$ satisfy the given conditions of the "intermediate" problem, $v_{tt} - v_{xx} = 0$, $v(0, t) = \sin \pi t/2$, $v(1, t) = 0$, $t \geq 0$. Let $u(x, t)$ satisfy the PDE and the conditions of the original problem. Then $w(x, t) = u - v$ satisfies the wave equation $w_{tt} - w_{xx} = 0$, and the boundary conditions

$$w(0, t) = u(0, t) - v(0, t) = \sin\frac{\pi t}{2} - \sin\frac{\pi t}{2} = 0$$

$$w(1, t) = u(1, t) - v(1, t) = 0 - 0 = 0$$

for $t \geq 0$. The initial conditions satisfied by w are $w(x, 0) = u(x, 0) - v(x, 0) = f(x) - v(x, 0)$, and $w_t(x, 0) = u_t(x, 0) - v_t(x, 0) = g(x) - v_t(x, 0)$.

(b). Let's look for $v(x, t)$ in the form $X(x) \sin(\pi t/2)$. The boundary conditions imply that $X(0) = 1$ and $X(1) = 0$, and for v to be a solution to the wave equation we must have $X'' = -(\pi/2)^2 X$. The function $X(x) = \cos(\pi x/2)$ fulfills these conditions. So $v = \cos(\pi x/2) \sin(\pi t/2)$.

(c). Note that $w(x, t) = u(x, t) - \cos(\pi x/2)\sin(\pi t/2)$ satisfies the conditions

$$
\begin{aligned}
w_{tt} - w_{xx} &= 0, & 0 < x < 1, \quad t > 0 \\
w(0, t) &= w(1, t) = 0, & t > 0 \\
w(x, 0) &= f(x), & 0 < x < 1 \\
w_t(x, 0) &= -(\pi/2)\cos(\pi x/2), & 0 < x < 1
\end{aligned}
$$

Following (9), we see that w has the form

$$
w(x, t) = \sum_{n=1}^{\infty} \sin n\pi x (A_n \cos n\pi t + B_n \sin n\pi t)
$$

where for $n \geq 1$

$$
A_n = 2 \int_0^1 f(x) \sin n\pi x\, ds
$$

$$
B_n = \frac{2}{n\pi} \int_0^1 \left(-\frac{\pi}{2}\cos\frac{\pi x}{2}\right) \sin n\pi x\, dx
$$

$$
= -\frac{1}{2n} \int_0^1 (\sin(n + 1/2)\pi x + \sin(n - 1/2)\pi x)\, dx = -\frac{4}{\pi(4n^2 - 1)}
$$

The solution u of the boundary/initial value problem [with $u_t(x, 0) = 0$] is

$$
u(x, t) = \sum_{n=1}^{\infty} \sin n\pi x (A_n \cos n\pi t + B_n \sin n\pi t) + \cos(\pi x/2)\sin(\pi t/2)
$$

where A_n and B_n are as given above.

6. The problem is $u_{tt} - u_{xx} = 6x$, $0 < x < 1$, $t > 0$; $u(0, t) = 0$, $u(1, t) = 0$, $t \geq 0$; $u(x, 0) = u_t(x, 0)$, $0 \leq x \leq 1$.

(a). First, we expand the "driving" term $F(x) = 6x$ of the PDE into a Fourier Sine Series on $0 \leq x \leq 1$. Now

$$
A_n = 2 \int_0^1 (6x) \sin n\pi x\, dx = \frac{12(-1)^{n+1}}{n\pi}
$$

and so $\text{FSS}[6x] = \sum_{n=1}^{\infty} 12(-1)^{n+1} \sin n\pi x / (n\pi)$. Now we shall determine $u(x, t)$ through its Fourier Sine Transform $\text{FSS}[u] = \sum_{n=1}^{\infty} U_n(t) \sin n\pi x$. After replacing u by $\sum U_n(t) \sin n\pi x$ in $u_{tt} - u_{xx} = 6x$, replacing $6x$ by $\text{FSS}[6x]$, and then matching coefficients, we have

$$
U_n'' + (n\pi c)^2 U_n = \frac{12(-1)^{n+1}}{n\pi}
$$

and from (16a), (16b), and (19) we see that $U_n(0) = 0$, $U_n'(0) = 0$. Solving the IVP for U_n, we find that

$$
U_n(t) = \frac{12(-1)^{n+1}}{n^3 \pi^3 c^2} (1 - \cos n\pi c t) \qquad n = 1, 2, \ldots.
$$

The formal solution to the given problem is

$$u(x, t) = \sum_{n=1}^{\infty} \sin(n\pi x) \frac{12(-1)^{n+1}}{n^3 \pi^3 c^2} (1 - \cos n\pi ct)$$

(b). The problem is $u_{tt} - u_{xx} = 0$, $0 < x < L$, $t > 0$; $u(0, t) = \sin(3\pi t/2L)$, $u(L, t) = 0$, $t \geq 0$; $u(x, 0) = u_t(x, 0) = 0$, $0 \leq x \leq L$. First, consider the intermediate problem

$$v_{tt} - v_{xx} = 0, \qquad 0 < x < L, \quad t > 0$$
$$v(0, t) = \sin(3\pi t/2L), \qquad v(L, t) = 0, \quad t > 0$$

To solve this problem, recall that $X(x)T(t)$ satisfies $v_{tt} - v_{xx} = 0$ if $X'' = \lambda X$ and $T'' = \lambda T$ for some constant λ. So one approach is to take $v(x, t) = X(x) \sin 3\pi t/(2L)$ where $X'' = -(3\pi/2L)^2 X$ and $X(0) = 1$, $X(L) = 0$. The function $X(x) = \cos(3\pi x/2L)$ has these properties, so the intermediate problem has the solution

$$v(x, t) = \cos \frac{3\pi x}{2L} \sin \frac{3\pi t}{2L}$$

Now put $w = u - v$, where u is the solution to the original problem and v is the above solution to the intermediate problem. Then we can verify that w must be a solution of the problem

$$w_{tt} - w_{xx} = 0, \qquad\qquad 0 < x < L, \quad t \geq 0$$
$$w(0, t) = w(L, t) = 0, \qquad\qquad t > 0$$
$$w(x, 0) = 0, \qquad\qquad 0 < x < L$$
$$w_t(x, 0) = -(3\pi/2L) \cos(3\pi x/2L), \qquad 0 < x < L$$

Use text formula (9) to find the series for w. Note that $f(x) = 0$, $g(x) = -(3\pi/2L) \cos(3\pi x/2L)$, so $A_n = 0$ for all n. To find the coefficients B_n, we must evaluate the integrals

$$B_n = \frac{2}{n\pi} \int_0^L \left(-\frac{3\pi}{2L} \cos \frac{3\pi x}{2L} \right) \sin \frac{n\pi x}{L} dx$$

$$= -\frac{3}{nL} \int_0^L \left\{ \sin \frac{(2n+3)\pi x}{2L} + \sin \frac{(2n-3)\pi x}{2L} \right\} dx = -\frac{24}{\pi(4n^2 - 9)}$$

The solution to the problem for w is

$$w(x, t) = -\frac{24}{\pi} \sum_{n=1}^{\infty} \frac{1}{4n^2 - 9} \sin \frac{n\pi x}{L} \sin \frac{n\pi t}{L}$$

so $u = w + v$ is the sought-for (formal) solution to the original problem.

7. We want to find a periodic function $\mu(t)$ to break the string modeled by $u_{tt} - c^2 u_{xx} = 0$, $0 < x < L$, $t > 0$; $u(0, t) = 0$, $u(L, t) = \mu(t)$, $t \geq 0$; $u(x, 0) = 0$, $u_t(x, 0) = 0$, $0 \leq x \leq L$. Consider the "shaking" function $\mu(t) = \cos \omega t$ for some constant ω. To solve the boundary/initial value problem for u, we first solve the "intermediate" problem

$$v_{tt} - c^2 v_{xx} = 0, \qquad 0 < x < L, \quad t > 0$$
$$v(0, t) = 0, \qquad t > 0$$
$$v(L, t) = \cos \omega t, \qquad t > 0$$

Now the separated function $X(x)T(t)$ solves the wave equation if $X'' = \lambda X$ and $T'' = c^2 \lambda T$ for some constant λ. Say we search for v in the form $v(x, t) = X(x) \cos \omega t$. Then clearly $X(x)$ must be a solution of $X'' = -(\omega/c)^2 X$ and satisfy the conditions $X(0) = 0$, $X(L) = 1$. Observe that if $\omega L/c \neq n\pi$, then

$$X(x) = \left(\sin \frac{\omega L}{c} \right)^{-1} \sin \frac{\omega x}{c}$$

has this property. The intermediate problem has the solution

$$v(x, t) = \left(\sin \frac{\omega L}{c} \right)^{-1} \sin\left(\frac{\omega x}{c}\right) \cos \omega t$$

Now we put $w = u - v$, where u solves the original problem and v the intermediate problem, and observe that w solves the problem

$$
\begin{aligned}
w_{tt} - w_{xx} &= 0, & 0 &< x < L, \quad t \geq 0 \\
w(0, t) &= w(L, t) = 0, & t &> 0 \\
w(x, 0) &= -(\sin(\omega L/c))^{-1} \sin(\omega x/c), & 0 &< x < L \\
w_t(x, 0) &= 0, & 0 &< x < L
\end{aligned}
$$

The solution of this problem for w is given by (9). Note that $B_n = 0$ for all n since $g(x) = 0$. To find the coefficients A_n, we must evaluate the integrals

$$\frac{2}{L} \int_0^L \sin \frac{\omega x}{c} \sin \frac{n\pi x}{L} dx = \frac{1}{L} \int_0^L \left\{ \cos\left(\frac{\omega}{c} - \frac{n\pi}{L} \right) x - \cos\left(\frac{\omega}{c} + \frac{n\pi}{L} \right) x \right\} dx$$

$$= \frac{(-1)^n 2 n\pi c^2}{(\omega L)^2 - (n\pi c)^2} \sin \frac{\omega L}{c} = \alpha_n$$

Then the coefficient A_n is

$$A_n = -\left(\sin \frac{\omega L}{c} \right)^{-1} \alpha_n = \frac{(-1)^{n+1} 2 n\pi c^2}{(\omega L)^2 - (n\pi c)^2}$$

The solution to the problem for w is

$$w(x, t) = \sum_{n=1}^{\infty} \frac{(-1)^{n+1} 2 n\pi c^2}{(\omega L)^2 - (n\pi c)^2} \sin \frac{n\pi x}{L} \cos \frac{n\pi c t}{L}$$

so $u = w + v$ is the solution of the original problem. If we allow ω to approach the fundamental frequency $\pi c/L$, then

$$\left(\frac{\omega L}{c} \right)^2 - \pi^2 \to 0$$

and both $v(x, t)$ and the first term of $w(x, t)$ become unbounded (i.e., the string "breaks").

10.8 The Heat Equation: Optimal Depth for a Wine Cellar

Suggestions for preparing a lecture

Topics: Discussion of the derivation of the heat equation (it may be simplest to do it for a homogeneous rod because then you only need the Fundamental Theorem of Calculus and not the Divergence Theorem). Then do the problem of temperature in a rod, and discuss various types of boundary conditions. Discuss properties of solutions of the heat equation and contrast these properties with those of the wave equation.

Making up a problem set

Problems 1(**a**), 1(**e**), 2, 6. Suggestion: Students will love you if you distribute a list of common Fourier trigonometric expansions on $0 \leq x \leq L$, e.g., FSS[1], FSS[x], FCS[x], for use in doing some of the problems in this set. Without a CAS or other reference at hand, students (and faculty) almost always make computational errors in calculating Fourier-Euler coefficients.

Comments

Wave motion has been the physical phenomenon supporting our discussion of PEs until now. The conduction of heat is shown in this section to be modeled by a linear PDE as well, but it is not the wave equation. The heat (or diffusion) PDE is derived from physical principles concerning the flow of thermal energy from warmer to cooler regions in a body. Natural boundary conditions are introduced that model heat sources or sinks or insulating material placed around the boundary of the body. Separation of variables is applied to construct a series solution for the flow of heat (measured by changing temperatures) in a long circular rod whose lateral walls are wrapped in thermal insulation. Then we show how to determine the optimal depth of a storage cellar to protect its contents from the temperature variations at the earth's surface. Finally, we compare and contrast properties of solutions of the heat equation with those of solutions of the wave equation. In particular, the heat operator smooths out any "corners" or discontinuities in the initial heat curve, while the wave equation simply propagates the initial kinks and jumps. Moreover, thermal energy moves "infinitely" fast, while a signal propagates with a finite speed. Some of the properties, incidentally, show that the wave and heat equations cannot be exact models of the corresponding real phenomena. We like to present the material of Sections 10.7 and 10.8 back-to-back because the models are so "physical" and lead to such contrasting conclusions.

---

1. We have shown that $X(x)T(t)$ is a solution of the heat equation: $u_t - Ku_{xx} = 0$ if $X'' = \lambda X$ and $T' = \lambda KT$ for the same constant λ. In parts (**a**)–(**d**) of this problem we require that $u(0, t) = u(L, t) = 0$. To save space, the details are only carried out for (**a**), (**e**), and (**g**). [*Hint*: See the subsection of 10.8, Temperature in a Rod.]

   (**a**). Let us consider the separated solution $= X(x)T(t)$. Inserting $X(x)T(t)$ into the PDE and applying the boundary conditions, we see that $X(x)$ and $T(t)$ must be solutions of

$$\frac{d^2 X}{d^2 x} - \lambda X = 0, \quad X(0) = 0, \quad X(L) = 0$$

$$\frac{dT}{dt} - \lambda KT = 0$$

As in (20) and (21) in the text, we have that $\lambda_n = -(n\pi/L)^2$, $X_n(x) = A_n \sin(n\pi x/L)$, $T_n(t) = B_n \exp[-K(n\pi/L)^2 t]$, $n = 1, 2, \ldots$, where A_n and B_n are any constants. The solutions of the PDE, $u_t - K u_{xx} = 0$ and the boundary conditions $u(0, t) = u(L, t) = 0$ are given by

$$u_n = X_n(x) T_n(t) = C_n \sin \frac{n\pi x}{L} \exp\left[-K\left(\frac{n\pi}{L}\right)^2 t\right], \qquad n = 1, 2, \ldots$$

where $C_n = A_n B_n$ is an arbitrary constant. Since the PDE and the boundary conditions are homogeneous in this problem,

$$u(x, t) = \sum_{n=1}^{\infty} C_n \sin \frac{n\pi x}{L} \exp\left[-K\left(\frac{n\pi}{L}\right)^2 t\right]$$

is also a solution of the PDE and the boundary conditions. Now we only need to determine the constants C_n so that $u(x, t)$ satisfies the initial condition. That is, we must choose C_n so that

$$u(x, 0) = \sum_{n=1}^{\infty} C_n \sin \frac{n\pi x}{L} = \sin \frac{2\pi x}{L}, \qquad 0 \leq x \leq L$$

So $C_1 = C_3 = \cdots = 0$ and $C_2 = 1$. The solution is

$$u(x, t) = \sin \frac{2\pi x}{L} \exp\left[-K\left(\frac{2\pi}{L}\right)^2 t\right]$$

This represents the temperature as one hump of a sine curve whose amplitude decays exponentially in time to zero.

(b). Since $f(x) = x$, the coefficients are

$$C_n = \frac{2}{L} \int_0^L x \sin \frac{n\pi x}{L} dx = \frac{2(-1)^{n+1} L}{n\pi}$$

The series solution is

$$u(x, t) = \frac{2L}{\pi} \sum_{n=1}^{\infty} \frac{(-1)^{n+1}}{n} \sin \frac{n\pi x}{L} \exp\left[-K\left(\frac{n\pi}{L}\right)^2 t\right]$$

In this case the initial temperature $u = x$, $0 \leq x \leq L$, is represented by a Fourier Sine Series; as time advances the amplitude of each of the sinusoidal terms in the series decays exponentially to zero.

(c). Since $f(x) = u_0$,

$$C_n = \frac{2}{L} \int_0^L u_0 \sin \frac{n\pi x}{L} dx = \frac{2u_0}{n\pi}[1 - (-1)^n]$$

The series solution is

$$u(x, t) = \frac{2u_0}{\pi} \sum_{n=1}^{\infty} \frac{1 - (-1)^n}{n} \sin \frac{n\pi x}{L} \exp\left[-K\left(\frac{n\pi}{L}\right)^2 t\right]$$

$$= \frac{4u_0}{\pi} \sum_{\text{odd } n} \frac{1}{n} \sin \frac{n\pi x}{L} \exp\left[-K\left(\frac{n\pi}{L}\right)^2 t\right]$$

Note that for all $t \geq 0$ the series solution for u is zero at $x = 0, L$. As time goes on, the temperatures everywhere decline to the zero values maintained at the endpoints.

(d). Since

$$f(x) = \begin{cases} u_0, & 0 \leq x \leq L/2 \\ 0, & L/2 \leq x \leq L \end{cases}$$

the coefficients are given by

$$C_n = \frac{2}{L} \int_0^{L/2} u_0 \sin \frac{n\pi x}{L} dx = \frac{2u_0}{n\pi}\left(1 - \cos \frac{n\pi}{2}\right) = \begin{cases} 2u_0/(n\pi), & n \text{ odd} \\ 2u_0/(n\pi)[1 - (-1)^{n/2}], & n \text{ even} \end{cases}$$

The series solution is

$$u(x, t) = \sum_{n=1}^{\infty} C_n \sin \frac{n\pi x}{L} \exp\left[-K\left(\frac{n\pi}{L}\right)^2 t\right]$$

The jump at $x = L/2$ from u_0 to 0 is immediately "smoothed out" as time increases from 0 in the series solution. All components decay exponentially to 0 as $t \to \infty$.

(e). Here the boundary conditions are $u(0, t) = 0$ and $u_x(L, t) = 0$ [perfect insulation at $x = L$]. Let's consider the separated solution $= X(x)T(t)$. Inserting $X(x)T(t)$ into the PDE and boundary conditions, we see that $X(x)$ and $T(t)$ must be solutions of

$$X''(x) - \lambda X(x) = 0, \quad X(0) = 0, \quad X'(L) = 0$$

$$T'(t) - \lambda K T(t) = 0$$

We have $\lambda_n = -[(2n-1)\pi/(2L)]^2$ and

$$X_n(x) = A_n \sin \frac{(2n-1)\pi x}{2L}, \quad T_n(t) = B_n \exp\left[-K\left(\frac{2n-1}{2L}\pi\right)^2 t\right], \quad n = 1, 2, \ldots$$

where A_n and B_n are any constants. The separated solutions of the PDE, $u_t - Ku_{xx} = 0$, and the boundary conditions $u(0, t) = u_x(L, t) = 0$ are given by

$$u_n = X_n(x)T_n(t) = C_n \sin \frac{(2n-1)\pi x}{2L} \exp\left[-K\left(\frac{2n-1}{2L}\pi\right)^2 t\right], \quad n = 1, 2, \ldots$$

where $C_n = A_n B_n$ is an arbitrary constant. Since the PDE and the boundary conditions are homogeneous in this problem,

$$u(x, t) = \sum_{n=1}^{\infty} C_n \sin \frac{(2n-1)\pi x}{2L} \exp\left[-K\left(\frac{2n-1}{2L}\pi\right)^2 t\right]$$

is also a solution of the PDE and the boundary conditions. Now we only need to determine constants C_n so that $u(x, t)$ satisfies the initial condition. That is, we must choose C_n so that

$$u(x, 0) = \sum_{n=1}^{\infty} C_n \sin \frac{(2n-1)\pi x}{2L} = \sin \frac{\pi x}{2L}, \quad 0 \le x \le L$$

We have $C_1 = 1, C_2 = C_3 = \cdots = 0$. The solution is

$$u(x, t) = \sin \frac{\pi x}{2L} \exp\left[-K\left(\frac{\pi}{2L}\right)^2 t\right]$$

and we have the initial temperature profile decaying exponentially to 0. Note that the condition $u_x(L, t) = 0$ means that a perfect insulator is clamped to the far end of the rod.

(f). Here we have $u(0, t) = 0$ and $u_x(L, t) = 0$. Proceeding as in part **(e)**, we see that λ_n, X_n, and T_n are as given in the solution to part **(e)**. Since $f(x) = x$, the coefficients are

$$C_n = \frac{2}{L} \int_0^L x \sin \frac{(2n-1)\pi x}{2L} dx = \frac{(-1)^{n+1} 8L}{(2n-1)^2 \pi^2}$$

The series solution is

$$u(x, t) = \frac{8L}{\pi^2} \sum_{n=1}^{\infty} \frac{(-1)^{n+1}}{(2n-1)^2} \sin \frac{(2n-1)\pi x}{2L} \exp\left[-K\left(\frac{(2n-1)}{2L}\pi\right)^2 t\right]$$

Note that this model represents perfect insulation at $x = L$, and indeed $\partial u(L, t)/\partial x = 0$ as we see by term-by-term differentiation of the series.

(g). Here we have perfect insulation at both ends: $u_x(0, t) = u_x(L, t) = 0$. Let us consider the separated solution $= X(x)T(t)$. Inserting $X(x)T(t)$ into the PDE and the boundary conditions, we see that $X(x)$ and $T(t)$ must be solutions of

$$\frac{d^2 X}{d^2 x} - \lambda X = 0, \quad X'(0) = 0, \quad X'(L) = 0$$

$$\frac{dT}{dt} - \lambda KT = 0$$

We have $\lambda_n = -(n\pi/L)^2$ and

$$X_n(x) = A_n \cos \frac{n\pi x}{L}, \qquad T_n(t) = B_n \exp[-K(\frac{n\pi}{L})^2 t], \quad n = 0, 1, 2, \ldots$$

where A_n and B_n are any constants. The separated solutions of the PDE: $u_t - Ku_{xx} = 0$ with boundary conditions $u_x(0, t) = u_x(L, t) = 0$ are given by

$$u_n = X_n(x)T_n(t) = C_n \cos\left(\frac{n\pi x}{L}\right) \exp\left[-K\left(\frac{n\pi}{L}\right)^2 t\right], \qquad n = 0, 1, 2, \ldots$$

where $C_n = A_n B_n$ is an arbitrary constant. Since the PDE and the boundary conditions are homogeneous in this problem,

$$u(x, t) = C_0 + \sum_{n=1}^{\infty} C_n \cos\left(\frac{n\pi x}{L}\right) \exp\left[-K\left(\frac{n\pi}{L}\right)^2 t\right]$$

is also a solution of PDE. Now we only need to determine the constants C_n so that $u(x, t)$ satisfies the initial condition. That is, we must choose C_n so that

$$u(x, 0) = \sum_{n=0}^{\infty} C_n \cos \frac{n\pi x}{L} = x, \quad 0 \le x \le L$$

Therefore, the coefficients C_n are given by

$$C_n = \frac{\langle f, \cos \frac{n\pi x}{L} \rangle}{\| \cos \frac{n\pi x}{L} \|^2} = \begin{cases} \frac{1}{L} \int_0^L x\,dx = \frac{L}{2}, & n = 0 \\ \frac{2}{L} \int_0^L x \cos \frac{n\pi x}{L}\,dx = \frac{2L}{n^2 \pi^2}[(-1)^n - 1], & n > 0 \end{cases}$$

The series solution is

$$u(x, t) = \frac{L}{2} - \frac{4L}{\pi^2} \sum_{\text{odd } n} \frac{1}{n^2} \cos \left(\frac{n\pi x}{L} \right) \exp \left[-K \left(\frac{n\pi}{L} \right)^2 t \right]$$

Note that since both ends of the rod and the cylindrical walls are perfectly insulated, no thermal energy can escape. This means that as time goes on the energy tends towards a level equilibrium. In fact, as $t \to +\infty$ the temperature tends to the uniform value of $L/2$.

2. **(a).** Let $u(x, t) = A(x) + v(x, t)$. The problem $u_t - Ku_{xx} = 0$, $0 < x < 1$, $t > 0$; $u(0, t) = 10$, $u(1, t) = 20$, $t \ge 0$; $u(x, 0) = 0$, $0 < x < 1$ becomes the two problems, one for $A(x)$ and the other for v:

$$A''(x) = 0, \qquad A(0) = 10, \qquad A(1) = 20$$

and

$$v_t - Kv_{xx} = 0, \qquad\qquad 0 < x < 1, \quad t > 0$$
$$v(0, t) = v(1, t) = 0, \quad t \ge 0; \qquad v(x, 0) = -A(x), \quad 0 \le x \le 1$$

It is not hard to show that $A(x) = 10(1 + x)$. So $v(x, 0) = -10(1 + x)$. We have

$$v(x, t) = -\sum_{n=1}^{\infty} \left[2 \int_0^1 10(1 + x) \sin n\pi x\,dx \right] \sin n\pi x \exp[-K(n\pi)^2 t]$$

$$= -\sum_{n=1}^{\infty} \frac{20[1 - 2(-1)^n]}{n\pi} \sin n\pi x \exp[-K(n\pi)^2 t]$$

The series solution is

$$u(x, t) = 10(1 + x) - \sum_{n=1}^{\infty} \frac{20[1 - 2(-1)^n]}{n\pi} \sin n\pi x \exp[-K(n\pi)^2 t]$$

(b). Recognizing that by geometric series, $\sum_1^{\infty} \exp(-kn\pi^2 t) = (1 + K\pi^2 t)^{-1} - 1$, we see that

$$|v(x, t)| \le \sum_{n=1}^{\infty} \left| \frac{20[1 - 2(-1)^n]}{n\pi} \right| \exp[-K(n\pi)^2 t] \le \frac{60}{\pi} \sum_{n=1}^{\infty} \exp(-Kn\pi^2 t)$$

$$= \frac{60}{\pi} \cdot \frac{e^{-K\pi^2 t}}{1 - e^{-K\pi^2 t}} \to 0 \text{ as } t \to +\infty$$

and the result holds.

3.   **(a).** It is easy to show by a direct calculation (given V and U as defined below) that $u(x,t) = V(x,t) + U(x,t)$ satisfies the PDE, $u_t - Ku_{xx} = 0$, $0 < x < 1$, $t > 0$, the boundary conditions $u(0,t) = g_1(t)$ and $u(1,t) = g_2(t)$, $t \geq 0$, and the initial condition $u(x,0) = f(x)$, $0 \leq x \leq 1$. Here $V = g_1(t) + [g_2(t) - g_1(t)]x$ and $U(x,t)$ satisfies the equations $U_t - KU_{xx} = -KV_t$, $0 < x < 1$, $t > 0$; $U(0,t) = U(1,t) = 0$, $t \geq 0$; $U(x,0) = f(x) - V(x,0)$, $0 \leq x \leq 1$.

  (b). If $g_1(t) = \sin t$ and $g_2(t) = 0$, $V(x,t) = (1-x)\sin t$. So $U(x,t)$ satisfies the equations $U_t - KU_{xx} = -KV_t = K(x-1)\cos t$, $0 < x < 1$, $t > 0$; $U(0,t) = U(1,t) = 0$, $t \geq 0$; $U(x,0) = f(x)$. Suppose that $U(x,t)$ and $-KV_t(x,t)$ are the sums of their Fourier Sine Series:

$$U(x,t) = \sum_{n=1}^{\infty} U_n(t) \sin n\pi x, \quad -KV_t(x,t) = K\sum_{1}^{\infty} C_n(t) \sin n\pi x$$

where

$$C_n(t) = 2\int_0^1 (x-1)\cos t \sin n\pi x \, dx = \frac{-2\cos t}{n\pi}$$

Insert these series into $U_t - KU_{xx} + KV_t = 0$:

$$\sum_{n=1}^{\infty} \left[U_n'(t) + KU_n(t)(n\pi)^2 - KC_n(t) \right] \sin n\pi x = 0$$

Then $U_n(t)$ satisfies the first-order linear IVP:

$$U_n'(t) + K(n\pi)^2 U_n + \frac{2K}{n\pi}\cos t = 0, \qquad U_n(0) = 2\int_0^1 f(x)\sin n\pi x \, dx$$

(because $U(x,0) = f(x) = \sum_{n=1}^{\infty} U_n(0)\sin n\pi x$). The solutions of the above ODE are

$$U_n(t) = e^{-K(n\pi)^2 t}\left[2\int_0^1 f(x)\sin n\pi x \, dx - \frac{2K}{n\pi}\int_0^t \cos\tau \, e^{K(n\pi)^2\tau} \, d\tau \right]$$

So

$$U(x,t) = \sum_{n=1}^{\infty} e^{-K(n\pi)^2 t}\left[2\int_0^1 f(x)\sin n\pi x \, dx - \frac{2K}{n\pi}\int_0^t \cos\tau \, e^{K(n\pi)^2\tau} \, d\tau \right]\sin n\pi x,$$

where the second integral can be evaluated explicitly, but we don't do so here.

  (c). By part **(a)** $u(x,t) = V(x,t) + U(x,t)$ where $V(x,t) = g(t) + [g_2(t) - g_1(t)]x$ and $U(x,t)$ satisfies a certain boundary/initial value problem. In this case $g_1 = \sin t$ and $g_2 = 0$, so $V = (1-x)\sin t$. The formula for $V(x,t)$ is derived in part **(b)**. The final formula turns out to be

$$u(x,t) = (1-x)\sin t + \sum_{n=1}^{\infty} e^{-K(n\pi)^2 t}\left[2\int_0^1 f(x)\sin n\pi x \, dx - \frac{2K}{n\pi}\int_0^t \cos\tau \, e^{K(n\pi)^2\tau} \, d\tau \right]\sin n\pi x$$

4. The problem is $u_t - Ku_{xx} = 3e^{-2t} + x$, $0 < x < 1$, $t > 0$; $u(0, t) = u(1, t)$, $t \geq 0$; $u(x, 0) = 0$, $0 \leq x \leq 1$. Let $u(x, t) = \sum_1^\infty U_n(t) \sin n\pi x$, $3e^{-2t} + x = \sum_1^\infty C_n(t) \sin n\pi x$, where

$$C_n(t) = 2 \int_0^1 (3e^{-2t} + x) \sin n\pi x \, dx = \frac{6e^{-2t} - 2(1 + 3e^{-2t})(-1)^n}{n\pi}$$

After inserting the series for u into the PDE, we see that $U_n(t)$ satisfies the first order linear IVP

$$U_n'(t) + K(n\pi)^2 + \frac{6e^{-2t} - 2(1 + 3e^{-2t})(-1)^n}{n\pi} = 0, \qquad U_n(0) = 0$$

(because $u(x, 0) = 0 = \sum_{n=1}^\infty U_n(0) \sin n\pi x$). The solution of the IVP is

$$U_n(t) = 2e^{-K(n\pi)^2 t} \int_0^t \frac{3e^{-2s} - (1 + 3e^{-2s})(-1)^n}{n\pi} e^{K(n\pi)^2 s} ds$$

$$= \frac{2(-1)^n}{K(n\pi)^3}[e^{-K(n\pi)^2 t} - 1] + \frac{6(-1)^n - 3}{n\pi(Kn^2\pi^2 - 2)}[e^{-K(n\pi)^2 t} - e^{-2t}]$$

So

$$U(x, t) = 2\sum_{n=1}^\infty \left\{ \frac{(-1)^n}{K(n\pi)^3}[e^{-K(n\pi)^2 t} - 1] + \frac{3(-1)^n - 3}{n\pi(Kn^2\pi^2 - 2)}[e^{-K(n\pi)^2 t} - e^{-2t}] \right\} \sin n\pi x$$

5. This problem relates to the wine cellar model described in the text.

(a). Let $u = U + T_0$, where $u(x, t)$ satisfies system (25) in the text. Then, we have

$$U_t - KU_{xx} = 0, \qquad 0 < x < \infty, \qquad -\infty < t < \infty$$
$$U(0, t) = A_0 \cos \omega t, \qquad -\infty < t < \infty,$$
$$|U(x, t)| < C, \qquad 0 \leq x < \infty, \qquad -\infty < t < \infty$$

since the surface temperature is given as $u(0, t) = T_0 + A_0 \cos \omega t$. As in the text, we have

$$U = \text{Re}\{A_0 e^{-\alpha x} e^{i(\omega t - \alpha x)}\} = A_0 e^{-\alpha x} \cos(\omega t - \alpha x), \qquad \alpha = \left(\frac{\omega}{2K}\right)^{1/2}$$

The temperature at depth x at time t is

$$u = T_0 + A_0 e^{-\alpha x} \cos(\omega t - \alpha x), \qquad \omega = \frac{2\pi}{3.15 \times 10^7}, \qquad \alpha = \left(\frac{\omega}{2K}\right)^{1/2}$$

The optimal depth x_0 reverses the "seasons" in the wine cellar in comparison to the natural seasons at the surface (see the text). We should have $\alpha x = \pi$:

$$x_0 = \frac{\pi}{\alpha} = \pi \left(\frac{2K}{\omega}\right)^{1/2} \approx 4.45 \text{ m}$$

If, however, ω corresponds to 1 day = 86400 seconds instead of 1 year, then, $\omega = 2\pi/86400$. The optimal depth is

$$x_0 = \frac{\pi}{\alpha} = \pi(0.004)^{1/2} \left(\frac{86400}{2\pi}\right)^{1/2} \approx 23.3 \text{ cm}$$

(b). From part **(a)**, we know that the temperature function at the optimal depth x_0 is $u = T_0 + A_0 e^{-\pi} \cos(\omega t - \pi)$ at time t.

(c). If the surface temperature wave has the form $T_0 + A_1 \cos \omega_1 t + A_2 \cos \omega_2 t$, let $u = T_0 + U + V$, where

$$
\begin{array}{llll}
U_t - K U_{xx} = 0, & V_t - K V_{xx} = 0, & 0 < x < \infty, & -\infty < t < \infty \\
U(0, t) = A_1 \cos \omega_1 t, & V(0, t) = A_2 \cos \omega_2 t, & -\infty < t < \infty, & \\
|U(x, t)| < C, & |V(x, t)| < C, & 0 \le x < \infty, & -\infty < t < \infty
\end{array}
$$

Similar to part **(a)**, we have

$$
U = \mathrm{Re}\{A_1 e^{-\alpha_1 x} e^{i(\omega_1 t - \alpha_1 x)}\} = A_1 e^{-\alpha_1 x} \cos(\omega_1 t - \alpha_1 x), \qquad \alpha_1 = \left(\frac{\omega_1}{2K}\right)^{1/2}
$$

$$
V = \mathrm{Re}\{A_2 e^{-\alpha_2 x} e^{i(\omega_2 t - \alpha_2 x)}\} = A_2 e^{-\alpha_2 x} \cos(\omega_2 t - \alpha_2 x), \qquad \alpha_2 = \left(\frac{\omega_2}{2K}\right)^{1/2}
$$

The temperature at depth x at time t is

$$
u = T_0 + A_1 e^{-\alpha_1 x} \cos(\omega_1 t - \alpha_1 x) + A_2 e^{-\alpha_2 x} \cos(\omega_2 t - \alpha_2 x)
$$

$$
\omega_1 = \frac{\omega_2}{365} = \frac{2\pi}{3.15 \times 10^7}, \qquad \alpha_1 = \frac{\alpha_2}{\sqrt{365}} = \left(\frac{\omega_1}{2K}\right)^{1/2}
$$

6.   **(a).** If $t = t_0 < 0$, then $-n^2 \pi^2 t_0 > 0$. So the term $e^{-n^2 \pi^2 t_0/4}/n^2 \to +\infty$ as $n \to \infty$ (use L'Hôpital's Rule). The series (25) diverges for every x, $0 < x < 2$, if $t = t_0 < 0$.

(b). The heat equation is valid only for t increasing from the initial time $t = 0$. If the initial data are not smooth, then only for $t > 0$ is the equation meaningful. The direction of time from angular to smooth temperature profiles determines time's advance.

(c). Consider $f(x) = \sin(\pi x/L)$. By text equation (23),

$$
C_n = \frac{2}{L} \int_0^L f(x) \sin \frac{n\pi x}{L} dx = \frac{2}{L} \int_0^L \sin \frac{\pi x}{L} \sin \frac{n\pi x}{L} dx = \begin{cases} 1 & n = 1 \\ 0 & n > 1 \end{cases}
$$

So

$$
u(x, t) = \sin \frac{\pi x}{L} \exp\left[-K \left(\frac{\pi}{L}\right)^2 t\right]
$$

which is a solution of the boundary/initial value problem (16) valid for all values of t.

7.   The problem is $u_t - u_{xx} = 0$, $0 < x < 1$, $t > 0$; $u(0, t) = te^{-t}, t \ge 0$; $u(1, t) = 0$, $t \ge 0$; $u(x, 0) = 0.01x(1 - x)$, $0 \le x \le 1$. By the Maximum Principle for the Heat Equation (Theorem 10.8.1), we know that $|u(x, t)| \le \max |u(\bar{x}, \bar{t})|$, where $(\bar{x}, \bar{t})$ is a point on the boundary of the region $R_t = \{(x, \bar{t}) : 0 < x < 1, \ 0 \le \bar{t} \le t\}$. Because $\max |u(\bar{x}, \bar{t})| = \max\{0, 0.0025, e^{-1}\} = 1/e$, we have that $|u(x, t)| \le 1/e$.

10.9 Laplace's Equation

Suggestions for preparing a lecture

Topics: The Laplacian and Laplace's equation, the Dirichlet problem, steady temperatures in a disk, harmonic functions and their properties.

Making up a problem set

Problems 1(**a**) or (**c**), 3.

Comments

Laplace's PDE and its solutions, called harmonic functions, have properties quite different from anything we have seen so far with the wave and heat equations and their solutions. We illustrate these points by solving a boundary value problem for steady temperatures in a disk, using circular harmonics in the process. Then, we discuss the averaging and maximal properties of harmonic functions and show that solutions of boundary value problems for Laplace's equation are unique (if there is a solution) and continuous in the data.

1. (**a**). The problem is $u_{xx} + u_{yy} = 0$, $0 < x < L$, $0 < y < M$; $u(0, y) = \alpha(y)$, $u(L, y) = 0$, $0 \le y \le M$; $u(x, 0) = u(x, M) = 0$, $0 \le x \le L$. The technique of separation of variables can be followed just as in previous sections to construct separated solutions $X_n(x)Y_n(y)$, where

$$\frac{-X_n''}{X_n} = \frac{Y_n''}{Y_n} = \lambda_n, \qquad Y_n(0) = Y_n(M) = 0$$

This implies that

$$\lambda_n = -(n\pi/M)^2, \quad Y_n = A_n \sin(n\pi y/m), \quad X_n = B_n \sinh(n\pi x/M) + C_n \cosh(n\pi x/M)$$

We also use the formula $\sinh(\alpha + \beta) = \sinh\alpha \cosh\beta + \cosh\alpha \sinh\beta$. Finally, we obtain

$$u_\alpha(x, y) = \sum_{n=1}^{\infty} A_n \sin\frac{n\pi y}{M} \sinh\frac{n\pi}{M}(L - x)$$

which satisfies the PDE and the boundary conditions. Since $u_\alpha(0, y) = \alpha(y)$, we need to choose the coefficients A_n so that the Fourier-Euler coefficient formula holds,

$$A_n \sinh\frac{n\pi L}{M} = \frac{2}{M}\int_0^M \alpha(y) \sin\frac{n\pi y}{M}\, dy$$

(**b**). The problem is $u_{xx} + u_{yy} = 0$, $0 < x < L$, $0 < y < M$; $u(0, y) = \alpha(y)$, $0 < y < M$; $u(L, y) = \beta(y)$, $0 \le y \le M$; $u(x, 0) = \gamma(x)$, $0 \le x \le L$; $u(x, M) = \delta(x)$, $0 \le x \le L$. [See the end of the solution for the formulas for B_n, C_n, D_n.] Similar to (**a**), we can show that

$$u_\beta(x, y) = \sum_{n=1}^{\infty} B_n \sin\frac{n\pi y}{M} \sinh\frac{n\pi}{M}x$$

is the solution of $u_{xx} + u_{yy} = 0$ with boundary conditions, $u(0, y) = 0$, $u(L, y) = \beta(y)$, $0 \le y \le M$; $u(x, 0) = u(x, M) = 0$, $0 \le x \le L$.

The function

$$u_\gamma(x, y) = \sum_{n=1}^{\infty} C_n \sin \frac{n\pi x}{L} \sinh \frac{n\pi}{L}(M - y)$$

is the solution of $u_{xx} + u_{yy} = 0$, $u(0, y) = u(L, y) = 0$, $0 \le y \le M$; $u(x, 0) = \gamma(x)$, $u(x, M) = 0$, $0 \le x \le L$.

The function

$$u_\delta(x, y) = \sum_{n=1}^{\infty} D_n \sin \frac{n\pi x}{L} \sinh \frac{n\pi}{L} y$$

is the solution of $u_{xx} + u_{yy} = 0$, $u(0, y) = u(L, y) = 0$, $0 \le y \le M$; $u(x, 0) = 0$, $u(x, M) = \delta(x)$, $0 \le x \le L$.

The coefficients are given by the Fourier-Euler formulas,

$$B_n \sinh \frac{n\pi L}{M} = \frac{2}{M} \int_0^M \beta(y) \sin \frac{n\pi y}{M} \, dy, \qquad C_n \sinh \frac{n\pi M}{L} = \frac{2}{L} \int_0^L \gamma(x) \sin \frac{n\pi x}{L} \, dx$$

$$D_n \sinh \frac{n\pi M}{L} = \frac{2}{L} \int_0^L \delta(x) \sin \frac{n\pi x}{L} \, dx$$

So $u = u_\alpha + u_\beta + u_\gamma + u_\delta$ is the solution of $u_{xx} + u_{yy} = 0$ in G with the boundary conditions

$$u(0, y) = \alpha(y), \qquad u(L, y) = \beta(y), \quad 0 \le y \le M$$
$$u(x, 0) = \gamma(x), \qquad u(x, M) = \delta(x), \quad 0 \le x \le L$$

(c). Replace the condition $u(0, y) = \alpha(y)$ in part **(a)** by an insulation condition $u_x(0, y) = \tilde{\alpha}(y)$, $0 \le y \le M$. We know that the solution of $u_{xx} + u_{yy} = 0$ with boundary conditions

$$u_x(0, y) = \tilde{\alpha}(y), \qquad u(L, y) = 0, \quad 0 \le y \le M$$
$$u(x, 0) = 0, \qquad u(x, M) = 0, \quad 0 \le x \le L$$

is

$$u_{\tilde{\alpha}}(x, y) = \sum_{n=1}^{\infty} \tilde{A}_n \sin \frac{n\pi y}{M} \sinh \frac{n\pi}{M}(L - x)$$

where

$$\tilde{A}_n \cosh \frac{n\pi L}{M} = \frac{2}{M} \int_0^M \tilde{\alpha}(y) \sin \frac{n\pi y}{M} \, dy$$

(d). This is as in part **(a)** but with the perfect insulation conditions $u_y(x, 0) = u_y(x, M) = 0$, $0 \le x \le L$. We find that the solution of $u_{xx} + u_{yy} = 0$ with boundary temperatures

$$u(0, y) = \alpha(y), \qquad u(L, y) = 0, \quad 0 \le y \le M$$
$$u_y(x, 0) = 0, \qquad u_y(x, M) = 0, \quad 0 \le y \le L$$

is

$$u_\alpha(x, y) = \sum_{n=0}^{\infty} A_n \cos \frac{(2n+1)\pi y}{2M} \sinh \frac{(2n+1)\pi}{2M}(L - x)$$

where

$$A_n \sinh \frac{(2n+1)\pi L}{2M} = \frac{2}{M} \int_0^M \alpha(y) \cos \frac{(2n+1)\pi y}{2M} \, dy$$

2. The problem is $u_{rr} + u_r/r + u_{\theta\theta}/r^2 = 0$, $\rho < r < R$, $|\theta| \leq \pi$; $u(\rho, \theta) = f(\theta)$, $u(R, \theta) = g(\theta)$, $|\theta| \leq \pi$. Then following text formulas (9a), (9b), and (10), but keeping both terms in (9a) and (9b), we have

$$u(r, \theta) = A_0 + B_0 \ln r + \sum_{n=1}^{\infty} \{r^n(A_n \cos n\theta + A_n^* \sin n\theta) + r^{-n}(B_n \cos n\theta + B_n^* \sin n\theta)\}$$

Note the boundary conditions $u(\rho, \theta) = f(\theta)$ and $u(R, \theta) = g(\theta)$. So

$$A_0 + B_0 \ln \rho = \frac{1}{2\pi} \int_{-\pi}^{\pi} f(\theta) \, d\theta, \qquad A_0 + B_0 \ln R = \frac{1}{2\pi} \int_{-\pi}^{\pi} g(\theta) \, d\theta$$

$$A_n\rho^n + B_n\rho^{-n} = \frac{1}{\pi} \int_{-\pi}^{\pi} f(\theta) \cos n\theta \, d\theta, \qquad A_n^*\rho^n + B_n^*\rho^{-n} = \frac{1}{\pi} \int_{-\pi}^{\pi} f(\theta) \sin n\theta \, d\theta$$

$$A_n R^n + B_n R^{-n} = \frac{1}{\pi} \int_{-\pi}^{\pi} g(\theta) \cos n\theta \, d\theta, \qquad A_n^* R^n + B_n^* R^{-n} = \frac{1}{\pi} \int_{-\pi}^{\pi} g(\theta) \sin n\theta \, d\theta$$

Formulas for A_n, B_n, A_n^* and B_n^* can be obtained by solving these equations for these coefficients.

3. The problem is $u_{rr} + u_r/r + u_{\theta\theta}/r^2 = 0$, $0 < r < 1$, $|\theta| \leq \pi$; $u(1, \theta) = f(\theta)$. We also want to find the largest and smallest temperatures in the disk.

(a). Since $f(\theta) = 3 \sin \theta$, we have from text formulas (10) and (11) that $A_0 = 0$,

$$A_n = \frac{1}{\pi} \int_{-\pi}^{\pi} 3 \sin \theta \cos n\theta \, d\theta = 0 \qquad B_n = \frac{1}{\pi} \int_{-\pi}^{\pi} 3 \sin \theta \sin n\theta \, d\theta = \begin{cases} 3, & n = 1 \\ 0, & n > 1 \end{cases}$$

The solution is $u(r, \theta) = 3r \sin \theta$. The maximum and minimum temperatures in the unit disk are ± 3, respectively.

(b). If $f(\theta) = \theta + \pi$ for $-\pi \leq \theta < 0$, $f(\theta) = -\theta + \pi$ for $0 \leq \theta < \pi$, then

$$A_n = \frac{1}{\pi} \int_{-\pi}^{\pi} f(\theta) \cos n\theta \, d\theta = \begin{cases} \pi, & n = 0 \\ 2[1 - (-1)^n]/\pi n^2, & n > 0 \end{cases} \qquad B_n = \frac{1}{\pi} \int_{-\pi}^{\pi} f(\theta) \sin n\theta \, d\theta = 0$$

The solution is

$$u(r, \theta) = \pi/2 + \frac{4}{\pi} \sum_{\text{odd } n} \frac{1}{n^2} r^n \cos n\theta$$

Since the max and min temperatures are achieved on the perimeter of the disk, we have max $u = \max f = x$ and min $u = \min f = 0$.

(c). If $f(\theta) = 0$ for $-\pi < \theta < 0$, $f(\theta) = 1$ for $0 < \theta < \pi$, we have

$$A_n = \frac{1}{\pi} \int_{-\pi}^{\pi} f(\theta) \cos n\theta \, d\theta = \begin{cases} 1, & n = 0 \\ 0, & n > 0 \end{cases}$$

$$B_n = \frac{1}{\pi} \int_{-\pi}^{\pi} f(\theta) \sin n\theta \, d\theta = \frac{1 - (-1)^n}{n\pi} = \begin{cases} 2n/\pi, & n \text{ odd} \\ 0, & n \text{ even} \end{cases}$$

The solution is

$$u(r, \theta) = \frac{1}{2} + \frac{2}{\pi} \sum_{odd \ n} \frac{1}{n} r^n \sin n\theta$$

Since the highest and lowest temperatures must be reached on the perimeter, we have max $u = \max f = 1$ and min $u = \min f = 0$.

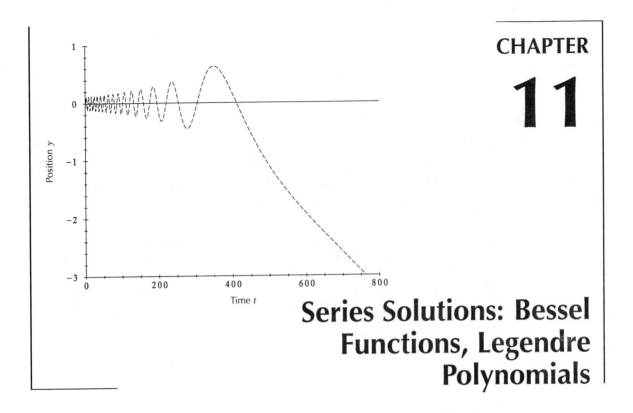

Series Solutions: Bessel Functions, Legendre Polynomials

Chapters 3 and 4 give the general theory, solution techniques, and properties of solutions for a scalar linear second-order ODE, but with little said about how to find solution formulas if the coefficients are nonconstant. The solutions of nonconstant-coefficient second-order linear ODEs cannot usually be found in closed form, but can be written as power series or "Frobenius" series, and that is the topic of this chapter. It is this kind of second-order ODE that often arises when separating variables in second-order PDEs written in polar, cylindrical, or spherical coordinates, and we will see some of these PDEs in Sections 11.1, 11.7, and 11.8. That is the reason we have placed this chapter here. On the other hand, aside from the PDE examples of steady temperature models, this chapter could be covered immediately following Chapter 4, and that is when we often cover it in our courses. In Section 11.1 (and Appendix B.2) we review properties of power series; of particular importance are the techniques of reindexing and the Identity Theorem. The notion of an ordinary point for a second-order linear ODE is introduced in Section 11.2, and the techniques needed for constructing convergent power series solutions in the neighborhood of an ordinary point are introduced via examples. The two best known (and widely applied) ODEs with nonconstant coefficients are the Legendre and the Bessel ODEs; these are treated in Sections 11.3, 11.6, and 11.7. Regular singular points are defined in Section 11.4 and discussed mostly in the context of the Euler ODEs. In Sections 11.5, 11.6, and 11.7 the methods of Frobenius are introduced in order to construct series solutions near regular singular points. These methods are straightforward, but involve much algebraic manipulation; a CAS may be appropriate here. To provide an alternative application of series techniques, we model the behavior of an aging spring in Section 11.1 by a second-order linear ODE with nonconstant coefficients and then return to that model in Sections 11.2 and 11.7.

The material of this section is now "classical" in the sense that it was largely completed by the end of the 19th century. It still is widely used in the applications of ODEs, especially when working with those ODEs that arise when variables are separated in a PDE.

11.1   Aging Springs and Steady Temperatures

Suggestions for preparing a lecture

Topics: A review of power series $\sum a_n x^n$, an example of how they are used to solve second-order linear ODEs, the idea of a recursion formula, and maybe the aging spring model to motivate the study of nonconstant coefficient linear ODEs. If your students are to do the problems by "hand" (rather than using a CAS), you might emphasize the need for care in doing the algebra.

Making up a problem set

No one of these problems stands out, except maybe for the group project on the aging spring (Problem 11). The project is not designed to enhance the students' ability to handle series, but rather to help them understand how one might go about modeling an aging spring and then using a numerical solver to explore the solutions of the corresponding ODEs. Other than that, we often assign a graphical problem (e.g., 2(**b**) or 7) and a small number of "crank" problems selected from Problems 1–10, a small number because the algebra can become tedious. As the students soon discover (and the instructors know from experience), the slightest algebraic mistake anywhere can lead to wildly wrong answers when trying to find a series expansion. This can be very discouraging to students, instructors, and paper graders—which is why we like to have students do just a few problems, but get them right.

Comments

Series expansions are central in analysis and, for at least 150 years, in ODEs as well. In this section we take the first steps in developing series techniques for solving a second-order linear ODE in y whose coefficients are polynomials or convergent power series in the independent variable, expressing the solution as a series in powers of the independent variable (usually x in this chapter, but occasionally t as in the aging spring model). The reason for this choice of variables is mainly historical: those were the variables used when the PDEs for heat flow in a rod and the displacement of a vibrating string were originally reduced to ODEs by the technique of separating variables (Chapter 10).

Series techniques are used throughout the chapter, so it is a good idea to brush up on these techniques; power series, Taylor series, the Ratio Test, geometric series, reindexing, and the Identity Theorem are given in Appendix B.2. A good CAS will handle much of the necessary bookkeeping when working with series and may even express the solution of an ODE directly in series form. However, it is good to do some of the problems "by hand" to solidify understanding of how series can be manipulated. We introduce the models of the aging spring and of steady-temperatures in a solid cylinder to show that series solutions can be used to solve "real" problems. We return to the aging spring problem in Sections 11.2 and 11.7 and the temperature problem in Section 11.8 and in Problem 4 of Section 11.3.

---

1.   Apply the Ratio Test (Appendix B.2, item 3) to each series to find the interval of convergence.

(a). The series is $\sum x^n/n!$ and the ratio of successive terms is

$$\left| \frac{x^{n+1}/(n+1)!}{x^n/n!} \right| = \frac{1}{n+1}|x| \to 0 \quad \text{as} \quad n \to \infty$$

so the interval of convergence is $-\infty < x < \infty$.

(b). Here the series is $\sum nx^n/2^n$ and the ratio of successive terms is

$$\left|\frac{(n+1)x^{n+1}/2^{n+1}}{nx^n/2^n}\right| = \frac{n+1}{2n}|x| \to \frac{1}{2}|x| \quad \text{as} \quad n \to \infty$$

so the series converges for $|x| < 2$.

(c). In this case the series is $\sum x^{2n+1}/(2n+1)!$ and the ratio of successive terms is

$$\left|\frac{x^{2n+3}/(2n+3)!}{x^{2n+1}/(2n+1)!}\right| = \frac{1}{(2n+3)(2n+2)}|x^2| \to 0 \quad \text{as} \quad n \to \infty$$

so the convergence interval is $-\infty < x < \infty$.

2. The graph of the function is the solid curve; the sums of the first 2, 3, and 4 nonzero terms of the polynomial approximations are shown, respectively, by the long-dashed, short-dashed, and long/short-dashed curves.

(a). The Taylor series about $x_0 = 0$ for e^x is $\sum_0^\infty x^n/n!$ since $D^n e^x = e^x$ and $e^0 = 1$. The Ratio Test shows that the series converges for all x. See Fig. 2(a).

(b). The Taylor series for $\sin x$ about $x_0 = 0$ is $\sum_{k=0}^\infty (-1)^k/[(2k+1)!]x^{2k+1}$ since $D^j \sin x = \pm \sin x$ if j is even, $\pm \cos x$ if j is odd, and $\sin 0 = 0$, $\cos 0 = 1$. The Ratio Test shows that the series converges for all x. See Fig. 2(b).

(c). We can find the Taylor series for $1/(1+x)$ about $x_0 = 0$ either from the Taylor formula for $x_0 = 0$, or more easily from the geometric series expansion $1/(1-r) = 1 + r + \cdots + r^n + \cdots$, where $r = -x$. So $1/(1+x) = \sum_{n=0}^\infty (-1)^n x^n$. The Ratio Test implies that the series converges for $|x| < 1$. See Fig. 2(c).

(d). Since $\sqrt{x} = (1 + (x-1))^{1/2}$, we may use the Binomial Theorem (Appendix B.2, item 9) to obtain the Taylor series about $x_0 = 1$:

$$(1 + (x-1))^{1/2} = 1 + \frac{1}{2}(x-1) + \frac{1}{2} \cdot (-\frac{1}{2}) \cdot \frac{1}{2!}(x-1)^2 + \cdots$$

$$+ \frac{1}{2}(-\frac{1}{2}) \cdots (\frac{1}{2} - n + 1)\frac{1}{n!}(x-1)^n + \cdots$$

Applying the Ratio Test, we have that the magnitude of the ratio of the $(n+1)$st and the nth term is $|(1 - 2n)(x-1)/(2n+2)|$ which tends to $|x - 1|$ as $n \to \infty$. So the interval of convergence is defined by $|x - 1| < 1$, that is, $0 < x < 2$. See Fig. 2(d).

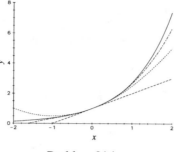

Problem 2(a).

Problem 2(b).

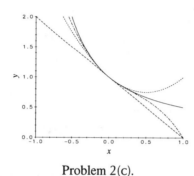

Problem 2(c).

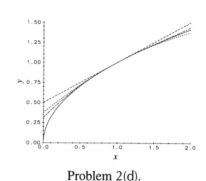

Problem 2(d).

3. In each case we have to lower or raise the power of x and (to compensate) raise or lower the bottom value of the range of the index by the same amount.

(a). The power of x must be lowered one step, so the lower limit of the range must be raised one step: $\sum_{n=0}^{\infty} 2(n+1)x^{n+1}/n! = \sum_{n=1}^{\infty} 2nx^n/(n-1)!$.

(b). The power of x should be raised two steps and the lower limit of the range accordingly lowered: $\sum_{n=2}^{\infty} n(n-1)a_n x^{n-2} = \sum_{n=0}^{\infty} (n+2)(n+1)a_{n+2}x^n$.

(c). The power of x must be lowered one unit and the lower range on the sum raised one unit: $\sum_{n=1}^{\infty}(-1)^{n-1}x^{n+1}/(n(n+1)) = \sum_{n=2}^{\infty}(-1)^n x^n/(n(n-1))$. Note that $(-1)^{n-2} = (-1)^n$, and we have selected the simpler form.

4. **(a).** $e^x + e^{-x} = \sum_{n=0}^{\infty} x^n/n! + \sum_{n=0}^{\infty}(-x)^n/n! = \sum_{k=0}^{\infty} 2x^{2k}/(2k)!$ for all x, since the odd-indexed terms cancel.

(b). $\sin x - \cos x = \sum_{k=0}^{\infty}(-1)^k x^{2k+1}/(2k+1)! + \sum_{k=0}^{\infty}(-1)^{k+1}x^{2k}/(2k)! = -1 + x + x^2/2 - x^3/6 - \cdots$, for all x.

(c). Since $\sin x \cos x = (\sin 2x)/2$, we have the Taylor series about $x_0 = 0$

$$\frac{1}{2}\sum_{k=0}^{\infty}(-1)^k (2x)^{2k+1}/(2k+1)!$$

for all x.

(d). According to 2**(c)** (or item 9 in Appendix B.2 with x replaced by $-x$), $1/(1+x) = \sum_{n=0}^{\infty}(-1)^n x^n$. Since $e^x = \sum_{n=0}^{\infty} x^n/n!$, we have that the series for the product (by the Product Theorem, item 6(ii) in Appendix B.2)

$$\frac{e^x}{1+x} = \sum_{n=0}^{\infty}\left(\sum_{k=0}^{n}(-1)^k \frac{1}{(n-k)!}\right)x^n$$

5. In each case we have $\sum_n[\cdot]x^n = 0$, so by the Identity Theorem (Appendix B.2, item 5), we must have $[\cdot] = 0$ for each n. This leads to the desired recursion formula.

   (a). By the Identity Theorem, $na_n - n + 2 = 0$ for $n = 1, 2, \ldots$, so $a_n = 1 - 2/n$.

   (b). Here we must have $(n+1)a_n = a_{n-1}$, or $a_n = a_{n-1}/(n+1)$ for $n = 1, 2, \ldots$. So

   $$a_n = \frac{1}{n+1}a_{n-1} = \frac{1}{n+1}\left(\frac{1}{n}a_{n-2}\right) = \cdots = \frac{1}{n+1}\cdot\frac{1}{n}\cdot\frac{1}{n-1}\cdots\frac{1}{2}a_0 = \frac{1}{(n+1)!}a_0$$

   (c). In this case, we have $(n+2)(n+1)a_{n+2} = a_n$, $n = 0, 1, 2, \ldots$. So

   $$a_{n+2} = \frac{1}{(n+2)(n+1)}a_n = \frac{1}{(n+2)(n+1)}\left(\frac{1}{n(n-1)}a_{n-2}\right) = \cdots$$

   $$= \frac{1}{n+2}\cdot\frac{1}{n+1}\cdot\frac{1}{n}\cdot\frac{1}{n-1}\cdots\frac{1}{2}a_0$$

   $$= \frac{1}{(n+2)!}a_0$$

   if n is even. If n is odd, we get $a_{n+2} = a_1/(n+2)!$. In any case a_n is a_0 or a_1 divided by $n!$ if $n \geq 2$.

6. We want to show that $y = \sum_0^{\infty} x^n/n!$ solves $y'' - y = 0$, $y(0) = 1$, $y'(0) = 1$. If $y = \sum_{n=0}^{\infty} x^n/n!$, then $y'' = \sum_{n=2}^{\infty} n(n-1)x^{n-2}/n! = \sum_{n=2}^{\infty} x^{n-2}/(n-2)!$ since the series converges for all x (by the Ratio Test), so can be differentiated term by term. (Note the first two terms in the series for y'' have been dropped since they are zero.) Then, $y'' - y = \sum_{n=2}^{\infty} x^{n-2}/(n-2)! - \sum_{n=0}^{\infty} x^n/n!$. Reindexing the first series, we have that $y'' - y = \sum_{n=0}^{\infty} x^n/n! - \sum_{n=0}^{\infty} x^n/n! = 0$ for all x. Since the exact solution is e^x, we see that $e^x = \sum_{n=0}^{\infty} x^n/n!$, which is hardly surprising since that is the Taylor series for e^x about $x_0 = 0$.

7. We want to use a series $y = \sum_0^{\infty} a_n x^n$ to solve the IVP, $y'' - 4y = 0$, $y(0) = 2$, $y'(0) = 2$.

   (a). Let $y = \sum_{n=0}^{\infty} a_n x^n$. Then $y'' = \sum_{n=2}^{\infty} n(n-1)a_n x^{n-2}$, which is $\sum_{n=0}^{\infty}(n+2)(n+1)a_{n+2}x^n$ after reindexing. So $y'' - 4y = \sum_{n=0}^{\infty}[(n+2)(n+1)a_{n+2} - 4a_n]x^n = 0$. By the Identity Theorem, $(n+2)(n+1)a_{n+2} = 4a_n$, $n = 0, 1, 2, \ldots$. Since y has the form $y = a_0 + a_1 x + a_2 x^2 + \cdots$ and we require that $y(0) = 2$ and $y'(0) = 2$, we might as well set $a_0 = 2$, $a_1 = 2$ to begin with. Then from the recursion identity, $a_{n+2} = 4a_n/(n+2)(n+1)$, we see (much as in 5**(c)**), that $a_n = 2^n a_0/n!$ if n is even, and $2^{n-1}a_1/n!$ if n is odd, where $n > 1$. The solution is

   $$y = \sum_{n=0}^{\infty}\frac{(2x)^{2n+1}}{(2n+1)!} + 2\sum_{n=0}^{\infty}\frac{(2x)^{2n}}{(2n)!} = \sum_{k=0}^{\infty}\frac{1}{k!}\left(\frac{3}{2}+\frac{1}{2}(-1)^k\right)(2x)^k$$

The exact solution is easily seen to be $y(x) = (3e^{2x} + e^{-2x})/2$, by the methods of Section 3.4. What we have found is the Taylor series for $(3e^{2x} + e^{-2x})/2$.

(b). See Fig. 7(b) where the solid curve is the graph of $y(x)$, and the long-dashed curve is the graph of the sum of the first two terms of the series; the graphs of the sums of the first three and four terms are hardly distinguishable from the graph of $y(x)$ to the accuracy of this plot.

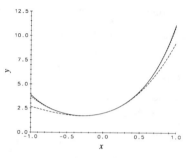

Problem 7(b).

8. The IVP is $y'' - y = \sum_0^\infty x^n$, $y(0) = 0$, $y'(0) = 0$. Since $y(0) = 0$ and $y'(0) = 0$, a solution of the form $y = a_0 + a_1 x + a_2 x^2 + \cdots$ must have $a_0 = a_1 = 0$. So $y = \sum_{n=2}^\infty a_n x^n$. We have $y'' = \sum_{n=2}^\infty n(n-1)a_n x^{n-2} = \sum_{n=0}^\infty (n+2)(n+1)a_{n+2}x^n$ (after reindexing), and $y'' - y = \sum_{n=0}^\infty x^n$ becomes $\sum_{n=0}^\infty [(n+2)(n+1)a_{n+2} - a_n - 1]x^n = 0$. The recursion identity is $a_{n+2} = (1 + a_n)/(n+2)(n+1)$, $n \geq 0$. Since $a_0 = a_1 = 0$, we have $a_2 = 1/2$, $a_3 = 1/6$, $a_4 = 3/4!$, $a_5 = 7/5!$, ... In general, $a_n = \sum_{k=1}^{[n/2]} (n - 2k)!/n!$, $n \geq 4$, where $[n/2]$ is the integer part of the number $n/2$. The solution is $y = \sum_2^\infty a_n x^n$, with a_n given above.

9. We want to find a_{100} if it is known that $\sum_0^\infty [(n+1)^2 a_{n+2} - n^2 a_{n+1} + (n-1)a_n]x^n = 0$. The recursion identity here is $(n+1)^2 a_{n+2} - n^2 a_{n+1} + (n-1)a_n = 0$, that is, $a_{n+2} = [n^2 a_{n+1} - (n-1)a_n]/(n+1)^2$, $n = 0, 1, 2, \ldots$, where $a_0 = a_1 = 1$. A computer may easily be programmed to calculate the coefficients using the identity and the initial data $a_0 = a_1 = 1$. In fact, $a_{100} \approx -1.0629 \times 10^{-6}$.

10. The IVP to be solved is $rR'' + R' + rR = 0$, $R(0) = 1$, $R'(0) = 0$. Let $R(r) = \sum_{n=0}^\infty b_n r^n$. Differentiating and reindexing, we have that $R'(r) = \sum_{n=0}^\infty (n+1)b_{n+1}r^n$ and $R''(r) = \sum_{n=0}^\infty (n+2)(n+1)b_{n+2}r^n$. Inserting into the ODE and switching the order of the second and third terms, we have

$$\sum_{n=0}^\infty [(n+2)(n+1)b_{n+2} + b_n]r^{n+1} + \sum_{n=0}^\infty (n+1)b_{n+1}r^n = 0$$

After reindexing the last summation by lowering the index range by 1 to obtain $b_1 +$

$\sum_{n=0}^{\infty}(n+2)b_{n+2}r^{n+1}$, we have

$$b_1 + \sum_{n=0}^{\infty}[(n+2)(n+1)b_{n+2} + b_n + (n+2)b_{n+2}]r^{n+1} = 0$$

So we have

$$(n+2)(n+1)b_{n+2} + b_n + (n+2)b_{n+2} = (n+2)^2 b_{n+2} + b_n = 0$$

which reduces to the recursion formula

$$b_{n+2} = -b_n/(n+2)^2$$

We know from the initial data that $b_0 = 1$ and $b_1 = 0$. From the recursion formula we see that $b_3 = b_5 = b_7 = b_9 = \cdots = 0$ since $b_1 = 0$. We also have $b_2 = -b_0/2^2$, $b_4 = b_0/(2^2 \cdot 4^2)$, and

$$b_6 = -\frac{1}{6^2} \cdot \frac{1}{4^2} \cdot \frac{1}{2^2}b_0, \cdots, b_{2n} = (-1)^n\frac{1}{[2 \cdot 4 \cdots (2n)]^2}b_0 = (-1)^n b_0/(2^n n!)^2$$

Since $b_0 = 1$,

$$R(r) = 1 - \frac{r^2}{2^2} + \frac{r^4}{(2 \cdot 4)^2} - \frac{r^6}{(2 \cdot 4 \cdot 6)^2} + \cdots.$$

11. Group project.

11.2 Series Solutions Near an Ordinary Point

Suggestions for preparing a lecture

Topics: Ordinary points, singular points, find the series solution about an ordinary point in a particular case (emphasize the need for algebraic care), compare the graph of the solution with the graphs of partial sums of the series solution.

Making up a problem set

Problems 3(**c**) (because it isn't too hard, or too easy), 4, 8. **Warning:** Before making any assignment, you might want to assess the level of algebraic complications in arriving at the solution of each proposed problem, so avoid requiring students to do too many problems with a lot of algebraic steps.

Comments

Ordinary points and singular points are defined for the ODE $y'' + P(x)y' + Q(x)y = 0$, and power series solutions are constructed in powers of $x - x_0$, where x_0 is an ordinary point. In all the examples and in the problems, x_0 is always taken to be zero for simplicity; in any case the variable change $z = x - x_0$ could be applied. Four of the famous "named" ODEs of classical analysis are introduced in the text and the problem set: the Legendre, Hermite (a particular case), Airy, and Mathieu ODEs. A series solution of the ODE for the aging spring is obtained and its usefulness examined. The Convergence Theorem for the series solutions about an ordinary point is presented, but not proved. In the text and in the problems when dealing with a specific power series we rely on the Ratio Test, rather than the Convergence Theorem, to determine intervals of convergence. Lots of interesting material here, but it may seem to be endless calculations with series. As in the first section, one algebraic mistake in working out the coefficients of a series solution, and the rest of the calculations are doomed. We think that every reader should carry through "by hand" the calculations of at least one of the series solutions, so learning the need for careful bookkeeping and methodical procedure, especially in reindexing and obtaining the recursion formula for the coefficients.

1. The definitions of ordinary point and singular point refer to the ODE $y'' + P(x)y' + Q(x)y = 0$.

    (a). The ODE is $y'' + k^2 y = 0$. Since $P(x) = 0$ and $Q(x) = k^2$ are real analytic at every point, every point x_0 is ordinary.

    (b). The ODE is $y'' + y/(1+x) = 0$. Here $P(x) = 0$, $Q(x) = (1+x)^{-1}$. Since $Q(x_0)$ is undefined at $x_0 = -1$, $Q(x)$ is not real analytic at x_0; $x_0 = -1$ is a singular point.

    (c). The ODE is $y'' + y'/x - y = 0$. Since $P(x) = 1/x$ is undefined at $x_0 = 0$, $P(x)$ is not real analytic at $x_0 = 0$, so 0 is a singular point.

2. The definitions of ordinary point and singular point refer to the ODE $y'' + P(x)y' + Q(x)y = 0$.

    (a). The ODE is $y'' + (\sin x)y' + (\cos x)y = 0$. Since $\sin x$ and $\cos x$ are real analytic for all x, every point is ordinary and there are no singular points.

    (b). The ODE is $y'' - (\ln|x|)y = 0$. The Taylor series for $\ln|x|$ at $x_0 \neq 0$ is $\ln|x_0| + \sum_{n=1}^{\infty} \frac{(-1)^{n+1}}{n}(\frac{x-x_0}{x_0})^n$ which converges for all x between 0 and $2x_0$ by the Ratio Test. Since $\ln 0$ is undefined, however, we see that $x_0 = 0$ is a singular point.

    (c). The ODE is $y'' + |x|y' + e^{-x}y = 0$. The coefficient function $|x|$ is not real analytic at $x = 0$ since it is not differentiable there. It is real analytic at all other values of x. The function e^{-x} is real analytic everywhere so $x_0 = 0$ is the only singular point.

    (d). The ODE is $(1-x)y'' + y' + (1-x^2)y = 0$. After normalizing the ODE, we see that $P(x) = (1-x)^{-1}$ and $Q(x) = (1-x^2)(1-x)^{-1} = 1 + x$. $P(x)$ is real analytic at every x except $x_0 = 1$ where $P(x)$ is not defined. For $x_0 \neq 1$, the Taylor series of $P(x)$ about x_0 is $\sum_{n=0}^{\infty} \frac{1}{(1-x_0)^{n+1}}(x - x_0)^n$, which converges for $x_0 - r < x < x_0 + r$, where $r = |1 - x_0|$. $Q(x)$ is real analytic for all x. The only singularity is $x_0 = 1$.

    (e). The ODE is $(1-x^2)y'' + xy' + y = 0$. After normalizing the ODE, we see that $P(x) = (1-x^2)^{-1}x$ and $Q(x) = (1-x^2)^{-1}$. Both $P(x)$ and $Q(x)$ are real analytic at every x_0, except $x_0 = \pm 1$ which are the two singularities. Rather than prove the real analyticity of $P(x)$ at $x_0 \neq \pm 1$ by constructing its Taylor series at x_0 and showing that it has a nontrivial interval of convergence, it is easier to cite a very general result: the quotient of two polynomials is real analytic at every x_0, unless x_0 is a root of the denominator. So $P(x)$ and $Q(x)$ are real analytic for all x, $x \neq \pm 1$. The only singular points are $x = \pm 1$.

    (f). The ODE is $x^2 y'' + xy' - y = 0$. After normalizing the ODE, we see that $P(x) = x^{-1}$ and $Q(x) = -x^{-2}$, which are real analytic at all $x_0 \neq 0$. So $x_0 = 0$ is the only singular point.

3. For **(a)**, **(b)** and **(c)**, the functions $P(x)$ and $Q(x)$ are polynomials, so real analytic for all x. For part **(d)**, the coefficient $x^2 + 1$ of y'' has no real roots; so $P(x)$ and $Q(x)$ are again real analytic for each x. In every case the initial conditions are $y(0) = 1$, $y'(0) = 1$.

(a). The ODE is $y'' + 4y = 0$. Here the most direct way is to note that the general solution is

$$y = C_1 \cos 2x + C_2 \sin 2x$$

$$= C_1 \sum_{k=0}^{\infty} [(-1)^k/(2k)!](2x)^{2k} + C_2 \sum_{k=0}^{\infty} [(-1)^k/(2k+1)!](2x)^{2k+1}$$

where we have used the Taylor series for $\cos 2x$ and $\sin 2x$ about $x_0 = 0$. By the Ratio Test, for the cosine series,

$$\lim_{n\to\infty} \left| \frac{(2x)^{2n+2}(2n)!}{(2x)^{2n}(2n+2)!} \right| = \lim_{n\to\infty} \left| \frac{(2x)^2}{(2n+2)(2n+1)} \right| = 0$$

The series converges for all x. The Ratio Test gives the same result for the series for $\sin(2x)$. The solution of the IVP is $y = \cos x + (\sin x)/2$. See Fig. 3(a).

(b). The ODE is $y'' - 4y = 0$. We have the general solution $y = C_1 e^{2x} + C_2 e^{-2x} = C_1 \sum_{n=0}^{\infty} 2^n x^n/n! + C_2 \sum_{n=0}^{\infty} (-1)^n 2^n x^n/n!$, where each series converges for all x by the Ratio Test. The solution of the IVP is $y = (3e^{2x} + e^{-2x})/4$. See Fig. 3(b).

(c). The ODE is $y'' + xy' + 3y = 0$. Let the general solution be $y = \sum_{n=0}^{\infty} a_n x^n$, so $y' = \sum_{n=1}^{\infty} n a_n x^{n-1}$, $y'' = \sum_{n=2}^{\infty} n(n-1) a_n x^{n-2}$. Then $y'' + xy' + 3y = 0$ becomes

$$\sum_{n=2}^{\infty} n(n-1) a_n x^{n-2} + \sum_{n=1}^{\infty} n a_n x^n + \sum_{n=0}^{\infty} 3 a_n x^n = 0$$

In order to include all three sums in a single summation, we must reindex the first sum and lower the range on the second from 1 to 0 (easy to do since $n = 0$ only gives the term 0 in the second sum):

$$\sum_{n=0}^{\infty} [(n+2)(n+1) a_{n+2} + n a_n + 3 a_n] x^n = 0$$

By the Identity Theorem, $a_{n+2} = -(n+3) a_n / [(n+2)(n+1)]$. If n is even [odd], we can push this recursion formula for the coefficients back to a_0 [a_1]. This is most easily done by considering the two cases separately, first n even ($n = 2k$) and then n odd ($n = 2k+1$).

$$a_{2k} = -\frac{2k+1}{(2k)(2k-1)} a_{2k-2} = \left(-\frac{2k+1}{(2k)(2k-1)} \right) \left(-\frac{2k-1}{(2k-2)(2k-3)} a_{2k-4} \right)$$

$$= \cdots = \frac{(-1)^k (2k+1)(2k-1)\cdots 3}{(2k)!} a_0$$

Since

$$(2k+1)(2k-1)\cdots 3 = \frac{(2k+1)(2k)(2k-1)(2k-2)\cdots 3\cdot 2}{(2k)(2k-2)\cdots 2} = \frac{(2k+1)!}{2^k k!}$$

we have that

$$a_{2k} = (-1)^k \frac{(2k+1)!}{(2k)!2^k k!} = (-1)^k \frac{2k+1}{2^k k!} a_0$$

In exactly the same way, it may be shown that

$$a_{2k+1} = -\frac{2k+2}{(2k+1)(2k)} a_{2k-1} = \cdots = (-1)^k \frac{(2k+2)(2k)\cdots 4}{(2k+1)!} a_1 = (-1)^k \frac{2^k (k+1)!}{(2k+1)!} a_1$$

We may write the general solution as

$$y = a_0 \sum_{k=0}^{\infty} (-1)^k \frac{(2k+1)}{2^k k!} x^{2k} + a_1 \sum_{k=0}^{\infty} (-1)^k \frac{2^k (k+1)!}{(2k+1)!} x^{2k+1}$$

where a_0 and a_1 are arbitrary constants. We may test the convergence of the series by the Ratio Test. For the first series we have that the magnitude of the ratio of successive terms is $|(2k+3)x^2/[2(k+1)(2k+1)]|$, which $\to 0$ as $k \to \infty$. The series converges for all x. It may be shown in much the same way that the second series also converges for all x. Since $y(0) = 1$ and $y'(0) = 1$, we set $a_0 = a_1 = 1$ to get the solution that meets the initial conditions. See Fig. 3(c).

(d). The ODE is $(x^2 + 1)y'' - 6y = 0$. Let $y = \sum_{n=0}^{\infty} a_n x^n$. The initial data imply that $a_0 = a_1 = 1$. We have that $y'' = \sum_{n=2}^{\infty} n(n-1)a_n x^{n-2} = \sum_{n=0}^{\infty} (n+2)(n+1)a_{n+2} x^n$. So $(x^2 + 1)y'' - 6y = \sum_{n=2}^{\infty} n(n-1)a_n x^n + \sum_{n=0}^{\infty} (n+2)(n+1)a_{n+2} x^n + \sum_{n=0}^{\infty} (-6a_n)x^n$. We have that

$$(2a_2 - 6a_0) + (6a_3 - 6a_1)x + \sum_{n=2}^{\infty} [(n+2)(n+1)a_{n+2} + n(n-1)a_n - 6a_n]x^n = 0.$$

By the Identity Theorem $a_2 = 3$, $a_3 = 1$ (since $a_0 = a_1 = 1$) and

$$a_{n+2} = -\frac{n^2 - n - 6}{(n+2)(n+1)} a_n = -\frac{n-3}{n+1} a_n$$

As in 3(c), we consider two cases: $n = 2k$ (even n) and $n = 2k + 1$ (odd n), where $n \geq 2$. We have that for $k \geq 2$

$$a_{2k} = -\frac{2k-5}{2k-1} a_{2k-2} = \cdots = (-1)^{k-1} \frac{(2k-5)(2k-7)\cdots(-1)}{(2k-1)(2k-3)\cdots 3} a_2$$

$$= (-1)^k \frac{3}{(2k-1)(2k-3)};$$

$$a_{2k+1} = -\frac{2k-4}{2k} a_{2k-1} = \cdots = (-1)^{k-1} \frac{(2k-4)(2k-6)\cdots 0}{2k(2k-2)\cdots 4} a_3$$

$$= 0$$

The solution is

$$y = 1 + x + 3x^2 + x^3 + 3 \sum_{k=2}^{\infty} \frac{(-1)^k}{(2k-1)(2k-3)} x^{2k}$$

To check for convergence we use the Ratio Test. The magnitude of the ratio of successive terms is $|\frac{2k-3}{2k+1}x^2|$ which $\to x^2$ as $k \to \infty$, so the series converges for $|x| < 1$. See Fig. 3(d). Note that the solution itself seems to be defined for values of x outside the interval of convergence of the series. This often happens; the series expansions may represent only a "part" of the solution.

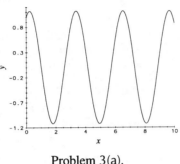

Problem 3(a).

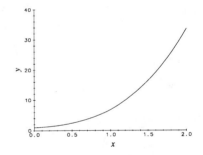

Problem 3(b).

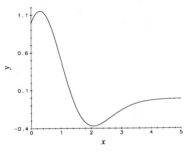

Problem 3(c).

Problem 3(d).

4. **(a).** Airy's ODE is $y'' - xy = 0$. Let $y = \sum_{n=0}^{\infty} a_n x^n$, $y'' = \sum_{n=2}^{\infty} n(n-1)a_n x^{n-2}$, $xy = \sum_{n=0}^{\infty} a_n x^{n+1}$. So

$$y'' - xy = \sum_{n=1}^{\infty} [(n+2)(n+1)a_{n+2} - a_{n-1}]x^n + 2a_2 = 0$$

where the term $2a_2$ has been split off from the summation since the new range is from 1 to ∞ rather than from 0 to ∞. By the Identity Theorem we see that $a_2 = 0$ and that

$$a_{n+2} = \frac{a_{n-1}}{(n+2)(n+1)}, \quad n \geq 1$$

Since $a_2 = 0$, we have $a_5 = a_8 = \cdots = a_{3k-1} = \cdots = 0$. From the above recursion formula we can express coefficients of the form a_{3k} [a_{3k+1}] as multiples of a_0 [a_1]. The general solution is

$$y = a_0 \left[1 + \sum_{k=1}^{\infty} \frac{x^{3k}}{(3k)(3k-1)(3k-3)(3k-4)\cdots 3 \cdot 2} \right]$$

$$+ a_1 \left[x + \sum_{k=1}^{\infty} \frac{x^{3k+1}}{(3k+1)(3k)(3k-2)(3k-3)\cdots 4 \cdot 3} \right]$$

where a_0 and a_1 are arbitrary. The Ratio Test may be used to show that each series converges for all x.

(b). See Fig. 4(b), where a numerical solver has been used to solve the IVP, $y'' - xy = 0$, $y(0) = 1$, $y'(0) = 0$.

(c). We have only plotted the sum of the first two terms and of the first four terms of the series expansion given in part **(a)** with $a_0 = 1$, $a_1 = 0$ in Fig. 4(c). Note how poor these approximations are for $x < -3$.

(d). From Fig. 4(b), we see that the solution of $y'' - xy = 0$, $y(0) = 1$, $y'(0) = 0$ is strictly increasing for $x > 0$ and oscillatory for $x < 0$. Is there a plausibility argument that this behavior should be expected? One way is to pretend that for all negative (or positive) x, the coefficient x is a constant with the same sign as x. The IVPs, $y'' + \alpha y = 0$ and $y'' - \alpha y = 0$, $y(0) = 1$, $y'(0) = 0$, α a positive constant, have the unique solutions $y = \cos \sqrt{\alpha} x$ and $y = (e^{\sqrt{\alpha} x} + e^{-\sqrt{\alpha} x})/2$, respectively. The function $y = \cos \sqrt{\alpha} x$, $x \leq 0$, is oscillatory, and $y = (e^{\sqrt{\alpha} x} + e^{-\sqrt{\alpha} x})/2$, $x \geq 0$, is strictly increasing. So we may expect that the solution of $y'' - xy = 0$, $y(0) = 1$, $y'(0) = 0$ is oscillatory for $x < 0$ and strictly increasing for $x > 0$.

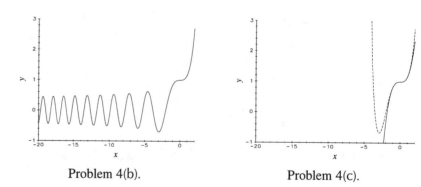

Problem 4(b).　　　　　　Problem 4(c).

5. The Matthieu equation is $y'' + (a + b \cos \omega x)y = 0$. Set $a = 1$, $b = 2$, $\omega = 1$, and $y(0) = 1$, $y'(0) = 0$. Since we only want the first few terms, we shall not attempt to find the general coefficient of the series solution. The initial data imply that $a_0 = 1$ and $a_1 = 0$ if $y = a_0 + a_1 x + a_2 x^2 + \cdots$. Since $\cos x$ has Taylor series $1 - x^2/2! + x^4/4! - x^6/6! + \cdots$ about $x_0 = 0$, we see that inserting the series for y and for $\cos x$ in the ODE and using item 6(ii) of Appendix B.2 to determine the first few terms of the product series for $(2 \cos x)y(x)$, we have that

$$0 = y'' + (1 + 2\cos x)y = (3 + 2a_2) + 6a_3 x + (12a_4 + 3a_2 - 1)x^2 + (3a_3 + 20a_5)x^3$$

$$+ (30a_6 + 3a_4 - a_2 + \frac{1}{12})x^4 + \cdots$$

So $a_2 = -3/2$, $a_3 = 0$, $a_4 = 13/48$, $a_5 = 0$, $a_6 = -23/288$, and the series for the solution is

$$y = 1 - \frac{3}{2}x^2 + \frac{13}{48}x^4 - \frac{23}{288}x^6 + \cdots$$

6. **(a).** The IVP is $y' = 1 + xy^2$, $y(0) = 0$. Let $y = a_0 + a_1 x + a_2 x^2 + \cdots$. Then $a_0 = 0$ since $y(0) = 0$, $y' = a_1 + 2a_2 x + 3a_3 x^2 + \cdots$, and $xy^2 = x(x^2)(a_1 + a_2 x + a_3 x^2 + a_4 x^3 + \cdots)^2 = x^3[a_1^2 + 2a_1 a_2 x + (a_2^2 + 2a_1 a_3)x^2 + 2(a_1 a_4 + a_2 a_3)x^3 + \cdots]$. So

$$0 = y' - 1 - xy^2 = (a_1 - 1) + 2a_2 x + 3a_3 x^2 + (4a_4 - a_1^2)x^3$$
$$+ (5a_5 - 2a_1 a_2)x^4 + (6a_6 - a_2^2 - 2a_1 a_3)x^5$$
$$+ (7a_7 - 2a_1 a_4 - 2a_2 a_3)x^6 + \cdots$$

By the Identity Theorem, $a_1 = 1$, $a_2 = a_3 = 0$, $a_4 = 1/4$, $a_5 = a_6 = 0$, $a_7 = 1/14$, $\cdots$, and

$$y = x + \frac{1}{4}x^4 + \frac{1}{14}x^7 + \cdots.$$

(b). The IVP is $y' = x^2 + y^2$, $y(0) = 0$. Since $y(0) = 0$, we have that $y = a_1 x + a_2 x^2 + \cdots$, $y' = a_1 + 2a_2 x + 3a_3 x^2 + \cdots$, and $y^2 = x^2(a_1 + a_2 x + \cdots)^2 = x^2(a_1^2 + 2a_1 a_2 x + (a_2^2 + 2a_1 a_3)x^2 + 2(a_1 a_4 + a_2 a_3)x^3 + (2a_1 a_5 + 2a_2 a_4 + a_3^2)x^4 + 2(a_1 a_6 + a_2 a_5 + a_3 a_4)x^5 + \cdots)$. We have that

$$0 = y' - x^2 - y^2 = a_1 + 2a_2 x + (3a_3 - 1 - a_1^2)x^2 + (4a_4 - 2a_1 a_2)x^3 + \cdots.$$

We see that $a_1 = a_2 = 0$, $a_3 = 1/3$, $a_4 = 0$. Proceeding in the same way to the coefficients of x^4, x^5, x^6, x^7, it may be shown that $a_6 = 1/54$, but the coefficients a_5, a_7 are zero. So

$$y = \frac{1}{3}x^3 + \frac{1}{54}x^6 + \cdots$$

7. The ODE is $y'' + (1 - e^{-x^2})y = 0$. Let $y = a_0 + a_1 x + a_2 x^2 + \cdots$. Since the Taylor series expansion of e^{-x^2} about $x_0 = 0$ is $1 - x^2 + x^4/2! - x^6/3! + \cdots$, we have that

$$y'' + y - e^{-x^2}y = 2a_2 + 6a_3 x + (12a_4 + a_0)x^2$$
$$+ (20a_5 + a_1)x^3 + (30a_6 + a_2 - \frac{a_0}{2})x^4 + \cdots$$

where we have used Appendix B.2, item 6(ii) to calculate the first few terms of the product series for $(1 - e^{-x^2})y(x)$. The coefficients a_0 and a_1 are arbitrary, while $a_2 = a_3 = 0$, $a_4 = -a_0/12$, $a_5 = -a_1/20$, $a_6 = a_0/60$. We have the general solution

$$y = a_0(1 - \frac{1}{12}x^4 + \frac{1}{60}x^6 + \cdots) + a_1(x - \frac{1}{20}x^5 + \cdots)$$

8. The IVP is $y'' + x^2 y' + 2xy = 0$, $y(0) = 1$, $y'(0) = 0$. Let $y = \sum_{n=0}^{\infty} a_n x^n$. Then $y'' = \sum_{n=2}^{\infty} n(n-1)a_n x^{n-2}$, so $x^2 y' = \sum_{n=1}^{\infty} na_n x^{n+1} = \sum_{n=2}^{\infty}(n-1)a_{n-1}x^n$, and $2xy =$

$\sum_{n=0}^{\infty} 2a_n x^{n+1} = \sum_{n=1}^{\infty} 2a_{n-1} x^n$. We have that

$$y'' + x^2 y' + 2xy = 2a_2 + (6a_3 + 2a_0)x + \sum_{n=2}^{\infty}[(n+2)(n+1)a_{n+2} + (n+1)a_{n-1}]x^n = 0$$

The initial conditions imply that $a_0 = 1$, $a_1 = 0$. The above expansion and the Identity Theorem imply that

$$a_2 = 0, \ a_3 = -\frac{1}{3}, \ a_{n+2} = -\frac{1}{n+2}a_{n-1}, \ n \geq 2$$

So $a_2 = 0$, $a_3 = -1/3$, $a_4 = a_5 = 0$, $a_6 = 1/18$, and so on. In general, $a_{3k+1} = a_{3k+2} = 0$, while $a_{3k} = (-1)^k[3k(3k-3)\cdots3]^{-1} = (-1)^k \cdot 1/(3^k k!)$. The series for $y(x)$ is then $\sum_{k=0}^{\infty}(-1)^k x^{3k}/(3^k k!)$. If $x = 1$ this is an alternating series whose terms decrease in magnitude as k increases. By the Alternating Series Test (Appendix B.2, item 10), $|y(1) - \sum_{k=0}^{4} 1/(3^k k!)| \leq 1/(3^5 5!) = 1/29160 < 10^{-4}$.

$$y(1) \approx 1 - \frac{1}{3} + \frac{1}{18} - \frac{1}{162} + \frac{1}{1944} \approx 0.7166$$

with an error less than 10^{-4}.

11.3   Legendre Polynomials

Suggestions for preparing a lecture

Topics: Legendre's equations of integer order, Legendre polynomials and a selection of their properties (the orthogonality property is important for Section 11.8).

Making up a problem set

Problems 1, 2, 3, 4. **Warning:** Problems 5, 6, and 7 are quite hard. If any of these three is assigned, students should be given suggestions on how to proceed.

Comments

The Legendre polynomials $\{P_n : n = 0, 1, 2, \ldots\}$ are the best known of the sets of orthogonal polynomials, and so we have chosen to give them and their ODEs a section of their own. This section, its problem set, and the full solutions of the problems should be regarded as an abbreviated compendium of properties of Legendre polynomials, other solutions of the Legendre ODEs, and other sets of orthogonal functions.

The Hermite ODEs and polynomials H_n and the Chebyshev ODEs and polynomials T_n make their appearance in the problem set. The reason the Chebyshev polynomials are labeled T_n and not C_n illustrates the difficulty in using the Latin alphabet to transliterate Cyrillic letters (Chebyshev was a well-known 19th century Russian mathematician). Sometimes one sees Tschebysheff, or Tchebyshev, or Chebyshov. Philip J. Davis has written an amusing and mathematically informative book based on his difficulties with these various transliterations [*The Thread: A Mathematical Gem*; Boston, Birkhaüser, 1983].

1. Assume that $u(x)$ is a solution of the ODE $y'' + a(x)y' + b(x)y = 0$ and that $u(x)$ is not the trivial solution $y = 0$, all x. Then a second solution $v(x)$, $|x| < 1$, independent of $u(x)$ can be found by solving the ODE

$$W[u, v](x) = uv' - u'v = e^{-\int^x a(s)\,ds}$$

In Legendre's ODEs of order 0 and 1, we have that $a(x) = -2x/(1 - x^2)$, $u_0(x) = 1$, and $u_1(x) = x$, respectively. So $v_0' = e^{\int^x 2s/(1-s^2)ds} = C_1/(1 - x^2)$, and (using an integral table)

$$v_0 = \frac{C_1}{2} \ln \frac{1+x}{1-x} + C_2, \quad (C_1 > 0)$$

Also, $xv_1' - v_1 = C_1/(1 - x^2)$, $v_1' - v_1/x = C_1/x(1 - x^2)$.

$$v_1 = e^{\int^x (1/s)\,ds}[C_3 + \int^x e^{-\int^s (1/\tau)\,d\tau}\frac{C_1}{s(1 - s^2)}\,ds]$$

$$= x[C_3 + C_1 \int^x \frac{C_1}{s^2(1 - s^2)}\,ds]$$

$$= x[C_3 + C_1(-\frac{1}{x} + \frac{1}{2} \ln \frac{1+x}{1-x})] \quad (C_1 > 0)$$

where we have used a table of integrals. So, second independent solutions for Legendre's equations of order 0 and 1 on the interval $|x| < 1$ are, respectively,

$$v_0 = \frac{C_1}{2} \ln \frac{1+x}{1-x} + C_2 \quad \text{and} \quad v_1 = x[C_3 + C_1(-\frac{1}{x} + \frac{1}{2} \ln \frac{1+x}{1-x})]$$

where C_1, C_2 and C_3 are constants and $C_1 > 0$. See Graph 1 and Graph 2, respectively, for graphs of v_0 and v_1, where $C_1 = 1$, $C_2 = 0$, $C_3 = 0$.

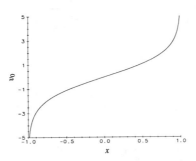

Problem 1, Graph 1.

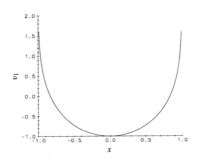

Problem 1, Graph 2.

2. Since $P_{2n+1}(x)$ contains only odd powers of x, $P_{2n+1}(0) = 0$. If we set $x = 0$ and replace n by $2n$ in (9), we see that the only nonzero term in the sum (with n replaced by $2n$) corresponds to the index $j = n$:

$$P_{2n}(0) = (-1)^n \frac{1}{2^{2n}} \frac{(2n)!}{(n!)^2}$$

From the identity $P_n(x) = xP_{n-1}(x) + (x^2 - 1)P'_{n-1}(x)/n$ (see Table 11.3.1), we see that $P_n(-1) = (-1)P_{n-1}(-1)$. Since $P_0(-1) = 1$, we conclude that $P_n(-1) = (-1)^n$.

3.   Suppose that $Q(x)$ is any polynomial of degree $k < n$. We want to show that the integral $\int_{-1}^{1} P_n(x)Q(x)\, dx = 0$. We prove by induction that there are constants $c_0, \ldots, c_k$ such that $Q(x) = \sum_{j=0}^{k} c_k P_k(x)$. This is certainly the case for $k = 0$ since $Q(x) = Q(0) \cdot 1 = Q(0) \cdot P_0(x)$. Suppppose the result is true for all polynomials Q of degree $\leq k$ (the induction hypothesis). Let Q be a polynomial of degree $k + 1$. Let $Q = a_0 + \cdots + a_{k+1}x^{k+1}$ and $P_{k+1} = b_0 + \cdots + b_{k+1}x^{k+1}$. Then $Q(x) = a_{k+1}P_{k+1}(x)b_{k+1}^{-1} + R(x)$, where $R(x)$ is a polynomial of degree k or less. $R(x) = \sum_{j=0}^{k} \alpha_j P_j(x)$ for some constants α_j by the induction hypothesis, so $Q(x)$ is a linear combination of $P_0(x), \ldots, P_{k+1}(x)$ and we are done. Note that the induction proof did not use the specific form of the Legendre polynomials, only that $P_n(x)$ is a polynomial of degree n.

Now to the problem itself: suppose $Q(x)$ is any polynomial of degree of $k < n$. Then $Q = c_0 P_0 + \cdots + c_k P_k$ and

$$\int_{-1}^{1} P_n(x)Q(x)\, dx = \sum_{j=0}^{k} c_j \int_{-1}^{1} P_n(x)P_j(x)\, dx = 0$$

by Theorem11.3.2.

4.   **(a).** We want to find the ODEs satisfied by $R(\rho)$ and $\Phi(\phi)$ if $R(\rho)\Phi(\phi)$ is a solution of the PDE $\sin\phi(\rho^2 u_\rho)_\rho + (\sin\phi u_\phi)_\phi = 0$. Let $u = R(\rho)\Phi(\phi)$. Then $u_\rho = R'(\rho)\Phi(\phi)$, $u_\phi = R(\rho)\Phi'(\phi)$, $u_{\rho\rho} = R''(\rho)\Phi(\phi)$ and $u_{\phi\phi} = R(\rho)\Phi''(\phi)$. The original PDE becomes

$$\sin\phi[2\rho R'(\rho)\Phi(\phi) + \rho^2 R''(\rho)\Phi(\phi)] + \cos\phi R(\rho)\Phi'(\phi) + \sin\phi R(\rho)\Phi''(\phi) = 0$$

After dividing by $R\Phi$ and separating variables:

$$\frac{\rho^2 R'' + 2\rho R'}{R} = -\frac{\Phi'' + \cot\phi\Phi'}{\Phi} = \lambda$$

where λ is a separation constant. Then,

$$\rho^2 R'' + 2\rho R' - \lambda R = 0, \quad 0 < \rho < 1$$

and

$$\Phi'' + \cot\phi\Phi' + \lambda\Phi = 0, \quad 0 < \phi < \pi$$

(b). Put $y = \Phi(x)$, where $x = \cos\phi$. By the Chain Rule, $\Phi'(\phi) = (dy/dx)(dx/d\phi) = -y'\sin\phi$. We have by the Chain Rule (again) that $\Phi''(\phi) = d\Phi'(\phi)/d\phi = -(\cos\phi)y' - [d(y')/dx][dx/d\phi]\sin\phi = -(\cos\phi)y' + (\sin^2\phi)y''$. So $\Phi'' + \cot\phi\Phi' + \lambda\Phi = 0$ becomes $y''\sin^2\phi - y'\cos\phi + \cot\phi(-y'\sin\phi) + \lambda y = 0$, [i.e., $(1 - x^2)y'' - 2xy' + p(p + 1)y = 0$, where $p = (-1 + \sqrt{1 + 4\lambda})/2$]. Suppose p is *not* a positive integer; from formulas (3a), (3b) we see that all nontrivial solutions of the Legendre equation are infinite series and (as shown in the text) these series converge only for $|x| < 1$. In fact, the solutions become unbounded as $|x| \to 1$, i.e., as $\phi \to 0, \pi$. This is a contradiction because $u(p, \phi)$ stands for the steady-state temperature in the unit ball $\rho \leq 1$ which, physically, remains bounded throughout the ball, that is, for $0 \leq \phi \leq \pi$, $\rho \leq 1$. However, if p is a positive integer, then the Legendre ODE has some polynomial solutions. These solutions are, of course, bounded for $|x| \leq 1$, that is, for $0 \leq \phi \leq \pi$.

5. We work with a family $\{R_n(x)\}$ of polynomials indexed by degree and we assume that for all $n, m, n \neq m$ we have

$$\int_a^b R_n(x) R_m(x) \rho(x) \, dx = 0$$

We say that $\{R_n : n = 0, 1, 2, \ldots\}$ is an orthogonal family with respect to the density ρ.

(a). Theorem 11.3.2 states that

$$< P_n, P_m >= \int_{-1}^{1} P_n(x) P_m(x) \, dx = \begin{cases} 0, & n \neq m \\ 2/(2n+1), & n = m \end{cases}$$

So, $a = -1$ and $b = 1$, $\rho(x) = 1$ for this orthogonal family.

(b). We want to show that the orthogonal family $\{R_n\}$ satisfies a three-term recursion formula. Let $R_n(x) = A_n x^n + \cdots + A_1 x + A_0$, $A_n \neq 0$, $n = 0, 1, 2, \ldots$, and $R_n^*(x) = R_n(x)/A_n$. The polynomials $R_n^*(x)$, $n = 0, 1, 2, \ldots$ are orthogonal on I to one another with respect to the density ρ since the family $R_n(x)$, $n = 0, 1, 2, \ldots$ have this property. Since $x R_n^*(x)$ is a polynomial of degree $n + 1$, there are constants $\lambda_1, \cdots, \lambda_n$ such that

$$x R_n^*(x) = R_{n+1}^*(x) + \sum_{j=0}^{n} \lambda_j R_j^*(x) \tag{i}$$

Multiplying by $R_s^*(x) \rho(x)$, $0 \leq s \leq n - 2$, and taking the integral, we have

$$\int_a^b x \rho(x) R_n^*(x) R_s^*(x) \, dx$$

$$= \int_a^b \rho(x) R_{n+1}^*(x) R_s^*(x) \, dx + \sum_{j=0}^{n} \lambda_j \int_a^b \rho(x) R_j^*(x) R_s^*(x) \, dx \tag{ii}$$

Because the degree of $x R_s^*(x)$ is less than n for $s = 0, 1, 2, \ldots, n - 2$, $x R_s^*(x)$ is a linear combination of $R_0(x), R_1(x), \cdots, R_{n-1}(x)$. Furthermore, $\int_a^b x \rho(x) R_n^*(x) R_s^*(x) \, dx = 0$, $s = 0, 1, 2, \ldots, n - 2$. Similarly, $\int_a^b \rho(x) R_{n+1}^*(x) R_s^*(x) \, dx = 0$, $s = 0, 1, 2, \ldots, n - 2$. When $s = 0, 1, 2, \ldots, n - 2$, (ii) becomes $\sum_{j=0}^{n} \lambda_j \int_a^b \rho(x) R_j^*(x) R_s^*(x) \, dx = 0$. Setting $s = 0$, the orthogonality property implies that $\lambda_0 \int_a^b \rho(x) [R_0^*(x)]^2 \, dx = 0$, i.e., $\lambda_0 = 0$. Similarly, setting $s = 1, 2, \ldots, n - 2$, respectively, we have $\lambda_j = 0$, $j = 1, 2, \ldots, n - 2$. So (i) becomes $x R_n^*(x) = R_{n+1}^*(x) + \lambda_n R_n^*(x) + \lambda_{n-1} R_{n-1}^*(x)$, $n = 2, 3, \ldots$. After replacing n by $n - 1$ and rearranging terms, we have the desired three-term recursion formula

$$R_n(x) = (a_n x + b_n) R_{n-1} + c_n R_{n-2}$$

$n = 2, 3, \ldots$, where $a_n = A_n/A_{n-1}$, $b_n = -A_n \lambda_{n-1}/A_{n-1}$ and $c_n = -A_n \lambda_{n-2}/A_{n-2}$.

(c). From formula (10) for Legendre polynomials and their recursion formula, we see that (10) has the form $R_n = (a_n x + b_n) R_{n-1} + c_n R_{n-2}$ if $a_n = 2 - 1/n$, $b_n = 0$, $c_n = -(1 - 1/n)$.

6.   **(a).** Hermite's equation of order n is $y'' - 2xy' + 2ny = 0$. If $y = \sum_{n=0}^{\infty} a_n x^n$, then $y'' - 2xy' + 2ny = 0$ becomes

$$\sum_{n=0}^{\infty} (k+2)(k+1)a_{k+2}x^k + \sum_{k=0}^{\infty} (-2ka_k)x^k + \sum_{k=0}^{\infty} 2na_k x^k = 0$$

where we have used k as an index to avoid confusion with the coefficient n in the last term. Combining the sums, after some calculation and use of the Identity Theorem, we get the recursion formula

$$a_{k+2} = \frac{2k - 2n}{(k+2)(k+1)} a_k, \qquad k \geq 0$$

The general solution of Hermite's equation is obtained from this recursion formula:

$$y = a_0 \left[1 - \frac{2n}{2!}x^2 - \frac{(4-2n)2n}{4!}x^4 - \frac{(8-2n)(4-2n)2n}{6!}x^6 - \cdots \right]$$

$$+ a_1 \left[x + \frac{2-2n}{3!}x^3 + \frac{(6-2n)(2-2n)}{5!}x^5 + \cdots \right]$$

For a fixed value of n one of the series in brackets terminates and polynomial solutions are obtained. The polynomial solution $H_n(x) = 2^n x^n + (\quad)x^{n-2} + \cdots$ has $a_0 \neq 0$ if n is even, $a_1 \neq 0$ if n is odd; H_n is an even [odd] function if n is even [odd].

(b). Next we show orthogonality over $\mathbb{R}$ with weight e^{-x^2}. Since $H_n'' - 2xH_n' + 2nH_n = 0$ and $H_m'' - 2xH_m' + 2mH_m = 0$, we have $[H_n'' H_m - H_m'' H_n] - 2x[H_n' H_m - H_m' H_n] + 2(n - m)H_n H_m = 0$. So we have

$$\frac{d}{dx}\left[e^{-x^2}(H_n' H_m - H_m' H_n) \right] + 2(n-m)e^{-x^2}H_n H_m = 0$$

that is,

$$e^{-x^2}[H_n' H_m - H_m' H_n]\Big|_{x=-\infty}^{x=+\infty} + 2(n-m)\int_{-\infty}^{+\infty} e^{-x^2} H_n H_m\, dx = 0$$

Since the first term on the left vanishes and $n \neq m$, we have that $\int_{-\infty}^{+\infty} e^{-x^2} H_n H_m\, dx = 0$. This means that the family of Hermite polynomials is orthogonal on the real line with respect to the density $\rho = e^{-x^2}$.

(c). Using the recursion formula $H_n = 2xH_{n-1} - 2(n-1)H_{n-2}$ and $H_0 = 1$, $H_1 = 2x$, we conclude that $H_2 = 4x^2 - 2$, $H_3 = 8x^3 - 12x$, $H_4 = 16x^4 - 48x^2 + 12$ and $H_5 = 32x^5 - 160x^3 + 120x$. See Fig. 6 for the graphs of H_0 and H_4 (solid), H_1 and H_5 (long dashes), H_2 (short dashes), H_3 (long/short dashes). The graphs suggest that the roots of H_n are all real and simple since each H_n seems to cross the x axis n times (the crossing points are the roots).

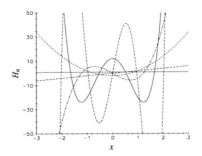

Problem 6(c).

7. **(a).** Chebyshev's equation is $(1 - x^2)y'' - xy' + n^2 y = 0$. Let $y = \sum_{k=0}^{\infty} a_k x^k$ and proceed much as in the preceding problems to write the ODE in the form $\sum_{k=0}^{\infty} [\cdots] x^k = 0$. In this case equating $[\cdots]$ to 0 gives the recursion formula,

$$a_{k+2} = \frac{k^2 - n^2}{(k+1)(k+2)} a_k$$

which can then be used to find the general solution

$$y = a_0 \left[1 - \frac{n^2}{2!} x^2 - \frac{(4 - n^2)n^2}{4!} x^4 - \frac{(16 - n^2)(4 - n^2)n^2}{6!} x^6 - \cdots \right]$$
$$+ a_1 \left[x + \frac{1 - n^2}{3!} x^3 + \frac{(9 - n^2)(1 - n^2)}{5!} x^5 + \cdots \right].$$

If n is an even [odd] integer, the first [second] series in brackets terminates and polynomial solutions are obtained. The polynomials are even functions if n is even, odd functions if n is odd. Let $T_0 = 1$ be the zeroth Chebyshev polynomial, and let the n-th Chebyshev polynomial $T_n(x)$ be the polynomial solution with leading coefficient 2^{n-1}.

(b). To show that the polynomials are orthogonal on $[-1, 1]$ with respect to the weight factor $(1 - x^2)^{-1/2}$, we proceed as follows. We have that $(1 - x^2)T_n'' - xT_n' + n^2 T_n = 0$ and $(1 - x^2)T_m'' - xT_m' + m^2 T_m = 0$. So, $(1 - x^2)(T_n'' T_m - T_m'' T_n) - x(T_n' T_m - T_m' T_n) + (n^2 - m^2)T_n T_m = 0$ [i.e., $(1 - x^2)^{1/2}(T_n'' T_m - T_m'' T_n) - x(1 - x^2)^{-1/2}(T_n' T_m - T_m' T_n) + (1 - x^2)^{-1/2}(n^2 - m^2)T_n T_m = 0$]. Therefore, $\frac{d}{dx}[(1 - x^2)^{1/2}(T_n' T_m - T_m' T_n)] = -(n^2 - m^2)(1 - x^2)^{-1/2} T_n T_m$ and $[(1 - x^2)^{1/2}(T_n' T_m - T_m' T_n)]_{-1}^1 = -(n^2 - m^2) \int_{-1}^1 (1 - x^2)^{-1/2} T_n T_m \, dx$. Because the left side of this equality vanishes for $n \neq m$, $T_n(x)$ and $T_m(x)$ are orthogonal on $I = [-1, 1]$ with respect to the density $\rho = (1 - x^2)^{-1/2}$.

(c). Use the given recursion formula $T_n = 2xT_{n-1} - T_{n-2}$, $T_0 = 1$, $T_1 = x$; therefore, $T_2 = 2x^2 - 1$, $T_3 = 4x^3 - 3x$, $T_4 = 8x^4 - 8x^2 + 1$, $T_5 = 16x^5 - 20x^3 + 5x$. See Fig. 7 for the graphs of T_0 and T_4 (solid), T_1 and T_5 (long dashes), T_2 (short dashes), and T_3 (long/short dashes). The graphs suggest that $T_n(x)$ has n simple real roots in $(-1, 1)$ since each T_n seems to cross the x-axis n times (each crossing point is a root).

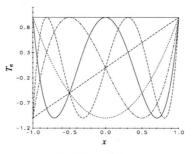

Problem 7(c).

Background Material: Proof of Bounds for Legendre Polynomial

Squaring the identity $P'_n = nP_{n-1} + xP'_{n-1}$, multiplying through by $(1-x^2)/n^2$, and adding it to the square of the identity $P_n = xP_{n-1} + [(x^2-1)/n]P'_{n-1}$, we obtain, for $|x| \leq 1$,

$$0 \leq \frac{1-x^2}{n^2}(P'_n)^2 + P_n^2 = P_{n-1}^2 + \frac{1-x^2}{n^2}(P'_{n-1})^2 < P_{n-1}^2 + \frac{1-x^2}{(n-1)^2}(P'_{n-1})^2 \qquad \text{(i)}$$

If we have (for $n = 1$) an upper bound for the right side of the inequalities in (i), that bound provides an upper bound for the second term in (i) for all $n \geq 1$. But the right side of (i) for $n = 1$ and for any x, $|x| \leq 1$, has value 1 since $P_0 = 1$ and $P'_0 = 0$. From all this, we conclude that

$$0 \leq \frac{1-x^2}{n^2}(P'_n(x))^2 + P_n^2(x) \leq 1, \qquad |x| < 1$$

so, $|P_n(x)| \leq 1$ for $|x| \leq 1$.

11.4 Regular Singular Points

Suggestions for preparing a lecture

Topics: Regular singular points, Euler ODEs.

Making up a problem set

Parts of 1 and 2, 3(**c**), 3(**e**), 5.

Comments

Now that we know how to construct power series solutions of a linear ODE in powers of $x - x_0$, where x_0 is an ordinary point, it is time to address the challenging problem of constructing series solutions if x_0 is singular. That can be done if x_0 is "regular singular" (isn't that an oxymoron?), and we start out by defining just what a regular singular point is. The simplest class of ODEs with regular singular points are the Euler ODEs, and most of the section is devoted to the construction of their solutions. The nice thing about an Euler ODE is that the intricate series techniques of the preceding three sections are not needed. Students (and the instructor) will have a welcome respite from the algebraic rigors of dealing with series. The indicial polynomial appears here, and should be emphasized because it plays a critical role in the analysis of the remaining sections of the chapter. There are some interesting pictures of solution behavior of an Euler ODE near its regular singularity at the origin. It should be noted that the basic Existence and Uniqueness Theorem (Theorem 3.2.1) doesn't work at the singularity, and an IVP may have no solution at all or else infinitely many (see Example 11.4.6).

1. The generic ODE here is $a_2 y'' + a_1 y' + a_0 y = 0$, which in normal form is $y'' + (a_1/a_2)y' + (a_0/a_2)y = 0$. The tests for a regular or irregular singularity at $x = x_0$ are concerned with the functions $(x - x_0)(a_1/a_2)$ and $(x - x_0)^2(a_0/a_2)$.

   (a). The ODE is $x^2 y'' + xy' + y = 0$. Here, $a_2 = x^2$, $a_1 = x$, $a_0 = 1$, while $x_0 = 0$. We have

   $$\lim_{x \to 0} \frac{xa_1}{a_2} = \lim_{x \to 0} \frac{x^2}{x^2} = 1, \quad \lim_{x \to 0} \frac{x^2 a_0}{a_2} = \lim_{x \to 0} \frac{x^2}{x^2} = 1$$

   and 0 is a regular singular point for $x^2 y'' + xy' + y = 0$.

   (b). The ODE is $xy'' + (1 - x)y' + xy = 0$. In this case $a_2 = x$, $a_1 = 1 - x$, $a_0 = x$ and $x_0 = 0$. Since

   $$\lim_{x \to 0} \frac{x(1 - x)}{x} = 1, \quad \lim_{x \to 0} \frac{x^2 \cdot x}{x} = 0,$$

   0 is a regular singular point of the ODE.

   (c). The ODE is $x(1 - x)y'' + (1 - 2x)y' - 4y = 0$. In this problem $x_0 = 1$ is the singular point to be tested. We have $a_2 = x(1 - x)$, $a_1 = (1 - 2x)$, $a_0 = -4$. We have

   $$\lim_{x \to 1} \frac{(x - 1)(1 - 2x)}{x(1 - x)} = 1, \quad \lim_{x \to 1} \frac{(x - 1)^2(-4)}{x(1 - x)} = \lim_{x \to 1} \frac{4(x - 1)}{x} = 0$$

   and $x_0 = 1$ is a regular singular point.

   (d). The ODE is $x^2 y'' + 2y'/x + 4y = 0$. Here $x_0 = 0$, $a_2 = x^2$, $a_1 = 2x^{-1}$, $a_0 = 4$ and $\lim_{x \to 0} x(2x^{-1})/x^2$ does not exist, so $x_0 = 0$ is an irregular point.

2. The normalized form of $a_2 y'' + a_1 y' + a_0 y = 0$ is $y'' + (a_1/a_2)y' + (a_0/a_2)y = 0$.

   (a). The ODE is $(1 - x^2)y'' - xy' + 2y = 0$. The singular points are $x = \pm 1$, and $a_2 = 1 - x^2$, $a_1 = -x$, $a_0 = 2$. We have that for -1

   $$\lim_{x \to -1} \frac{(x + 1)(-x)}{1 - x^2} = \lim_{x \to -1} \frac{-x}{1 - x} = \frac{1}{2}, \quad \lim_{x \to -1} \frac{(x + 1)^2(2)}{1 - x^2} = \lim_{x \to -1} \frac{(x + 1)(2)}{1 - x} = 0$$

   and -1 is a regular singular point. A similar analysis shows that $+1$ is also a regular singular point.

   (b). The ODE is $y'' + x^2 y'/(1 - x^2) + (1 + x)^2 y = 0$. Here $x = 1$ is the singular point, $a_2 = 1$, $a_1 = x^2/(1 - x)^2$, $a_0 = (1 + x)^2$. We have that

   $$\lim_{x \to 1} \frac{(x - 1)x^2/(1 - x)^2}{1}$$

   does not exist, and 1 is an irregular singular point.

(c). The ODE is $x(1 - x^2)^3 y'' + (1 - x^2)^2 y' + 2(1 + x)y = 0$. There are three singular points $x_0 = 0, +1$, and -1; $a_2 = x(1 - x^2)^3$, $a_1 = (1 - x^2)^2$, and $a_0 = 2(1 + x)$. We leave it to the reader to show that the limits for the three values of x_0

$$\lim_{x \to x_0} \frac{(x - x_0)a_1(x)}{a_2(x)} \text{ and } \lim_{x \to x_0} \frac{(x - x_0)^2 a_0(x)}{a_2(x)}$$

all exist, except for one:

$$\lim_{x \to 1} \frac{(x - 1)^2 a_0(x)}{a_2(x)} = \lim_{x \to 1} \frac{(x - 1)^2 2(1 + x)}{x(1 - x^2)^3} = \lim_{x \to 1} \frac{2}{x(1 + x)^2(1 - x)}$$

which does not exist. So 0 and -1 are regular singularities, while $+1$ is irregular.

(d). The ODE is $x^3(1 - x^2)y'' - x(x + 1)y' + (1 - x)y = 0$. The singularities are $x_0 = 0$, $+1, -1$, while $a_2 = x^3(1 - x^2)$, $a_1 = -x(x + 1)$ $a_0 = 1 - x$. The appropriate limits all exist for $x_0 = \pm 1$, which then are regular singular points. However, $x_0 = 0$ is irregular since

$$\lim_{x \to 0} \frac{x(-x)(x + 1)}{x^3(1 - x^2)} = \lim_{x \to 0} \frac{-x - 1}{x(1 - x)^2}$$

does not exist.

3. Each of the equations may be written as an Euler equation $x^2 y'' + p_0 x y' + q_0 y = 0$, and the solutions are determined by the roots r_1 and r_2 of the indicial polynomial $r^2 + (p_0 - 1)r + q_0$. The three types of solutions are given in (13) in the text. In each case we list the indicial polynomial, its roots, and then the general solution; c_1 and c_2 denote arbitrary constants. It is assumed throughout that $x > 0$, and absolute value signs around x are not needed. See the appropriate figures for plots of some solutions of each equation. For negative x, replace x by $|x|$ in all formulas.

(a). $r^2 - r - 6$; $r_1 = -2, r_2 = 3$; $y = c_1 x^{-2} + c_2 x^3$.

(b). $r^2 - 4$; $r_1 = -2, r_2 = 2$; $y = c_1 x^{-2} + c_2 x^2$.

(c). $r^2 + 9$; $r_1 = 3i, r_2 = -3i$; $y = c_1 \cos(3 \ln x) + c_2 \sin(3 \ln x)$.

(d). $r^2 - r/2 - 1/2$; $r_1 = -1/2, r_2 = 1$; $y = c_1 x^{-1/2} + c_2 x$.

(e). $r^2 - 2r + 5$; $r_1 = 1 + 2i = \bar{r}_2$; $y = x[c_1 \cos(2 \ln x) + c_2 \sin(2 \ln x)]$.

(f). $r^2 + 6r + 9$; $r_1 = r_2 = -3$; $y = x^{-3}(c_1 + c_2 \ln x)$.

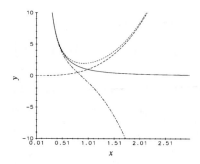

Problem 3(a).

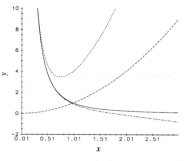

Problem 3(b).

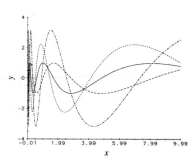

Problem 3(c).

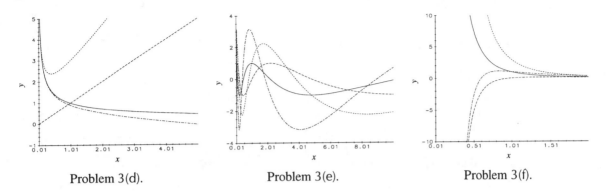

Problem 3(d).   Problem 3(e).   Problem 3(f).

4. The operators on the left side are Euler operators whose null spaces may be found by the method of this section. A particular solution in each case is found by Undetermined Coefficients.

(a). The indicial polynomial of the Euler operator is $r^2 - 5r + 6$ with roots $r_1 = 2, r_3 = 3$. A particular solution will have the form $y_p = a + bx$. So $-4xb + 6a + 6bx = 2x + 5$ and $a = 5/6, b = 1$. The general solution is $y = c_1 x^3 + c_2 x^2 + x + 5/6$, where c_1 and c_2 are any constants.

(b). The indicial polynomial of the Euler operator is $r^2 - r - 6$ with roots $r_1 = -2, r_3 = 3$. A particular solution has the form $y_p = a + bx + cx^2$. So $2cx^2 - 6a - 6bx - 6cx^2 = 2x^2$, and $a = b = 0, c = -1/2$. The general solution is $y = c_1 x^{-2} + c_2 x^3 - x^2/2$, where c_1 amd c_2 are any constants and $x > 0$.

5. The solutions of $x^2 y'' + p_0 xy' + q_0 y = 0$ are given by (13), and we see that every solution tends to 0 as $x \to +\infty$ if and only if r_1 and r_2 are negative or else have negative real parts. This is clear if r_1 and r_2 are real and distinct. If $r_1 = r_2 < 0$, then the term $x^{r_1} \ln x = \ln(x)/x^{-r_1} \to 0$ as $x \to +\infty$ by L'Hôpital's Rule. If $r_1 = a + bi = \bar{r}_2$ and $a < 0, b \neq 0$, then $|x^a \cos(b \ln x)| \leq |x|^a \to 0$ as $x \to +\infty$, and similarly for $x^a \sin(b \ln x)$. The roots r_1 and r_2 of the indicial polynomial $r^2 + (p_0 - 1)r + q_0$ have negative real parts if and only if $p_0 > 1$ and $q_0 > 0$, as can be seen by using the quadratic formula for the roots.

6. (a). The function $y = x^{r_1}$ is a solution of the nth order Euler ODE $x^n y^{(n)} + a_{n-1} x^{n-1} y^{(n-1)} + \cdots + a_1 xy' + a_0 y = 0$ if and only if $x^n [r_1 (r_1 - 1) \cdots (r_1 - n + 1)]x^{r_1 - n} + x^{n-1}[r_1 (r_1 - 1) \cdots (r_1 - n + 2)]a_{n-1} x^{r_1 - n+1} + \cdots + r_1 a_1 x^{r_1} + a_0 x^{r_1} = 0$, that is, if and only if r_1 is a root of $p(r) = r(r-1) \cdots (r-n+1) + r(r-1) \cdots (r-n+2)a_{n-1} + \cdots + ra_1 + a_0$.

(b). The indicial polynomial for $x^3 y''' + 4x^2 y'' - 2y = 0$ is $r(r-1)(r-2) + 4r(r-1) - 2 = r^3 + r^2 - 2r - 2 = (r+1)(r^2 - 2)$, and the roots are -1 and $\pm\sqrt{2}$. The functions x^{-1}, $x^{-\sqrt{2}}, x^{\sqrt{2}}$ are three independent solutions for $x > 0$.

11.5   Series Solutions Near Regular Singular Points, I

Suggestions for preparing a lecture

Topics: Indicial polynomial, Frobenius series, one example. You might want to construct a Frobenius series "procedure" like that on page 620 to help students get through a Frobenius series problem.

Making up a problem set

Problems 1(a) or 4, 3.

Comments

In the next three sections of the chapter we take up the techniques for constructing a pair of independent solutions of a homogeneous second-order linear ODE with nonconstant coefficients near a regular singularity. In Section 11.5 we address the simplest case when the roots of the indicial polynomial do *not* differ by an integer. We also show that even if the roots do differ by an integer we can always construct a solution corresponding to the larger root. The process by which Frobenius series expansions of solutions are constructed is similar to the methods used in Section 11.2 to find power series expansions of solutions near an ordinary point, but with the twist that the powers of the independent variable need not be integers. As before, the successful construction of the series requires patience, algebraic "smarts," and careful bookkeeping. Once again, a CAS may be helpful. We have noticed over the years that readers have a lot of trouble with this material. But those who finally "see" how to do it are quite proud of their accomplishment.

---

1. The detailed solution process for finding y_1 and y_2 is given only for **(a)**. In the other cases only the indicial polynomial, its roots r_1 and r_2, the recursion formula and the corresponding independent solutions, $y_1 = x^{r_1} \sum_{n=0}^{\infty} a_n(r_1)x^n$ and $y_2 = x^{r_2} \sum_{n=0}^{\infty} a_n(r_2)x^n$, are given. In each case $r_1 - r_2$ is not an integer, and the Method of Forbenius given in this section produces the two independent solutions y_1 and y_2.

    (a). The ODE is $9x^2 y'' + 3x(x+3)y' - (4x+1)y = 0$. The equation in standard form is $x^2 y'' + x(1 + x/3)y' + (-1/9 - 4x/9)y = 0$. So $p_0 = 1$ and $q_0 = -1/9$. The indicial polynomial is $r^2 + (p_0 - 1)r + q_0 = r^2 - 1/9$, and its roots are $r_1 = 1/3$, $r_2 = -1/3$. Let $y = \sum_{n=0}^{\infty} a_n x^{n+r}$ be a solution. Then

$$y' = \sum_{n=0}^{\infty} (n+r)a_n x^{n+r-1}, \quad y'' = \sum_{n=0}^{\infty} (n+r)(n+r-1)a_n x^{n+r-2}$$

Inserting these series into the ODE,

$$9x^2 y'' + 3x(x+3)y' - (4x+1)y = 0$$

reindexing, and rearranging, we have that

$$\sum_{n=0}^{\infty} [9(n+r)(n+r-1) + 9(n+r) - 1]a_n x^{n+r} + \sum_{n=1}^{\infty} [3(n+r-1) - 4]a_{n-1} x^{n+r} = 0$$

Note that the term corresponding to $n = 0$ in the first summation above is zero because $r^2 = 1/9$. The index in that sum actually runs from $n = 1$ to ∞, and the two summations can be combined because they have the same index range and involve the same power of x. Applying the Identity Theorem to the coefficient of x^{n+r}, we obtain the recursion relation:

$$[9(n+r)(n+r-1) + 9(n+r) - 1]a_n = -[3(n+r-1) - 4]a_{n-1}$$

and, solving this relation for a_n,

$$a_n = -\frac{3(n+r-1)-4}{9(n+r)^2-1}a_{n-1}$$

So for $r = 1/3$,

$$a_n = -(n-2)a_{n-1}/[n(3n+2)], \quad n \geq 1$$

Setting $a_0 = 1$, we have that $a_1 = 1/5$, while $a_n = 0$, $n \geq 2$. We have a solution corresponding to $r = 1/3$:

$$y_1 = x^{1/3} + \frac{1}{5}x^{4/3}$$

For $r = -1/3$,

$$a_n = -(3n-8)a_{n-1}/[3n(3n-2)]$$
$$= a_0(-1)^n(3n-8)\cdots(-2)(-5)/[3^n n!(3n-2)(3n-5)\cdots 1]$$
$$= (-1)^n \frac{10}{3^n n!(3n-2)(3n-5)}a_0, \quad n \geq 2$$

So a second independent solution (corresponding to $r = -1/3$) is

$$y_2 = x^{-1/3}\sum_{n=0}^{\infty}a_n x^n$$

where $a_0 = 1$ and a_n for $r = -1/3$ is given above. The series converge for all x but the general solution $y = c_1 y_1 + c_2 y_2$ is not defined at $x = 0$, because of the factor $x^{-1/3}$ in y_2.

(b). The ODE is $4x^2 y'' + x(2x+9)y' + y = 0$. The indicial polynomial is $r^2 + 5r/4 + 1/4$, and its roots are $r_1 = -1/4$, $r_2 = -1$. The recursion formula is

$$a_n(r) = -2(n+r-1)a_{n-1}/[(n+r)(4n+4r+5) + 1]$$

We can "solve" the formula to write $a_n(-1/4)$ and $a_n(-1)$ as multiples of a_0, but the formulas are not simple and we do not do so. We have that

$$y_1 = x^{-1/4}\sum_{n=0}^{\infty}a_n(-1/4)x^n \quad \text{and} \quad y_2 = x^{-1}\sum_{n=0}^{\infty}a_n(-1)x^n$$

where we set $a_0 = 1$ and find $a_1, a_2, \cdots$ from the recursion formula.

(c). The ODE is $x^2 y'' + x(1-x)y' - 2y = 0$. The indicial polynomial is $r^2 - 2$ and its roots are $r_1 = \sqrt{2}$, $r_2 = -\sqrt{2}$. The recursion formula is

$$a_n(r) = (n+r-1)a_{n-1}/[(n+r)^2 - 2]$$

So

$$a_n(\pm\sqrt{2}) = \frac{(n \pm \sqrt{2} - 1)(n \pm \sqrt{2} - 2) \cdots (\pm\sqrt{2})a_0}{n!(n \pm 2\sqrt{2})(n - 1 \pm 2\sqrt{2}) \cdots (1 \pm 2\sqrt{2})}$$

A pair of independent solutions is

$$y_1 = x^{\sqrt{2}} \sum_{n=0}^{\infty} a_n(\sqrt{2})x^n \quad \text{and} \quad y_2 = x^{-\sqrt{2}} \sum_{n=0}^{\infty} a_n(-\sqrt{2})x^n$$

where $a_0 = 1$ and $a_n(\pm\sqrt{2})$ is given above.

(d). The ODE is $x^2(1 - x^2)y'' + x(x - 1)y' + 8y/9 = 0$. The standard form is

$$x^2 y'' + (x - 1)y'/(1 - x^2) + 8y/[9(1 - x^2)] = 0$$

and so

$$x^2 y'' + x(-1/(1 + x))y' + [8/(9(1 - x^2))]y = 0$$

Since

$$-1/(1 + x) = -1 + x - x^2 + \cdots$$

and

$$1/(1 - x^2) = 1 + x^2 + x^4 + \cdots$$

(using geometric series with $|x| < 1$), we see that $p_0 = -1$, $q_0 = 8/9$. Hence, the indicial polynomial is

$$f(r) = r^2 - 2r + 8/9$$

and its roots are $r_1 = 4/3$, $r_2 = 2/3$. To find the recursion formula, use the ODE in its original form and conclude after working with the series that

$$a_n(r) = \frac{-(n + r - 1)a_{n-1} + (n + r - 2)(n + r - 3)a_{n-2}}{(n + r)(n + r - 2) + 8/9}, \qquad n \geq 2$$

$$a_1(r) = -\frac{ra_0}{(r + 1)(r - 1) + 8/9}$$

Then

$$y_1 = x^{4/3} \sum_{n=0}^{\infty} a_n(4/3)x^n \quad \text{and} \quad y_2 = x^{2/3} \sum_{n=0}^{\infty} a_n(2/3)x^n$$

where $a_0 = 1$ and $a_n(r)$ is given by the above for $n = 1, 2, \cdots$.

2. The ODE is $x^2 y'' + x(1 - x)y' - 2y = \ln(1 + x)$. See the solution for Problem 1**(c)** for a spanning set for the solution space of the homogeneous equation. For a particular solution of the nonhomogeneous equation, we expand $\ln(1 + x)$ in its Taylor series about $x_0 = 0$,

$$\ln(1 + x) = \sum_{n=1}^{\infty} (-1)^{n+1} x^n / n$$

assume a solution of the form $y = \sum_{n=0}^{\infty} a_n x^n$, insert this series on the left of the ODE, and then match coefficients to determine the a_n's. In this problem we find the first four nonvanishing terms:

$$-2 a_0 - a_1 x + (2a_2 - a_1)x^2 + (7a_3 - 2a_2)x^3 + (14a_4 - 3a_3)x^4 + \cdots$$
$$= x - x^2/2 + x^3/3 - x^4/4 + \cdots$$

So $a_0 = 0$, $a_1 = -1$, $a_2 = -3/4$, $a_3 = -1/6$, $a_4 = -3/56$ and a particular solution is

$$y_p = -x - 3x^2/4 - x^3/6 - 3x^4/56 - \cdots$$

Then $y = c_1 y_1 + c_2 y_2 + y_p$, where c_1 and c_2 are arbitrary constants and y_1 and y_2 are as in the solution for Problem 1(c), is the general solution of the nonhomogeneous problem.

3. The ODE is $3x^2 y'' + 5xy' - e^x y = 0$. The Taylor series for e^x about $x_0 = 0$ is $1 + x + x^2/2! + x^3/3! + \cdots$. So $p_0 = 5/3$, $q_0 = -1/3$ for the ODE in nonstandard form

$$x^2 y'' + 5xy'/3 - e^x y/3 = x^2 y'' + 5xy'/3 + (-1/3 - x/3 - \cdots)y = 0$$

The indicial polynomial is $r^2 + 2r/3 - 1/3$ and $r_1 = 1/3$, $r_2 = -1$. There are independent solutions of the form

$$y_1 = x^{1/3} + a_1 x^{4/3} + a_2 x^{7/3} + \cdots, \qquad y_2 = x^{-1} + b_1 + b_2 x + \cdots$$

where we have set $a_0 = b_0 = 1$ in each series. Although we could use Frobenius Theorem I to determine the coefficients, it is probably simpler to make a direct substitution of y_1 and then y_2 into the ODE. For y_1 we have that

$$3x^2 y_1'' + 5xy_1' - e^x y_1 = (7a_1 - 1)x^{4/3} + (20a_2 - a_1 - 1/2)x^{7/3}$$
$$+ (39a_3 - a_2 - a_1/2 - 1/6)x^{10/3} + \cdots = 0$$

So $a_1 = 1/7$, $a_2 = 9/280$, $a_3 = 227/32760$ and

$$y_1 = x^{1/3} + x^{4/3}/7 + 9x^{7/3}/280 + 227x^{10/3}/32760 + \cdots$$

In the same way we see that

$$y_2 = x^{-1} - 1 - x/8 - 11x^2/360 + \cdots$$

4. The ODE is $xy'' - y' - 4x^3 y = 0$. The standard form for the ODE is

$$x^2 y'' - xy' - 4x^4 y = 0$$

So $p_0 = -1$, $q_0 = 0$, the indicial polynomial is $r^2 - 2r = 0$, and $r_1 = 2$, $r_2 = 0$. According to the methods of this section, there is a family of solutions of the form $y_1 = x^2[a_0 + a_1 x + \cdots]$, where a_0 is arbitrary, but since $r_1 - r_2$ is an integer the methods are inadequate to find a second independent solution (see Section 11.7 for a way to find the second solution in this case). However, we only need to find y_1 in this problem. It is already clear that $y_1(0) = y_1'(0) = 0$ regardless of the value of a_0 because of the x^2 factor. This does not contradict the Uniqueness Theorem (Theorem 3.2.1) since the normalized equation $y'' - x^{-1}y' - 4x^2 y = 0$ has a singularity at $x = 0$ and the hypotheses of the theorem are not satisfied. With y_1 as above, we have that

$$xy_1'' - y_1' - 4x^3 y_1 = \sum_{n=0}^{\infty} n(n+2)a_n x^{n+1} - \sum_{n=0}^{\infty} 4a_n x^{n+5} = 0$$

After lowering n by 4 in the last sum and raising the lower limit of the range from 0 to 4 to compensate, we have

$$3a_1 x^2 + 8a_2 x^3 + 15a_3 x^4 + \sum_{n=4}^{\infty} [n(n+2)a_n - 4a_{n-4}]x^{n+1} = 0$$

where the first three terms of the first sum above have been pulled outside the sum. We see that a_0 is arbitrary, $a_1 = a_2 = a_3 = 0$, $a_4 = a_0/6$, and, in general,

$$a_{4n} = \frac{a_0}{n!2^n(2n+1)(2n-1)\cdots 1} = \frac{a_0}{(2n+1)!}$$

while $a_{4n+k} = 0$, $k = 1, 2, 3$. So

$$y_1 = a_0 \sum_{k=0}^{\infty} x^{4k}/(2k+1)!$$

where a_0 is any constant. Each of these solutions satisfies the initial conditions $y(0) = y'(0) = 0$.

5. Suppose $x^3 y'' + y = 0$ has a solution $y = \sum_{n=0}^{\infty} a_n x^{n+r}$ for some constant r and constants a_n, where the series $\sum_{n=0}^{\infty} a_n x^n$ converges in a nontrivial interval I containing 0. Then we obtain a power series after a certain amount of reindexing

$$x^3 y'' + y = a_0 x^r + \sum_{n=1}^{\infty} [(n+r-1)(n+r-2)a_{n-1} + a_n]x^{n+r} = 0$$

So $a_0 = 0$, and

$$a_n = -(n+r-1)(n+r-2)a_{n-1}$$

This implies that $a_1 = a_2 = \cdots = a_n = \cdots = 0$, and the only solution in power series form is the trivial solution ($y = 0$).

6. (a). Laguerre's ODE of order p is $xy'' + (1-x)y' + py = 0$. The normalized form of Laguerre's equation is $y'' + x^{-1}(1-x)y' + px^{-1}y = 0$. Since $\lim_{x\to 0}(1-x)$ and $\lim_{x\to 0}(px)$ both exist, 0 is a regular singular point.

(b). The indicial polynomial of $x^2 y'' + x(1-x)y' + pxy = 0$ is r^2 since $p_0 = 1$ and $q_0 = 0$. So $r_1 = r_2 = 0$ and the methods of this section may be used to find a nontrivial solution of the form $y_1 = \sum_{n=0}^{\infty} a_n x^n$, but a second independent solution can only be found by the methods of Section 11.7. We have that

$$xy'' + (1-x)y' + py = \sum_{n=0}^{\infty} [(n+1)^2 a_{n+1} + (p-n)a_n]x^n = 0$$

The recursion formula is

$$a_{n+1} = (n-p)a_n/(n+1)^2$$

and setting $a_0 = 1$, we have that

$$y_1(x) = 1 - px - p(1-p)x^2/4 + \cdots + [(-p)(1-p)\cdots(n-1-p)]x^n/(n!)^2 + \cdots$$

(c). Suppose p is a nonnegative integer, $p = n - 1$ say. Then the factor $n - 1 - p = 0$ appears as a factor of x^k for all $k \geq n$. So y_1 is a polynomial of degree p.

(d). Let L_n and L_m be the Laguerre polynomials of degree n and m, respectively, $n \neq m$. We have

$$xL_n'' + (1-x)L_n' + nL_n = 0, \quad xL_m'' + (1-x)L_m' + mL_m = 0$$

If the first equation is multiplied by L_m and the second by L_n, subtraction yields

$$x(L_n''L_m - L_m''L_n) + (1-x)(L_mL_n' - L_nL_m') + (n-m)L_nL_m = 0$$

and so

$$e^{-x}x(L_n''L_m - L_m''L_n) + e^{-x}(1-x)(L_mL_n' - L_m'L_n) + (n-m)e^{-x}L_nL_m = 0$$

$$[e^{-x}x(L_mL_n' - L_m'L_n)]' + (n-m)e^{-x}L_nL_m = 0$$

Integrating both sides from 0 to $+\infty$, we have that

$$[e^{-x}x(L_mL_n' - L_m'L_n)]\Big|_0^{+\infty} = (m-n)\int_0^{+\infty} e^{-x}L_nL_m\,dx$$

Since the left side of this equality is 0 and since $n \neq m$, we have that $\int_0^{+\infty} e^{-x}L_nL_m\,dx = 0$. This means $\{L_p : p = 0, 1, 2, \ldots\}$ is orthogonal on the interval $(0, \infty)$ with density $\rho(x) = e^{-x}$. Using $a_0 = 1$ and the recursion formula from part **(b)**, we see that the first five Laguerre polynomials are $L_0(x) = 1$, $L_1(x) = 1 - x$, $L_2(x) = 1 - 2x + x^2/2$, $L_3(x) = 1 - 3x + 3x^2/2 - x^3/6$, and $L_4(x) = 1 - 4x + 3x^2 - 2x^3/3 + x^4/24$.

7. **(a).** The Legendre ODE of order p is $(1-x^2)y'' - 2xy' + p(p+1)y = 0$. First, we let $x = s + 1$ so that the regular singularity at $x = 1$ is translated to the origin. Legendre's ODE for $z(s) = y(x)$ then becomes

$$\left[1 - (s+1)^2\right]\frac{d^2z}{ds^2} - 2(s+1)\frac{dz}{ds} + p(p+1)z = 0$$

We need to write the ODE in the standard form $s^2y'' + sp(s)z' + q(s)z = 0$. Since the coefficient of z'' is $1 - (s+1)^2 = s(-s-2)$, we multiply the ODE by s and divide by $-s - 2$ to obtain

$$s^2\frac{d^2z}{ds^2} + \frac{2(s+1)s}{s+2}\frac{dz}{ds} + \frac{p(p+1)(-s)}{s+2}z = 0$$

Since the Taylor series for $2(s+1)/(s+2)$ and $p(p+1)(-s)/(s+2)$ are, respectively, $1 + (\cdot)s + \cdots$, and $0 - p(p+1)s/2 + \cdots$, we see that $p_0 = 1$, $q_0 = 0$, the indicial polynomial is r^2, and the indicial roots are $r_1 = r_2 = 0$.

(b). Frobenius series techniques may be used to find a solution $z_1(s)$ of

$$(s+2)s\frac{d^2z}{ds^2} + 2(s+1)\frac{dz}{ds} - p(p+1)z = 0$$

The recursion formula turns out to be

$$a_{n+1} = \frac{p(p+1) - n(n+1)}{2(n+1)^2} a_n$$

Setting $a_0 = 1$ and replacing s by $x - 1$ we have that

$$y_1(x) = 1 + \frac{p(p+1)}{2}(x-1) + \frac{p(p+1)[p(p+1) - 2]}{16}(x-1)^2 + \cdots$$

$$+ \frac{1}{2^n (n!)^2} p(p+1)[p(p+1) - 2] \cdots [p(p+1) - n(n+1)](x-1)^n + \cdots$$

which is a series solution in powers of $(x - 1)$, and converges for $|x - 1| < 2$.

11.6 Bessel Functions

Suggestions for preparing a lecture

Topics: Bessel functions of integer order, of any order, the gamma function, properties of Bessel functions. Emphasize the graphical properties, the roots, and orthogonality.

Making up a problem set

Problems 1, 4, 6 (part **(b)** is hard), 7, 10.

Comments

The classic examples of ODEs with a regular singularity are the Bessel equations, and this section and the next are mostly about these equations and their solutions. Along the way we define and give a few of the properties of another classic function, Euler's gamma function (or the interpolated factorial). The construction of the Bessel functions of the first kind follows the Frobenius steps outlined in Section 11.5. The gamma function has unusual properties and is fun to work with. Bessel functions of the first kind resemble decaying sinusoids in many ways. For later use in solving PDEs in Section 11.8, we discuss the positive zeros λ_n of Bessel functions J_p, the resemblance between $J_p(\lambda_n x)$ and $\sin(n\pi x)$, and the orthogonality properties.

1.   Using (6), we have that

$$J_3(x) = \left(\frac{x}{2}\right)^3 \left[\frac{1}{3!} - \frac{1}{4!}\left(\frac{x}{2}\right)^2 + \frac{1}{2!5!}\left(\frac{x}{2}\right)^4 - \frac{1}{3!6!}\left(\frac{x}{2}\right)^6 + \cdots\right]$$

$$= x^3/48 - x^5/768 + x^7/30720 - x^9/2211840 + \cdots$$

2.   **(a).** Since

$$J_3(x) = \sum_{k=0}^{\infty} \frac{(-1)^k}{k!(k+3)!}\left(\frac{x}{2}\right)^{2k+3}$$

we see (as in Problem 1) that the first four terms in the series add up to a polynomial $f(x)$ of degree 9:

$$f(x) = \left(\frac{x}{2}\right)^3 \left[\frac{1}{3!} - \frac{1}{4!}\left(\frac{x}{2}\right)^2 + \frac{1}{2!5!}\left(\frac{x}{2}\right)^4 - \frac{1}{3!6!}\left(\frac{x}{2}\right)^6\right]$$

Since the magnitude of the difference between the sum of a convergent alternating series of terms of decreasing magnitude and a partial sum does not exceed the magnitude of the first term omitted (Appendix B.2, item 10), we have

$$|J_3(x) - f(x)| \leq \left|\frac{x}{2}\right|^3 \frac{1}{4!7!}\left|\frac{x}{2}\right|^8 = \frac{1}{4!7!}\left|\frac{x}{2}\right|^{11}, \quad |x| \leq 4$$

This yields that $|J_3(x) - f(x)| \leq 2^{11}/(4!7!) \leq 0.02$ for $0 \leq x \leq 4$. See Fig. 2(a) for the graph of $f(x)$, $0 \leq x \leq 4$.

(b). For $|x| \leq 4$ the series $\frac{1}{3!} - \frac{1}{4!}\left(\frac{x}{2}\right)^2 + \cdots$ is a convergent alternating series of terms of decreasing magnitudes. If $a_0 - a_1 + a_2 - \cdots$ is an alternating series with these properties, then the sum S of the series is bounded by:

$$(a_0 - a_1) + (a_2 - a_3) + \cdots + (a_{2k} - a_{2k+1}) \leq S \leq a_0 - (a_1 - a_2) - \cdots - (a_{2k-1} - a_{2k})$$

We have $f(x) \leq J_3(x)$ for $|x| \leq 4$. But analysis of $f(x)$ shows that $f(x)$ is positive for $0 < x \leq 4$. The first positive zero of $J_3(x)$ is larger than 4.

(c). The second derivative of $J_3(x)$ is positive for positive x close to zero, so the graph is concave upward for those values of x.

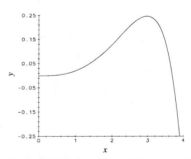

Section 11.6, Problem **2(a)**

3. According to Problem 6, Section 3.7, a second independent solution y_2 of Bessel's equation of order p satisfies the first-order linear equation $J_p y_2' - J_p' y_2 = x^{-1}$, where the term on the right side is $\exp[-\int^x a(s)\,ds]$ and $a(x) = x^{-1}$ [the normalized form $y'' + ay' + by = 0$ of Bessel's equation is $y'' + x^{-1}y' + (1 - p^2 x^{-2})y = 0$]. So $y_2' - J_p' y_2/J_p = 1/(xJ_p)$ and the integrating factor of the first-order linear ODE is $1/J_p$. We have $(y_2/J_p)' = 1/(xJ_p^2)$ and we can take $y_2 = J_p(x)\int^x (sJ_p^2(s))^{-1}\,ds$. The general solution of Bessel's equation of order p is $y = c_1 J_p(x) + c_2 J_p(x)\int^x (sJ_p^2(s))^{-1}\,ds$, c_1 and c_2 any constants.

4. The magnitude of the ratio of the $(k+1)^{\text{st}}$ and k^{th} terms in the series (15) for $J_p(x)$ is

$$\frac{x^2}{4} \cdot \frac{\Gamma(k+p+1)}{(k+1)\Gamma(k+p+2)} = \frac{x^2}{4} \cdot \frac{1}{(k+1)(k+p+1)}$$

where we have used formula (12) to express $\Gamma(k+p+2)$ as a multiple of $\Gamma(k+p+1)$. Since the above ratio $\to 0$ as $k \to \infty$ for each x, the series for $J_p(x)$ converges for all x.

5. See Figs. 5 for plots of J_0, J_1, J_2, J_3 (Graph 1) and $J_{1/3}$, $J_{4/3}$, $J_{7/3}$ (Graph 2).

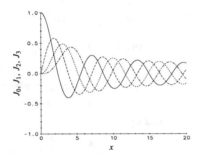
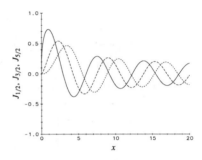

Problem 5, Graph 1.   Problem 5, Graph 2.

6. **(a).** As to the zeros of J_0, J_1, J_2 and $J_{1/2}$, $J_{3/2}$, $J_{5/2}$, see Figures 11.6.1 and 11.6.5. The two figures lead us to guess that the positive zeros of J_p and J_{p+1} interlace, where $p = 0$, 1, 2, 1/2, 3/2, or 5/2. That is, between two adjacent zeros of J_p there is a zero of J_{p+1}, and between two adjacent zeros of J_{p+1} there is a zero of J_p.

(b). Here's how to show that the zeros of J_p and J_{p+1} interlace. First, we show that J_p and J_{p+1} do not have any zeros in common. By (19), $(x^{-p}J_p(x))' = -x^{-p}J_{p+1}(x)$. If $J_p(x_0) = J_{p+1}(x_0) = 0$, then $x_0^{-p}J_p'(x_0) = 0$. So if $x_0 > 0$, $J_p'(x_0) = 0$. But we cannot have $J_p(x_0) = J_p'(x_0) = 0$, $x_0 > 0$, since that violates the Uniqueness Theorem (only $y(x) = 0$ has the property that $y(x_0) = y'(x_0) = 0$ at a regular point of a second-order linear ODE). Now suppose a and b, $a < b$ are successive positive zeros of $J_p(x)$. Using this fact and Eq. (19), we have that $J_p'(a) = -J_{p+1}(a)$ and $J_p'(b) = -J_{p+1}(b)$. Since a and b are successive zeros of $J_p(x)$, $J_p'(a)$ and $J_p'(b)$ have opposite signs. So $J_{p+1}(a)$ and $J_{p+1}(b)$ have opposite signs, and $J_{p+1}(x)$ must have a zero inside the interval (a, b). The same kind of argument using (18) with p replaced by $p + 1$ shows that between successive zeros of $J_{p+1}(x)$ there is a zero of $J_p(x)$. So the positive zeros of $J_p(x)$ and $J_{p+1}(x)$ interlace.

7. **(a).** From (11), we have $\Gamma(1/2) = \int_0^\infty x^{-1/2} e^{-x}\, dx$. Let's show that $\Gamma(1/2) = \sqrt{\pi}$. Let $x = u^2$. Then $\Gamma(1/2) = \int_0^\infty 2 e^{-u^2}\, du$. To evaluate this integral, let $I = \int_0^\infty e^{-u^2}\, du$. Then

$$I^2 = \int_0^\infty e^{-u^2}\, du \int_0^\infty e^{-v^2}\, dv = \int_0^\infty \left(\int_0^\infty e^{-(u^2+v^2)}\, du \right) dv$$

$$= \int_0^{\pi/2} \int_0^\infty \left(r e^{-r^2}\, dr \right) d\theta = -\frac{1}{2} \int_0^{\pi/2} \{e^{-r^2}\} \Big|_0^\infty d\theta = \pi/4$$

where we have introduced polar coordinates in the first quadrant of the uv plane to carry out the indicated double integral. So $I^2 = \pi/4$, $I = \sqrt{\pi}/2$, and $\Gamma(1/2) = 2I = \sqrt{\pi}$.

(b). Now let's show that

$$\Gamma\left(n + \frac{1}{2}\right) = \frac{(2n)!\sqrt{\pi}}{2^{2n}n!}$$

Certainly, by **(a)** we have that

$$\Gamma\left(n + \frac{1}{2}\right) = \frac{(2n)!}{2^{2n}n!}\sqrt{\pi}$$

when $n = 0$. Suppose the formula is true for n (the Induction Hypothesis). We shall show that it is true for $n + 1$. We have by (12) that

$$\Gamma\left(n + \frac{3}{2}\right) = (n + \frac{1}{2})\Gamma\left(n + \frac{1}{2}\right) = (n + \frac{1}{2})\frac{(2n)!}{2^{2n}n!}\sqrt{\pi}$$

where we have used the Induction Hypothesis. So

$$\Gamma\left(n + \frac{3}{2}\right) = \frac{(2n+1)!}{2^{2n+1}n!}\sqrt{\pi} = \frac{(2n+2)(2n+1)!}{2^{2n+2}(n+1)!}\sqrt{\pi} = \frac{(2n+2)!}{2^{2n+2}(n+1)!}\sqrt{\pi}$$

and the formula is verified for the case $n + 1$. By Induction, the formula holds for $n = 0, 1, 2, \ldots$.

(c). We have

$$\Gamma(3/2) = \Gamma(1/2 + 1) = \Gamma(1/2)/2 = \sqrt{\pi}/2$$

and

$$\Gamma(5/2) = \Gamma(3/2 + 1) = 3\Gamma(3/2)/2 = 3\sqrt{\pi}/4$$

Similarly,

$$\Gamma(-3/2) = \Gamma(-3/2 + 1)/(-3/2) = -2\Gamma(-1/2)/3 = 4\sqrt{\pi}/3$$

and

$$\Gamma(-5/2) = \Gamma(-5/2 + 1)/(-5/2) = -2\Gamma(-3/2)/5 = -8\sqrt{\pi}/15$$

8. Using (16) we have

$$J_2 = \frac{2}{x}J_1 - J_0, \quad J_3 = \frac{4}{x}J_2 - J_1, \quad J_4 = \frac{6}{x}J_3 - J_2$$

9. **(a).** Set $p = 0$ in equation (19) and obtain that $J_0' = -J_1$. So $\int^x J_1(t)\,dt = -J_0(x) + C$. Equation (18) for $p = 1$ yields that $(xJ_1)' = xJ_0$, and this becomes $\int^x tJ_0(t)\,dt = xJ_1(x) + C$.

(b). The identity $\int^x J_0(t)\sin t\,dt = xJ_0(x)\sin x - xJ_1(x)\cos x + C$ holds if and only if

$$J_0(x)\sin x = (xJ_0(x))'\sin x + xJ_0(x)\cos x - (xJ_1(x))'\cos x + xJ_1(x)\sin x$$

which we obtain from the identity by differentiation. The differentiated identity holds if and only if

$$J_0(x) \sin x = [(x J_0(x))' + x J_1(x)] \sin x$$

[since part (a) implies $x J_0(x) = (x J_1(x))'$] which in turn holds if and only if

$$J_0 = x(J_0' + J_1) + J_0$$

It follows from part (a) that $J_0' + J_1 = 0$. So the identity is true. In a similar way, we can show the second identity in (b), but we omit the details.

(c). $\int^x t^2 J_0 \, dt = \int^x t(t J_0) \, dt$ [using (a)] $= \int^x t(t J_1)' \, dt$ [integration by parts] $= x(x J_1) - \int^x t J_1 \, dt = x^2 J_1 - \int^x t(-J_0)' \, dt = x^2 J_1 + x J_0 - \int^x J_0 \, dt$.

(d). We have that

$$\int x^5 J_2(x) \, dx = \int x^2 (x^3 J_2) \, dx = \int x^2 (x^3 J_3)' \, dx \qquad \text{[use (18)]}$$

$$= x^2 (x^3 J_3) - 2 \int x(x^3 J_3) \, dx \qquad \text{[integration by parts]}$$

$$= x^5 J_3 - 2 \int x^4 J_3 \, dx = x^5 J_3 - 2 \int (x^4 J_4)' \, dx \qquad \text{[use (18) again]}$$

$$= x^5 J_3 - 2 x^4 J_4 + C$$

10. Let's show that $J_{-1/2} = \left(\frac{2}{\pi x}\right)^{1/2} \cos x$ satisfies Bessel's equation of order 1/2.

$$J_{-1/2}' = -\left(\frac{2}{\pi}\right)^{1/2} \frac{\sin x}{x^{1/2}} - \left(\frac{2}{\pi}\right)^{1/2} \frac{\cos x}{2 x^{3/2}} = -J_{1/2} - \frac{1}{2} x^{-1} J_{-1/2}$$

$$J_{-1/2}'' = -\left(\frac{2}{\pi}\right)^{1/2} \frac{\cos x}{x^{1/2}} + \left(\frac{2}{\pi}\right)^{1/2} \frac{\sin x}{2 x^{3/2}} + \frac{1}{2} \left(\frac{2}{\pi}\right)^{1/2} \frac{\sin x}{x^{3/2}} + \frac{3}{4} \left(\frac{2}{\pi}\right)^{1/2} \frac{\cos x}{x^{5/2}}$$

$$= -J_{-1/2} + x^{-1} J_{1/2} + \frac{3}{4} x^{-2} J_{-1/2}$$

So $x^2 J_{-1/2}'' + x J_{-1/2} + (x^2 - 1/4) J_{-1/2} = -x^2 J_{-1/2} + x J_{1/2} + 3 J_{-1/2}/4 - x J_{1/2} - J_{-1/2}/2 + (x^2 - 1/4) J_{-1/2} = 0$, and $J_{-1/2}(x)$ solves Bessel's equation of order 1/2. Since $\sin x$ is not a constant multiple of $\cos x$, then $J_{-1/2} = \left(\frac{2}{\pi x}\right)^{1/2} \cos x$ and $J_{1/2} = \left(\frac{2}{\pi x}\right)^{1/2} \sin x$ form an independent set of solutions of Bessel's equation of order $1/2$, $x > 0$.

11. See the figure for the graph of $J_1(x)$, $0 \le x \le 30$. From the graph the 8th and 9th positive zeros λ_8, λ_9 appear to be about 25.9 and 29, respectively. The difference is slightly less than π.

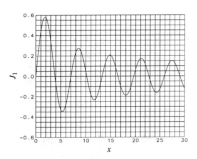

Problem 11.

12.   (a). The modified Bessel equation of order p is $t^2 y'' + ty' - (t^2 + p^2)y = 0$. Let $x = -it$; then $y(t) = z(-it)$, where $z(x)$ is any solution of Bessel's equation of order p, has the property that $dy/dt = -iz'$ and $d^2 y/dt^2 = -z''$. So $x^2 z'' + xz' + (x^2 - p^2)z = 0$ becomes

$$t^2 \frac{d^2 y}{dt^2} + t\frac{dy}{dt} - (t^2 + p^2)y = 0$$

(b). Let $p = 0$ in part (a) and observe that for the resulting equation the indicial polynomial is r^2 and the indicial roots are $r_1 = r_2 = 0$. So there is a solution of the form $y = I_0(t) = 1 + \sum_{n=1}^{\infty} a_n t^n$. We have that

$$ty'' + y' - ty = a_1 t + \sum_{n=2}^{\infty} (n^2 a_n - a_{n-2})t^n = 0$$

where $a_0 = 1$. The recursion relation is $a_n = a_{n-2}/n^2$. So $a_1 = a_3 = a_5 = \cdots = a_{2n+1} = \cdots = 0$, while

$$a_{2n} = \frac{1}{(2n)^2} a_{2n-2} = \cdots = \frac{1}{2^{2n}(n!)^2} a_0 = \frac{1}{2^{2n}(n!)^2}$$

Then $I_0(t) = \sum_{n=0}^{\infty} (t/2)^{2n}/(n!)^2$ is a solution to the modified Bessel equation of order 0.

(c). The indicial polynomial for the modified Bessel equation of order p, $t^2 y'' + ty' - (t^2 + p^2)y = 0$, is $r^2 - p^2$, and the characteristic exponents are $r_1 = p$, $r_2 = -p$. There is a solution $I_p(t) = t^p \sum_{n=0}^{\infty} a_n t^n$. We have that

$$t^2 y'' + ty' - (t^2 + p^2)y = (2p + 1)a_1 t^p + \sum_{n=2}^{\infty} [n(n + 2p)a_n - a_{n-2}]t^{p+n} = 0$$

So $a_1 = 0$ and

$$a_n = a_{n-2}/[n(n + 2p)] \quad \text{for } n \geq 2$$

where a_0 is arbitrary. This implies that $a_3 = \cdots = a_{2n+1} = \cdots = 0$, while for $a_0 \neq 0$, we have

$$a_{2n} = \frac{1}{2n(2n + 2p)} a_{2n-2} = \cdots = \frac{a_0}{2^{2n} n!(n + p)(n + p - 1)\cdots(p + 1)}$$

Now set $a_0 = [2^p \Gamma(p+1)]^{-1}$ and use the fact that $\Gamma(n+p+1) = (n+p)\Gamma(n+p) = \cdots = (n+p)(n+p-1)\cdots(p+1)\Gamma(p+1)$ to obtain the solution

$$I_p(t) = \left(\frac{t}{2}\right)^p \sum_{n=0}^{\infty} \frac{1}{n!\Gamma(n+p+1)} \left(\frac{t}{2}\right)^{2n}$$

Background Material: Verification of Bessel Recursion Relations

Let's verify formula (18). Multiply each side of (15) by x^p, differentiate term-by-term, and use the fact that $\Gamma(k+p+1) = \Gamma(k+p)(k+p)$:

$$[x^p J_p]' = \frac{d}{dx}\left\{ \sum_{k=0}^{\infty} \frac{(-1)^k 2^p}{k!\Gamma(k+p+1)} \left(\frac{x}{2}\right)^{2k+2p} \right\}$$

$$= \sum_{k=0}^{\infty} \frac{(-1)^k (k+p) 2^p}{k!\Gamma(k+p+1)} \left(\frac{x}{2}\right)^{2k+2p-1}$$

$$= x^p \left(\frac{x}{2}\right)^{p-1} \sum_{k=0}^{\infty} \frac{(-1)^k}{k!\Gamma(k+p)} \left(\frac{x}{2}\right)^{2k} = x^p J_{p-1}$$

The verification of (19) is similar. To verify (16), multiply each side of (18) by x^{-p}, each side of (19) by x^p, and subtract the two:

$$x^{-p}[x^p J_p]' - x^p[x^{-p} J_p]' = J_{p-1} + J_{p+1}$$

$$x^{-p}[px^{p-1} J_p + x^p J_p'] - x^p[-px^{-p-1} J_p + x^{-p} J_p'] = J_{p-1} + J_{p+1}$$

from which (16) follows after a little algebraic rearrangement. Formula (17) is derived similarly. Note that (16)–(19) also hold for the functions J_{-p} if p is positive and not an integer.

11.7 Series Solutions Near Regular Singular Points, II

Suggestions for preparing a lecture

Topics: Work through one example where the roots of the indicial polynomial are equal or differ by an integer, explaining in the latter case why a problem may arise. Explain the logarithm term by analogy with an Euler equation whose indicial polynomial has a double root (Section 1.4). Talk briefly about Bessel functions of the second kind, show the graphs of some of them, and emphasize that these functions tend to $-\infty$ as $x \to 0^+$. Discuss the aging spring model in terms of Bessel functions.

Making up a problem set

Problems 1(a) or (b) or (c), 3(a) (b) or 4(a) (d). Group Problem 6 is a good way to wrap up the main themes of the chapter.

Comments

The cases where the roots of the indicial polynomial are equal or differ by an integer are addressed in this section. Two examples are worked out in detail so that the reader can see the step-by-step process for obtaining a second solution that involves a logarithm, once the first solution has been obtained by the methods of Section 11.5. Alternatively, the Wronskian Reduction Method outlined in Problem 6 of Section 3.7 also leads to a logarithmic term in the second solution. Bessel functions Y_p of the second kind are introduced and (briefly) discussed. The important point here is that $Y_p(x) \to -\infty$ as $x \to 0^+$. Finally, applications of all of this are made to the two problems introduced at the start of the chapter: the model of the aging spring, and the ODE arising when the PDE boundary problem model for steady temperatures in a cylinder has its variables separated. Both involve ODEs that reduce to a Bessel equation when new variables are introduced. This, incidentally, is another reason for the importance of the Bessel equations: many seemingly intractable ODEs can be reduced to a Bessel equation by a variable change.

---

1. It is straightforward to show that 0 is a regular singular point of each ODE; the details are omitted.

(a). The ODE is $xy'' + (1+x)y' + y = 0$. The equation in standard form is $x^2 y'' + x(1+x)y' + xy = 0$. So, $p_0 = 1$, $q_0 = 0$ and the indicial polynomial is r^2 with roots $r_1 = r_2 = 0$. We are in Case II of Frobenius Theorem II. First, there is a solution of the form $y_1 = \sum_{n=0}^{\infty} a_n x^n$ with $a_0 = 1$. Using the techniques of Section 11.5, we see that the recursion formula in this case is

$$a_{n+1} = -a_n/(n+1), \quad n = 0, 1, 2, \cdots$$

and a solution is

$$y_1 = \sum_{n=0}^{\infty} (-1)^n x^n/n! = e^{-x}$$

We have by the Wronskian Reduction Method of Problem 6, Section 3.7, that $e^{-x}y_2' + e^{-x}y_2 = e^{-x}/x$, where the right-hand side is $\exp[-\int^x a(s)ds]$, $a(s) = s^{-1} + 1$, for the normalized ODE, $y'' + x^{-1}(1+x)y' + x^{-1}y = 0$. So, $y_2' + y_2 = 1/x$, or $(y_2 e^x)' = e^x/x$. We have as a second independent solution

$$y_2 = e^{-x} \int^x (e^s/s)ds$$

Since the integral can't be expressed in terms of elementary functions, we use power series techniques. Expanding e^s in its Taylor series about $x_0 = 0$, dividing each term by s, and integrating term by term, we have that

$$y_2 = e^{-x}[\ln x + x + x^2/4 + \cdots + x^n/(n!n) + \cdots] = y_1(x)[\ln x + x + x^2/4 + \cdots + x^n/(n!n) + \cdots]$$

(b). The ODE is $x^2 y'' + x(x-1)y' + (1-x)y = 0$. Here $p_0 = -1$, $q_0 = 1$, and the indicial polynomial is $r^2 - 2r + 1$ with roots $r_1 = r_2 = 1$. We are in Case II of Frobenius's

Theorem II. Using the methods of Section 11.5, we see that there is a solution of the form

$$y_1 = \sum_{n=0}^{\infty} a_n x^{n+1}$$

where $a_0 = 1$. In fact, the recursion formula is

$$a_n = (1 - n)a_{n-1}/n^2$$

and we see that $a_n = 0$ for $n \geq 1$, so $y_1 = x$ is a solution. We have by the Wronskian Reduction Method of Problem 6, Section 3.7, that $xy_2' - y_2 = xe^{-x}$ since $a(x) = 1 - x^{-1}$ in the normalized ODE and $\exp[-\int^x a(s)ds] = xe^{-x}$. So $y_2' - y_2/x = e^{-x}$, $(y_2/x)' = e^{-x}/x$ and a second solution is

$$y_2 = x \int^x s^{-1}e^{-s} \, ds = x[\ln x - x + x^2/4 + \cdots + (-1)^n x^n/(n!n) + \cdots]$$

where the factor $[\cdots]$ is obtained by writing the Maclaurin series for e^{-s}, dividing each term by s, and integrating.

(c). The ODE is $xy'' - xy' + y = 0$. Here the ODE in standard form is $x^2 y'' - x^2 y' + xy = 0$, and the indicial polynomial is $r^2 - r$ and $r_1 = 1$, $r_2 = 0$. We are in Case III of Frobenius Theorem II. We may use the methods of Section 11.5 to find a solution $y_1 = \sum_{n=0}^{\infty} a_n x^{n+1}$, $a_0 = 1$. The recursion formula is

$$a_{n+1} = \frac{(1-n)a_n}{n(n+1)}, \quad n \geq 0$$

So $a_n = 0$ for $n \geq 1$ and

$$y_1 = x$$

is a solution. We have by the Wronskian Reduction Method of Problem 6, Section 3.7 that $xy_2' - y_2 = e^x$, $y_2' - y_2/x = e^x/x$, $(y_2/x)' = e^x/x^2$ and

$$y_2 = x \int^x e^s/s^2 ds = x[-1/x + \ln x + \cdots + x^{n-1}/(n!(n-1)) + \cdots]$$

where we replaced e^s by its Maclaurin series, divided each term by s^2, and then integrated term by term.

(d). The ODE is $xy'' - x^2 y' + y = 0$. The standard form is $x^2 y'' - x^3 y' + xy = 0$, and the indicial polynomial is $r^2 - r$ with roots $r_1 = 1$, $r_2 = 0$. The methods of Section 11.5 give a solution $y_1 = \sum_{n=0}^{\infty} a_n x^{n+1}$, $a_0 = 1$, where the recursion formula is

$$(n+1)(n+2)a_{n+1} + a_n - na_{n-1} = 0, \quad n \geq 1$$

and $2a_1 + a_0 = 0$. Three-term recursion formulas are not easy to solve. We only find the first few terms in the series for y_1

$$y_1(x) = x - x^2/2 + x^3/4 - 5x^4/48 + \cdots$$

A second independent solution has the form

$$y_2 = \alpha y_1 \ln x + \sum_{n=0}^{\infty} d_n x^n$$

where $d_0 = 1$. Now let the operator $L = xD^2 - x^2 D + 1$. Then $L[y_2]$ becomes

$$(\alpha \ln x) L[y_1] + \alpha(2y_1' - y_1/x - xy_1) + L\left[\sum_{n=0}^{\infty} d_n x^n\right]$$

$$= 2\alpha(1 - x + 3x^2/4 - 5x^3/12 + \cdots) - \alpha(1 - x/2 + x^2/4 - 5x^3/48 + \cdots)$$

$$- \alpha(x^2 - x^3/2 + x^4/4 - 5x^5/48 + \cdots) + d_0$$

$$+ \sum_{n=1}^{\infty} [(n^2 + n)d_{n+1} + d_n - (n-1)d_{n-1}]x^n = 0$$

where we have used $L[y_1] = 0$ and the series expansion for y_1 found above. Combining coefficients of like powers of x and setting $d_0 = 1$, we have

$$(\alpha + 1) + (-3\alpha/2 + 2d_2 + d_1)x + (\alpha/4 + 6d_3 + d_2 - d_1)x^2 + \cdots = 0$$

So, $\alpha = -1$ and we may take d_1 to be any convenient value (say $d_1 = 0$). We have $d_2 = -3/4$, $d_3 = 1/6$, and so on. We have that

$$y_2 = -y_1(\ln x + 1 - 3x^2/4 + x^3/6 + \cdots)$$

2. (a). Let $y(x) = e^{ax}w(bx)$, where w is a solution of Bessel's equation of order p, $s^2 w'' + sw' + (s^2 - p^2)w = 0$. Then, by the Chain Rule, $y'(x) = ay(x) + be^{ax}w'$, $y''(x) = ay'(x) + abe^{ax}w' + b^2 e^{ax}w'' = a^2 y(x) + 2abe^{ax}w' + b^2 e^{ax}w''$. Using the fact that $b^2 x^2 w'' + bxw' + (b^2 x^2 - p^2)w = 0$, it is a straightforward, but lengthy, calculation to show that $x^2 y'' + x(1 - 2ax)y' + [(a^2 + b^2)x^2 - ax - p^2]y = 0$. So, the Bessel function solutions of the w-equation can be used to solve the y-equation.

    (b). Comparing the given equation $x^2 y'' + x(1 - 2x)y' + (2x^2 - x - 1)y = 0$ with the y-equation in (a), we have $a = 1$, $b = 1$, $p = 1$. So, $y = e^x w(x)$ where $w = c_1 J_1(x) + c_2 Y_1(x)$.

3. (a). Let $y(x)$ be a solution of Bessel's equation of order p: $x^2 y''(x) + xy'(x) + (x^2 - p^2)y(x) = 0$. Let $w(z) = z^{-c}y(az^b)$. Then

$$w' = -\frac{cw}{z} + abz^{b-c-1}y'$$

$$w'' = \frac{c}{z^2}w - \frac{c}{z}(-\frac{c}{z}w + abz^{b-c-1}y') + ab(b-c-1)z^{b-c-2}y' + a^2b^2 z^{2b-c-2}y''$$

A lengthy calculation then shows that w satisfies the equation $z^2 w'' + (2c + 1)zw' + [a^2 b^2 z^{2b} + (c^2 - p^2 b^2)]w = 0$. Solutions of the w-equation can be expressed in terms of $z^{-c}J_p(az^b)$ and $z^{-c}Y_p(az^b)$ [or $z^{-c}J_{-p}(az^b)$ if p is not an integer].

    (b). Comparing Airy's equation $d^2 y/dz^2 + zy = 0$ (note the replacement of $-x$ by z) with the w-equation in (a), we see, after multiplying Airy's equation by z^2, that $c = -1/2$,

$b = 3/2$, $a^2 = 4/9$, $p = 1/3$. So, $y(x) = |x|^{1/2}[c_1 J_{1/3}(2|x|^{3/2}/3) + c_2 J_{-1/3}(2|x|^{3/2}/3)]$. Since $\{J_{1/3}, J_{-1/3}\}$ is an independent set, $\{1/3(J_{1/3} + I_{1/3}), \frac{-1}{\sqrt{3}}(J_{1/3} - J_{-1/3}\}$ is also an independent set, and so is $\{Ai(x), Bi((x)\}$.

(c). Comparing the given equation with the w-equation of **(a)**, we have that $c = -1/2$, $b = 2$, $a = 1/2$, $p = 1/(4\sqrt{2})$. So, $y = x^{1/2}[c_1 J_p(x^2/2) + c_2 J_{-p}(x^2/2)]$, $p = 1/(4\sqrt{2})$, $x > 0$.

(d). Comparing the given equation (after multiplying by x^2) with the w-equation of **(a)**, we have $c = -1/2$, $b = 3$, $a = 1/3$, $p = 1/6$. So, $y = x^{1/2}[c_1 J_{1/6}(x^3/3) + c_2 J_{-1/6}(x^3/3)]$, $x > 0$.

4. See Fig. 5. Observe that $y \to 1$ as $t \to +\infty$ since $J_0(0) = 1$, $6e^{-t} \to 0$ as $t \to +\infty$ and we are looking at $J_0(6e^{-t}$. This corresponds to the spring tending to a unit compression.

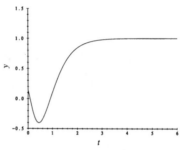

Section 11.7, Problem **5**

5.   Group problem.

6.   Group problem.

11.8 **Steady Temperatures in Spheres and Cylinders**

Suggestions for preparing a lecture

Topics: Singular Sturm-Liouville problems, Legendre basis, Bessel basis, either the temperature problem in the sphere or the problem in a cylinder (but not both).

Making up a problem set

Problems 1**(a)(b)**, 3**(a)**, 6**(a)**.

Comments

Finally, we find out why Bessel functions and Legendre polynomials were introduced in Chapter 11. It turns out that the wave, heat, and potential PDEs can be separated into individual coordinates only by using cylindrical or spherical coordinates when the bounding surfaces have those respective shapes. Moreover, one or more of the resulting ODEs will almost always be a Bessel equation, a Legendre equation, or some other nonconstant-coefficient second order linear ODE like those studied in Chapter 11. We have already seen in Section 10.9 that an Euler ODE comes up in solving a heat problem in a disk. In this section, we concentrate on just two models, steady temperatures in a cylinder immersed in ice water (except at one end), and steady temperatures in a spherical solid with prescribed temperatures on the boundary. Extensive use is made of the results in Sections 11.3, 4, 6, 7 on Legendre, Euler, and Bessel equations.

1. We want to find a formula for the steady temperature in a unit ball with the given surface temperature $f(\phi)$. We follow the steps given in the text.

   (a). Because $f = \cos 2\phi - \sin^2 \phi = 3\cos^2 \phi - 2$, we have $g(x) = f(\arccos x) = 3x^2 - 2$. Using formulas (12) and (15), we have

   $$u = \sum_{n=0}^{\infty} A_n \rho^n P_n(\cos \phi)$$

   $$A_n = \frac{2n+1}{2} \int_{-1}^{1} g(x) P_n(x)\, dx = \begin{cases} -1 & n = 0 \\ 2 & n = 2 \\ 0 & \text{otherwise} \end{cases}$$

   So, the steady-state temperature in a ball of unit radius is $u(\rho, \phi, \theta) = -1 + 2P_2(\cos \phi)\rho^2$.

   (b). Because $f = \begin{cases} 1, & 0 < \phi < \pi/2 \\ 0, & \pi/2 < \phi < \pi \end{cases}$, $g(x) = f(\arccos x) = \begin{cases} 1, & 0 < x < 1 \\ 0, & -1 < x < 0 \end{cases}$

   So

   $$A_n = \frac{2n+1}{2} \int_{-1}^{1} g(x) P_n(x)\, dx = \frac{2n+1}{2} \int_{0}^{1} P_n(x)\, dx$$

   Because $xP_n' - P_{n-1}' = nP_n$ [see one of the identities in Table 11.3.1], we have (integrating by parts),

   $$n \int_{0}^{1} P_n(x)\, dx = \int_{0}^{1} (xP_n' - P_{n-1}')\, dx = xP_n(x)\Big|_{0}^{1} - \int_{0}^{1} P_n(x)\, dx - P_{n-1}\Big|_{0}^{1}$$

   $$= P_n(1) - P_{n-1}(1) + P_{n-1}(0) - \int_{0}^{1} P_n(x)\, dx = P_{n-1}(0) - \int_{0}^{1} P_n(x)\, dx$$

   since $P_k(1) = 1$ for all k. So $\int_{0}^{1} P_n(x)\, dx = P_{n-1}(0)/(n+1)$, $n \geq 1$. So

   $$A_n = \begin{cases} 1/2 & n = 0 \\ \dfrac{2n+1}{2n+2} P_{n-1}(0) & n > 0 \end{cases}$$

   Since

   $$P_k(0) = \begin{cases} 0, & k \text{ odd} \\ (-1)^{k/2} \dfrac{k!}{2^k (k!)^2}, & k \text{ even} \end{cases}$$

we only need to consider odd values of n, $n = 2k + 1$. So, the steady-state temperature in a ball of unit radius is

$$u(\rho, \phi, \theta) = \frac{1}{2} + \sum_{k=0}^{\infty} \frac{4k+3}{4k+4} P_{2k}(0) P_{2k+1}(\cos\phi) \rho^{2k+1}$$

where $P_{2k}(0)$ can be obtained from the formula given above.

(c). Because $f = |\cos\phi|$, $g(x) = f(\arccos x) = |x|$. So

$$A_n = \frac{2n+1}{2} \int_{-1}^{1} g(x) P_n(x)\, dx = \frac{2n+1}{2} \int_{-1}^{1} |x| P_n(x)\, dx$$

$$= \frac{2n+1}{2} \left(\int_{0}^{1} x P_n(x)\, dx - \int_{-1}^{0} x P_n(x)\, dx \right)$$

Note from Table 11.3.1 that $n P_n = x P_n' - P_{n-1}'$, so we have

$$n \int_0^1 x P_n\, dx = \int_0^1 (x^2 P_n' - x P_{n-1}')\, dx = x^2 P_n \Big|_0^1 - 2\int_0^1 x P_n\, dx - x P_{n-1}\Big|_0^1 + \int_0^1 P_{n-1}\, dx$$

$$= P_n(1) - P_{n-1}(1) - 2\int_0^1 x P_n\, dx + \int_0^1 P_{n-1}\, dx$$

$$= -2\int_0^1 x P_n\, dx + \int_0^1 P_{n-1}\, dx$$

since $P_k(1) = 1$ for all k. Since $\int_0^1 P_k(x)\, dx = P_{k-1}(0)/(k+1)$, we have

$$\int_0^1 x P_n\, dx = \frac{1}{n+2} \int_0^1 P_{n-1}\, dx = \frac{P_{n-2}(0)}{n(n+2)}$$

Similarly, we can obtain

$$\int_{-1}^0 x P_n\, dx = \frac{1}{n+2} \int_{-1}^0 P_{n-1}\, dx = \frac{(-1)^{n-1} P_{n-2}(0)}{n(n+2)}$$

It is easy to calculate the coefficients A_0 and A_1 directly: $A_0 = 1/2$, $A_1 = 0$. For $n \geq 2$, we have that

$$A_n = \frac{2n+1}{2} \left(\frac{P_{n-2}(0)}{n(n+2)} + \frac{(-1)^{n-2} P_{n-2}(0)}{n(n+2)} \right) = \frac{(2n+1)(1+(-1)^{n-2})}{n(n+2)}$$

So, the steady-state temperature in a ball of unit radius is

$$u(\rho, \phi, \theta) = \frac{1}{2} + \sum_{n=2}^{\infty} \frac{(2n+1)(1+(-1)^{n-2}) P_{n-2}(0)}{n(n+2)} P_n(\cos\phi)\rho^n$$

2. We want to find steady temperatures in the hollow ball described by $\rho_1 < \rho < \rho_2$ where the temperature on the inner and outer walls are given, respectively, by $f(\phi)$ and $g(\phi)$. Similar to the process in the text, we have

$$u(\rho, \phi) = \sum_{n=1}^{\infty} [A_n \rho^n P_n(\cos\phi) + B_n \rho^{-n} P_n(\cos\phi)]$$

Note that the boundary conditions are $u(\rho_1, \phi) = f(\phi)$, $u(\rho_2, \phi) = g(\phi)$. So

$$A_n \rho_1^n + B_n \rho_1^{-n} = \frac{2n+1}{2} \int_{-1}^{1} \tilde{f}(x) P_n(x)\, dx \qquad A_n \rho_2^n + B_n \rho_2^{-n} = \frac{2n+1}{2} \int_{-1}^{1} \tilde{g}(x) P_n(x)\, dx$$

where $\tilde{f}(x) = f(\arccos x)$, $\tilde{g}(x) = g(\arccos x)$. The above equations may be solved for A_n and B_n and the formulas for A_n and B_n inserted into the series for u.

3. **(a).** We are to solve boundary problem (16) in a cylinder if $f(r) = 1$, $0 \geq r \geq 1$. If $f(r) = 1$, $0 \leq r \leq 1$, then with x_n, $n = 1, 2, ...$, denoting the consecutive positive zeros of $J_0(x)$, we have from (20) that

$$A_n = \frac{2}{[J_1(x_n)]^2 \sinh(x_n a)} \int_0^1 r J_0(x_n r)\, dr = \frac{2}{[J_1(x_n)]^2 \sinh(x_n a) k_n^2} \int_0^1 (r J_1(x_n r))'\, dr$$

$$= \frac{2}{[J_1(x_n)]^2 \sinh(x_n a)) x_n^2} (r J_1(x_n r)) \Big|_0^1 = \frac{2}{J_1(x_n) \sinh(x_n a)) x_n^2}$$

where we have used the Bessel identity, $(x J_1(x))' = x J_0(x)$. So

$$u(r, z) = \sum_{n=1}^{\infty} \frac{2}{x_n^2} \cdot \frac{\sinh x_n (a - z)}{\sinh(x_n a)} \cdot \frac{J_0(x_n r)}{J_1(x_n)}$$

(b). Now assume perfect insulation on the sidewalls of the cylinder: $u_r(1, \theta, z) = 0$ for $0 \leq z \leq a$, $-\pi \leq \theta \leq \pi$. If the sidewall temperature condition $u(1, \theta, z) = 0$ is replaced by the perfect-insulation condition $u_r(1, \theta, z) = 0$, boundary problem (17) becomes

$$R'' + \frac{1}{r} R' - \lambda R = 0, \qquad R(0^+) < \infty, \qquad R'(1) = 0$$

The corresponding eigenfunctions are given by $R_n(r) = J_0(x_n^* r)$, $n = 1, 2, \ldots$, where x_n^* are the consecutive positive zeros of $J_1(x)$, $n = 1, 2, \ldots$ (because $J_0' = -J_1$). Similar to the procedure in the text, we have the following solution:

$$u(r, z) = \sum_{n=1}^{\infty} A_n \sinh x_n^* (a - z) J_0(x_n^* r)$$

where

$$A_n = \frac{2}{(x_n^*)^2 [J_0(x_n^*)]^2 \sinh(x_n^* a)} \int_0^1 r J_0(x_n^* r)\, dr, \qquad J_1(x_n^*) = 0, \quad n = 1, 2, \ldots$$

4. We want to show that $\int_0^1 r J_0^2(ar)\, dr = [J_0^2(a) + J_1^2(a)]/2$. Multiplying Bessel's equation of order 0 by $2J_0'(r)$ and rewriting it, we can obtain $[r^2 (J_0')^2]' + r^2 (J_0^2)' = 0$. Integrating from 0 to x_n using integration by parts, we have

$$0 = \int_0^{x_n} ([r^2 (J_0')^2]' + r^2 (J_0^2)')\, dr = [r^2 [J_0'(r)]^2 + r^2 (J_0^2)] \Big|_0^{x_n} - 2 \int_0^{x_n} r J_0^2\, dr$$

$$= [r^2 J_1^2(r) + r^2 J_0^2(r)] \Big|_0^{x_n} - 2 \int_0^{x_n} r J_0^2(r)\, dr$$

Setting $r = t/x_n$ and at the last step using formula (19) in Section 11.6, we have

$$\int_0^1 r J_0^2(x_n r)\, dr = \int_0^{x_n} x_n^{-2} t J_0^2(t)\, dt = \frac{1}{2}[J_1^2(x_n) + J_0^2(x_n)]$$

5. In a Neumann problem $\nabla^2 u = 0$ inside a region G and $\partial u/\partial n$ is a prescribed funtion of position on the boundary of G.

(a). Suppose that u_1 and u_2 are solutions and set $w = u_1 - u_2$. Then $\nabla^2 w = 0$ in G and $\partial w/\partial n = 0$ on ∂G. Let G^* consist of G together with its boundary. Since $\nabla(w^2) = 2w\nabla w$ and $\nabla w \cdot n = \partial w/\partial n = 0$, using the planar Divergence Theorem (Theorem B.5.13) on w^2, we have

$$\int_{\partial G} \nabla(w^2) \cdot n\, ds = \int_{G^*} \text{div}(\nabla w^2)\, dA = 0$$

On the other hand, since $\nabla^2 w = 0$,

$$\int_{G^*} \text{div}(\nabla w^2)\, dA = \int_{G^*} 2[w\nabla^2 w + ||\nabla w||^2]\, dA = 2\int_{G^*} ||\nabla w^2||\, dA = 0$$

So $\nabla w = i\partial w/\partial x + j\partial w/\partial y$ is the zero vector throughout G^*. That means that $w(x, y)$ does not change as x and y change. So $w = $ constant on G^*, and it follows that $u_1 = u_2 +$ constant.

(b). Applying the planar Divergence Theorem to u, we have

$$\int_{G^*} \text{div}(\nabla u)\, dA = \int_{G^*} (\nabla^2 u)\, dA = \int_{\partial G} \nabla u \cdot n\, ds = \int_{\partial G} \partial u/\partial n\, ds = \int_{\partial G} f(s)\, ds = 0$$

(c). Suppose G is the unit disk in the plane and $f(\theta) = \sin\theta$ on the perimeter ∂G, while $u = 0$ if $\theta = 0$ and $r = 1$. As in formula (10) in Section 10.9, we have the following solution form:

$$u(r, \theta) = \frac{A_0}{2} + \sum_{n=1}^{\infty} r^n (A_n \cos n\theta + B_n \sin n\theta)$$

where A_n and B_n are arbitrary. We will determine A_n and B_n by imposing the boundary condition $\partial u/\partial n = \sin\theta$ on ∂G. That is

$$\frac{\partial u}{\partial r}(1, \theta) = \sin\theta = \sum_{n=1}^{\infty} [n A_n \cos n\theta + n B_n \sin n\theta]$$

We know that $A_n = 0$, $n = 1, 2, \ldots$, $B_1 = 1$, $B_2 = 0$, $n = 2, 3, \ldots$ and $u(r, \theta) = A_0/2 + r\sin\theta$. Since $u = 0$ when $\theta = 0$, $r = 1$, we must have $A_0 = 0$, so $u = r\sin\theta$.

6. In each case we want to write the eigenvalue problem in the form $Ly = \lambda y$.

(a). The ODE is $x^4 y'' + 2x^3 y' = \lambda y$, where $y(1) = y(2) = 0$. L has the symmetric form $L[y] = (x^2 y')'/x^{-2}$, and $\text{Dom}(L) = \{y \text{ in } C^2[1, 2] : y(1) = y(2) = 0\}$. The eigenvalues of L are nonpositive. First try $\lambda = 0$. The eigenfunctions must have the form $y = A/x + B$, and the conditions imply that $A + B = 0$, $A/2 + B = 0$, and hence $A = B = 0$. So $\lambda = 0$ is not an eigenvalue for L. Next, try $\lambda = -k^2$, $k > 0$. To solve the equation $L[y] = -k^2 y$ we make the independent variable substitution $s = 1/x$. Since $dy/dx = (dy/ds)(-1/x^2)$, we see that

$x^2 y' = -dy/ds$, and so $d(x^2 y')/dx = -(d^2 y/ds^2)(-1/x^2)$. The transformed equation is $d^2 y/ds^2 = -k^2 y$; the general solution of the original ODE is $y = A\cos(k/x) + B\sin(k/x)$. The endpoint conditions imply that $A\cos k + B\sin k = 0$, $A\cos(k/2) + B\sin(k/2) = 0$. The determinant of this linear system is

$$\Delta = \sin\frac{k}{2}\cos k - \cos\frac{k}{2}\sin k = \sin\left(\frac{k}{2} - k\right) = -\sin\frac{k}{2}$$

Note that $\Delta = 0$ if and only if $k/2 = n\pi$, $n = 1, 2, \ldots$. So $k_n = 2n\pi$, $n = 1, 2, \ldots$, and the eigenvalues of L are $\lambda_n = -k_n^2 = -4n^2\pi^2$, $n = 1, 2, \ldots$. To find the corresponding eigenspaces, substitute k_n into the linear system above to obtain $A\cos 2n\pi + B\sin 2n\pi = 0$, $A\cos n\pi + B\sin n\pi = 0$. So $A = 0$ and B is arbitrary. The eigenspace V_{λ_n} corresponding to λ_n is spanned by $y_n = \sin(2n\pi/x)$. The eigenspaces of L are orthogonal in $PC[1, 2]$ under the weighted scalar product $\langle f, g \rangle = \int_1^2 x^{-2} fg\, dx$, and the set $\Phi = \{\sin(2n\pi/x) : n = 1, 2, \ldots\}$ is a basis for $PC[1, 2]$ under that same scalar product.

(b). The ODE is $xy'' - y' = \lambda x^3 y$ with boundary data $y(0) = y(a) = 0$. The operator associated with the Sturm-Liouville problem can be written in the symmetric form $L[y] = (y'/x)'/x$, with $\rho = x$ and $\text{Dom}(L) = \{y \text{ in } C^2[0, a] : y(0) = y(a) = 0\}$. The Nonpositivity of Eigenvalue Theorem (Theorem 10.6.3) shows that the eigenvalue problem $L[y] = \lambda y$ has no solutions with $\lambda > 0$. Trying $\lambda = 0$, we see that the corresponding eigenfunctions have the form $y = C_1 x + C_2$. The endpoint conditions imply that $C_1 = C_2 = 0$, and hence $\lambda = 0$ is not an eigenvalue. Next try $\lambda = -4k^2$, $k > 0$. The change of independent variable $s = x^2$, $x \leq 0$, gives the equation $L[y] = \lambda y$ the form $d^2 y/ds^2 = -k^2 y$. Eigenfunctions in this case have the form $y = A\cos(kx^2) + B\sin(kx^2)$. The endpoint conditions imply that $A = 0$, and $\sin(ka^2) = 0$. So $ka^2 = n\pi$, $n = 1, 2, \ldots$ and hence $\lambda_n = -(2n\pi/a^2)^2$, $n = 1, 2, \ldots$ are the eigenvalues of L, and the corresponding eigenspace E_n is spanned by the eigenfunction $y_n = \sin(2n\pi x^2/a^2)$. The eigenspaces are orthogonal in $PC[0, a]$ under the weighted scalar product $\langle f, g \rangle = \int_0^a xfg\, dx$. The orthogonal set $\Phi = \{\sin(2n\pi x^2/a^2) : n = 1, 2, \ldots\}$ is a basis for $PC[0, a]$ under the same scalar product.

(c). The ODE is $[(xy')' - p^2 y/x]/x = \lambda y$, where y is in $C^2[0, R_0]$ and $y(R_0) = 0$. The operator $L[y] = [(y'/x)' - p^2 y/x]/x$ with $\text{Dom}(L) = \{y \text{ in } C^2[0, R_0] : y(R_0) = 0\}$ is already in symmetric form. The Nonpositivity of Eigenvalue Theorem (Theorem 10.6.3) shows that no eigenvalue of L can be positive. First, try $\lambda = 0$. Then the eigenvalue equation $L[y] = 0$ takes the form of an Euler equation: $x^2 y'' + xy' - p^2 y = 0$. Any corresponding eigenfunctions have the form $y = Ax^p + Bx^{-p}$. But $B = 0$ if this function is to be in $C^2[0, R_0]$, and $A = 0$ if $y(R_0) = 0$. Hence, $\lambda = 0$ is not an eigenvalue. Next try $\lambda = -k^2$, $k > 0$. Then the eigenvalue equation $L[y] = 0$ takes the form of an ODE which can be reduced to a Bessel equation of order p: $x^2 y'' + xy' + (k^2 x^2 - p^2)y = 0$ by replacing kx by x. A consequence of Problem 3(a), Section 11.7, is that this equation has only one solution that is bounded near the origin; namely, $y = J_p(kx)$. The boundary condition $y(R_0) = 0$ implies that $J_p(kR_0) = 0$. Let $x_1^{(p)}, x_2^{(p)}, \ldots$ be the consecutive positive zeros of J_p, then $kR_0 = x_n^{(p)}$, $n = 1, 2, \ldots$, and hence $\lambda_n = -(x_n^{(p)}/R_0)^2$, $n = 1, 2, \ldots$ are the eigenvalues of L, and the corresponding eigenspace V_{λ_n} is spanned by the eigenfunction $y_n = J_p(x_n^{(p)}x/R_0)$. The eigenspaces are orthogonal in $PC[0, a]$ under the weighted scalar product $\langle f, g \rangle = \int_0^{R_0} xfg\,dx$. Moreover, the orthogonal set $\Phi = \{J_p(x_n^{(p)}x/R_0) : n = 1, 2, \ldots\}$ is a basis for $PC[0, R_0]$ under the same scalar product cited above.

APPENDIX A

Basic Theory of Initial Value Problems

Appendices A.1–A.4 discuss and verify the results of the Existence and Uniqueness Theorem (Theorem 2.1.1), the Extension Theorem (Theorem 2.2.1) and the Sensitivity/Continuity part of the Fundamental Theorem (Theorem 2.3.3). Appendix A.1 looks at Uniqueness, Appendix A.2 at Existence, Appendix A.3 at Extension, and Appendix A.4 at Sensitivity/Continuity.

A.1 Uniqueness

1. **(a).** The function $f = |y|$ is not differentiable at $y = 0$ since $\lim_{h \to 0}(f(h) - f(0))/h = \lim_{h \to 0}((|h| - 0)/h) = \pm 1$, depending upon whether h goes to zero through positive or negative values. So f doesn't satisfy the hypotheses of Theorem 2.1.1.

   (b). Since one of the hypotheses does not hold in any region containing a portion of the t-axis, the Uniqueness Principle cannot be used to show that the problem $y' = |y|$, $y(t_0) = 0$, has a unique solution. The problem has the trivial solution $y(t) = 0$, all t. To show that there are no other solutions, we proceed as follows. Note that if $y_0 \neq 0$, then the problem $y' = |y|$, $y(t_0) = y_0$, meets the conditions of the Uniqueness Principle in the region $y > 0$ and also in the region $y < 0$. The unique solution is $y = y_0 e^{\pm(t - t_0)}$, $-\infty < t < +\infty$, where "+" is used if $y_0 > 0$, and "−" is used if $y_0 < 0$. Any solution $y_1(t)$ of $y' = |y|$, $y(t_0) = 0$ (except $y = 0$) must intersect at least one of these exponential solutions at some t-value. But if it does so, then by uniqueness in the region above and below the t-axis, it would have to coincide with that exponential solution for all t, so could never vanish. This contradiction shows that $y(t) = 0$, all t, is the unique solution of the given problem.

   (c). The proof of the Uniqueness Principle in the textbook actually works when f satisfies a Lipschitz condition. From the Triangle Inequality (Section 7.2) we have that $||y_1| - |y_2|| \leq |y_1 - y_2|$, for all y_1, y_2. So $f(y) = |y|$ satisfies a Lipschitz condition [see (4) in the text] with $L = 1$; it follows that the IVP of part **(b)** has a unique solution.

2. **(a).** The ODE is $y' = |y|^{m/n}$, $y(0) = 0$; m and n are relatively prime positive integers. Certainly, $y(t) = 0$, all t, is a solution. To show that there are no other solutions, note first that if $m/n = 1$, then by Problem 1 there are no other solutions. Suppose $m > n$ and let $m/n = 1 + \alpha$, where $\alpha > 0$. We are finished if we can show that the conditions of the

Uniqueness Principle hold. The function

$$f = \begin{cases} y^{1+\alpha}, & y \geq 0 \\ (-y)^{1+\alpha}, & y < 0 \end{cases}$$

is certainly continuous everywhere and is continuously differentiable for $y \neq 0$ with derivative

$$f' = \begin{cases} (1+\alpha)y^{\alpha}, & y > 0 \\ (-1)(1+\alpha)(-y)^{\alpha}, & y < 0 \end{cases} = (1+\alpha)|y|^{\alpha}\mathrm{sgn}(y)$$

It is differentiable with derivative 0 at $y = 0$ since $\lim_{h \to 0}(f(h) - f(0))/h = \lim_{h \to 0}(|h|^{1+\alpha} - 0)/h = \lim_{h \to 0}(\pm|h|^{\alpha}) = 0$. Since $\lim_{y \to 0}(1+\alpha)y^{\alpha}\mathrm{sgn}(y)$ is also 0, we see that $f(y)$ is continuously differentiable everywhere, and the conditions of the Uniqueness Principle hold.

(b). If $m < n$, then $f = |y|^{m/n}$ is no longer differentiable at $y = 0$. The conditions of the Uniqueness Principle do not hold on intervals containing 0. The IVP $y' = |y|^{m/n}$, $y(0) = 0$, has many solutions. One solution is $y(t) = 0$, all t, and there are infinitely many other solutions. To see this, we separate variables and write the ODE as $y^{-m/n}y' = 1$, or in derivative form as $(a^{-1}y^a)' = 1$, where $a = 1 - m/n$. Integrating, we obtain $a^{-1}y^a = t - C$ where C is an arbitrary constant. For $t \geq C$ the function $z(t) = (at - aC)^{1/a}$ is a solution of the ODE $y' = y^{m/n}$. Notice that $\lim_{t \to C^+} z'(t) = 0$, for any value of C. The functions

$$y(t) = \begin{cases} (at - aC)^{1/a}, & t \geq C \\ 0, & t < C \end{cases}$$

are all solutions of the IVP $y' = y^{m/n}$, $y(0) = 0$, if C is any positive constant.

A.2 The Picard Process for Solving an Initial Value Problem

1. We want to find the Picard iterates for $y' = -y$, $y(0) = 1$.

(a). Using iteration scheme (5), we have the Picard iterates $y_0(t) = 1$, all t

$$y_1(t) = 1 + \int_0^t (-y_0(s))\,ds = 1 + \int_0^t (-1)\,ds = 1 - t$$

$$y_2(t) = 1 + \int_0^t (-y_1(s))\,ds = 1 + \int_0^t -(1-s)\,ds = 1 - t + t^2/2$$

$$y_3(t) = 1 + \int_0^t (-y_2(s))\,ds = 1 + \int_0^t -(1 - s + s^2/2)\,ds = 1 - t + t^2/2 - t^3/6$$

These iterates are defined for all t.

(b). See Fig. 1(b) for the graphs of $y_0 - y_3$. The dashed line is the solution curve for the IVP.

(c). Following the pattern of y_0, y_1, y_2, y_3, we guess that the n-th Picard iterate is given by $y_n(t) = \sum_{j=0}^{n}(-1)^j t^j / j!$. This is proved by induction on n: The formula holds for $n = 1$ (see part **(a)**). Suppose it holds for n (the induction hypothesis). We shall show it holds for $n + 1$. From the iteration scheme (5) we have

$$y_{n+1} = 1 + \int_0^t (-y_n(s)) \, ds = 1 + \int_0^t \left(-\sum_{j=0}^{n} \frac{(-1)^j s^j}{j!} \right) ds = 1 - \sum_{j=0}^{n} \int_0^t \frac{(-1)^j s^j}{j!} \, ds$$

$$= 1 + \sum_{j=0}^{n} \frac{(-1)^{j+1} t^{j+1}}{(j+1)!} = \sum_{j=0}^{n+1} \frac{(-1)^j t^j}{j!}$$

where we have used the induction hypothesis and reindexed a sum. So the formula holds for $n + 1$, if it holds for n. By induction, the formula holds for all n, and all t.

(d). The exact solution is $y = e^{-t}$, $-\infty < t < \infty$, as we see by the technique of Section 1.3. See the dashed line in Fig. 1(b).

(e). The Taylor series for e^{-t} is $\sum_{j=0}^{\infty}(-1)^j t^j / j!$ about $t_0 = 0$ since if $f(t) = e^{-t}$ then $f^{(j)}(0) = (-1)^j$, and the Taylor series for $f(t)$ is $\sum_{j=0}^{\infty} f^{(j)} t^j / j!$. So $y_n(t)$ is just the partial sum $\sum_{j=0}^{n}(-1)^j t^j / j!$ of the Taylor series for e^{-t}. In this problem, Picard iteration constructs the Taylor series for the solution by adding one term to the series at each iteration.

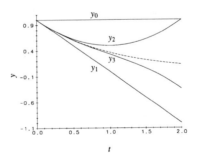

Problem 1(b).

2. The IVP is $y' = 2ty$, $y(0) = 2$.

(a). The iteration scheme (5) for this problem becomes $y_0(t) = 2$, $y_{n+1}(t) = 2 + \int_0^t 2sy_n(s) \, ds$, so

$$y_1(t) = 2 + \int_0^t 2s(2) \, ds = 2 + 2t^2,$$

$$y_2(t) = 2 + \int_0^t 2s(2 + 2s^2) \, ds = 2 + 2t^2 + t^4,$$

$$y_3(t) = 2 + \int_0^t 2s(2 + 2s^2 + s^4) \, ds = 2 + 2t^2 + t^4 + t^6/3$$

All are defined for all t.

(b). See Fig. 2(b) for the graphs of $y_0 - y_3$. The dashed line is the solution curve for the IVP.

(c). To "guess" a formula for $y_n(t)$, we first solve the problem exactly by separating variables to obtain $y^{-1}\,dy = 2t\,dt$, so $\ln|y| = t^2 + C$, or $y = 2e^{t^2}$, where the initial condition $y(0) = 2$ was used to find C. Since the Taylor series for e^t is $\sum_{j=0}^{\infty} t^j/j!$, the Taylor series for e^{t^2} is $\sum_{j=0}^{\infty} t^{2j}/j!$, and the partial sum through term n for $2e^{t^2}$ is $2\sum_{j=0}^{n} t^{2j}/j!$. So we suspect that $y_n(t)$ is given by the latter sum. To prove this for all n by induction, first note that the formula is correct for $n = 1$ since $y_1(t) = 2 + 2t^2$ from part **(a)**. Suppose that $y_n(t)$ is given by the sum above. We use the iteration scheme (5):

$$y_{n+1}(t) = 2 + \int_0^t 2sy_n(s)\,ds = 2 + \int_0^t 2s\left(2\sum_{j=0}^{n}\frac{s^{2j}}{j!}\right)ds = 2 + 4\sum_{j=0}^{n}\int_0^t \frac{s^{j2j+1}}{j!}\,ds$$

$$= 2 + 4\sum_{j=0}^{n}\frac{t^{2j+2}}{(2j+2)j!} = 2\sum_{j=0}^{n+1}\frac{t^{2j}}{j!}$$

where we have reindexed to obtain the last sum. So the formula holds for $n+1$, if it does for n, and the induction proof is complete. All iterates are defined for all t.

(d). The exact solution is $y = 2e^{t^2}$, $-\infty < t < \infty$. [See part **(c)**]. See the dashed line in Fig. 2(b).

(e). See part **(c)**.

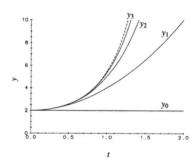

Problem 2(b).

3. The IVP is $y' = y^2$, $y(0) = 1$.

(a). Using the iteration scheme (5) to calculate the Picard iterates, we have

$$y_0(t) = 1,$$

$$y_1(t) = 1 + \int_0^t y_0^2(s)\,ds = 1 + \int_0^t 1\,ds = 1 + t$$

$$y_2(t) = 1 + \int_0^t y_1^2(s)\,ds = 1 + \int_0^t (1+s)^2 ds = 1 + t + t^2 + t^3/3$$

$$y_3(t) = 1 + \int_0^t y_2^2(s)\, ds = 1 + \int_0^t (1 + s + s^2 + s^3/3)^2\, ds$$

$$= 1 + t + t^2 + t^3 + 2t^4/3 + t^5/3 + t^6/9 + t^7/63$$

All are defined for all t.

(b). See Fig. 3(b) for the graphs of y_0–y_3. The exact solution (shown by the dashed line) is $y = (1 - t)^{-1}$ where $t < 1$.

(c). Certainly, $y_1(t) = 1 + t$ is a polynomial of degree 1. Assume that $y_n(t)$ is a polynomial of degree $2^n - 1$ (the induction hypothesis). To show that $y_{n+1}(t)$ is a polynomial of degree $2^{n+1} - 1$, we note that

$$y_{n+1}(t) = 1 + \int_0^t y_n^2(s)\, ds = 1 + \int_0^t (c_0 + \cdots + c_k s^{2^n - 1})^2\, ds$$

$$= 1 + \int_0^t (c_0^2 + \cdots + c_k^2 s^{2^{n+1} - 2})\, ds = \cdots + c_k^2 (2^{n+1} - 1)^{-1} t^{2^{n+1} - 1}$$

where we have set $y_n(t) = c_0 + \cdots + c_k t^{2^n - 1}$. So $y_{n+1}(t)$ is a polynomial of the desired degree and the proof is complete.

(d). Since polynomials are defined for all values of their variables, every Picard iterate is defined for all t, even though the exact solution is defined only for $t < 1$. On the other hand, the Taylor series for the exact solution $y = (1 - t)^{-1}$ is the geometric series $\sum_{j=0}^{\infty} t^j$, which by the Ratio Test converges only for $-1 < t < 1$.

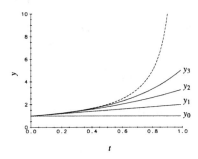

Problem 3(b).

4. Using the wrong "starting point" $y_0(t) = 0$ in the Picard iteration algorithm (5) for approximating the solution of $y' = -y$, $y(0) = 1$, we have that $y_0(t) = 0$, $y_1(t) = 1 - \int_0^t y_0(s)\, ds = 1 + 0 = 1$, $y_2(t) = 1 - \int_0^t y_1(s)\, ds = 1 - \int_0^t 1\, ds = 1 - t$, $y_3(t) = 1 - \int_0^t y_2(s)\, ds = 1 - \int_0^t (1 - s)\, ds = 1 - t + t^2/2$. By comparison with the solution to Problem 1, we see that the Picard iterates with the wrong starting point coincide with those with the "right" starting point $y_0(t) = 1$, but lag behind by 1 step. The sequence of iterates still converges to the exact solution of the problem, $y = e^{-t}$, even if the first step is wrong.

5.   In this case we have $y' = -y$, $y(0) = 1$, but we take $y_0 = \sin t$ as the zeroth Picard iterate:

$$y_0(t) = \sin t$$

$$y_1(t) = 1 + \int_0^t (-y_0(s))\,ds = 1 + \int_0^t (-\sin s)\,ds = \cos t,$$

$$y_2(t) = 1 + \int_0^t (-y_1(s))\,ds = 1 + \int_0^t -\cos s\,ds = 1 - \sin t$$

If we continue this process, being careful to separate the terms as above, we see that each $y_n(t)$ is the sum of two terms, the first being the partial sum $\sum_{j=0}^{n-1} (-1)^j t^j/j!$ of the Taylor series of e^{-t}, while the magnitude of the second term is the magnitude of the difference between $\pm \sin t$ [or $\pm \cos t$] and its corresponding Taylor series through the term t^{n-1}. Since the latter Taylor series tends to $\sin t$ [or $\cos t$] as $n \to \infty$, these terms tend to zero as $n \to \infty$ and the sequence of iterates converges to e^{-t}, even with the absurd choice of $y_0(t) = \sin t$.

6.   The IVP is $y' = 1 + y^2$, $y(0) = 0$.

(a). Using iteration scheme (5) to find the Picard iterates, we have that

$$y_0(t) = 0$$

$$y_1(t) = 0 + \int_0^t (1 + y_0^2(s))\,ds = \int_0^t 1\,ds = t$$

$$y_2(t) = \int_0^t (1 + y_1^2(s))\,ds = \int_0^t (1 + s^2)\,ds = t + t^3/3$$

$$y_3(t) = \int_0^t (1 + y_2^2(s))\,ds = \int_0^t (1 + s^2 + 2s^4/3 + s^6/9)\,ds = t + t^3/3 + 2t^5/15 + t^7/63$$

(b). We use induction to show that the iterate $y_n(t)$ is a polynomial of degree $2^n - 1$. The iterate $y_1(t) = t$ is a polynomial of degree 1. Suppose that $y_n(t)$ is a polynomial of degree $2^n - 1$ (the induction hypothesis). We must show that $y_{n+1}(t)$ is a polynomial of degree $2^{n+1} - 1$. We have that

$$y_{n+1}(t) = \int_0^t \left(1 + y_n^2(s)\right)\,ds = \int_0^t \left[1 + (c_0 + \cdots + c_k s^{2^n-1})^2\right]\,ds$$

$$= \cdots + \frac{c_k^2 t^{2^{n+1}-1}}{2^{n+1} - 1}$$

just as in 3(c). So $y_{n+1}(t)$ has the desired degree and the induction is complete.

(c). Separating variables and using a table of integrals, we have that $(1 + y^2)^{-1}\,dy = dt$, and hence $\arctan y = t + C$, or $y = \tan t$, where the initial condition has been used to evaluate C. The exact solution is defined only for $|t| < \pi/2$.

(d). Because the IVP $y' = 1 + y^2$, $y(0) = 0$ satisfies the conditions of Theorem A.2.1, the sequence of Picard iterates $\{y_n(t)\}_1^\infty$ converges to the exact solution for $|t| < \pi/2$, but not otherwise since the exact solution is not even defined for $|t| = \pi/2$.

7.   The IVP is $y' = \sin(t^3 + ty^5)$, $y(0) = 1$. We have $y_0(t) = 1$, $y_1(t) = 1 + \int_0^t \sin(s^3 + s)\,ds$, which cannot be expressed in terms of elementary functions. In general, the use of the Picard scheme is limited by the difficulty of carrying out the integration of any but the simplest of functions. This does not mean that the integrals fail to exist, only that they cannot be written in terms of elementary functions.

8. The terms $y_0 = 3$, $y_1 = 3 - 27t$, $y_n = 3 - \int_0^t y_{n-1}^3(s)\, ds$, ... are Picard iterates for the IVP $y' = -y^3$, $y(0) = 3$. The exact solution may be obtained by separating variables: $-y^{-3}y' = 1$, so $y^{-2}/2 = t + C$, or $y = (2t + 2C)^{-1/2}$, or $y(t) = (2t + 1/9)^{-1/2}$, where the initial condition has been used to evaluate C. Note that the exact solution is defined only for $t > -1/18$. The sequence $\{y_n(t)\}_{n=1}^\infty$ converges uniformly to $y(t) = (2t + 1/9)^{-1}$ on some closed interval in $t > -1/18$ which contains $t = 0$.

9. **(a).** Let ε be any positive number. We must find a positive integer N, which may depend upon ε, so that for all integers $n \geq N$, $|(\sin nt)/n| < \varepsilon$, for all t in some interval I. Since $|(\sin nt)/n| \leq 1/n$, we see that we may take for N any integer larger than $1/\varepsilon$. Then if $n \geq N$, we have

$$|\sin nt/n| \leq \frac{1}{n} \leq \frac{1}{N} < \frac{1}{1/\varepsilon} = \varepsilon$$

So $\{(\sin nt)/n\}_1^\infty$ converges uniformly to the zero function on any interval I.

(b). Let ε be any positive number. We must find a positive integer N such that for all integers $n \geq N$, $|t^n| < \varepsilon$ for all t, $0 \leq t \leq 0.5$. Since for t in this interval we have $|t^n| = t^n \leq (0.5)^n$, we may "solve" the equality $(1/2)^k = \varepsilon$, for k in order to find a suitable integer N: $2^{-k} = \varepsilon$, $2^k = 1/\varepsilon$, $k = -(\ln \varepsilon)/\ln 2$. So if N is any positive integer larger than $-(\ln \varepsilon)/\ln 2$, and if $n \geq N$, then $n > -(\ln \varepsilon)/\ln 2$, $n \ln 2 > -\ln \varepsilon = \ln(1/\varepsilon)$, $2^n > 1/\varepsilon$, $2^{-n} < \varepsilon$, and we have $t^n < \varepsilon$ if $0 \leq t \leq 1/2$. Hence, $\{t^n\}_1^\infty$ converges uniformly to 0 on the interval $0 \leq t \leq 1/2$. The same kind of analysis shows the uniform convergence of $\{t^n\}_1^\infty$ to 0 on the interval $0 \leq t \leq b$, where $0 < b < 1$. However, if $0 \leq t \leq 1$, then the sequence converges to $f(t)$, where $f(t) = 0$ if $0 \leq t < 1$ and $f(1) = 1$, but the convergence cannot be uniform on that interval. To show nonuniformity, we only need to find a value of $\varepsilon > 0$ for which there is no positive integer N such that $|t^n - f(t)| < \varepsilon$ for $0 \leq t \leq 1$. For example, let $\varepsilon = 0.1$. Since t^n is a continuous function of t for each n and since $0^n = 0$, $1^n = 1$, there is a value of t, $0 < t < 1$ for which $t^n = 1/2$ (in fact, $t = 2^{-1/n}$). On the other hand, for this value of t it is certainly not the case that $|t^n - f(t)| = |1/2 - 0| < 0.1$, no matter how large we take n. No suitable N exists and the convergence of $\{t^n\}_1^\infty$ to $f(t)$ for $0 \leq t \leq 1$ is *not* uniform.

A.3 Extension of Solutions

1. **(a).** The variables in the ODE $y' = -y^3$ may be separated and the ODE solved (using the initial condition $y(0) = 1$) to obtain $y = (2t + 1)^{-1/2}$. This solution is defined for $t > -1/2$.

As t decreases to $-1/2$, $y(t)$ increases to $+\infty$, while as $t \to +\infty$, $y(t)$ decreases to 0.

(b). Separating variables in the ODE $y' = -2ty$, we have $e^y\,dy = -t\,dt$, so $e^y = -t^2/2 + C$. From the initial condition we find that $e^2 = C$, so we have the solution $y(t) = \ln|e^2 - t^2/2|$, which is defined for $|t| < 2^{1/2}e$. As $t \to \pm 2^{1/2}e$, $y(t)$ tends to $-\infty$.

2. Separate variables in the ODE $y' = 2ty^2$; we have $y^{-2}\,dy = 2t\,dt$, $-y^{-1} = -C + t^2$, $y(t) = (C - t^2)^{-1}$. Requiring that y be positive, we see that $|t| < C^{1/2}$ and $C > 0$. As $t \to -(C^{1/2})^+$ and as $t \to (C^{1/2})^-$, $y(t) \to +\infty$. There are, of course, other examples [e.g., Problem 1**(b)**].

3. The IVP is $y' + p(t)y = q(t)$, $y(t_0) = y_0$, where p and q are continuous on an interval I. The function $P(t) = \int_{t_0}^t p(s)\,ds$ is continuous for all t in I since $p(t)$ is continuous for all t in I. So $e^{P(t)}$ and $\int_{t_0}^t e^{P(s)}q(s)\,ds$ are continuous for all t in I, the latter since $q(t)$ is continuous for all t in I. If $p(t)$ and $q(t)$ are continuous for all t in I, the solution formula (3) in Section 1.3 tells us that $y(t) = e^{P(t_0)}y_0 e^{-P(t)} + e^{-P(t)}\int_{t_0}^t e^{P(s)}q(s)\,ds$, so $y(t)$ is defined for all t. There is no "finite escape time" inside I.

A.4 Sensitivity of Solutions to the Data

1. The two IVPs to be compared are $y' = y/10$, $y(0) = a$, and $y' = y/10 + ye^{-y^2}$, $y(0) = \tilde{a}$, with respective solutions $y(t)$ and $\tilde{y}(t)$. Since the number M in the Perturbation Estimate is an upper bound on the "perturbation" term ye^{y^2}, we find the maximum value M of that term by equating its derivative to 0 and finding $y = 2^{-1/2}$. So we can take $M = 2^{-1/2}e^{-1/2}$: for simplicity we take $M = 1/2$, which is slightly larger. The number L in the estimate (Theorem A.4.2) is an upper bound for $|\partial(y/10)/\partial y| = 1/10$, so $L = 1/10$. Since $t_0 = 0$, the inequality (11) becomes

$$|y(t) - \tilde{y}(t)| \le |a - \tilde{a}|\,e^{t/10} + 5\left(e^{t/10} - 1\right), \qquad 0 \le t$$

The coefficient 5 is not unique.

2. Suppose that the differentiable function $y(t)$ satisfies the inequality $|y'(t)| \le |y(t)|$. We want to show that either $y(t) \ne 0$ for all t, or else $y(t) = 0$ for all t. Suppose $y(t_0) = 0$ for some t_0. We have that $y(t) = \int_{t_0}^t y'(s)\,ds$ which implies that (using Theorem B.5.9)

$$|y(t)| \le \left|\int_{t_0}^t y'(s)\,ds\right| \le \int_{t_0}^t |y'(s)|\,ds \le \int_{t_0}^t |y(s)|\,ds$$

where we assume $t \ge t_0$ and use the assumption that $|y'| \le |y|$. Let $z(t) = |y(t)|$. Then the inequality $z(t) \le \int_{t_0}^t z(s)\,ds$ has the form of inequality (2) in the Basic Estimate (Theorem A.4.2) with $A = B = 0$, $L = 1$, $a = t_0$. By (3) and the definition of $z(t)$, we have $0 \le |y(t)| \le 0$. So $y(t) = 0$ for $t \ge t_0$ if $y(t_0) = 0$. A similar proof applies for $t \le t_0$ as well.

3. Now $y(t)$ and $\bar{y}(t)$ are both solutions of the ODE $y' = f(t, y)$, so $y'(t) = f(t, y(t))$ and $\bar{y}'(t) = f(t, \bar{y}(t))$, for all $a \le t \le b$. Integrating these equations over the interval $a \le s \le t$, we see that

$$y(t) = y(a) + \int_a^t f(s, y(s))\, ds, \quad \text{and } \bar{y}(t) = \bar{y}(a) + \int_a^t f(s, \bar{y}(s))\, ds$$

for all $a \le t \le b$. Without loss of generality, suppose that $y(a) > \bar{y}(a)$. Then from the Existence and Uniqueness Theorem (Theorem 2.1.1) we see that $y(t) > \bar{y}(t)$, so $|y(t) - \bar{y}(t)| = y(t) - \bar{y}(t)$, for all $a \le t \le b$. Subtracting the second integral equation from the first, we see that

$$y(t) - \bar{y}(t) = y(a) - \bar{y}(a) + \int_a^t \{f(s, y(s)) - f(s, \bar{y}(s))\}\, ds$$

and from the Mean Value Theorem (Theorem B.5.4) it follows that

$$y(t) - \bar{y}(t) = y(a) - \bar{y}(a) + \int_a^t f_y(s, c(s))(y(s) - \bar{y}(s))\, ds \tag{i}$$

where c is a value (which may depend on s) with $\bar{y}(s) < c(s) < y(s)$.

We are now ready to address the bulleted items in turn:

- If $K = 0$, then $f_y(t, y) \ge 0$ for all (t, y) in R. We have

$$\int_a^t f_y(s, c(s))(y(s) - \bar{y}(s))\, ds \ge 0, \quad \text{for all } a \le t \le b$$

so from (i) above we see that

$$y(t) - \bar{y}(t) \ge y(a) - \bar{y}(a), \quad \text{for all } a \le t \le b$$

Since $|y(t) - \bar{y}(t)| = y(t) - \bar{y}(t)$, for all $a \le t \le b$, the desired inequality follows. If $L = 0$, then $f_y(t, y) \le 0$, for all (t, y) in R, and the asserted inequality follows in a similar fashion.

- Now assume that $0 < K \le f_y(t, y) \le L$ for all (t, y) in R. We have

$$K \int_a^t (y(s) - \bar{y}(s))\, ds \le \int_a^t f_y(s, c(s))(y(s) - \bar{y}(s))\, ds \le L \int_a^t (y(s) - \bar{y}(s))\, ds$$

for all $a \le t \le b$. If we put $z(t) = y(t) - \bar{y}(t)$, then from (i) above we have that

$$K \int_a^t z(s)\, ds \le z(t) - z(a) \le L \int_a^t z(s)\, ds, \quad \text{for all } a \le t \le b \tag{ii}$$

Following the proof of the Basic Estimate (Theorem A.4.1), we put $w(t) = \int_a^t z(s)\, ds$, and observe that $w'(t) = z(t)$ and so the inequality above becomes

$$Kw \le w' - z(a) \le Lw \tag{iii}$$

Working with the top half of inequality (iii), we have that

$$w' - Lw - z(a) \le 0, \quad \text{for } a \le t \le b$$

which, after multiplying through by e^{-Lt} becomes

$$\left(we^{-Lt} + \frac{z(a)}{L}e^{-Lt}\right)' \leq 0, \quad \text{for } a \leq t \leq b$$

The function $[w - z(a)/L]e^{-Lt}$ is a nonincreasing function, and so

$$\left(w + \frac{z(a)}{L}\right)e^{-Lt} \leq \left(w(a) + \frac{z(a)}{L}\right)e^{-La} \qquad \text{(iv)}$$

or [since $w(a) = 0$]

$$Lw + z(a) \leq z(a)e^{L(t-a)}, \quad a \leq t \leq b$$

But from (iii), $z(t) \leq Lw + z(a)$, and so the desired inequality holds:

$$z(t) \leq z(a)e^{L(t-a)}, \quad a \leq t \leq b$$

To show the other half of the asserted inequality, write the bottom part of inequality (iii) as

$$\left(we^{-Kt} + \frac{z(a)}{K}e^{-Kt}\right)' \geq 0, \quad \text{for all } a \leq t \leq b$$

The function $(w + z(a)/K)e^{-Kt}$ is nondecreasing, so

$$(w(t) + z(a)/K)\,e^{-Kt} \geq (w(a) + z(a)/K)\,e^{-Ka} \qquad \text{(v)}$$

or [since $w(a) = 0$]

$$Kw + z(a) \geq z(a)e^{K(t-a)}, \quad a \leq t \leq b$$

But from (iii), $z(t) \geq Kw + z(a)$, so the desired inequality holds:

$$z(a)e^{K(t-a)} \leq z(t), \quad a \leq t \leq b$$

The above argument fails when $K < 0$ or $L < 0$ because, when the inequalities (iv) or (v) are multiplied by K or L, the resulting inequality goes the wrong way.

- We treat only the case where $L < 0$; the case where $K < 0$ is treated similarly. From the top half of (ii) we have that

$$z(t) \leq z(a) + L\int_a^t z(s)\,ds, \quad a \leq t \leq b$$

where $z(t) = y(t) - \bar{y}(t) \geq 0$, for all $a \leq t \leq b$. Iterating the above inequality, we have that

$$z(t) \leq z(a) + L\int_a^t \left\{z(a) + L\int_a^s z(r)\,dr\right\} ds$$

$$\leq z(a) + z(a)L(t - a) + L^2\int_a^t \left\{\int_a^s z(r)\,dr\right\} ds$$

Changing the order of integration in the iterated integral, we obtain

$$z(t) \leq z(a) + z(a)L(t-a) + L^2 \int_a^t z(r) \left\{ \int_r^t ds \right\} dr$$

$$\leq z(a) + z(a)L(t-a) + L^2 \int_a^t (t-r)z(r) \, dr$$

Continuing the iteration in this manner we obtain after n iterations

$$z(t) \leq z(a) + z(a)L(t-a) + z(a)\frac{L^2(t-a)^2}{2!} + \cdots$$

$$+ L^{n+1} \int_a^t \frac{(t-r)^n}{n!} z(r) \, dr$$

Since the last term in the estimate above goes to zero for any choice of t in the interval $a \leq t \leq b$, it follows that

$$z(t) \leq z(a) \left\{ 1 + L(t-a) + \frac{L^2(t-a)^2}{2!} + \cdots + \frac{L^n(t-a)^n}{n!} + \cdots \right\}$$

$$\leq z(a)e^{L(t-a)}$$

as was asserted.

- Observe that $f_y(t, y) = (-1/10)y/(1+y)$, and so

$$-\frac{1}{10} \leq f_y(t, y) \leq 0, \quad \text{for } t \geq 0, \ y \geq 0$$

We have the estimates

$$|y(0) - \bar{y}(0)|e^{-t/10} \leq |y(t) - \bar{y}(t)| \leq |y(a) - \bar{y}(a)|$$

for $t \geq 0$, and any two solutions $y(t)$, $\bar{y}(t)$ of the ODE $y' = (1/10)(-y + \ln(1+y))$ which lie in the closed region $t \geq 0$, $y \geq 0$. See Fig. 3, Graph 1 for the graphs of the two solutions of the ODE with data $y(0) = 4$, and $\bar{y}(0) = 1$. It is evident that $|y(t) - \bar{y}(t)| < 3$, for all $0 \leq t \leq 40$. Direct measurement shows that $|y(10) - \bar{y}(10)| \geq 3e^{-1} \approx 1.104$, and that $|y(20) - \bar{y}(20)| \geq 3e^{-2} \approx 0.406$, and that $|y(40) - \bar{y}(40)| \geq 3e^{-4} \approx 0.055$.

- Since $f(t, y) = \sin t - (1/10)(0.5 \cos y + y)$, we see that

$$f_y(t, y) = (-1 + 0.5 \sin y)/10$$

so

$$-0.15 \leq f_y(t, y) \leq -0.05, \quad \text{for all } t \text{ and } y$$

It follows from the estimate proven above that

$$10e^{-0.15t} \leq |y(t) - \bar{y}(t)| \leq 10e^{-0.05t}, \quad \text{for all } t$$

for the given solutions $y(t)$ and $\bar{y}(t)$. See Fig. 3, Graph 2.

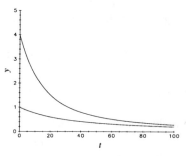

Problem 3, Graph 1.

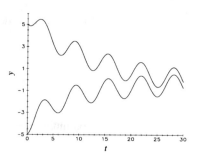

Problem 3, Graph 2.

4. Group problem.

Notes

Notes

Notes

Notes

Notes

Notes

Notes

Notes

Notes

Notes

Notes

Notes

Notes